# Student Reference Manual for Electronic Instrumentation Laboratories

## Stanley Wolf

*Department of Electrical Engineering and Computer Science*
*California State University, Long Beach*

## Richard F. M. Smith

*Department of Electrical Engineering and Computer Science*
*California State Polytechnic University, Pomona*

PRENTICE HALL, Englewood Cliffs, New Jersey 07632

*Library of Congress Cataloging-in-Publication Data*

WOLF, STANLEY
    Student reference manual for electronic instrumentation
    laboratories / Stanley Wolf, Richard F.M. Smith.

            p.    cm.
    Bibliography: p.
    Includes index.
    ISBN 0-13-855776-4
    1. Electronic instruments.   2. Electronic measurements.
I. Smith, Richard F. M.   II. Title.
TK7878.4.W65   1990                          89-8713
621.381'54—dc20                              CIP

*To Carol,*
*Jennifer, and Ross*

*To Donna,*
*Steve, and Mark*

Editorial/production supervision and
    interior design: **KATHLEEN SCHIAPARELLI**
Cover design: **BEN SANTORA**
Manufacturing buyer: **DONNA DOUGLASS**

 © 1990 by Prentice-Hall, Inc.
A Paramount Communications Company
Englewood Cliffs, New Jersey 07632

Printed in the United States of America

10   9   8   7   6   5

ISBN 0-13-855776-4

Prentice-Hall International (UK) Limited, *London*
Prentice-Hall of Australia Pty. Limited, *Sydney*
Prentice-Hall Canada Inc., *Toronto*
Prentice-Hall Hispanoamericana, S.A., *Mexico*
Prentice-Hall of India Private Limited, *New Delhi*
Prentice-Hall of Japan, Inc., *Tokyo*
Simon & Schuster Asia Pte. Ltd., *Singapore*
Editora Prentice-Hall do Brasil, Ltda., *Rio de Janeiro*

# Contents

# *Preface*

This book is a unique reference manual for the engineering student, designed to provide practical information and technical background data for the laboratory courses of the electrical engineering curriculum. The need for such a book has arisen from the growing tendency to omit the coverage of practical information from engineering courses and textbooks. With additional material being continually crammed into the curriculum, other material is forced to be deleted. Since theoretical concepts are perceived (probably correctly) as being more fundamental and less ephemeral than technology-based topics, the latter are being de-emphasized (in both the course syllabus and textbook).

And yet, students are still being held responsible for acquiring such practical knowledge, essentially on their own. To some degree, this learning is supposed to occur through the "hands on" experience of the laboratory courses. Unfortunately, the skimpy, and ad hoc character of the supporting information that accompanies many such courses makes the learning process less than optimal. If a text was available that contained practical reference material relevant to the laboratory topics, such learning would be significantly enhanced. We have written this text to serve this role.

The book is an outgrowth of an earlier book by S. Wolf, *Guide to Electronic Measurements and Laboratory Practice*. Much of the new material, mostly added by R. Smith, has been added to make the new book suitable for a larger number of laboratory courses. The book is unique in that its presentation style is user-friendly (it can be used in a self-tutorial manner) and its material is suitable for many laboratory courses. An outline of the book's material as it relates to various laboratory courses is as follows:

- *Basic Instruments and Measurements Laboratory:* Chapter 1, Language of Electrical Measurements; Chapter 2, Experimental Data and Errors; Chapter 3, Electrical Laboratory Practice (Electrical Safety, Grounds, Circuit Protection Devices, Input Impedance, Output Impedance, and Loading); Chapter 4, Analog DC and AC Meters; Chapter 5, Digital Electronic Meters; Chapter 6, The Oscilloscope; Chapter 10, Resistors and the Measurement of Resistance; Chapter 11, Measurement of Capacitance, Inductance, and Impedance; Chapter 12, DC Signal Sources (Batteries and Power Supplies).

- *Analog Electronics Laboratory:* Chapter 4, Analog AC Ammeters and Voltmeters; Chapter 5, Digital Electronic Meters; Chapter 6, The Oscilloscope, including Digital Oscilloscopes and the Curve Tracer; Chapter 8, Time and Frequency Measurements; Chapter 12, DC Signal Sources; Chapter 13, AC Signal Sources.

- *Digital Electronics Laboratory:* Chapter 1, Language of Digital Systems; Chapter 5, Digital Electronic Meters; Chapter 18, Data Transmission in Digital Instrument Systems.

- *Power Laboratory:* Chapter 9, Power and Energy Measurements; Chapter 10, Use of the Megohmmeter.

- *Industrial Process Control and Data Acquisition Laboratory:* Chapter 14, Electrical Transducers; Chapter 15, Electronic Amplifiers; Chapter 16, Interference Signal and their Elimination or Reduction; Chapter 17, Introduction to Instrumentation Systems.

- *Biomedical Instrumentation Laboratory:* Chapter 3, Electrical Safety; Chapter 7, Potentiometers and Recorders; Chapter 14, Electrical Transducers; Chapter 15, Instrumentation Amplifiers; Chapter 16, Interference Signals and their Elimination or Reduction.

- *Digital Data Communications in Computer-Controlled Instrumentation Systems:* Chapter 18, Data Transmission in Digital Instrument Systems.

Because of the scope of the information provided, students need only purchase a single text that will serve as their reference for most undergraduate electrical engineering laboratory courses. In an endeavor to provide the book at minimum cost, a soft-cover format was chosen for its production.

The overall thrust of the book is to teach students to become proficient *users* of electronic measuring instruments and for them to gain a practical understanding of electrical laboratory practices. In this regard, the book will set out to explain how to select instruments for various measurement applications, how to evaluate their capabilities, how to connect them together, and how to operate them properly. In addition, descriptions of the terminology, apparatus, and measurement techniques unique to the electrical laboratory environment are provided. In summary, the book is meant to serve as a self-contained vehicle for carrying the reader through most electronic measurement tasks.

The presentation is kept at a basic level, so that the book is easy to understand, once the theory that precedes the laboratory work has been presented to the student. At this level we can afford the luxury of developing the material in ample detail. Furthermore, many important subjects that tend to be overlooked in more advanced texts can be covered. For example, we place considerable emphasis on the discussion of concepts that are usually regarded by instrumentation experts as self-evident (e.g., grounds and grounding, electrical safety, ground loops, and impedance matching). Such concepts often remain as puzzles to the beginner unless they are explicitly explained.

Several chapters are allotted to a description of the most common components and quantities encountered in electrical laboratory work. These chapters present practical information dealing with the construction, appearance, and uses of such items as resistors, capacitors, inductors, transformers, relays, batteries, power supplies, cables, switches, connectors, fuses, transducers, and amplifiers.

The continued importance and versatility of the oscilloscope provided the impetus for including an even more comprehensive treatment of this measurement tool. The new material also includes a discussion of digital and storage scopes, as well as an introduction to the curve tracer.

Digital measurements language is introduced in Chapter 1. Chapter 5, in its entirety, is allocated to digital meters, analog-to-digital, and digital-to-analog conversion methods. Aspects of data communications between digital instruments and computers are carefully explored in Chapter 18. The discussion includes a basic but thorough introduction to the two predominant standards for communication between instruments and computers: the *serial-asynchronous communication link with ASCII formatted data (RS-232C standard)* and the *IEEE-488 standard bus*.

In recognition of the trend toward greater employment of *instrumentation systems,* the chapters of the latter part of the book (Chapter 14–18) are organized as a group that examines topics relevant to measurement system implementation and use. Electrical transducers, the sources of the signals in many systems, are dealt with in Chapter 14. An expanded treatment of temperature transducers has been added. Chapter 15 covers electronic amplifiers and their use in measurement applications. New material on differential amplifiers, CMRR, operational amplifiers, and instrumentation amplifiers is included. Chapter 16 deals entirely with electrical interference signals and their suppression. Sources of both external and internal noise are considered. A careful discussion of ground-loop and common-mode interference, along with techniques available to minimize their effects, is presented. Chapter 17 explores various instrumentation system configurations as well as interfacing considerations in analog and analog-to-digital systems. Included are such topics as analog signal conditioning, analog signal transmission, multiplexers, and sample-and-hold circuits. As described earlier, the material of Chapter 18 examines techniques for interfacing instruments and computers.

Many people contributed their talent and knowledge to the preparation of this book. The authors offer their appreciation to all who contributed to it, directly and indirectly. We especially thank our wives and families for their continued support and encouragement in this project. A special note of thanks is also extended to Michael Muller for his assistance in the preparation of the Solution Manual.

## A NOTE TO THE STUDENT

This book has been designed to provide you with a great deal of important, practical information on electronic instruments and electronic measurement techniques generally not covered in the formal lecture portions of the engineering courses you take. It is, however, information assumed to be part of the knowledge that electrical engineering students acquire while pursuing their BSEE degree. Ironically, our book is one of the few places where much of this kind of information is assembled in one place.

In the event that you are asked to study the material in the text without the benefit of an instructor's guidance, we have tried to present the material in as easy and as straightforward a manner as possible. As a result, if you have completed the basic college physics sequence and a course on electric circuits, the material should be readily digestible.

As you move to more advanced laboratory courses (or even after graduation), you will appreciate being able to refer back to sections on specific instrument operating instructions (e.g., the oscilloscope, the curve tracer, the power supply, or the VOM) to help connect together test setups and to make valid measurements. The sections dealing with the characteristics of various circuit components (resistors, capacitors, inductors, batteries, relays, circuit protection devices, and electrical transducers) should be a handy reference when designing circuits and systems that call for them. Finally, the information on grounding, ground loops, and interference sources (and techniques to suppress them) will probably be consulted by you when problems arise from these all-too-common circuit glitches. In summary, we have tried to bring together a great deal of information that will be of use in many of your undergraduate electrical engineering laboratory courses, and probably even beyond.

*Stanley Wolf*

*Richard F.M. Smith*

# 1

# *Language of Electrical Measurements*

Electrical measurements and instruments are described with the help of various symbols, conventions, and terms, many of which are unique to electrical science. One should be acquainted with the most common of these terms and symbols before studying the details of electrical instrument operation and measurement techniques. Familiarity with the "language" paves a solid path for continuing study in any subject.

In keeping with this principle, our discussion begins with a chapter that introduces some of the most general concepts associated with electrical measurements. These particular concepts are part of the vocabulary used to describe all phases of electrical measurement work. Because of the basic nature of these concepts, some readers may already be acquainted with them and should feel free to bypass familiar material. However, if questions arise during later study, the information in this chapter can be used as a reference.

## CHARGE, VOLTAGE, AND CURRENT

The concepts of electric charge, electric current, and voltage are introduced first. Electrical measurements almost always involve determining one or more of these quantities.

1

## Electric Charge

Electrical phenomena arise from the nature of the particles that constitute matter. For example, atoms are composed largely of electrically charged particles. The nucleus of an atom is a central core consisting of protons (which have a positive charge) and neutrons. The nucleus is surrounded by a swarm of electrons. The electron has an electric charge which is equal in magnitude but opposite in polarity to the charge of a proton. Therefore, an electrically neutral atom must contain an equal number of electrons and protons.

If one or more electrons are removed from an atom, it is no longer neutral. There are now fewer electrons than protons, and the atom has a net positive charge. If electrons are removed from many neutral atoms of a substance and are then removed from the boundaries of the body itself, the entire body acquires a net positive charge. Similarly, if extra electrons are somehow injected into a body of electrically neutral matter, the body possesses a net negative charge.

The unit used to describe an amount of charge is the *coulomb* (abbreviated C). One coulomb is equivalent to the total electric charge possessed by $6.2 \times 10^{18}$ electrons; therefore, one electron has a charge of $1.6 \times 10^{-19}$ C.[1]

A body exhibiting a net charge will experience a force when placed in the neighborhood of other charged bodies. The magnitude of such an electrostatic force between two charged bodies is found from *Coulomb's law*, published in 1785.

$$F = \frac{kQ_1Q_2}{d^2} \tag{1-1}$$

where $Q_1$ is the charge, in coulombs, on one body, and $Q_2$ is the charge on the other. $F$ is the force in *newtons*,[2] $d$ is the distance in meters separating the charged bodies, $k = (4\pi\epsilon_0)^{-1}$ is a constant of value $9 \times 10^9$ newton-meters$^2$/coulomb$^2$ (N-m$^2$/C$^2$), and $\epsilon_0$ is the permittivity of free space, $8.85 \times 10^{-12}$ C$^2$/N-m$^2$.

If charges possess identical polarities (i.e., both positive or both negative), the force between them is *repulsive*. If the charges are of opposite polarity, the force is *attractive* (Fig. 1-1).

### Example 1-1

If all the electrons are removed from a penny and are transferred 1.6 km to a second penny, what is the charge on each penny and what is the force of attraction? There are approximately $8.41 \times 10^{23}$ electrons in a penny.

### Solution

(a) $Q = (8.41 \times 10^{23}$ electrons$) \times (1.6 \times 10^{-19}$ coulombs/electron$)$
    $= 1.344$ C on each penny.

---

[1]The value of the charge possessed by a single electron was discovered about a century after the coulomb unit was established. This is why the relationship between a coulomb and the charge of one electron is not a simple number.

[2]A *newton* is the force required to accelerate a mass of 1 kilogram at a rate of 1 meter per second each second. The force of 1 N is equal to the force of 0.2248 lb.

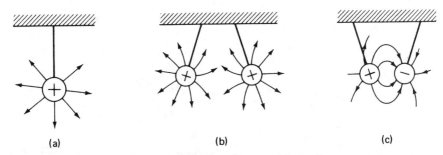

**Figure 1-1**  Forces between charged bodies: (a) electric field lines around charged body; (b) like charges repel; (c) opposite charges attract.

(b) $F = \dfrac{kQ_1Q_2}{d^2} = \dfrac{\left(9 \times 10^9 \dfrac{N-m^2}{C^2}\right)(1.344 \text{ C})^2}{\left(1.6 \times 10^3 \text{ m}\right)^2} = 6.35 \times 10^3 \text{N}$

This force is probably large enough to pull a car with its brakes locked and on dry pavement, even at this distance of 1.6 km.

## Voltage

The concept of *voltage* is related to the concepts of potential energy and work. That is, when electric charges are moved against the force of an electric field, work must be done to move them. This work involves an expenditure of energy. Since the law of conservation of energy says that energy cannot be created or destroyed, the energy used to move charges against an electric field must be converted to another form. This conversion is similar to the energy conversion involved in lifting a weight against the force of gravity. The energy expended in lifting a weight from the floor to a tabletop is stored by the weight in its location on the tabletop. The stored energy is called *potential energy* because it has the potential to be released and reconverted to the energy (kinetic) associated with a moving mass. This would occur if the weight were dropped from the table [Fig. 1-2(a)].

If an electric charge is infinitely far away from other electric charges, it will not feel any force of repulsion or attraction due to them. At that point, the *electrostatic potential* of the charge is defined to be zero. If the charge is then brought closer to other electric charges, its electrostatic potential (and potential energy) will change. That is, if the charge is moved closer to a charge of the same polarity, it must be moved against the force of the electric field, and this will increase its *potential energy*. (If the charge is moved closer to charges of opposite polarity, it is moved *with* the force of the electric field, and thus will lose potential energy.) The *electrostatic potential* of any point in space is thus defined as the energy, per unit charge, that would be required to bring the charge to that point from a point of zero electrostatic potential. If a charged body is moved from one point in an electrical system to another, the two points that locate the positions of a charged particle

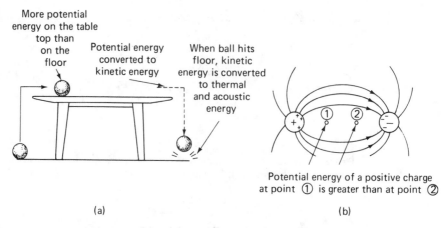

**Figure 1-2** Potential difference in gravitational and electric fields: (a) potential energy of a ball on a table; (b) potential energy of a charge in an electric field.

before and after being moved can be characterized by the *difference of* (electrostatic) *potential* between them. This *potential difference* is more commonly referred to as *voltage*, and it indicates how much energy would be acquired or lost (per unit of charge) by a particle as it was moved within the electric field [Fig. 1-2(b)].

Two points in a system (e.g., points 1 and 2) are said to have a difference of potential of 1 *volt* if 1 *joule*[3] of energy is required to move 1 coulomb of charge from the one point to the other. This is written mathematically as

$$\text{potential difference}_{1-2} = \text{volts} = \frac{\text{joules}}{\text{coulomb}} \qquad (1\text{-}2)$$

The unit of potential difference is the *volt*.

Note once again that it is the difference of potential between *two* points that is being measured by the value of the voltage. However, in many practical systems, one particular potential level is chosen as the reference level and assigned an arbitrary value of zero. The potential of all other points of the system are compared to this level. In such systems we can speak of single points in the system as having voltage values, because the zero value of the reference level is assumed to be the second level to which other potential levels are compared.

The planet earth is the most commonly used zero reference[4] (and is also called *ground*). This means that the potential of the earth at any point to which an electric circuit is connected is generally considered to be zero.

[3]One *joule* (J) is the work done when a force of 1 newton must be used to move an object a distance of 1 meter.

[4]We defined the point at which an electric charge is infinitely far from other charges as a point of zero electrical potential (i.e., it does not feel any repulsive or attractive electric forces). Since the Earth is electrically neutral and so very large, any human-made charge will not appreciably affect this neutrality. Hence, for all practical purposes, the earth can also be defined as having an electrical potential of zero.

There are, however, exceptions to this. Hospital grounding systems and computer grounding systems are two common examples. Both of these grounding systems are adversely affected by very small changes in earth ground or potential. For example, two copper rods driven into the ground at different locations may not be at the same potential. Since a finite resistance exists between the two locations, any potential difference between the two rods will set up a current flow between them. Although the potential differences are usually small, the microcurrents that they cause cannot be ignored in medical electronic applications where humans or animals can be subjected to such currents. Computers are also susceptible to grounding problems if careful consideration is not given to the grounding system.

However, in circuits not connected to the earth (such as airplanes, automobiles, and ships), another surface or point (such as the airplane fuselage) may be assigned a zero potential level for convenience.

## Electric Current

*Electric current* is defined as the number of charges moving past a given point in a circuit in 1 second. This definition is written mathematically for a steady current as

$$i = \frac{q}{t} \tag{1-3}$$

where $i$ is the current and $q$ is the net charge that moved past the point in $t$ seconds. The unit of current is the ampere (A), and 1 ampere indicates that 1 coulomb of charge is transported past a point in 1 second. (Since the charge on a single electron is about $1.6 \times 10^{-19}$ C, a current of 1 ampere corresponds to a flow of about $6 \times 10^{18}$ electrons per second.) Smaller currents are more conveniently described by using the milliampere (mA, $10^{-3}$), the microampere ($\mu$A, $10^{-6}$), or picoampere (pA, $10^{-9}$).

The moving charges that make up current can take such forms as the motion of electrons in a vacuum or solid or the motion of ions in liquids or gases. Most of the currents found in electric circuits involve the motion of electrons in solids or vacuums. However, in devices such as batteries or in certain transducers, the current may also involve the motion of positive and negative ions. Nevertheless, in this section we describe only the phenomenon of current flow in a solid conductor since it is the type of current most frequently encountered in measurement circuits.

Electrical conductors contain essentially free electrons that can move about quite easily within the boundaries of the conductor. When an electric field is applied to the conductor, these electrons move in response to the applied electric field. If the conductor is a wire (as shown in Fig. 1-3) and the electric field is applied in the direction shown, the motion of the free electrons in the wire will be from left to right. The current resulting from the charge motion is said to flow from right to left. The total number of electrons that move past some cross-sectional area of the wire per unit of time yields the magnitude of the current.

When a voltage is applied across a conductor, a current flows almost instantaneously throughout the conductor. The rapidity with which current appears in

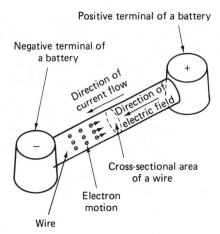

Positive terminal of a battery

Negative terminal of a battery

Direction of current flow

Direction of electric field

Cross-sectional area of a wire

Electron motion

Wire

**Figure 1-3**  Current flow in a wire.

the entire conductor is due to the velocity with which the electric field propagates within the conductor. (This velocity is effectively the speed of light—approximately $3 \times 10^8$ m/s.)

The electric field acts on each electron in the conductor. The electrostatic force on the electrons due to the field causes them to acquire a component of average velocity in the direction of the field. Although this component has a rather small value (on the order of 0.001 cm/s), the magnitude of the current can still be quite large because there are about $10^{23}$ free electrons per cubic centimeter of conductor material.

## ELECTRICAL UNITS

To be able to speak quantitatively about any group of quantities, we must devise a set of *units* which describes a fixed amount of each quantity. The International System of Units (SI), which includes the units used to describe electrical quantities, is the unit system used in this book. Table 1-1 gives those SI units most commonly used in connection with electrical measurements.

The SI system was formerly known as the meter-kilogram-second-ampere (MKSA) system because these four quantities are used to define all the other units used by the system. Prior to 1960, when the SI units were adopted as the standard, other systems were also acceptable for use. Therefore, they still may be found in some older publications. These other systems include the CGS (centimeter-gram-second) and the English Gravitational (foot-pound-second) systems. If these other systems are encountered elsewhere, conversion factors may be used to convert the units into SI units.

**TABLE 1-1** SI ELECTRICAL UNITS

| Quantity | Unit | Abbreviation |
|---|---|---|
| Length | meter | m |
| Mass | kilogram | kg |
| Time | second | s |
| Current | ampere | A |
| Temperature | degree Kelvin | °K |
| Voltage | volt | V |
| Resistance | ohm | $\Omega$ |
| Capacitance | farad | F |
| Inductance | henry | H |
| Energy | joule | J |
| Power | watt | W |
| Frequency | hertz | Hz |
| Charge | coulomb | C |
| Force | newton | N |
| Magnetic flux | weber | Wb |
| Magnetic flux density | webers/meter$^2$ | Wb/m$^2$ |

## SINE WAVES, FREQUENCY, AND PHASE

The instantaneous values of electrical signals can be graphed as they vary with time. Such graphs are known as the *waveforms* of the signals. Signal waveforms are analyzed and measured in many electrical applications.

Generally speaking, if the value of a signal waveform remains constant with time, the signal is referred to as a *direct-current* (dc) *signal*. An example of a dc signal is the voltage supplied by a battery. If a signal is time varying and has positive and negative instantaneous values, the waveform is known as an *alternating-current* (ac) *waveform*. If the variation is continuously repeated (regardless of the shape of the repetition), the waveform is called a *periodic waveform*.

The most common periodic waveform encountered in electrical systems is the *sinusoid*. In describing its characteristics we can also introduce the most important characteristics that are used to define other periodic waveforms. Figure 1-4 shows an example of a sinusoid. The mathematical expression for this waveform is

$$v = V_o \sin \omega t = V_o \sin (2\pi f t) \qquad (1\text{-}4)$$

The amplitude of the sine wave (and also other waveforms) denotes the maximum value of the function and is given by $V_o$ in Eq. (1-4). The frequency, $f$, of the sine wave (and of other periodic waveforms) is defined as the number of cycles traversed in 1 second. Accordingly, the frequency is measured in cycles per second, or hertz (Hz). The time duration (in seconds) of one cycle of a waveform is called the *period, T*. The frequency and period of the waveform are related by

$$f = \frac{1}{T} \qquad (1\text{-}5)$$

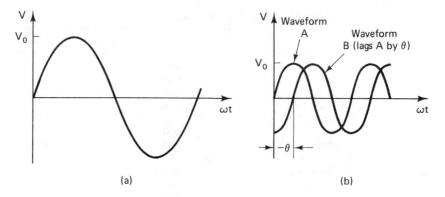

(a)                                                (b)

**Figure 1-4**   (a) Sine wave; (b) waveform.

In addition, one cycle of a waveform is defined as spanning $2\pi$ radians. Thus, if $2\pi$ is multiplied by the frequency, we obtain the angular (radian) frequency ($\omega$) of the sine wave:

$$\omega = 2\pi f = \frac{2\pi}{T} \qquad (1\text{-}6)$$

The units of $\omega$ are radians per second.

In Fig. 1-4(b), two sine waves of equal frequencies are plotted on a single time axis. The equations for both of these waveforms cannot be identical because each waveform possesses a different instantaneous value from the other at any given time. The manner in which the equations of the two waveforms differ is in the value of their *phase angles*. The concept of *phase angle* involves the comparison of two relative quantities. For example, if we define waveform $A$ as having a phase angle of zero, its equation is written as

$$v = V_o \sin \omega t \qquad (1\text{-}7)$$

Then the waveform $B$ will have a phase angle $\theta$, which indicates how much the waveforms are displaced from each other in time. If waveform $B$ has a zero value (for a positive slope) that occurs *later* in time than the zero value (for a positive slope) of waveform $A$, then waveform $B$ is said to *lag* waveform $A$, and vice versa. For example, in Fig. 1-4(b) we can say that waveform $B$ lags waveform $A$ by $\theta$ degrees. The equation of waveform $B$ is then written as

$$v = V_o \sin(\omega t - \theta) \qquad (1\text{-}8)$$

where the minus sign of $\theta$ indicates that the waveform of Eq. (1-8) lags the waveform of Eq. (1-7) by an angle $\theta$. A very common error results when waveforms $A$ and $B$ are observed on an oscilloscope. Students often erroneously assume that $B$ leads $A$ since it appears to the right of $A$.

## AVERAGE AND ROOT-MEAN-SQUARE VALUES

If the signals applied to a circuit are exclusively dc signals, it is relatively easy to calculate such quantities as the number of amperes flowing in the circuit or the energy dissipated by the components of the circuit over a period of time. In addition, one measurement of the dc waveform at any time will reveal all that must be known about the quantity it represents. However, the magnitudes of electrical quantities usually vary with time rather than keep constant values. If a signal is time varying, its waveform is no longer as simple as the dc waveform shown in Fig 1-5(a). Instead, the waveform has some time-varying shape. The variation may be periodic [as in Fig. 1-5(b) and (c)] or it may be a more random variation.

When waveforms possess time-varying shapes, it is no longer sufficient to measure the value of the quantity they represent at only one instant of time. It is not possible from one measurement to determine all that must be known about the signal. However, if the shape of a time-varying waveform can be determined, it is possible to calculate some characteristic values of the waveform shape (such as its average value). These values can be used to compare the effectiveness of various waveforms with other waveforms, and they can also be used to predict the effects that a particular signal waveform will have on the circuit to which it is applied.

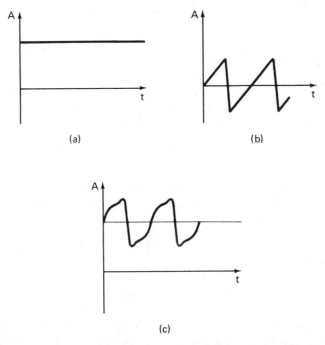

**Figure 1-5** Signal waveforms: (a) dc waveform; (b) time-varying periodic waveform; (c) time-varying periodic waveform superimposed on a dc level.

The two most commonly used characteristic values of time-varying waveforms are their average and their root-mean-square values. We will see how and why both of these values are determined.

### Average Value

The meaning of the *average value* of a waveform can best be understood if we use a current waveform as an example. The average value of a time-varying current waveform over the period, $T$, is the value that a dc current would have to have if it delivered an equal amount of charge in the same period, $T$. Mathematically, the average value of any periodic waveform is found by dividing the area under the curve of the waveform in one period, $T$, by the time of the period. This can be written as

$$A_{av} = \frac{\text{area under the curve}}{\text{length of the period (seconds)}} \tag{1-9}$$

where $A_{av}$ is the average value of the waveform.

For readers who are familiar with calculus, this expression is written more generally as

$$A_{av} = \frac{1}{T} \int_0^T f(t)\, dt \tag{1-10}$$

where $T$ is the length of the period of the curve and $f(t)$ is the equation of the shape of the waveform.[5]

**Example 1-2**

Find the average value of the curves given in Fig. 1-6.

**Solution.**  The average value of the curve of Fig. 1-6(a) is found from Eq. (1-9):

(a)
$$A_{av} = \frac{(40 \times 1) + (0) - (10 \times 2)}{4} = \frac{20}{4} = 5$$

(Note that the area under the curve from $t = 2$ to $t = 4$ is a negative area.)

(b) In this case the curve is a sinusoid and the area under the curve is found by using Eq. (1-10).

$$A_{av} = \frac{1}{T} \int_0^T A_o \sin \frac{2\pi t}{T}\, dt = \frac{A_o}{T} \int_0^T \sin \frac{2\pi}{T} t\, dt$$

$$= -\frac{TA_o}{2\pi T} \left[ \cos \frac{2\pi t}{T} \right]_0^T$$

[5]Note that it is not absolutely necessary to know calculus to continue using this text; but the problem of calculating average values is an example of one area where calculus or numerical methods are almost indispensable in order to obtain analytical results. For those who are not familiar with calculus, the average values of several common waveforms encountered while making measurements are presented in Fig. 1-7.

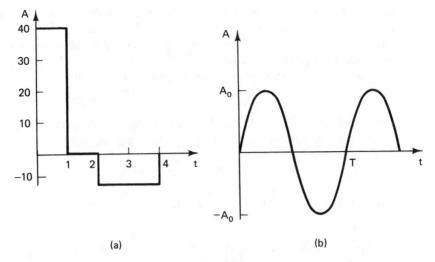

Figure 1-6

$$= \frac{A_o}{2\pi}[(1) - (+1)] = 0$$

$$= 0$$

This important result shows that the average value of a purely sinusoidal waveform is zero!

## Root-Mean-Square Values

The second common characteristic value of a time-varying waveform is its *root-mean-square* (rms) value. In fact, the rms value is used more often than the average value to describe electrical signal waveforms. The major reason for this is that the average value of symmetrical[6] periodic waveforms is zero. (Example 1-2 showed that this was indeed the case for a sinusoidal waveform.) A value of zero certainly does not provide much useful information about the properties of a signal. In contrast, the rms value of any waveform is not zero.

The rms value of a waveform refers to its *power delivering* capability. In connection with this interpretation, the rms value is sometimes called the *effective value*. This name is used because the rms value is equal to the value of a dc waveform, which would deliver the same power if it replaced the time-varying waveform.

To determine the rms value of a waveform, we first square the magnitude of the waveform at each instant. (This makes the value of the magnitude positive even when the original waveform has negative values.) Then the average (or mean) value of the squared magnitudes is found. Finally, the square root of this average value is

---

[6]By *symmetrical,* we mean in this context that a periodic waveform has equal positive and negative areas.

taken to get the result. Because of the sequence of calculations that is followed, the result is given the name root-mean-square. The situation that led to an average value of zero for some waveforms is avoided because the squaring process makes the entire quantity positive before the mean is taken.

Mathematically, the rms value of a waveform is written as

$$A_{rms} = \sqrt{\langle f(t)^2 \rangle} = \sqrt{\text{average}[f(t)^2]} \tag{1-11}$$

where the symbol $\langle \cdot \rangle$ means that the average of the quantity within the brackets is taken. For a given waveform, $f(t)$, the rms value is found by using the expression

$$A_{rms} = \sqrt{\frac{1}{T}\int_0^T [f(t)]^2 \, dt} \tag{1-12}$$

where $T$ is the length of one period of the waveform (in seconds).

**Example 1-3**

Find the rms value of the sinusoidal waveform of Fig. 1-6(b).

**Solution.**

$$A_{rms} = \sqrt{\frac{1}{T}\int_0^T [f(t)]^2 \, dt} = \sqrt{\frac{1}{T}\int_0^T A_o^2 \left(\sin\frac{2\pi}{T}t\right)^2 dt}$$

$$= \sqrt{\frac{A_o^2}{T}\left[\frac{t}{2} - \frac{T\sin}{4\pi}\left(\frac{4\pi}{T}t\right)\right]_0^T} = \frac{A_o}{\sqrt{2}}$$

Therefore, the rms value of sinusoidal waveforms is

$$A_{rms} = \frac{A_o}{\sqrt{2}} = 0.707 \, A_o \tag{1-13}$$

When referring to sinusoidal signals, it is most common to describe them in terms of their rms values. For example, the 115-V, 60-Hz voltage delivered by electric power companies to domestic consumers is really a sinusoidal waveform whose amplitude is about 163 V and whose rms value is therefore 115 V.

Figure 1-7 shows six time-varying waveforms commonly encountered in electrical measurement work. The average and rms values of each waveform are given in relation to their peak amplitudes.

## LANGUAGE OF DIGITAL MEASUREMENT SYSTEMS

Instrumentation systems can be divided into two broad categories: analog systems and digital systems. In *analog systems*, the measurement information is processed and displayed in analog form. The measured quantity is an *analog* quantity (i.e., a quantity whose value can vary in a *continuous* manner). In *digital systems*, the measurement information is processed and displayed in *digital* form. However, the

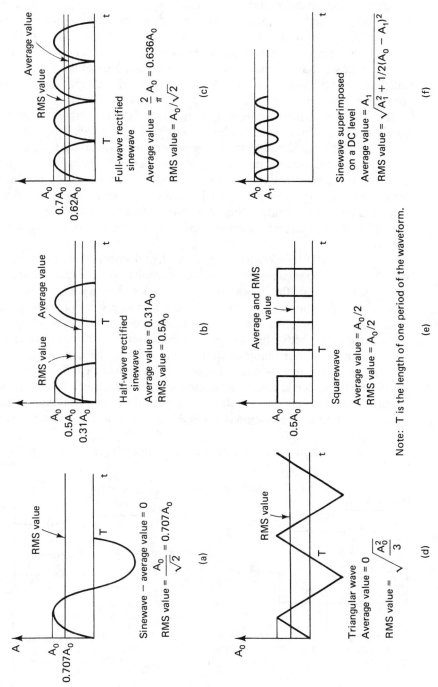

**Figure 1-7** Average and rms values of some waveforms commonly encountered in electrical measurements.

**(a)**

RMS value

Sinewave — average value = 0

RMS value = $\dfrac{A_0}{\sqrt{2}} = 0.707 A_0$

$A_0$
$0.707 A_0$

**(b)**

RMS value   Average value

Half-wave rectified sinewave

Average value = $0.31 A_0$

RMS value = $0.5 A_0$

$A_0$
$0.5 A_0$
$0.31 A_0$

**(c)**

Average value

RMS value

Full-wave rectified sinewave

Average value = $\dfrac{2}{\pi} A_0 = 0.636 A_0$

RMS value = $A_0/\sqrt{2}$

$A_0$
$0.7 A_0$
$0.62 A_0$

**(d)**

RMS value

Triangular wave

Average value = 0

RMS value = $\sqrt{\dfrac{A_0^2}{3}}$

$A_0$

**(e)**

Average and RMS value

Squarewave

Average value = $A_0/2$

RMS value = $A_0/2$

$A_0$
$0.5 A_0$

**(f)**

Sinewave superimposed on a DC level

Average value = $A_1$

RMS value = $\sqrt{A_1^2 + 1/2(A_0 - A_1)^2}$

$A_0$
$A_1$

Note: T is the length of one period of the waveform.

13

majority of the manufacturers of electronic systems now combine both digital and analog components in their products to minimize cost and maximize both reliability and versatility. A purely *analog* system handles only continuous functions, while a purely digital system operates on *discrete* or *discontinuous* pulses. As an example, assume that we wish to display the velocity of an automobile. One of the ways in which this can be done is to use a drag generator, whose input signal is transmitted through a cable connected to a gear in the transmission. This type of velocity display system is a *continuous* or *analog* system, since all the variables in the process are *continuous*. The same automobile velocity can be displayed digitally by using a pulse generator connected to the transmission and a digital display of the output of the pulse generator on the dashboard. A more versatile system might use the digital signal for velocity in combination with other signals, so that other variables in digital form could be displayed on the dashboard as well (e.g., mpg, distance traveled, rpm, and distance to empty). Simultaneously, the same digital signal could be used to control the velocity of the automobile (an analog function) through the cruise control.

In this section we portray a general digital instrumentation system in block diagram form. To facilitate the portrayal, we shall introduce many of the terms used in the description of the operation of digital electronic systems. The scope of the discussion will be limited, however, to those terms necessary for understanding the operation and interfacing of digital instruments. We cannot hope to be a primer on digital electronic theory and design. Interested readers are referred to some excellent references on these subjects in the list at the end of the chapter.

In later chapters, a more detailed discussion of digital instruments, digital measurement systems, and their components is presented. When studying those chapters, readers may find it helpful to refresh their understanding of some of the basic digital terms. This section is designed to provide a convenient reference source for this purpose.

A block diagram of a basic digital instrument is shown in Fig. 1-8. It can be

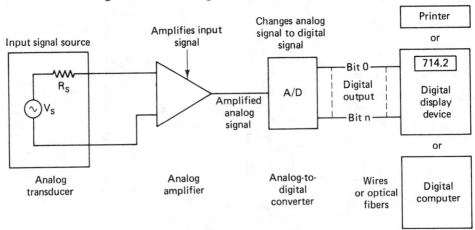

**Figure 1-8**  Block diagram of basic digital instrument.

seen that an analog signal is received by the digital instrument from a circuit or transducer under test. Typically, the instrument first subjects the analog signal to amplification. Next, the amplified signal is converted into digital form by an analog-to-digital (A/D) conversion circuit. Finally, the digital signal is either displayed on a digital display device or is made available for transmission to other digital instruments (such as recorders or computers) for further processing and display. Note that each of the blocks of this diagram is slated for further discussion later in the book. More specifically, (1) transducers and their operation are described in Chapter 14, (2) amplifiers in Chapter 15 (including the instrumentation amplifiers used in many systems), (3) analog-to-digital conversion techniques and digital display devices in Chapter 5, (4) digital recording devices in Chapter 7, and (5) digital systems, including the proper *interfacing* that must exist between various types of digital instruments and devices, in Chapter 18.

## Digital Data Nomenclature

The output of digital measuring instruments is data represented in a digital format. To introduce the concept of the digital format, consider that the data in this format may be used to activate digital display devices (as shown in Fig. 1-8), thereby indicating the output in the form of a numerical display. For display devices to be designed such that they can respond to the digital data in an appropriate manner, it is first necessary to be able to identify and interpret the particular digital format in which the data are presented.

Most digital data formats are based on signal levels that are restricted to binary values, or states (i.e., only one of two possible values).[7] The two states are represented by the symbols 1 and 0, which are known as the *binary digits* or by the terms TRUE or FALSE. Binary digits are often referred to as *bits* (a contraction of binary digits). TRUE can refer to either 1 or 0 and FALSE would then refer to 0 or 1. The use of TRUE and FALSE is very useful in analyzing propositional logic and active low devices. The digits used in the decimal number system $0, 1, 2, 3, \ldots,$ 9 are known as *decimal digits*. Groups of *bits* are called *bytes, words,* or *strings* with delineators.

In most digital systems the electrical quantity that represents the two binary states is either a voltage or a current possessing one of two discrete amplitudes. For example, the two voltage amplitudes representing the binary digits in a system might be $+5.0$ V $= 1$ and $0.0$ V $= 0$, or $+3.4$ V $= 1$ and $+0.4$ V $= 0$. Actually, in practical digital systems, the voltage values representing 1 and 0 are really voltage ranges. That is, any voltage from 2.4 to 5.0 V would be interpreted as a 1 (nominal voltage 3.4 V $= 1$) and any voltage in the range 0.0 to 0.4 (nominal voltage 0.4 V $= 0$) would be interpreted as a 0 (Fig. 1-9). The ranges are different for each logic type such as TTL, CMOS, I²L, ECL. By using transistors or operational amplifiers, any voltage range can be created and used.

---

[7]A tristate device is also used in many microprocessor-based systems. This device has the two binary states and a third state that is a high impedance state.

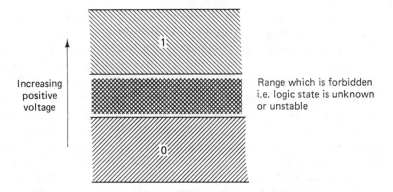

**Figure 1-9**   Logic-level signal range. A voltage level of any value within the 1 range or 0 range is considered as 1 or 0.

     To represent a value of measured data in digital form, a group of such binary digits (bits) must be used. For example, a value such as 25 expressed in decimal form (i.e., using the decimal digits, 2 and 5) could be represented in binary form (i.e., using only bits, 0 and 1) as 11001. Note that the binary number 11001 is composed of 5 bits.

     Electronic digital systems or instruments are typically designed to function by handling data formatted in groups containing a specific number of bits. For example, each decimal digit or alphabetic character may be represented by a group made up of a unique combination of bits. Such groups are commonly known as *digital words*. Digital words usually contain groups of 4, 8, 16, or 32 bits, but sometimes words of other lengths are used. For example, the word for the number 9 in an 8-bit format might be 10001001, while in a 4-bit word format it might be 1001. Eight-bit digital words, because of their widespread use, have acquired their own designation and have come to be known as *bytes*. Note that the left-most bit of a digital word is known as the *most significant bit* (MSB) and the right-most bit is the *least significant bit* (LSB). A graphical representation of a digital signal, showing the variations in logic state versus time, is known as a *digital waveform*. For example, Fig. 1-10 shows

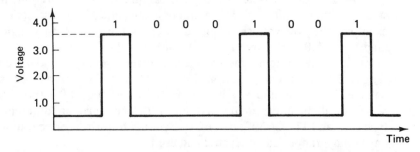

**Figure 1-10**   Representation of the decimal digit, 9, as an 8-bit digital word (or *byte*). Note that "1" = +3.4 V and "0" = +0.4 V.

a digital waveform of the decimal digit 9, represented as an 8-bit binary word, or byte, with logic levels 1 = +3.4 V and 0 = +0.4 V.

It is customary to describe systems in which the binary digit 1 is represented by the more positive of the two bit levels as systems that employ *positive* logic [Fig. 1-11(a)]. In systems where the binary digit 0 is represented by the more positive level, the system is said to employ negative logic [Fig. 1-11(b)]. That is, two examples of *positive* binary logic levels are 1 = +5 V and 0 = −5 V, and 1 = +3.4 V and 0 = +0.4 V. Two examples of *negative* binary logic levels are 1 = −5 V and 0 = +5 V, and 1 = +0.4 V and 0 = +3.4 V.

Digital systems are designed to transfer digital words from one part of the system to another. Such transmission can be done in either a serial or a parallel fashion. In *serial* transmission, one bit at a time of the digital word is sent from one part of the system to the other, and only one signal path is required. In *parallel* transmission, all the bits of the word are transmitted simultaneously, and this requires that there be available an individual signal path (i.e., a wire or conductor) for each bit. Thus a digital system that uses 8-bit words and parallel transmission would require eight parallel signal paths between those parts of the system between which words are transmitted. Sometimes, digital information is transmitted serially in one part of a system and in parallel in another part. In such cases, parallel-to-serial and serial-to-parallel conversion must be used to make the transmission formats compatible.

## Digital Codes

In the preceding section we introduced the concept of expressing a decimal number (e.g., 25) in a binary format (e.g., 11001). There are many other digital formats (which also use binary digits but in modified groupings) in which data can be expressed or *coded*. Each of these specific formats, or *digital codes*, can be likened to a separate digital alphabet. Let us describe three of the most commonly used of such *digital codes*: binary, binary-coded decimal (BCD), and the American Standard Code for Information Interchange (ASCII).

The *binary* format (which we will refer to as straight binary code) consists of numbers expressed by combinations of the two states mentioned earlier, "0" and

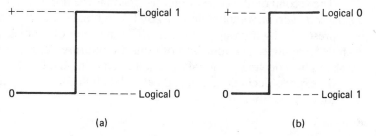

(a)                                         (b)

**Figure 1-11**   Digital systems utilize one of two logic conventions: (a) positive logic; (b) negative logic.

"1." In straight binary code, the decimal number "0" is also represented by "0," and the decimal number "1" is represented by "1." Larger numbers are generated by using the binary counting system, which is founded upon a *base*, or radix, of 2. For example, the binary number $11001_2$ (where the subscript 2 identifies the number as being represented in the binary counting system) is equivalent (reading left to right) to

$$1101_2 = (1 \times 2^4) + (1 \times 2^3) + (0 \times 2^2) + (0 \times 2^1) + (1 \times 2^0)$$

where

$$2^4 = 16, \quad 2^3 = 8, \quad 2^2 = 4, \quad 2^1 = 2, \quad \text{and} \quad 2^0 = 1$$
$$\text{(in the decimal counting system)}$$

Thus

$$11001_2 = 16_{10} + 8_{10} + 0 + 0 + 1_{10} = 25_{10}$$

Table 1-2 shows examples of other simple decimal numbers converted into straight binary form.

**TABLE 1-2    CONVERTING SELECTED DECIMALS**

| | | | |
|---|---|---|---|
| 0 | 0000 | 10 | 1010 |
| 1 | 0001 | 15 | 1111 |
| 2 | 0010 | 16 | 10000 |
| 3 | 0011 | 17 | 10001 |
| 4 | 0100 | 32 | 100000 |
| 5 | 0101 | 64 | 1000000 |
| 6 | 0110 | 128 | 10000000 |
| 7 | 0111 | | |
| 8 | 1000 | | |
| 9 | 1001 | | |

We can see that a group of 4 bits can represent any of 16 decimal numbers (0 through 15), and that it takes progressively larger numbers of bits to represent larger decimal numbers in straight binary code.

The second digital code of our discussion is the popular digital code known as *binary-coded decimal* (abbreviated as BCD). In binary, we saw that any number can be represented by a group of bits. In BCD, a group of 4 bits is used to represent each decimal digit. The four bit positions are interpreted as in binary code. Table 1-3 shows how decimal numerals are expressed using the BCD code.

To express decimal numbers greater than 9 in BCD, a separate 4-bit group is used for each number.

**Example 1-4**

Express the following decimal numbers in binary-coded decimal (BCD): (a) $83_{10}$; (b) $495_{10}$.

**Solution.**

(a) $83_{10} = \overset{8}{1000} \quad \overset{3}{0011}$ in BCD.

(b) $495_{10} = \overset{4}{0100} \quad \overset{9}{1001} \quad \overset{5}{0101}$ in BCD.

**TABLE 1-3**  DECIMAL DIGITS EXPRESSED IN BINARY-CODED DECIMAL (BCD) CODE

| BCD 4-bit set | Decimal digit | BCD 4-bit set | Decimal digit | BCB 4-bit set | Decimal digit |
|---|---|---|---|---|---|
| 0000 | 0 | 0101 | 5 | 1010 | undefined |
| 0001 | 1 | 0110 | 6 | 1011 | undefined |
| 0010 | 2 | 0111 | 7 | 1100 | undefined |
| 0011 | 3 | 1000 | 8 | 1101 | undefined |
| 0100 | 4 | 1001 | 9 | 1110 | undefined |
|  |  |  |  | 1111 | undefined |

Clearly, BCD is a convenient but somewhat wasteful code for representing digital numbers. More bits are required to represent a decimal number than in *binary code*. For example, the digital number $83_{10}$ requires 8 bits in BCD code (1000 0011) but only 7 bits in straight binary code ($1010011_2$). However, the convenience and relative ease of interpreting a number expressed in BCD has made it one of the most widely used of the digital codes (i.e., it is easier to interpret 1000 0011 as being equal to $83_{10}$ than it is to determine the decimal value of $1010011_2$).

At this point it is useful to define the terms *encode* and *decode*. We pointed out that most electronic digital systems are based on voltage or current signals that assume one of the two binary states "logic 0" and "logic 1." The systems must therefore perform any arithmetic operations using binary arithmetic. Human beings, however, are accustomed to working with decimal-formatted numbers. Thus we frequently want to be able to convert decimal-formatted numbers into digital formats. The process of converting decimal-formatted numbers (and other symbols, including alphabetic characters) into the various digital formats is known as *encoding*. The electronic digital systems can then operate on such encoded digital numbers. When it is required to obtain the output in decimal form, the digital-formatted numbers are *decoded* into the decimal format.[8] Specific digital electronic devices called encoders and decoders are designed to perform such operations. For example, a *decimal-to-BCD encoder* would convert decimal-formatted numbers that were fed into its inputs to numbers in the BCD format at its output. A BCD-to-decimal decoder, on the other hand, would reverse the process and present information in decimal format at its output. In Chapter 5 we show how decoders are used to operate the decimal display of a digital instrument.

[8]Note that, in general, *to encode* means to change information from one code to another, whereas *to decode* means to reverse the previous encoding. Occasionally, these more general definitions are meant.

As noted in the footnote, it is not necessary to restrict the encoding and decoding process to decimal digits. It is also important to encode other symbols, such as the characters of the English alphabet; symbols such as $+$ , $-$ , $*$, $\#$, $\$$, $\%$, $!$, and @; and certain keyboard operations (for use with teletypewriters, etc.) such as "carriage return," "back space," "shift," and "space." A popular "alphanumeric" digital code exists for encoding these symbols (including the alphabet, the decimal numbers, 28 symbols, and a variety of keyboard control functions). It is known as the *American Standard Code for Information Interchange (ASCII)*. It was originally developed to allow encoding of numbers, alphabetic characters, and symbols on a teletypewriter. Seven bits of binary information are used to encode each of the characters. Table 1-4 is an ASCII code table. It lists the 128 seven-bit ASCII code words, including the 26 alphabetical characters, the 18 decimal digits, and the 28 symbols. The remaining 64 ASCII code words shown are keyboard operations (with the abbreviations given in the lower table) or lowercase alphabetic characters. Other alphanumeric digital codes include the *EBDCIC* (Extended Binary Decimal Interchange Code) and *Selectric* code. Both are used predominantly on IMB machines and are somewhat similar to ASCII. There are numerous codes used by industry. When digital data is in a continuous string (for example from a shaft encoder) the *reflected* code is used, since only 1 bit changes from one number to the next.

### Combinational and Sequential Logic Circuits; Timing Diagrams

The electronic devices that operate on digital signals and transmit them from one part of a digital system to another are known as *gates*. Gates, which are the building blocks of digital circuitry, are usually constructed from transistors, diodes, resistors, and so on, connected in specific circuit configurations to perform the required digital logic operations. Digital logic circuits built with such gates are designed to respond to digital input signals according to the rules of *Boolean algebra*. The topics of digital systems design and digital electronics are concerned with a study of Boolean algebra and its application to the analysis and design of digital electronic systems. For our discussion it is sufficient to mention that there are two broad classes of digital logic circuits: combinational logic circuits and sequential logic circuits or algorithmic state machines.

In *combinational logic circuits*, if a given combination of digital logic levels is simultaneously applied to the inputs of the digital circuit, a specific response will occur directly at the outputs of the circuit. In *sequential logic circuits*, the element of time is introduced into the system. For a desired output, not only must the correct combination of digital signals with proper logic levels appear at the digital circuit inputs, but the logic levels must appear at the inputs in the right order with respect to time (much like the procedure that must be followed when a combination of numbers is used to open a safe). A vending machine is a typical example of a state machine.

Most digital instruments and systems utilize the sequential (time-dependent)

**TABLE 1-4** ASCII CODE[a]

| Bits 3–6 | | 0000 | 0001 | 0010 | 0011 | 0100 | 0101 | 0110 | 0111 | 1000 | 1001 | 1010 | 1011 | 1100 | 1101 | 1110 | 1111 |
|---|---|---|---|---|---|---|---|---|---|---|---|---|---|---|---|---|---|
| Bits 0–2 | | 0 | 1 | 2 | 3 | 4 | 5 | 6 | 7 | 8 | 9 | A | B | C | D | E | F |
| 000 | 0 | NUL | SOH | STX | ETX | EOT | ENQ | ACK | BEL | BS | HT | LF | VT | FF | CR | SO | SI |
| 001 | 1 | DLE | DC1 | DC2 | DC3 | DC4 | NAK | SYN | ETB | CAN | EM | SUB | ESC | FS | GS | RS | US |
| 010 | 2 | SP | ! | ” | # | $ | % | & | ' | ( | ) | * | + | , | - | . | / |
| 011 | 3 | 0 | 1 | 2 | 3 | 4 | 5 | 6 | 7 | 8 | 9 | : | ; | < | = | > | ? |
| 100 | 4 | @ | A | B | C | D | E | F | G | H | I | J | K | L | M | N | O |
| 101 | 5 | P | Q | R | S | T | U | V | W | X | Y | Z | [ | \ | ] | ∧ | — |
| 110 | 6 | ` | a | b | c | d | e | f | g | h | i | j | k | l | m | n | o |
| 111 | 7 | p | q | r | s | t | u | v | w | x | y | z | { | ¦ | } | ~ | DEL |

[a] Symbols and abbreviations are as follows:

| Row/column | Symbol | Name | Control character | Function |
|---|---|---|---|---|
| 2/0 | SP | Space (normally non-printing) | NUL | Null |
| 2/1 | ! | Exclamation point | SOH | Start of heading |
| 2/2 | ” | Quotation marks | STX | Start of text |
| 2/3 | # | Number sign | ETX | End of text |
| 2/4 | $ | Dollar sign | EOT | End of transmission |
| 2/5 | % | Percent | ENQ | Enquiry |
| 2/6 | & | Ampersand | ACK | Acknowledge |
| 2/7 | ' | Apostrophe (closing single quotation mark; acute accent) | BEL | Bell (audible or attention signal) |
| | | | BS | Backspace |
| 2/8 | ( | Opening parenthesis | HT | Horizontal tabulation |
| 2/9 | ) | Closing parenthesis | | (punch card skip) |
| 2/A | * | Asterisk | LF | Line feed |
| 2/B | + | Plus | VT | Vertical tabulation |
| 2/C | , | Comma (cedilla) | FF | Form feed |
| 2/D | - | Hyphen (minus) | CR | Carriage return |
| 2/E | . | Period (decimal point) | SO | Shift out |
| 2/F | / | Slant | SI | Shift in |
| 3/A | : | Colon | DLE | Data link escape |
| 3/B | ; | Semicolon | DC1 | Device control 1 |
| 3/C | < | Less than | DC2 | Device control 2 |
| 3/D | = | Equals | DC3 | Device control 3 |
| 3/E | > | Greater than | DC4 | Device control 4 (Stop) |
| 3/F | ? | Question mark | NAK | Negative acknowledge |
| 4/0 | @ | Commercial at | SYN | Synchronous idle |
| 5/B | [ | Opening bracket | ETB | End of transmission block |
| 5/C | \ | Reverse slant | CAN | Cancel |
| 5/D | ] | Closing bracket | EM | End of medium |
| 5/E | ∧ | Circumflex | SUB | Substitute |
| 5/F | — | Underline | ESC | Escape |
| 6/0 | ` | Grave accent (opening single quotation mark) | FS | File separator |
| | | | GS | Group separator |
| 7/B | { | Opening brace | RS | Record separator |
| 7/C | ¦ | Vertical line | US | Unit separator |
| 7/D | } | Closing brace | DEL | Delete |
| 7/E | ~ | Overline (tilde, general accent) | | |

logic circuits. Thus it needs to be emphasized that the proper functioning of sequential circuits is critically dependent on the timing relationships between activities occurring in the digital system. For example, if an 8-bit word is to be transferred serially over a data line in a sequential system, specific timing relationships within the system must be maintained. That is, if the data bits do not appear on the line at exactly the right times, they are likely to be irrevocably lost and will no longer be transferred to the following steps in the operation.

The tools employed to illustrate the specific timing relationships required by sequential digital circuits are the *timing diagram, state diagram*, or the *algorithmic state machine* (ASM) *chart*. These show various signals in the digital circuit as a function of time or state. When constructing timing diagrams, several variables are usually plotted against the same time scale so that the times at which variables change with respect to each other can easily be observed.

For example, it may take a finite time for the output signal of a gate to react to a change in its input signal. This indicates that the change in the gate output signal is delayed with respect to the input signal change. Figure 1-12 is a simple timing diagram that demonstrates the delay associated with the output signal of a gate known as an *inverter*. It shows that the output signal change is delayed by time $\epsilon_1$ when the input signal changes from 0 to 1, and by $\epsilon_2$ when the input signal changes from 1 to 0.

Sequential digital circuits are themselves divided into two further subclasses of circuits, *clocked circuits* and *unclocked circuits*. Clocked sequential circuits are usually referred to as *synchronous circuits*, and unclocked sequential circuits are known as *asynchronous circuits*.

In all *synchronous circuits*, a master oscillator (known as the *clock*) is used to provide a train of regular timing (or *clock*) pulses. The clock pulses of the system control the operation of all the elements of the system. The only times during which events in the system are permitted to occur are during one of these clock pulses.

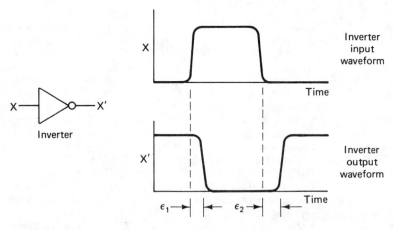

**Figure 1-12**   Timing diagram to illustrate the propagation delay in an inverter.

(For example, the transfer of data from one point in the circuit to another can occur only within the duration of a clock pulse.) The circuit is inert during all other times. A personal computer is an example of a synchronous machine.

In *asynchronous sequential circuits*, events can occur immediately after the previous event has been completed. That is, it is not necessary to await the arrival of another clock pulse before initiating the next event. (An example of an asynchronous system is a Touch-Tone telephone circuit. The user can key-in another number as soon as the previous number has been registered.) There must, however, be some method in asynchronous systems to inform the transmitting and receiving digital devices that the transmission of a digital word has been properly started and completed. A discussion of the various techniques employed to perform this task properly belongs to the subject of *interfacing* between digital circuits and instruments. Therefore, a more complete discussion of the subject is given in Chapter 18, which deals with digital instruments and their interfacing requirements.

## PROBLEMS

1. How many coulombs are represented by the following numbers of electrons?
   (a) $51.6 \times 10^{18}$
   (b) $1.0 \times 10^{19}$
   (c) $7.55 \times 10^{15}$

2. Calculate the force (in pounds) that exists between a positive charge of 0.4 C and a negative charge of 1.2C, 30 cm apart.

3. A copper penny has a mass of 3.1 g and contains about $2.9 \times 10^{22}$ atoms. Let us assume that one electron is removed from each atom and these electrons are all removed from the penny to a distance such that the force of attraction between the positively charged penny and the removed group of electrons is 0.7 N. How far apart must the penny and the group of electrons be in such a case?

4. (a) If 50 J of energy is required to move 5 C of charge from infinity to some point in space designated as point A, what is the potential difference between infinity and point A?
   (b) If 24 additional joules are required to move the 5 C from point $A$ to point $B$, what is the potential difference between point $A$ and point $B$?

5. If the potential difference between two points in an electric circuit is 40 V, how much work is required to move 15 C of charge from one point to the other?

6. List three different possible locations at which the electrostatic potential of a point can be defined as having a value of zero.

7. A body attached to earth ground by an electrical conductor is at zero electrical potential. A charge of 25 C is added to the body. What is the potential of the body now? Explain.

8. What is meant by the term "free electron" in reference to electrons in material bodies? Explain why some materials are conductors of electricity whereas others are not.

9. If $40.847 \times 10^{18}$ electrons pass through a wire in 13 s, find the current that was flowing in the wire during that time.

10. How many electrons pass through a conductor in 10 min if a steady current of 15 mA flows during that time period?

**11.** If the current flowing in a conductor is 5 mA, how much time is required for 0.0064 C to pass through the conductor?

**12.** For the waveform shown in Fig. P1-1,

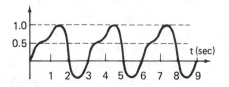

Figure P1-1

(a) Find the period $T$.

(b) How many cycles of the waveform are shown?

(c) What is the frequency of the waveform?

(d) What is its amplitude?

**13.** (a) Draw two sine waves and label one of them *current* and the other *voltage*.

(b) Draw the voltage waveform with an amplitude of twice the current waveform but with the same frequency.

(c) Draw the waveforms on the same set of axes and have the voltage waveform lead the current waveform by 45°.

**14.** Repeat Problem 13, but draw the current waveform leading the voltage waveform by 180°.

**15.** Convert the following numbers of degrees to radians.

(a) 30°

(b) 90°

(c) 240°

**16.** Find the angular velocity in radians per second of the waveforms that have the following frequencies.

(a) 60 Hz

(b) 500 Hz

(c) 0.05 MHz

**17.** Find the amplitudes and frequencies of the following waveforms.

(a) $40 \cos 377t$

(b) $-63 \sin (14t)$

(c) $0.015 \sin (800t)$

**18.** Find the average value of the current waveform shown in Fig. P1-2.

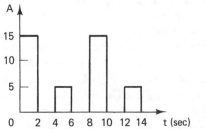

Figure P1-2

**19. (a)** Find the average value of the waveform shown in Fig. P1-3.

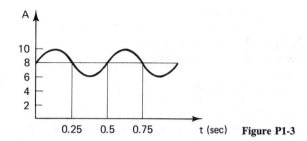

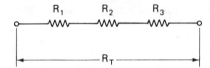

Figure P1-3

**(b)** Find the average value of this waveform over the first half-cycle.
**(c)** Find the rms value of the entire waveform, and
**(d)** Find the rms value of the waveform over the first half of one cycle of the waveform.

**20.** A light bulb is used in a 20-V dc system. What rms value of ac voltage is necessary to have the bulb light as brightly as it does when powered by the dc system?

**21.** If two or more resistors are connected as shown in Fig. P1-4, they are said to be connected in *series*. The total series connection has a resistance given by

$$R_T = R_1 + R_2 + \cdots + R_n$$

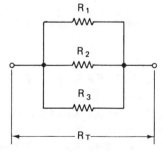

Figure P1-4

If the values of the resistors in Fig. P1-4 are $R_1 = 35 \ \Omega$, $R_2 = 720 \ \Omega$, and $R_3 = 15 \ \Omega$, what is $R_T$?

**22.** If two or more resistors are connected as shown in Fig. P1-5, the connection is called a *parallel* connection. The resistance of the total parallel connection is given by

$$\frac{1}{R_T} = \frac{1}{R_1} + \frac{1}{R_2} + \cdots + \frac{1}{R_n}$$

Figure P1-5

If the values of the resistors in Fig. P1-5 are $R_1 = 35 \ \Omega$, $R_2 = 720 \ \Omega$, and $R_3 = 5 \ \Omega$, what is the value of $R_T$?

**23.** Find $R_T$ for the connection of resistors shown in Fig. P1-6.

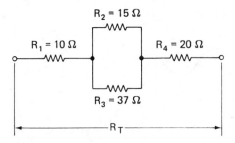

**Figure P1-6**

**24.** Express the following decimal numbers in **(a)** binary form, **(b)** binary-coded-decimal form, and **(c)** ASCII form.
   **(1)** 60
   **(2)** 111
   **(3)** 511

**25.** Express the following binary numbers in decimal form.
   **(a)** 10101
   **(b)** 1001010
   **(c)** 11011110

**26.** Express the following BCD-formatted numbers in decimal form.
   **(a)** 1001 0001 0111 1000
   **(b)** 0111 1000 0001 0010
   **(c)** 1001 0111 0111 0101

**27.** Why is the binary system so useful for electronic data processing applications?

**28.** Define the following terms.
   **(a)** Bit
   **(b)** Byte
   **(c)** To encode
   **(d)** Clock
   **(e)** Positive logic

**29.** Explain the difference between combinational digital circuits and sequential digital circuits.

**30.** Explain the difference between synchronous and asynchronous sequential digital circuits.

## REFERENCES

1. Halliday, D., and Resnick, R., *Physics*. New York: John Wiley., 1962.
2. Hostetter, G., *Fundamentals of Network Analysis*. New York: Harper & Row, 1980.
3. Hayt, W., and Kemmerly, J., *Engineering Circuit Analysis*. New York: McGraw-Hill, 1986.
4. Fletcher, W., *An Engineering Approach to Digital Design*. Englewood Cliffs, N.J.: Prentice Hall, 1980.
5. Gothmann, W., *Digital Electronics*. Englewood Cliffs, N.J.: Prentice Hall, 1977.

6. Olenick, R., Apostol, T., and Goodstein, D., *Beyond the Mechanical Universe*. New York: Cambridge University Press, 1986.
7. Mano, M., *Digital Design*. Englewood Cliffs, N.J.: Prentice Hall, 1984.
8. Tocci, R., *Digital Systems Principles and Applications*. Englewood Cliffs, N.J.: Prentice Hall, 1985.

# 2

# *Experimental Data and Errors*

Measurements play an important role in substantiating the laws of science. They are also essential for studying, developing, and monitoring many devices and processes. However, the process of measurement itself involves many steps before it yields a useful set of information. To study the methods that will produce effective measurement results, let us consider the measurement process as a sequence of five operations. These operations can be listed as follows:

1. Designing an efficient measurement setup. This step includes a proper choice of available equipment and a correct interconnection of the separate components and instruments.
2. Intelligent operation of the measurement apparatus.
3. Recording the data in a manner that is clear and complete. The recorded information should provide an unambiguous record for future interpretation.
4. Estimating the accuracy of the measurement and magnitudes of possible attendant errors.
5. Preparing a report that describes the measurement and its results for those who may be interested in using them.

All five of these items must be successfully completed before a measurement is truly useful.

The first two items of this list are the legitimate concerns of the remainder of our book. The latter three are discussed to some extent in this chapter. The material presented here is only a short introduction to these topics; however, for many measurement applications, this level of presentation should be adequate. If some situations are encountered where more details are required, the reader can consult other references dealing more extensively with the subject. Some of these are listed at the end of the chapter.

## MEASUREMENT RECORDING AND REPORTING

The *original data sheet* is a most important document. Mistakes can be made in transferring information, and therefore copies cannot have the validity of an original. If disputes arise, the original data sheet is the basis from which they are resolved (even in courts of law). Thus it is an excellent practice to label, record, and annotate data carefully as they are taken. A short statement at the head of the data sheet should explain the purpose of the test and list the variables to be measured. Items such as the date, wiring diagrams used, equipment serial numbers and models, and unusual instrument behavior should all be included. The measurement data themselves should be neatly tabulated and properly identified. (All this should emphasize the fact that jotting down data on scrap papers and trusting the memory to record data are not acceptable procedures for recording data. Such bad habits will certainly lead to the eventual loss of valuable pieces of information.) In general, the records of the experiment on the data sheet should be complete enough to specify exactly what was done and, if need be, to provide a guide for duplicating the work at a later date.

The report presented at the end of a measurement should also be carefully prepared. Its objective is to explain what was done and how it was accomplished. It should give the results that were obtained, as well as an explanation of their significance. In addition to containing all pertinent information and conclusions, the report must be clearly written with proper attention to spelling and grammatical structure. To aid in organizing the report and avoid omitting important information, an outline and rough draft should always be used. The rough draft can later be polished to produce a concise and readable document.

The form of the report should consist of three sections:

1. Abstract of results and conclusions
2. Essential details of the procedure, analysis, data, and error estimates
3. Supporting information, calculations, and references

In industrial and scientific practices, the abstract is likely to be read by higher-level managers and other users who are scanning reports for possible information contained in the report body. The details, on the other hand, will usually be read by those needing specific information contained in the report or by others wanting to duplicate the measurement in some form. The latter groups will be interested in the details on the data sheets, the analysis of the level of accuracy, and the calculations and results that support the conclusions and recommendations. For these readers, the references from which source material and information were obtained should also be provided.

The results and conclusions of the report form its most important parts. The measurement was made to determine certain information and to answer some specific questions. The results indicate how well these goals were met.

## GRAPHICAL PRESENTATION OF DATA

Graphical presentation is an efficient and convenient way of portraying and analyzing data. Graphs are used to help visualize analytic expressions, to interpolate data, and to discuss errors. This section provides a brief introduction to proper methods of graphing data.

Graphs should always contain a title, the date the data was taken, and adequately labeled and scaled axes. An effective way to detect experimental errors is to plot the theoretical data before an experiment is performed. Then, as the experiment is performed, the actual data are compared with the predicted data. That is, the data are plotted at the same time they are taken, not afterward. In many instances, the experiment can be stopped and examined as soon as any major discrepancies between the predicted and the actual data are noted. In addition, any unexpected data points can be rechecked before the experiment is dismantled.

When deciding whether to plot the independent or dependent variable along the ordinate (vertical axis) or the abscissa (horizontal axis), the usual convention is to plot the dependent variable (i.e., the controlled parameter, such as the input voltage) on the abscissa, and the independent variable (i.e., the measured response to the controlled parameter) on the ordinate. Base your final decision on what you are trying to show the reader. Consider what will best illustrate the information in an unambiguous manner. Clarity of the presentation is very important. The data may be presented, for instance, in the form of a bar graph, lines, or a pie chart, depending upon what you want to be emphasized.

The data points on the graph are typically shown as small circles, the diameter of which can be proportional to the estimated error of the readings. Sometimes, however, I-bars are used to plot data points. In such cases, the top and bottom of the bar should indicate the approximate errors of that data point. When it is necessary to plot more than one line on a graph, symbols, such as squares, diamonds, or triangles, can be used to differentiate the data points associated with each line.

Once the data are plotted, a smooth curve needs to be drawn that best fits the data. When the data points appear to lie in a straight line, the best fit is usually a line that has data points lying as closely as possible on either side of the line. A useful technique for locating the "best-fit" line is the *method of least squares*. A detailed description of how to apply this method is found in References 1 and 2 cited at the end of the chapter. There are many computer programs that will automatically plot the data and perform statistical analysis on the data such as curve fitting.

## Types of Graph Paper

In addition to Cartesian graph paper (in which both the abscissa and ordinate have linear scales), there are several other graph paper types commonly used, including semilog, log-log, and polar.

*Semilog* paper has one logarithmic axis and one linear axis (Fig. 2-1). It is especially valuable when one is faced with the task of graphing exponential functions or functions that have one widely ranging parameter. Such functions are "expanded" in the low range and "compressed" in the high range by the logarithmic scale. An example is the output voltage of a wide band-pass filter versus frequency.

When an exponential function, $y = Ae^{x\alpha}$ (with $A$ and $\alpha$ being constants), is graphed so that $y$ is plotted on the log axis and $x$ on the linear axis, a straight line is the result. The value of $\alpha$ can be determined from the "slope" of the line, and $A$ is found from the value of $y$ at $x = 0$ (Fig. 2-1). The slope of the line on semilog paper, $(\Delta y/\Delta x)_s$, is, however, not the same as the slope in Cartesian coordinates, $(\Delta y/\Delta x)_c$.

Thus, to determine $\alpha$ from the slope of the line on semilog paper $(\Delta y/\Delta x)_s$, the following procedure is used. Since $\alpha$ is the amount by which the variable $x$ must change in order to create a change in $y$ by a factor of $e = 2.718$, the appropriate $\Delta y$ to use in $(\Delta y/\Delta x)_s$ is the linear distance on the $y$ axis that corresponds to $e$ on the log scale. An easy way to find the appropriate $\Delta y$ distance is to note the distance from 1.0 to 2.718 (or 0.1 to 0.2178) on the log scale (see Fig. 2-1). Then $\alpha$ will be equal to the corresponding distance $\Delta x$ on the linear scale. The sign of $\alpha$ is positive or negative, depending on the sign of the slope of the line (as defined in the ordinary manner). An example is the characteristic curve of a thermistor, which is

$$R_T = R_0\, e^{\,\beta(1/T - 1/T_0)} \tag{2-1}$$

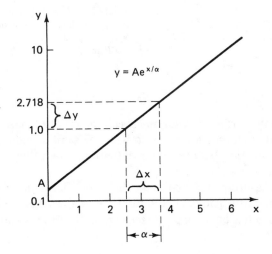

**Figure 2-1** Semilog graph with a two-cycle logarithmic axis, illustrating methods of determining $A$ and $\alpha$.

The value of β can be found mathematically by taking the natural logarithm of both sides of the equation and solving for β.

$$\beta = \frac{\ln\left(\dfrac{R_T}{R_0}\right)}{\left(\dfrac{1}{T} - \dfrac{1}{T_0}\right)} \tag{2-2}$$

Use caution when extrapolating logarithmic data beyond actual experimental data. For example, β in Equation 2-1 is constant only for a narrow range of temperatures.

Semilog paper is available with the number of cycles in the logarithmic axis ranging from 1 to 5.

Graph paper on which both axes are scaled logarithmically is called *log-log* paper (see Fig. 2-2). For functions in which both independent and dependent parameters vary over a wide range, log-log paper serves best.

If the functional form $y = Bx^\beta$ is plotted on log-log paper, and $B$ and $\beta$ are constants, a straight line will result. The value of $B$ is determined from the value of $x = 1$ (see Fig. 2-2). To determine β, the "slope" of the log-log plotted line is used. In this case, however, β is found from the ratio of the total linear length along the $y$ axis for a decade change in $x$ ($\Delta y$ in Fig. 2-2), divided by the linear length of one decade change in $y$ ($\Delta z$ in Fig. 2-2). The sign of β is found from the sign of the slope of the line in the ordinary manner.

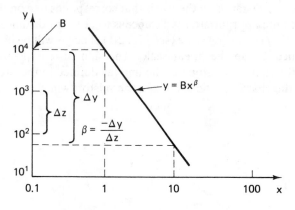

**Figure 2-2** Log-log graph (with four-cycle logarithmic variation on both $x$ and $y$ axis) illustrating methods of determining $B$ and β.

## PRECISION AND ACCURACY

In measurement analysis the terms *accuracy* and *precision* are often misunderstood and used incorrectly. Although they are taken to have the same meaning in every-day speech, there is a distinction between their definitions when they are used in descriptions of experimental measurements.

The *accuracy* of a measurement specifies the difference between the measured and the true value of a quantity. The deviation from the *true value* is the indication of how accurately a reading has been made. *Precision*, on the other hand, specifies the repeatability of a set of readings, each made independently with the same instrument. An estimate of precision is determined by the deviation of a reading from the *mean* (average) *value*. To illustrate the difference between accuracy and precision more graphically, consider an instrument that has a defect in its operation. The instrument may be giving a result that is highly repeatable from measurement to measurement, yet far from the true value. The data obtained from this instrument would be highly precise but quite *inaccurate*. It may also occur that two instruments produce readings which are equally precise, but which differ in accuracy because of differences in the instrument design. Such examples emphasize that *precision does not guarantee accuracy*, although accuracy requires precision.

The concept of *accuracy*, when applied to instruments that display a reading by the use of a scale and pointer, usually refers to their full-scale reading (unless otherwise specified). When a meter is said to be accurate to 1 percent, this means that a reading taken anywhere along one of its scales will not be in error by more than 1 percent of the full-scale value.

### Example 2-1

A voltmeter is specified as being accurate to 1 percent of its full-scale reading. If the 100-V scale is used to measure voltages of (a) 80 V and (b) 12 V, how accurate will the readings be (assuming all other errors besides the meter reading error are negligible)?

**Solution.** Since the meter is accurate to within 1 percent of its full-scale value, any reading taken will be accurate to $1\% \times 100$ V $= 1$ V. Thus the error of the 80-V reading will be $80 \pm 1$ V. The possible percent error is

$$\text{percent error} = \frac{\text{true value} - \text{measured value}}{\text{true value}} \times 100\%$$

$$= \frac{80 - 79}{80} \times 100\% \simeq 1.25\%$$

The error of the meter while the 12-V measurement is made can still be $\pm 1$ V. Then the possible percent error is

$$\text{percent error} = \frac{12 - 11}{12} \times 100\% \simeq 8\%$$

Example 2-1 shows that the use of a small segment of a meter's scale to make a reading can result in a larger measurement error than if a greater segment of the scale is utilized. In Example 2-1, a meter with a smaller full-scale voltage than 100 V could have been used to reduce the error of the 12-V measurement.

Digital accuracy is usually specified as plus or minus a percentage of the reading plus a number of counts of the least significant digit [e.g., $\pm 0.05$ percent full scale (F.S.) $\pm 1$ count]. The accuracy may change as different ranges are selected. There is also usually a specific temperature range over which the manufacturer guarantees the stated accuracy. Some manufacturers state the accuracy of their instruments by using the following format: The accuracy (at $25° \text{ C} \pm 5° \text{ C}$) is 1 percent of full scale $\pm 1$ LSD on the 200 and 2000 pf ranges.

## RESOLUTION AND SENSITIVITY

*Resolution* is the significance of the least significant digit (LSD). For example, the range of a capacitance meter might be given as 199 pf, with a resolution of 0.1 pf. The meter range would be from 000.0 to 199.9 pf, and it would be referred to as a $3\frac{1}{2}$ digit meter. (The $\frac{1}{2}$ indicates that the most significant digit (MSD) can only be either a 0 or 1.) A $\frac{3}{4}$ (e.g., a $3\frac{3}{4}$ meter) would indicate that the MSD can only be either 0, 1, 2, or 3. Defining resolution as the smallest incremental quantity that can be measured with certainty is not followed by all instrument manufacturers. Adding resolution to a meter may or may not be of any value depending on the sensitivity and the end use of the meter. For example, it is possible to indicate the speed of an automobile to three decimal places (e.g., 55.5 mph), but the average driver would not consider the additional resolution useful. Consider the case of providing an instrument with even a higher resolution capability. At some point (e.g., if a six-decimal place resolution is provided) there is even doubt that the last one or two digits have any real physical meaning. In many instruments with such resolution, the final digits would probably be responding to random noise rather than to actual changes in the input.

    *Sensitivity* is the smallest incremental change that the meter can detect. This does not mean that the smallest detectable change has to be displayed for the user. For example, if a meter had the capability to detect a 1 rpm change in a motor that is driving a varying load, the two least significant digits of the display would nevertheless normally be forced to indicate zero, so that a person reading the meter would see stable digits.

    Sensitivity is sometimes alternatively expressed as the *ratio* of the incremental change in the output for an incremental change in the input. *Threshold* is another performance characteristic of a meter that is also related to sensitivity. It is the *smallest change* in the input that will result in a change in the output. Some instruments exhibit enough *hysteresis* to cause them to indicate different output values for the same input value, depending upon the direction in which the input value was approached. (The second law of thermodynamics prohibits perfect reversibility, and, hence, all instruments have some degree of hysteresis.) It is usually expressed as a percent of full scale output.

    Recently, some manufacturers of indicating instruments have been changing the method of specifying scale errors as a result of design improvements in the meters. Scale errors are sometimes being stated as a *percentage of a reading* rather than as a percentage of the full-scale value. Instruments designed to satisfy this type of accuracy rating will eliminate some of the errors possible in the older-style instruments.

    Determining the errors of instrumentation systems are more complicated than determining the error of individual instruments. To illustrate this idea, consider an instrumentation system used to test the strength of steel lifting cables. Before manufacturers sell such cables, they must test them to make sure that their actual breaking strength (in tension) is greater than the value stated in the manufacturers' sales literature. To measure the breaking strength, the cable is installed in a fixture that stretches it (i.e., applies tension to it). The instrumentation system consists of a tensiometer (a device that converts a tension signal into a voltage), a signal conditioner that amplifies the voltage signal from the tensiometer, an analog-to-digital converter, and a digital meter (see Fig. 2-3). The accuracy of the digital tensile force reading is affected by all the components in the system, including the power supplies. In addition, the test is performed outdoors, which means humidity and temperature also influence the accuracy of the output reading. To complicate matters further, each device is manufactured by a different corporation, which sets its own accuracy standards and presentation format.

A statistical approach is used to estimate the system error in the digital output reading. To determine the worst case error, all the individual errors are added (although it should be noted that an error this magnitude seldom occurs because the individual errors are random in nature). A test fixture is also usually constructed for the specific purpose of calibrating the instrumentation system in addition to allowing each individual component to be calibrated.

## Powers of 10 and Their Abbreviations

Both very large and very small numbers are often used in measuring and expressing electrical quantities. It is usually more convenient and more precise to express these numbers in terms of their powers of 10 rather than to write the whole numbers. By

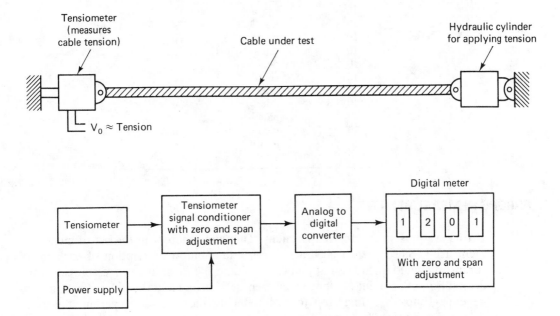

**Figure 2-3**   Fixture for testing steel cables.

utilizing this procedure, we can clearly state the exact number of significant digits of a quantity. In addition, we can avoid the use of many zeros when dealing with both large and small numbers. The following example shows how to express numbers by using the powers-of-10 notation.

**Example 2-2**

$$1,390,000 = 1.39 \times 10^6$$

$$0.000032 = 3.2 \times 10^{-5}$$

Certain standard prefixes and symbols are used to denote particular multipliers. They are shown in Table 2-1. Examples of how these abbreviations are used in connection with electrical units are given below.

**Example 2-3**

$$10,000 \text{ ohms} = 10 \text{ kilohms} = 10 \text{ k}\Omega$$

$$1.0 \times 10^{-6} \text{ farad} = 1 \text{ microfarad} = 1 \text{ } \mu\text{F}$$

$$\tfrac{1}{1000} \text{ ampere} = 1 \text{ milliampere} = 1 \text{ mA}$$

**TABLE 2-1**    POWERS OF 10

| Multiplier | Prefix | Abbreviation |
|------------|--------|--------------|
| $10^{12}$  | tera   | T            |
| $10^9$     | giga   | G            |
| $10^6$     | mega   | M            |
| $10^3$     | kilo   | k            |
| $10^2$     | hecto  | h            |
| 10         | deka   | da           |
| $10^{-1}$  | deci   | d            |
| $10^{-2}$  | centi  | c            |
| $10^{-3}$  | milli  | m            |
| $10^{-6}$  | micro  | $\mu$        |
| $10^{-9}$  | nano   | n            |
| $10^{-12}$ | pico   | p            |
| $10^{-15}$ | femto  | f            |
| $10^{-18}$ | atto   | a            |

## ERRORS IN MEASUREMENT

Errors are present in every experiment. They are inherent in the act of measurement itself. Since perfect accuracy is not attainable, a description of each measurement should include an attempt to evaluate the magnitudes and sources of its errors. From this point of view, an awareness of errors and their classification into general groups is a first step toward reducing them. If an experiment is well designed and carefully performed, the errors can often be reduced to a level where their effects are smaller than some acceptable maximum. Figure 2-4 classifies the

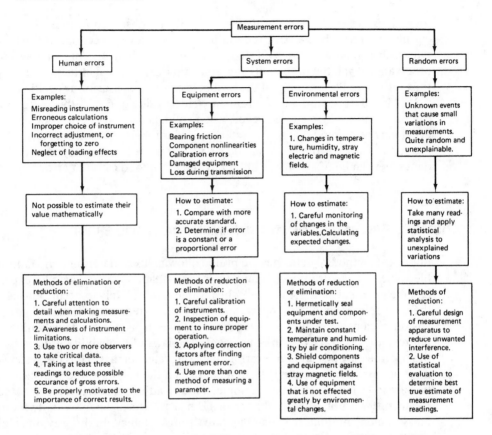

**Figure 2-4** Measurement errors: how to estimate, reduce, or eliminate them.

main categories of errors and describes some causes and methods of correcting them.

Errors that are inherent to the specific type of instrument being used for making a measurement are described in more detail in the relevant sections of chapters devoted to such instruments.

Sometimes a specific reading taken during a measurement is rather far from the mean value. If faulty functioning of the measurement instruments is suspected as the cause of such unusual data, the value can be rejected. However, even such data should be retained on the data sheet (although they should be labeled as suspect data). Nevertheless, even when all the items involved in a measurement setup appear to be operating properly, unusual data may still be observed. We can use a guide to help decide when it is permissible to reject some suspect data. This guide is derived by using the methods of statistical evaluation of errors that will be covered in the following sections. It is stated as follows: Individual measurement readings taken when all the instruments of a measurement system appear to be operating properly may be rejected when their deviation from the average value is four times larger than the *probable error* of one observation (the procedure for calculating the probable error is described in the following section). It can be shown that such a random error will not occur in more than one out of one hundred observations, and the probability that some unusual external influence was at play is thus very high. For example, if the probable error in measuring a specific voltage of 5.21 V is ±0.21 V, and a measurement yields a value of 6.2 V, this piece of datum can probably be rejected. However, when an unusually large error does occur, this event may be a signal that some systematic error is being committed. An attempt to locate the cause of the error should be undertaken. Keeping rejected data in the data sheet can be of assistance in finding the extent and cause of error.

## STATISTICAL EVALUATION OF MEASUREMENT DATA AND ERRORS

Statistical methods can be very helpful in allowing us to determine the most proba-ble value of a quantity from a limited group of data. That is, given an experiment and the resulting data, we can tell which value is most likely to occur. Furthermore, the probable error of one observation and the extent of uncertainty in the best answer obtained can also be determined. However, a statistical evaluation cannot improve the accuracy of a measurement. The laws of probability utilized by statis-tics operate only on random errors and not on system errors. Thus the system errors must be small compared to the random errors if the results of the statistical evalua-tion are to be meaningful. For example, if a zero adjustment is incorrect, a statistical treatment will not remove this error. But a statistical analysis of two different measurement methods may demonstrate the discrepancy. In this way, the measure-ment of precision can lead to a detection of inaccuracy. We will now show how the following quantities can be calculated using statistics:

1. Average or mean value of a set of measurements
2. Deviation from the average value
3. Average value of the deviations
4. Standard deviation (related to the concept of rms)
5. Probability of error size in one observation

   1. *Average or mean value.* The most likely value of a measured quantity is found from the arithmetic average or mean (both words have the same definition) value of the set of readings taken. Of course, the more readings taken, the better will be the results. The average value is calculated from

$$a_{av} = \frac{a_1 + a_2 + \cdots + a_n}{n} \qquad (2\text{-}3)$$

where

$$a_{av} = \text{average value}$$
$$a_1, a_2, a_3, \ldots = \text{value of each reading}$$
$$n = \text{number of readings}$$

**2.** *Deviation from the average value.* This number indicates the departure of each measurement from the average value. The value of the deviation may be either positive or negative.

**3.** *Average value of the deviations.* This value will yield the precision of the measurement. If there is a large *average deviation*, it is an indication that the data taken varied widely and the measurement was not very precise. The average value of the deviations is found by taking the absolute magnitudes (i.e., disregarding any minus signs) of the deviations and computing their mean.

**4.** *Standard deviation and variance.* The average deviation of a set of measurements is only one of the methods of determining the dispersion of a set of readings. However, the average deviation is not mathematically as convenient for manipulating statistical properties as the *standard deviation* (also known as the root-mean-square of rms deviation). Although the difference between the average and the standard deviation cannot be completely appreciated at our level of presentation, the fact remains that the standard deviation is a much more useful statistical quantity. As such, it is used almost exclusively in expressing dispersions of data. The standard deviation is found from the formula

$$\sigma = \sqrt{\frac{d_1^2 + d_2^2 + d_3^2 + \cdots + d_n^2}{n-1}}$$

where

$$\sigma = \text{standard deviation}$$
$$d_1, d_2, d_3, \ldots = \text{deviation from the average value}$$
$$n - 1 = \text{one less than the number of measurements taken}$$

The variance $V$ is the value of the standard deviation $\sigma$ squared:

$$V = \sigma^2$$

**5.** *Probable size of error and Gaussian distribution.* If a *random* set of errors about some average value is examined, we find that their frequency of occurrence relative to their size is described by a curve (Fig. 2-5) known as a Gaussian curve (or bell-shaped curve). Gauss was the first to discover the relationship expressed by this curve. It shows that the occurrence of small random deviations from the mean value are much more probable than large deviations. In fact, it shows that large deviations are extremely unlikely. The curve also indicates that random errors are equally likely to be positive or negative. If we use the standard deviation as a measure of error, we can use the curve to determine what the probability of an error greater

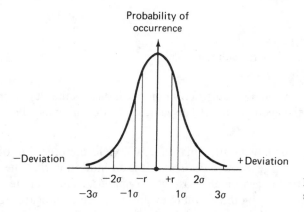

**Figure 2-5**  Error size in terms of standard deviations.

than a certain $\sigma$ value will be for each observation. Table 2-2 shows the probability of an error occurring greater than a specific $\sigma$ value for each observation.

**6.** *Probable error.* From Table 2-2 we can also calculate the probable error that will occur if only one measurement is taken. Since a random error can be either positive or negative, an error greater than $|0.675\sigma|$ is probable in 50 percent of the observations. Hence, the probable error of one measurement is

$$r = \pm 0.675\sigma$$

**Example 2-4**

Given the following set of current measurements taken from an ammeter, find their (a) average value, (b) average deviation, (c) standard deviation, and (d) probable error.

Data:    153 mA    162 mA    157 mA    161 mA    155 mA

**Solution.**

(a)  Average value: $I_{av} = \dfrac{I_1 + I_2 + I_3 + I_4 + I_5}{5} = \dfrac{785}{5} = 157$ mA

(b)  Average deviation: $D = \dfrac{|d_1| + |d_2| + |d_3| + |d_4| + |d_5|}{5}$

$$= \frac{4 + 5 + 0 + 4 + 2}{5} = 3 \text{ mA}$$

(c)  Standard deviation: $\sigma = \sqrt{\dfrac{d_1^2 + d_2^2 + d_3^2 + d_4^2 + d_5^2}{4}} = \sqrt{15} \approx 3.8$ mA

(d)  Probable error: $r = 0.675\sigma \approx 2.6$ mA

**TABLE 2-2**

| Error ($\pm$) (standard deviations) | Probability of error being greater than given $+\sigma$ or $-\sigma$ value in one observation |
|---|---|
| 0.675 | 0.250 |
| 1.0 | 0.159 |
| 2.0 | 0.023 |
| 3.0 | 0.0015 |

Random errors associated with components in a system (such as the steel cable tester described earlier) can be treated statistically. Specifically, the variance of the sum of the errors $\sigma^2$ output equals the sum of the individual variances.

$$\sigma^2_{output} = \sigma^2_{transducer} + \sigma^2_{signal\ conditioner} + \sigma^2_{A/D} + \sigma^2_{meter} \qquad (2\text{-}4)$$

When the ratio of the standard deviation of a component to the gain of a component is compared to the ratio of the system standard deviation to the system gain, the result is

$$\frac{\sigma_{component}}{k_{component}} = \sqrt{n}\ \frac{\sigma_{system}}{k_{system}} \qquad (2\text{-}5)$$

where $k$ = gain
$n$ = the number of components or subsystems

For example, when there are $n$ transducers that are used to supply an output signal, for example a truck scale that has four load cells, the random error in individual load cells can be twice the system error. That is, the random error of a load cell should not exceed $\sqrt{n}$ times the allowable system error if ideal error populations that are centered around the mean are assumed. This means that the individual component errors of the cable tension tester or a four load cell scale can also be twice the permissible system error because there are four components (subsystems). When systematic errors are present, this method cannot be used. If a signal exists in the presence of noise, sampling techniques can increase the sign-to-noise ratio. This occurs because random noise will tend to cancel out, and only the nonrandom signal will remain as the number of samples is increased.

## THE DECIBEL

Occasionally, when measuring a quantity such as power, it may not be as important to know the absolute value of the quantity as it is to know its value relative to some other quantity. For instance, we might want to know the ratio of the power developed in one part of a circuit to the power in another part. Also, it is often easier to measure this ratio rather than the absolute value itself. For example, one prime quantity of interest in an amplifier is the *power gain* (defined as the ratio of the power at its output to the power at its input, $P_{out}/P_{in}$). A logarithmic scale can be used to describe such ratios quite conveniently and the one that is commonly employed to do it is called the *decibel* scale ("bel" after Alexander Graham Bell.) The ratio in decibels of two power values $P_2$ and $P_1$ is defined as

$$G = 10 \log_{10}\left(\frac{P_2}{P_1}\right) \qquad \text{decibels} \qquad (2\text{-}6)$$

and the result is expressed as the difference in decibels (dB). For example, when there is a difference of 1 dB in power, $P_2/P_1 = 1.26$. A 3-dB difference corresponds to $P_2/P_1 = 2$. A 10-dB difference indicates that the power ratio is also 10. Figure 2-6 is a chart that shows how a ratio of powers corresponds to positive decibel (dB) values (power *gain*). If the power of $P_2$ is smaller than that of $P_1$, there is a power *loss* and the dB value is negative. The use of decibel scales is especially advantageous when we want to display graphically the change in the value of ratios that go from a very small to a very large quantity. For example, the ratio of $P_{out}$ to $P_{in}$ of an amplifier may vary from much less than one to many thousand as the frequency of its input signal changes. To describe such a change on a graph, a logarithmic scale is more useful and compact than a linear scale.

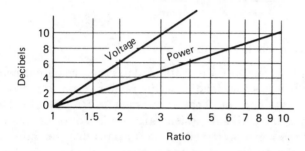

**Figure 2-6**  Chart showing ratio of decibels to power and voltage.

In addition to being used to describe power ratios, the decibel has also come to be used in describing voltage and current ratios or gains. Voltage and current are more frequently measured than power, and their ratios can also be expressed logarithmically. Strictly speaking, the decibel scale for expressing the ratios of voltages is valid only if the input and load impedances of the circuit being measured are equal ($R_i = R_L$). (See Chapter 3 for a discussion of input and output impedances.) In this case, from the definition of gain in decibels,

$$G = 10 \log_{10}\left(\frac{P_o}{P_i}\right) = 10 \log_{10}\left(\frac{V_o^2}{R_L} \times \frac{R_i}{V_i^2}\right) = 20 \log_{10}\left(\frac{V_o}{V_i}\right) \qquad (2\text{-}7)$$

However, common usage has changed the correct meaning as defined above to the definition that $G_v = 20 \log_{10} (V_o/V_i)$ regardless of whether $R_i$ and $R_L$ are equal. Under such circumstances, it is not always possible to convert the voltage gain to a power gain. $G_v$ is called the *voltage gain* in decibels.

When a certain power level is chosen as a standard reference, the ratio of a measured power to this level can also be described in decibels. For example, in telephone circuit use, the milliwatt ($1 \times 10^{-3}$ W) is used as the standard power reference level. Thus a 10-mW signal is referred to as +10 dBm (i.e., +10 dB referred to the milliwatt standard). Since 600 Ω is the standard impedance level of telephone circuits, voltage can also be expressed in decibels. Many voltmeters have calibrated scales in decibels for use in such applications. A reading of 0 dBm on such a scale corresponds to a voltage of 0.775 V (this number is derived from the voltage existing at 1 mW and 600 Ω).

# PROBLEMS

1. Describe in your own words the difference between the terms *accuracy* and *precision* as they are used in reference to experimental measurements.

2. What are the three general classes of measurement errors?

3. List five different specific errors that frequently occur in the process of making measurements.

4. A voltmeter whose accuracy is guaranteed to within 2 percent of its full-scale reading is used on its 0–50 V scale. The voltage measured by the meter is 15 and 42 V. Calculate the possible percentage error of both readings.

5. A 0- to 50-mA ammeter has an accuracy of 0.5 percent. Between what limits may the actual current be when the meter indicates 13 mA?

6. Make the following conversions.
   (a) 0.35A to milliamperes and microamperes
   (b) 0.041 mV to microvolts and volts
   (c) 400,000 $\Omega$ to megohms
   (d) 73 $\mu$V to millivolts and volts

7. State the number of significant digits in each of the following numbers.
   (a) 0.35                                (b) 0.041
   (c) 400,000                             (d) 73
   (e) 14.5                                (f) .000029

8. Four resistors are connected in series. The values of the resistors are 14.5 $\Omega$, 5.32 $\Omega$, 64.377 $\Omega$, and 0.43 $\Omega$, with an uncertainty of one unit in the last digit of each value. Calculate the resistance of the total connection. (See Chapter 1, Problem 20 if you need to know the formula for calculating the resistance of a series connection.)

9. Subtract $296 \pm 4$ from $635 \pm 4$, and express the uncertainty that exists in the answer as a percentage of the answer.

10. Ten measurements of current in a circuit branch yield values of 50.2, 50.6, 49.7, 51.1, 50.3, 49.9, 50.4, 49.6, 50.3, and 51.0 mA. Assume that only random errors were present in the measurement system. Calculate
    (a) The average value
    (b) The standard deviation of the readings
    (c) The probable error of the readings

11. The following voltage values are listed on a data sheet as the values obtained from measuring a certain voltage: 21.45, 21.74, 21.66, 19.07, 21.53, and 21.19 V. By examining the numbers, calculate
    (a) The average value
    (b) The probable error
    If only random errors are present, how does one treat the 19.07 value?

12. Calculate the decibel power gain for the following input and output powers.
    (a) $p_o = 75$ mW, $p_i = 5$ mW          (c) $p_o = 50$ mW, $p_i = 25$ mW
    (b) $p_o = 45$ $\mu$W, $p_i = 10$ $\mu$W          (d) $V_o = 0.707$ V, $V_i = 1$ V

13. Find the magnitude power gain corresponding to a decibel power gain of $+40$ dB.

14. Find the magnitude power gain corresponding to a decibel power gain of $-13$ dB.

**15.** The input power to a circuit is 15,000 W at a voltage of 1000 V. The power output is 500 W and the output impedance is 30$\Omega$.
   **(a)** Find the power gain in decibels
   **(b)** Find the voltage gain in decibels (use the relation $P = V^2/R$ to find $P_o$)
   **(c)** Explain why the results of parts (a) and (b) agree and disagree
**16.** An amplifier rated at 30 W output is connected to a speaker whose impedance is 10 $\Omega$.
   **(a)** If the power gain of the amplifier is +32 dB, what is the input power required to obtain the full output from the amplifier?
   **(b)** If the voltage gain of the amplifier is 40 dB, what is the required input voltage if the amplifier is to produce its rated output?

# REFERENCES

1. Goodwin, H. M., *Elements of the Precision of Measurements and Graphical Methods*. New York: McGraw-Hill, 1913.
2. Burrows, W. H., *Graphical Techniques for Engineering Computations*, chap. 2 and 6. New York: Chemical, 1965.
3. Coombs, C., ed., *Basic Electronic Instrument Handbook*. New York: McGraw-Hill, 1972.
4. Tuve, G. L., and Domholdt, L. C., *Engineering Experimentation*, chap. 2. New York: McGraw-Hill, 1966.
5. Cooper, W. D., and Helfrick, A. *Electronic Instrumentation and Measurement Techniques*, chap. 1 and 2. Englewood Cliffs, N.J.: Prentice Hall, Inc., 1985.
6. Young, H. D., *Statistical Treatment of Experimental Data*. New York: McGraw-Hill, 1962.
7. Richards, J. W., *Interpretation of Technical Data*. Princeton, N.J.: D. Van Nostrand, 1967.
8. Lannon, J. M., *Technical Writing*, 4th edi. Glenview, Ill.: Scott, Foresman, 1988.
9. *In Tech*, "Understanding Digital Panel Meter Specifications," p. 55, Feb. 1988.
10. R. J. Larson, and Marx, M. L., *An Introduction to Mathematical Statistics and its Applications*, p. 178. Englewood Cliffs, N.J.: Prentice Hall, 1986.

# 3

# *Electrical Laboratory Practice*

In addition to electrical measuring instruments and the body of terms and symbols used to describe them, there are a number of laboratory techniques and associated concepts that are uniquely related to electrical measurements. Sometimes these techniques involve the direct application of a physical principle (such as impedance matching); at other times they are more in the realm of an art (as the methods used to eliminate external interference and *ground loops*).

Usually, a student is exposed to such laboratory techniques indirectly and learns about them through a kind of osmosis. Often, this type of learning is not completely effective. There is still a need for reinforcing the student's intuitive knowledge by a physical understanding of the processes that underlie the techniques. The objective of this chapter is to provide this reinforcement by examining some of these techniques from a physical point of view.

The techniques discussed in this chapter include electrical safety and electric shocks, the laying out of circuits, and the use of cables, switches, and circuit protection devices. Additional concepts of a similar nature, including impedance matching, input impedance, and the loading effects of measuring devices, are also covered. External interference signals and their elimination from measurement systems are discussed in Chapters 15 and 16.

Even an introductory discussion will often put us into a position of using concepts that have not yet been defined in the text. Unfortunately, this problem cannot be avoided. If the material were to be placed toward the end of the book after all the necessary concepts had been developed, there would be a greater likelihood of it being overlooked by the reader. Since the information contained in this chapter is quite essential for making proper electrical measurements, and since such information is often hard to find in other texts, it is important that the reader be aware of its presence in this book. Thus, in the interests of maximum exposure, the chapter appears near the beginning. Once the reader has finished the remainder of the book, he should be able to reread this chapter and understand it with greater clarity.

## SAFETY

When working in the electrical laboratory or when using electrical equipment, observing proper safety precautions is as important as making accurate measurements. Potentially lethal hazards exist in the electrical laboratory environment, and failure to follow careful safety procedures can make you or a fellow worker the victim of a serious accident. The best way to avoid accidents is to recognize their causes and carefully adhere to well-established safety procedures. A full awareness of the dangers and the possible consequences of accidents helps develop the proper motivation for following such procedures.

The most common and serious hazard of the electrical laboratory is electric shock. Other hazards that should also be recognized include dangerous chemicals, moving machinery, and soldering irons.

### Electric Shock

When electric current is passed through the human body, the effect that it causes is called *electric shock*. Electric shock can accidentally occur owing to poor equipment design, electrical faults, human error, or a combination of unfortunate circumstances. The lethal aspect of electric shock is a function of the amount of current that is forced through the human conducting path and time. It is not necessarily dependent on the value of the applied voltage. A shock from 100 V may turn out to be as deadly as a shock from 1000 V.

The severity of an electric shock varies somewhat with the age, sex, and physical condition of the victim. But, in general, the level of current required to kill any human is remarkably small. For this reason, extreme care must always be exercised to prevent electric shock from occurring.

The threshold of perception of current in most humans is about 1 mA. The sensation caused by this current level takes the form of an unpleasant tingling or heating at the point of contact. Currents above 1 mA but below 5 mA are felt more strongly, but usually do not produce severe pain. However, current levels of 1 to 5 mA can still be dangerous because of the startling reaction they may cause. For example, a shock from such currents might lead one to jump back against a hot oven or a moving piece of machinery or to fall off a ladder, thereby causing an injury. (Note that 5 mA is the maximum current allowed to leak from home appliance circuits to their causes and still be able to pass the Underwriters Laboratory specifications.)

At levels above 10 mA, current passing through the body begins to cause involuntary muscular contractions. Owing to these spasms, the victim loses the ability to control muscles. Even though the pain is severe, the victim is unable to release the grip on the electrical conductor being held. For this reason, such a current level is called "can't let go" current. If it is sustained, "can't let go" current can lead to fatigue, collapse, and even death.

If the current level flowing in the body exceeds 100 mA, it begins to interfere with the coordinated motion of the heart.[1] This *fibrillation* prevents the heart from pumping blood, and death will occur in minutes unless the fibrillation is stopped. Above 300 mA, the heart's muscular contractions are so severe that fibrillation is prevented. If the shock is halted quickly enough, the heart will probably resume a normal rhythm. In such cases, breathing may have stopped, and artificial respiration may have to be applied. If proper first aid is provided, the shock may not be fatal, even though severe burns may have resulted. (In fact, a method of administering large current pulses to fibrillating hearts is used to restore them to their normal rhythm.)

From this discussion we can see that the current that passes through the skin and then through the body is most lethal in the range between 100 and 300 mA. (Note that 100 mA is about one-tenth of the current flowing in a 100-W lamp.) Figure 3-1 summarizes the effects of current on the human body if it enters along a path through the skin.

By using indirect methods (i.e., by applying electric currents to laboratory animals), investigators have determined that current can be fatal in 1 out of 20 cases if it exceeds the value given in Equation 3-1:

$$i(\text{mA}) = 116/\sqrt{t(\text{seconds})} \qquad\qquad (3\text{-}1)$$

For example, this equation predicts that a current of 100 mA must flow for approximately 1.3 seconds before it is probabilistically fatal. Therefore, just brushing against an energized conductor may be unpleasant but probably not fatal. Direct current voltages are extremely dangerous because dc current penetrates body muscles and nerves even more easily than 60 Hz current. This results in deeper and more severe flesh burns. Large electric switch gear (the industrial equivalent of the circuit breaker panel of a residential building) is very dangerous because of the large amount of energy that can be delivered by its conductors in case of a *fault* (e.g., by the accidental shorting of two conductors or of one conductor to ground). Electricians have been severely injured after being literally blown 20 feet as the result of

---

[1] Note that there is also an important relationship between the frequency of the applied current and the minimum current required for fibrillation. As bad luck would have it (or as Murphy's law dictates), the condition of maximum human susceptibility to electric current effects occurs at about 60 Hz, the common power-line frequency. At higher frequencies the susceptibility to fibrillation decreases rapidly because such high-frequency ac current changes direction faster than the heart tissue can respond.

Effects of 60 Hz Electric Shock (current) on an average human through the body trunk

| Current intensity — 1 second contact | Effect |
|---|---|
| 1 milliampere | Threshold of perception |
| 5 milliamperes | Accepted as maximum harmless current intensity |
| 10–20 milliamperes | "Let-go" current before sustained muscular contraction |
| 50 milliamperes | Pain. Possible fainting, exhaustion, mechanical injury, heart and respiratory functions continue. |
| 100–300 milliamperes | Ventricular fibrillation will start but respiratory center remains intact |
| 6 amperes | Sustained myocardial contraction followed by normal heart rhythm. Temporary respiratory paralysis. Burns if current density is high. |

**Figure 3-1** Effects of various current levels on the human body. (Courtesy of Hewelett-Packard Co.)

arcing faults. Currents from such faults can be large enough to vaporize the fluids in a limb, resulting in a vapor explosion within the limb.

The voltage required for a fatal current level to flow in the human conducting path can vary. Its value depends on the skin resistance at the point of contact. Wet skin may have a resistance as low as 1 kΩ, whereas dry skin may have as much as 500 kΩ resistance. (Once the current passes into the body, the resistance is much less, owing to the conductivity of body fluids.) Thus, a 100-V potential applied to wet skin can be fatal. In fact, even 50 V under certain circumstances can be as deadly as 5000 V. Furthermore, skin resistance falls rapidly as current passes through the point of contact because the current breaks down the protective, dry, outer-skin layer. This makes it important to break the contact with the live conductor as soon as possible. Since the voltage at the point of contact usually remains constant, and since the resistance decreases, the current can soon rise to a lethal level.

In many medical applications, however, electronic sensing devices (catheters) are inserted directly into the body (along arteries or veins) to monitor such physical phenomena as blood flow. These devices are in intimate contact with the body fluids, and, therefore, only a very small resistance between the sensing device and the patient exists. Under such circumstances it would not be unusual to expect very small currents (entering the body through a catheter) to lead to electric shock effects. Clinical studies have in fact shown that currents as small as 180 μA (0.18 mA) from electrodes placed directly on each side of the heart can trigger ventricular fibrillation. Thus electric currents much smaller than 100 to 300 mA can be expected to be lethally hazardous to patients if they are connected to electrical medical equipment. Since leakage currents in the hospital environment may inadvertently reach such dangerous levels, special care and safety measures must be utilized to protect medical patients from electric shock accidents. Some of the techniques used to protect medical patients against such shock hazards are discussed in the sections of Chapter 16 that deal with the elimination of external electric interference signals and ground loops in electronic measurement systems.

The best method for protecting oneself from the hazard of shock when using ordinary (i.e., nonmedical) electrical equipment and instruments is to rely on proper grounding of the equipment employed. The details of how and why to ground equipment properly are given in a later section entitled "Equipment Grounding for Safety." In addition to good grounding techniques, one should avoid handling equipment that has exposed wires or conductors. Always try to shut off power when touching any circuits. Furthermore, always wear shoes to insulate yourself further from ground. Avoid coming in contact with such grounds as metal plumbing while handling the wires or instruments. If "hot" equipment must be repaired, use only one hand, keeping the other far away from any part of the circuit. Do not wear metal, rings, bracelets, or wristwatches when working with electrical systems.

### First Aid for Electric Shock

The first step in aiding a victim of electric shock is to try and shut off the power to the conductor with which the victim is in contact. If the attempt is not successful and the victim is still receiving a shock, break the contact of the victim and the source of electricity without endangering yourself. Do this by using an insulator (such as a piece of dry wood, rope, cloth, or leather) to pull or separate the victim and the live conductor. Do not touch the victim with bare hands as long as the victim is electrified. (Even momentary contact with the victim can be fatal if the current level is sufficiently high.) The contact must be broken quickly because skin resistance falls rapidly with time and a fatal current of 100 to 300 mA can be reached if the shock is allowed to continue long enough.

If breathing has stopped and the individual is unconscious, start giving artificial respiration immediately. Do not stop until a medical authority pronounces the victim beyond further help. This may take up to 8 hours. Symptoms of rigor mortis and the lack of a detectable pulse should be disregarded, because these are sometimes results of the shock. They are not necessarily proof that the victim has expired.

### Other Hazards of the Electrical Laboratory

When using power tools such as drills or saws, care must also be taken to prevent serious injury. Power tools should not be operated unless instructions on how to operate them have been received. In addition, loose clothing or long hair which could become caught in moving machinery should not be worn when such equipment is being operated. Finally, always wear goggles or safety glasses when drilling or cutting with power tools.

The soldering iron is another instrument that can cause accidents if used carelessly. Unattended hot irons can burn unsuspecting workers and may set fire to the surroundings. To prevent soldering-iron accidents, always replace the iron into its holder when not soldering. Also, make sure to turn off soldering irons after use.

When using cleaning solvents (such as trichlorethylene) or corrosive chemicals (such as acids in semiconductor laboratories), care must be exercised in their use and disposal. Well-ventilated fume hoods must be used when working with these chemicals to dispose of the corrosive or poisonous chemical fumes. Gloves, special clothing, and goggles should be used to protect against chemical splattering and contamination. When the corrosive chemicals are dumped into sinks, a large volume of water should be allowed to flow after them to dilute their harmful properties. In case of acid spills, flush the sink with copious amounts of water.

### Safety Rules

**1.** Never work alone. Be sure there are others in the laboratory to summon and provide aid in case of accidents.

**2.** Use only instruments and power tools provided with three-wire power cords. (See the section "Equipment Grounding for Safety.")

**3.** Always shut off power before handling wiring.

**4.** Check all power cords for sign of damage. Replace or repair damaged cords and leads.

**5.** Always wear shoes. Keep shoes dry. Avoid standing on metal or damp concrete. (All these precautions prevent you from becoming a low-impedance path to ground.) Do not wear metal, rings, etc.

**6.** Never handle electrical instruments when your skin is wet (the moisture decreases your skin's resistance and allows a greater current to flow through you).

**7.** Hot soldering irons should not be left unwatched. Keep hot soldering irons in holders when not soldering.

**8.** Never wear loose clothing around machinery. Always wear safety goggles when using chemicals or power tools.

**9.** Always connect a cable or lead to the point of high potential as the last step. This is, do not connect the lead to the "hot" side of a circuit first, or you will end up holding a "hot" connector in your hand.

## GROUNDS

### Importance of Grounds

The concepts of *ground* and *grounding* are basic and integral concepts utilized in the design of electrical measurement systems. For proper operation of such systems, these concepts should be well understood. However, grounds are often not clearly defined during a student's training. As a result, he may end up working with measurement systems that are not properly grounded. If such situations lead to erroneous measurements, a consequent confusion may also develop as to why the error exists. To keep such problems from arising, a discussion concerning some of the basic principles of grounds is presented in the following sections.

## Grounding

Since all measurements of potential difference (voltage) are relative, the voltage level of any point in a circuit must always be compared to some reference level. This means that there must be a voltage level at one point that is defined as the reference voltage. Usually, this reference level is assigned a voltage value of zero and is known as the *circuit ground* or *common point* of the system.

To provide one common and convenient reference potential for the majority of measurements, the potential of the earth (the planet) was chosen as zero. When a conductor or a circuit is connected at some point to the earth by a low-impedance electrical connection, that point will be at essentially the same potential as the earth (zero). The conductor or circuit is then said to be *earth grounded, earthed*, or *grounded*. (In the electrical systems of buildings, this ground point is usually referred to as the *service entrance ground*.) In actuality, all such earth grounds, however, are not necessarily at exactly the same potential. Even grounds within a building may not be at the same potential. If indeed two such grounds develop a difference in potential, a large circulating ground current may be established. Such a ground current can be very destructive to equipment connected to the ground line of this electrical system (especially computer and data communication equipment, which operate at high frequencies). Grounding problems of this kind are therefore usually the result of having more than one service entrance ground installed in an industrial complex.

The resistance to earth ground for a grounding system can be determined by using the method shown in Fig. 10-6. The National Electric Code requires that the maximum resistance of a grounding electrode shall not exceed 25 ohms. To reduce the resistance to ground in electrical utility substations, a copper ground mat is installed under the entire substation. Many computer installations use raised floors that have a ground plane directly under the entire floor. Such a ground plane, in contrast to a ground cable, provides a low resistance ground path for high frequency currents. For buildings, steel columns and concrete reinforcing bars make excellent ground electrodes.

Now note that the *circuit ground* mentioned at the outset of the discussion may be an earth ground, or it may simply be a point in the circuit to which all other voltages are referred without being connected to earth ground. For example, a flashlight may operate from a 6-V battery. The *circuit ground* of its system is not connected to earth and may or may not be at zero potential with respect to earth. However, the positive terminal of the battery is always at 6 V relative to the flashlight circuit ground. Other examples of such non-earth grounded circuits include the automobile and the airplane. For the electrical systems of these machines, the *circuit ground* may be the metal body of the automobile or the fuselage of the airplane. In such cases, the chassis takes the place of the earth in serving as a zero potential level. When the chassis acts as the zero reference potential, the system is said to be *chassis grounded*. Note that the chassis voltage may be many volts above that of earth ground and yet provide a zero reference level for its own internal circuits. When a circuit is connected to a chassis that is deliberately disconnected from earth ground, the circuit is said to be *floating*. (This condition is often specifically created when a power supply or oscilloscope is used for making certain types of voltage measurements.) A potential can exist between the chassis and earth grounds, and if a conducting path is connected between them, a current will flow. The conducting path could be a human being; thus *floating* equipment must be handled as if it were at some higher potential to avoid electric shock.

This discussion emphasized the fact that the term *ground* can have distinctly different meanings, all related closely enough to occasionally cause confusion. A *circuit ground* is the most general definition because it can be an *earth ground*, a *chassis ground*, or neither (it may just be a convenient point in the circuit to which other voltages can be referred). One must clearly determine which meaning is the relevant one each time the term is encountered. The symbol ⏚ is often used interchangeably to refer to all types of grounds. Sometimes, however, if there is no earth ground connection, the symbol ⌁ is used to denote a floating chassis ground.

Connections to earth ground are constructed by burying or driving conductors into the earth. Such connections are effective if they provide a very low resistance path to earth ground.

In the common three-wire wall outlet from which electric power is obtained in the home, there are two wires connected to ground (Fig. 3-2). Wire 2 is connected to ground and carries the return current from the load back to ground. It is called the *neutral* and the color used for it is white. Wire 3 is a non-current–carrying wire under normal operating conditions. That is, its purpose is to supply a low resistance path back to the service entrance panel. The resistance must be low enough so that if a fault occurs, it will actuate the protective device of the circuit. If the electrical path to ground exhibited a high resistance, an arcing fault could be established that would literally melt anything in the vicinity of the fault, including metal. In house-

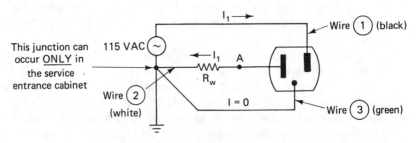

**Figure 3-2**   Three-wire wall outlet.

hold wiring, wire 3 is usually a bare copper wire, but it sometimes has green insulation, especially in three-wire appliances. (Wire 3, as we will describe in more detail in the section on "Equipment Grounding for Safety," is used as a protective measure.) Wire 1 is not connected to ground but is connected to the terminal of higher potential of the ac source; its color designation is any color other than white or green. It is usually black, red, or blue.

A current flows in wires 1 and 2 when an appliance or instrument is connected to the outlet. Since all wire contains some resistance per foot of length, point $A$ on wire 2 will not be at zero potential, even though it is connected to ground. The potential of the wire at point $A$ will instead be

$$V_A = R_w I_1 \tag{3-2}$$

where $R_w$ is the total resistance of the path from point $A$ to ground. (The sum of the resistance of the wire and the resistance of the contact to ground determines $R_w$.) This means that, although wire 2 is connected to ground, it is not really a true point of zero reference.

A common example of a failure to observe proper grounding techniques involves the use of measuring equipment connected to earth ground through the third (ground) wire of the three-wire power cord [e.g., an oscilloscope that has one input terminal connected to the chassis (and the chassis is connected to earth ground)]. For example, consider the voltage measurement being attempted in Fig. 3-3. Since one of the input terminals of the scope is grounded, an attempt to measure the nongrounded voltage between points $A$ and $B$ of the circuit results in short-circuiting point $B$ to ground. This short circuit effectively eliminates the remainder of the elements from the circuit. Thus the voltage value measured by the scope is erroneous because the circuit has been drastically altered by the connection of the scope.

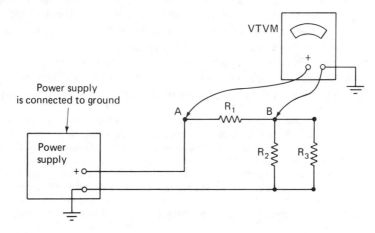

**Figure 3-3** Example of incorrect grounding technique. By connecting the VTVM into the circuit as shown, point $B$ becomes grounded.

A method that allows the scope to be used for the measurement of non-grounded voltages is to *float* the scope. Two ways of doing this include (1) the use of a battery-powered oscilloscope and (2) the use of an oscilloscope possessing a differential input. On the other hand, use of the (*A-B*) mode of a dual-trace oscilloscope to float a scope can be potentially hazardous, as is described further in the section "How to Operate an Oscilloscope" in Chapter 6.

## Equipment Grounding for Safety

Ultimately, the most important reason for grounding electrical equipment is to provide additional protection against electrical shocks. Electrical instruments and household appliances are built so that their equipment cases (also called the *chassis*) are electrically isolated from the wires that carry power to their circuits. The isolation is provided by the insulation of the wires and the chassis is thereby prevented from becoming electrically "hot."

If the chassis of the equipment somehow comes into contact with an exposed part of one of the current-carrying wires (possibly due to wear or damage of the wire insulation), it will possess the same potential above the earth ground as the wire with which it is in contact. If there is no good electrical connection from the chassis to earth ground, the chassis will remain "hot." The unaware user may simultaneously touch the "hot" chassis and earth ground and be subjected to an electric shock.[2] Such an accident can occur if the appliance or equipment utilizes a two-wire power cord. In such cases, both wires carry current when the equipment is in normal operation. There is no wire available for grounding the chassis in case of accidental electrical contact [Fig. 3-4(b)].

On the other hand, if there is a good connection from the chassis to ground and an exposed wire touches the chassis, the current can flow directly to ground through a very low resistance path. This low-resistance path usually offers less impedance to current flow than the conducting path through the appliance. A large current flow in the circuit is the result. Such a large current surge should cause the circuit fuse to burn out or the circuit breaker to open.

Another device used to protect against the hazard of electric shock is the *ground-fault circuit interrupter* (GFCI). A *ground fault* is a current leak that occurs when either energizing conductor (*hot* or *neutral*) contacts the frame or chassis of an

---

[2]Note that leakage currents are always present between the circuitry and chassis of electrical equipment even when the equipment is operating properly and the power wires are accordingly isolated from the chassis. Such leakage currents exist because the materials used to isolate the current-carrying conductors from the chassis are never perfect insulators and because capacitive coupling exists between the current-carrying conductors and the chassis. Thus even if no apparent ohmic conducting paths between the circuitry and the chassis are in evidence, a potential difference (voltage) with respect to ground always exists on the chassis. Of course, the leakage currents flowing to the chassis are supposed to be kept so small in well-designed, normally operating units that the resulting voltage on the chassis is harmless.

The Underwriters Laboratory (UL) performs testing on many consumer appliances to verify that such leakage current levels are indeed below dangerous levels.

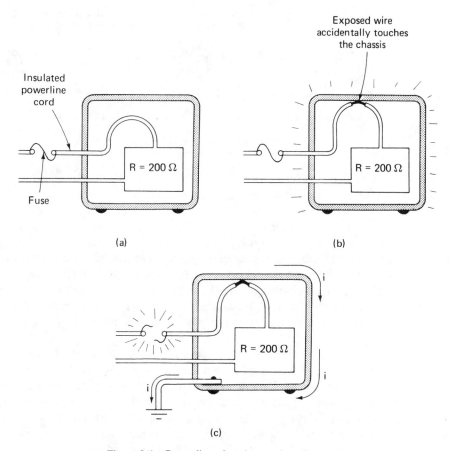

**Figure 3-4**  Grounding of equipment for safety.

electrically powered device. As previously discussed, if the *hot* wire faults to a *grounded* chassis and the resistance of path to ground is low, the resultant heavy current flow is likely to burn out a fuse or trip a circuit breaker and thus prevent damage to equipment. However, if the path to ground is through a higher resistance (such as the human body) the current flow may not be great enough to trip a breaker, but it may be sufficient to kill.

Furthermore, if a *neutral* conductor faults to ground, the chassis may still be a source of dangerous fault currents. This is because in a neutral wire-to-chassis ground fault, the fault current will probably not be interrupted by the common overload devices (fuse, etc.). Thus the problem may remain undetected.

*Ground-fault circuit interrupters* (GFCIs) are designed to automatically shut off power to a piece of equipment in less than one-fortieth of a second if a difference (i.e., an imbalance) between the hot conductor current and neutral conductor current is detected. The existence of such an imbalance implies the presence of a ground fault (possibly through the conductive path of a human being).

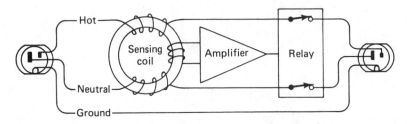

**Figure 3-5** Ground-fault circuit interrupter.

The sensing device of the GFCI is a magnetic toroid (sensing coil) around which several turns of the hot and neutral conductors are wrapped (Fig. 3-5). If the currents in the hot and neutral conductors are equal, each will produce an equal magnetic field in the toroid. These fields will cancel one another and no net magnetic field will exist in the toroid. If the currents in the hot and neutral conductors are not equal (imbalanced), a net magnetic field will be present in the toroid. This field will be detected by a third *sensing winding*, which is also wrapped about the toroid. The signal resulting from the detected magnetic field is amplified and used to activate relays which are connected in the power circuit supplying the equipment. Activation of the relay causes the power being supplied to the equipment to be interrupted.

Ground-fault circuit interrupters were originally developed to protect against electric shocks in and around swimming pools, but they are now installed wherever users of electrical equipment are operating on well-grounded work areas (such as damp ground or steel scaffoldings). Modern electrical code regulations also require that GFCIs be installed in the circuits that furnish power to the bathrooms, kitchens, and exterior power outlets of new buildings.

Ground-fault currents of only 0.0001 percent of the load current can be detected by GFCIs, and they are suitable for use in both two-wire and three-wire equipment. However, they do not protect against shock from line-to-line contacts, nor are they meant to replace fuses or circuit breakers. They are meant merely to complement other safety devices as an added measure of protection.

The leading cause of electrocutions in homes is caused by appliances that fall into water. Such accidents are therefore not caused by a faulty appliance. For example, severe shock, which is often fatal, can result if a hairdryer is dropped into an occupied bathtub. The same results can occur if a plugged-in coffee pot is immersed in a sink. Over 100 deaths occur each year as a result of such accidents. A relatively new device that can prevent such accidents, and which is smaller and less expensive to manufacture than a GFCI, is the *immersion detection circuit interrupter* (IDCI). This device can be built directly into the plug of such appliances as hair dryers and coffee pots. A schematic of the IDCI is shown in Figure 3-6. A hair dryer cord with an IDCI contains two power wires and a third sense wire for detecting immersion. Current flowing through the sense wire causes a positive voltage on the gate of the silicon-controlled rectifier (SCR), which then energizes the solenoid.

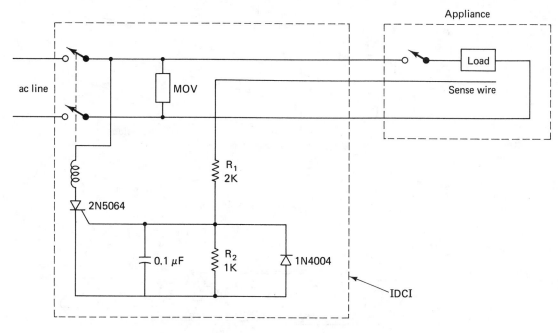

**Figure 3-6**  Immersion detection circuit interrupter schematic.

The armature of the solenoid trips the spring-loaded switch, which is manually resettable. For appliances that are permanently damaged when immersed, the switch is not resettable. These devices meet the same UL current-versus-time sensitivity requirements as GFCIs (e.g., 5.6 s with 6 mA or 26 ms with 264 mA flowing in the sense wire). The IDCI devices actually exceed the UL requirements by tripping when less than 1 mA of current flows in the sense wire. The devices are also nonpolarized (i.e., they trip regardless of which power lead is hot and which is ground). Furthermore, they have another advantage over GFCIs, which may not trip in an insulated bathtub (because the extensive use of plastic prevents the ground currents that trip a GFCI).

## CIRCUIT PROTECTION DEVICES

Large current surges from overloads or short circuits sometimes accidentally occur in electric circuits. Such current surges can lead to component destruction, electric shocks, or fires if not stopped in time. To guard electrical systems against damage from such unexpected overloads, certain protective devices are used. The most common of these are the fuse and the circuit breaker. They function by rapidly interrupting the flow of current in a circuit if it exceeds a specified value.

A fuse is basically a length of fine wire or thin metal that is designed to heat up and melt if its maximum current rating is exceeded (Fig. 3-7). It is placed in series with the circuit it is meant to protect. Fuses must be installed on the hot side because

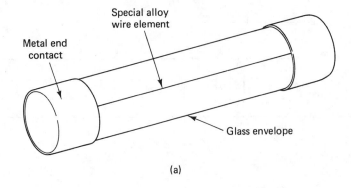

Special alloy
wire element

Metal end
contact

Glass envelope

(a)

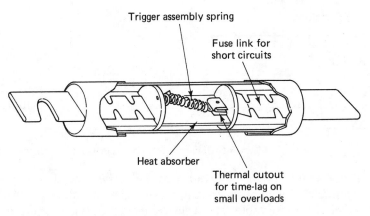

Trigger assembly spring

Fuse link for
short circuits

Heat absorber

Thermal cutout
for time-lag on
small overloads

Busmann slow-blow or dual-element fuse

(b)

**Figure 3-7**    Fuses. (Courtesy of Bussmann Division, Cooper Industries.)

if the fuse were placed in series with the low side of the line, the electrified circuit would remain at the potential of the hot wire, even if the fuse burned out. Therefore, a shock hazard would still exist. By melting when the current flowing in the circuit exceeds its capability, the fuse destroys a portion of the conducting path. This halts the current flowing in the rest of the circuit (Fig. 3-8). Usually, this break must occur on the rising edge of the first cycle of the fault current to prevent damage. For example, if the short-circuit let-through current of a power source is 200,000 amperes, this current would do considerable damage to equipment if it was allowed to flow for any significant duration. If the fuse blows on the rising edge of the first cycle, however, this large current is prevented from flowing through the circuit.

      The energy is absorbed by the melting fuse element and its surrounding gas or silica sand (as is the case for large industrial or power plant fuses). The total energy

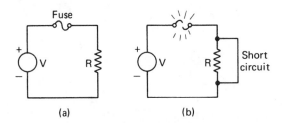

(a)  (b)  **Figure 3-8**  Fusing a circuit.

*let-through* by the fuse is a function of $I^2t$, where $I$ is the peak let-through current and $t$ is the total clearing time. UL-listed, current limiting fuses must have performance characteristics that comply with the $I^2t$ specifications of UL. In such applications, so-called *fast fuses* are used. Some circuits, however, are designed to produce or withstand short, high current pulses without damages. Such circuits still need to be protected against current surges that are too large or too long in duration. In these cases, delayed-action or *slow-blow* fuses may be used as the protective element. A slow-blow fuse resists melting if its current rating is exceeded for a short period of time. However, if the overload is too large or persists too long, the fuse also eventually melts and opens.

The blowing of a fuse is an indication that there is a malfunction within the circuit the fuse is guarding. Before a blown fuse is replaced with a new one, the fault should be located and repaired. If high ambient temperature conditions exist, the fuse size must be increased, otherwise the fuse will blow at a lower current than desired. A common example of this situation involves the fuses located in the engine compartment of an automobile.

The *circuit breaker* is a protective device that also opens a circuit if an overload is applied to it. However, unlike the fuse, the breaker is not destroyed by the overload unless the *current-interrupting rating* of the circuit breaker is exceeded. This maximum overload current may be as low as 5000 amperes for a common household-type circuit breaker or 50,000 for an industrial breaker. Fuses are used for larger currents. Circuit breakers for industrial plants are also coordinated. That is, the trip settings of the circuit breakers are selected so that the circuit breaker closest to a fault is the one that is tripped.

Breakers generally consist of a switch that is held closed by a catch. To open the circuit, the catch is released. Two common breaker-release mechanisms are the electromagnet and the bimetallic strip (Fig. 3-9). When the current exceeds the critical value in the electromagnetic coil, its magnetic field draws in the metallic bar, and the breaker catch is released. In the bimetallic strip type of mechanism, the current passing through the breaker heats the strip and causes it to bend. If the strip is heated by an excessive current, it bends backward so far that it causes the catch to spring open. In the thermal-magnetic breaker, both mechanisms are used. Normal overloads cause the bimetallic strip to release the breaker catch, whereas short circuits cause the electromagnet to activate the release mechanism. When the cause of the excess current has been located and repaired, the circuit breaker can be reset to its conducting position by a switch or pushbutton. Because of the switch, circuit

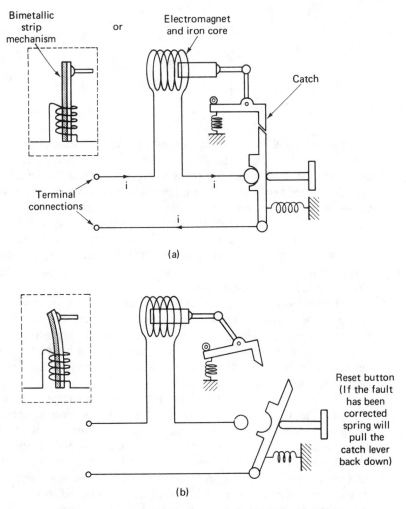

**Figure 3-9** Simplified sketch of two of the mechanisms used in circuit breakers. (a) Bimetallic and electromagnetic breaker release mechanisms are shown in their "closed" positions. (b) If excess current passes through the mechanism: (1) heat from the current bends the bimetallic strip backward, causing the catch to release, or (2) current passing through coil draws in the iron rod, thereby releasing the catch.

breakers can also be used as ON–OFF switches. They are also available with motor-operated trip coils. This allows such breakers to be operated (reset or opened) from a remote location. Most large circuit breakers are operated remotely for safety reasons. On the other hand, whenever a manually operated circuit breaker is opened or closed, this should be done while standing to one side and with

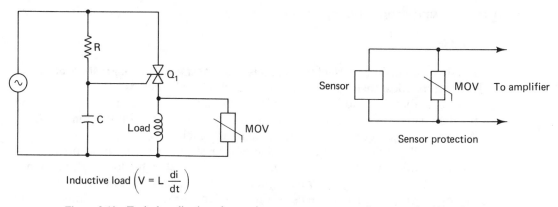

Inductive load $\left( V = L \dfrac{di}{dt} \right)$

**Figure 3-10** Typical application of a transient surge suppressor such as a metal oxide varistor (MOV).

your back to the breaker. This safety precaution should be followed because, if a breaker is accidentally closed on a short circuit, it can explode.

Circuit breakers and fuses protect circuits from overcurrent but do not protect them from overvoltage. Devices that protect circuits from overvoltage are generally known as *transient suppressors*. There are many devices available to the engineer for providing overvoltage protection. Some of these include zener diodes, silicon carbide varistors, and metal-oxide varistors (MOVs). Such devices are usually placed across the incoming power line or inductive load (Fig. 3-10). Because of the diversity of characteristics and nonstandardized manufacturer specifications, transient suppressors are not easy to compare. Metal oxide varistors (MOVs) and zener diodes are the most widely used. A *crowbar* is another transient suppressor device and consists of a zener diode and SCR, and such devices are available as packaged units. The device combination within the crowbar provides a very low resistance path to ground until the voltage drops below the SCR cutoff voltage.

Transient suppressors are rated by their breakdown voltage and the energy (joules) that they can dissipate. If the voltage transient lasts for more than several milliseconds, however, the low resistance path will allow large currents to flow, resulting in the destruction of the protective device.

Transient suppressors are widely used in industrial applications in which SCRs are used to drive inductive loads. The large voltage transient that occurs when the load is turned OFF can cause the SCR to fail. When an SCR fails, it usually exhibits a short circuit between its terminals. The result of such voltage transient failure is that the inductive load can be inadvertently turned back ON. Transient suppressors are therefore used to protect the SCR and prevent such accidents. The best protection results by placing a suppressor directly across the inductive load and at the entry of the electronic circuit. Voltage surge suppressors are also placed across voltage sensitive sensors to prevent damage from transients caused by starting motors and other line induced voltage transients.

## CABLES, CONNECTORS, SWITCHES, AND RELAYS

### Cables

A large proportion of electrical signals are transmitted through solid electrical conductors. Most such signal-carrying conductors are in the form of wires or cables. A *wire* is a single conductor. A *cable* is a configuration of insulated wire bound together with a plastic sleeve or ties, and is color coded for identification purposes. A single large power wire is also referred to as a cable.

Usually, the best electrical conductor is most suitable as a carrier of electrical signals. In other words, the better the electrical conductor, the lower the resistance losses that occur during the transmission of electrical signals. Therefore, the conductors in wires and cables are usually made of copper or aluminum. Gold, tin, and nickel are used extensively as conductors in semiconductors, relays, and sensors. The earth and the ocean are used as conductors for large dc transmission systems. Wire manufactured in the United States is sized according to the American Wire Gauge convention. It is based on a constant ratio, which is 1.123 between diameters of successive gauge numbers. For example, the diameter of #6 AWG wire is 1.123 times larger than the diameter of a #7 AWG wire. Gauge numbers are only used for wire numbers between #40 and #0000. Wire numbers larger than 0000 are expressed in circular mils, because the wires are stranded rather than solid.

The properties of the wire shown in Table 3-1 are based upon the National Electric Code® resistance and current carrying capacity. [National Electrical Code® (NEC®) is a trademark of the National Fire Protection Association, Inc.,

**TABLE 3-1**   AMERICAN WIRE GAUGE (AWG) SIZES OF COPPER WIRE

| Application | AWG number | Area (circular[a] mils) | Ohms per 1000 ft (20°C) | Maximum allowable current (A) (90°C) |
|---|---|---|---|---|
| | 500 MCM | $25 \times 10^{10}$ | 0.0216 | 427 |
| | 0000 | 211,600 | 0.049 | 253 |
| Power distribution | 00 | 133,100 | 0.078 | 186 |
| | 1 | 83,690 | 0.124 | 137 |
| | 4 | 41,470 | 0.240 | 89 |
| House main power carriers | ⌈6 | 26,240 | 0.395 | 65 |
| | ⌊8 | 16,510 | 0.620 | 48 |
| Lighting, outlets, | ⌈12 | 6,530 | 1.588 | 20 |
| general home use | ⌊14 | 4,110 | 2.52 | 15 |
| Television, radio | ⌈20 | 1021.5 | 10.1 | —[b] |
| | ⌊22 | 642.4 | 16.1 | —[b] |
| Telephone instruments | 28 | 159.8 | 64.9 | —[b] |
| | 35 | 31.5 | 329.0 | —[b] |
| | 40 | 9.9 | 1049.0 | —[b] |

[a] 1 circular mil = 1 CM = (diameter of wire in mils)$^2$ = $d^2$
[b] Current rating must be calculated per NEC® section 310-15.

Batterymarch Park, Quincy, Mass.] The resistance (in ohms) given in the table refers to the dc resistance value and not to the ac impedance. The current rating of conductors depends upon the temperature rating of the insulation used. The maximum current of a wire is established so that the temperature rise of the conductor will not reduce the life expectancy of the insulating material. (For example, the maximum allowable current given in Table 3-1 is based on the condition that there are three conductors in a conduit and that the conductor has an insulation that is rated 90°C).

Wires and cables are also usually surrounded by some type of electrical insulator. The insulation prevents the current they carry from leaking away to any conducting materials with which the cable makes contact. The materials chosen to make this insulation have a high insulation resistance, high mechanical strength, and durability. Furthermore, they are designed to be able to operate over a fairly wide temperature range and to withstand oil and corrosive chemicals without deterioration. The most popular insulating materials are PVC (polyvinyl chloride), Teflon, polythylene, and rubber. PVC and polythylene are the most widely used materials of this group. Teflon is very inert but quite costly, and it is therefore limited to use in cables that most undergo extreme environmental conditions.

The most common types of wires and cables (Fig. 3-11) include the following ones:

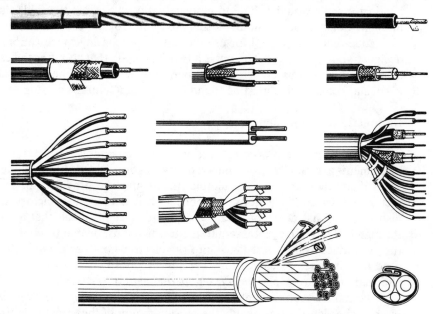

**Figure 3-11**  Typical electronic wires and cables. From top to bottom: hookup wire, test prod wire, shielded cable, special audio and sound cable, coaxial cable, multiple conductor cable, twin-lead cord, shielded multiple conductor cable, TV cable, multipurpose cable with shielded pairs. (Courtesy of Belden Manufacturing Corp.)

**1.** *Hookup wire.* Generally consists of a multistranded single conductor surrounded by PVC or polyethylene. Used for connecting elements in ordinary low-frequency circuits.

**2.** *Test-prod wire.* Very flexible wire surrounded by rubber insulation. Used in test leads of measuring instruments. High flexibility is desirable so that wire does not break with repeated bending. Rubber insulation provides high insulation resistance as well as flexibility.

**3.** *Shielded cable.* Consists of an inner signal-carrying conductor and braided metal sheath surrounding the inner conductor. The inner conductor and surrounding sheath are separated from one another by a flexible insulating layer. The outer layer of wire is also an insulator. This type of cable is used as cable for carrying low-level signals. The braided sheath is effective in reducing the pickup of interference signals by the inner signal-carrying conductor.

**4.** *Multiple-conductor cables.* Consist of many conductors bundled together in one sheath. Can have any number of conductors as well as different types in the same bundle. In Fig. 3-11 there are several multiple-conductor cables shown.

**5.** *Coaxial cable.* Similar to the shielded cable in construction, but used to carry high-frequency and pulse-type signals. At high frequencies, ordinary single-conductor cable would radiate too much energy away from the cable during transmission. Coaxial cables eliminate this problem. Coaxial cable is usually identified by a code that was evolved for use in military purchasing specifications. A typical designation is RG-mn/U. The *mn* is a two-digit code that specifies the cable. There is no pattern to this choice of numbers and letters, so the specifications of a particular coaxial cable must be found in the manufacturer's catalog.

### Connections and Connectors

As described in the preceding section, the components of electric circuits and instruments are usually interconnected by wires and cables. At the points where the wires themselves are joined, suitable means must be provided for making satisfactory electrical connections. A connection is deemed satisfactory when it furnishes a path that does not alter the characteristics of signals that are transmitted through it. Thus one general requirement which must be met by connections is that they introduce as little resistance into the electrical path as possible. The methods used to join wires or cables together can be classified according to whether the connections are to be *permanent, semipermanent,* or *separable.*

Permanent connections are usually made by *soldering, welding, crimping,* or *wrapping.* Soldering is probably the most common of these methods. In soldering, two metal surfaces are united when a solder junction is formed between them. *Solder* itself is a metal or metal composition that melts at a fairly low temperature ($\approx 400°C$) and "wets" the surfaces to which it is applied. Upon cooling, solder forms a low-resistance, permanent connection between these surfaces. Soldered connections can be made rapidly between wires, and multiple solder connections sometimes can be made simultaneously. This means that solder connections lend themselves to mass soldering and automated techniques. It should also be mentioned that solder connections are not strictly permanent, that is, they can actually be disconnected and remade a limited number of times.

The tools and equipment necessary for making solder connections are a soldering iron, solder, tinned wires, and terminals. The quality of the connections depends to some extent on the skill of the operator.

The second type of permanent connection, the *weld*, is the strongest and most permanent type of electrical connection. In welding, a direct contact is made by heating and fusing the metals of the cables being joined. A very strong connection results. However, welding requires special equipment and is suitable only for solid-wire and single-lead connections. Since other methods yield suitable connections which can be fabricated more easily, welded connections are limited in their use to special applications (i.e., for connections that must be able to withstand high temperatures).

*Crimping* is the third method of permanently joining wires. In crimping, two metals are pressed together with a special *crimping tool*. The high pressure forces the metals into intimate contact with one another and forms a low-resistance contact through deformation. The reliability of well-made crimped connections is high. Crimping is the most common method used to join solderless wire terminals to wires.

The final permanent method for joining wires and terminals is called *wire wrapping*. In this technique a solid wire is tightly wrapped around terminals that have sharp edges. Special tools (either power or manually activated) perform each wire-wrap operation in a fraction of a second.

*Semipermanent* connections between elements are usually made by screwing on various types of terminals to *binding posts* or *terminal blocks*. Some various screw-on *terminals* (also called lugs) are shown in Fig. 3-12. Such terminals are attached to wires by crimping or soldering. The wire with the attached terminal can then be connected to a binding post or terminal block with a screw clamp.

The typical *binding post* is shown in Fig. 3-13. The nuts of binding posts can be made of metal or insulator material. Such posts are designed to accommodate bare wire, banana and phone plugs, as well as spade and hook wire terminals (Fig. 3-13).

Two types of terminal blocks are shown in Fig. 3-14. The *insulating-barrier type* has insulating materials between each connecting strip to isolate it from the neighboring strips. Terminal lugs can be attached to the terminal block by screwing them into place. The lug type of terminal strip is designed to accept solder connections. Terminal block connections are used most often in low-voltage, low-power applications that need infrequent detachable connections.

Whenever circuits or instruments are designed to be readily joined or connected to other electrical components, some form of *separable connection* is required. The class of components used to provide separable connections are called *connectors*. Connectors usually have two mating halves. One of the halves is called

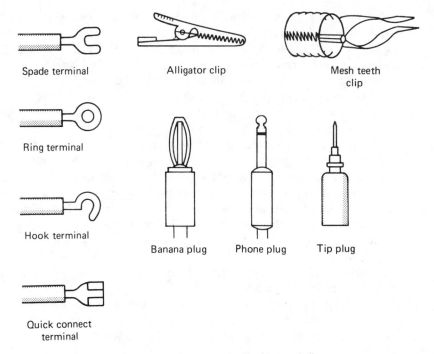

**Figure 3-12** Typical terminals, plugs, and clips.

*plugs, pins*, or *male ends*. The corresponding halves are known as *receptacles, jacks, sockets*, or *female ends*. Usually, the receptacle is mounted on the more permanent member of the equipment being connected (i.e., the chassis, box, or other fixed part.) The plug is usually connected to the cable or movable part.

The simplest plugs are single-pin plugs such as the banana plug, the phone plug, or the tip plug. These plugs are shown in Fig. 3-12 (In some applications, double-tip banana plugs are also available.) Other simple plug-like terminations are the alligator clip and the claw (or meshed teeth) type of clip. These last two clips are ordinarily found on the probes of test leads, instrument accessories, and on cords used for making quick connections.

For cables that have more than one conductor, a multipin connector is used. Since each pin is connected to a specific conductor of the cable, all conductors can be correctly connected together each time the multipin connector is mated. To ensure that multipin connectors can be mated only in the proper way, the pins can be aligned in special patterns. This built-in method of providing correct orientation of the connectors is called *polarization*.

There are many different types of multipin connectors. In the home, the common power cord is a multipin connector which has either a two- or three-pin plug and socket. Figure 3-15 shows some other examples of the many types of multipin connectors that are available. The circular connectors shown are usually used for connecting two cables together. The rectangular connectors are more often

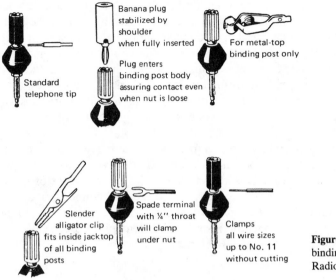

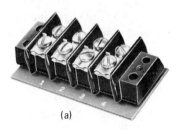

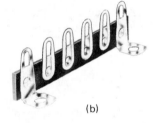

**Figure 3-13** Making connections to a binding post. (Courtesy of General Radio Corp.)

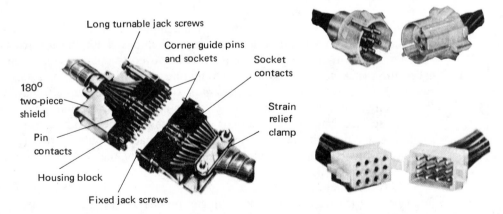

**Figure 3-14** Terminal blocks: (a) insulated barrier type; (b) lug type. (Courtesy of TRW Cinch Connector Div.)

**Figure 3-15** Examples of multipin connectors. (Courtesy of American Pamcor, Inc.)

employed for making connections between a cable and chassis. The contacts of the multipin connectors are usually mounted in an insert in such a way that they can be self-aligning. The insert is then fitted in a connector housing. An insulator such as phenolic or melamine is used to form the insert. The contacts of connectors are made of nickel or gold-plated brass or bronze, and they have a springlike action incorporated into their design. The spring tension provides for low-resistance contacts and a sure connection. The housings of the two mating halves are most often joined by screw threads or a so-called bayonet-type design. A few other types use jackscrews or latches.

For coaxial cables, special connectors must be utilized. These connectors are usually designed so that their impedances are matched to the coaxial cables for which they form the connection. This allows them to furnish a low-distortion path for the signals they carry. The most common type of "coax" connectors are called BNC types. These have a *bayonet type* of connector (Fig. 3-16). Their threaded equivalents are called TNC connectors. Some other widely used coax connector classifications are the N, HN, C, and UHF types.

### Switches

A switch is a device for turning on, turning off, or directing electric current. The most common types of switches used in electrical instruments and measurement systems follow:

1. Toggle switch
2. Pushbutton switch
3. Rotary switch
4. Slide switch
5. Snap switch
6. Mercury switch

We will examine the operation and some applications of each of these types. First, however, let us define some terms used when describing the construction of switches. The arm or part of the switch that is moved to open or close a circuit is called the *switch pole*. If a switch has only one pole, it is known as a *single-pole switch*. If it has two poles it is a *double-pole switch*. Switches can also have three, four, or any other number of poles (e.g., triple-pole, four-pole, multipole, etc.).

If each contact alternately opens and closes only one circuit, the switch is a single-throw type. On the other hand, if the contact is double-acting (i.e., if it breaks one circuit while simultaneously closing another), the switch is referred to as a *double-throw* type.

A switch therefore can be a single-pole, single-throw (SPST); single-pole, double-throw (SPDT); double-pole, single-throw (DPST); double-pole, double-throw (DPDT); or any other combination of multipole and either single- or double-

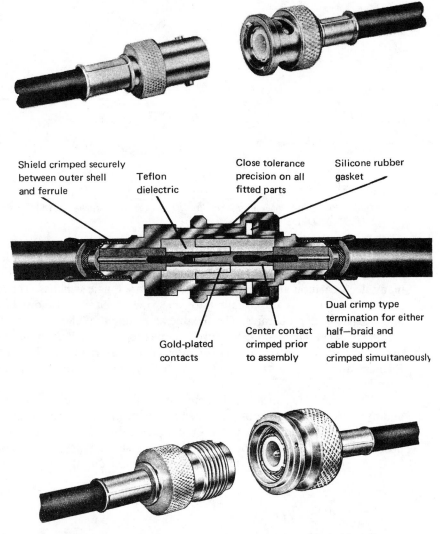

Shield crimped securely between outer shell and ferrule

Teflon dielectric

Close tolerance precision on all fitted parts

Silicone rubber gasket

Dual crimp type termination for either half—braid and cable support crimped simultaneously

Gold-plated contacts

Center contact crimped prior to assembly

**Figure 3-16** Examples of coaxial connectors. (Courtesy of American Pamcor, Inc.)

throw types. DPDT switches may also be make-before-break. That is, both contacts are closed simultaneously as the pole moves between contacts. Figure 3-17 shows the various symbols for these switch types.

Relays are switches that are operated by magnetic action, and they employ the same type of notation to describe their contacts.

The *snap-action toggle switch* is a switch in which a projecting arm or knob, moving through a small arc, causes the contacts of a circuit to open or close suddenly. The fact that the contact is made and broken suddenly reduces arcing and

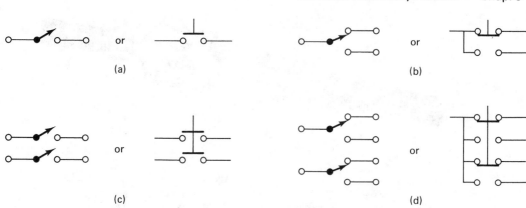

**Figure 3-17** Circuit diagram symbols for various types of switches: (a) SPST; (b) SPDT; (c) DPST; (d) DPDT.

yields a sure contact. For these reasons toggle switches are used in a wide range of switching applications. Lighting switches in most houses and the ON–OFF switch on many electrical tools and industrial instruments are toggle switches. The reset switch on circuit breakers is also usually a toggle switch. Figure 3-18 shows how a simple SPST toggle switch operates. The flexible activating arm allows the switch to snap quickly between ON and OFF. More complex toggle switches can have more than two positions and may open and close more than one circuit branch simultaneously. Because not all toggle switches are snap-action, the manufacturer's specifications must be reviewed.

The *pushbutton switch* is designed to open or close a circuit when depressed and to return to a normal position when released. In some pushbutton switches the contacts remain open or closed after the pushbutton has returned to its normal position (alternate-action type). In the alternate-action types, the pushbutton must be depressed twice to return to the original position. In other types, the contact is

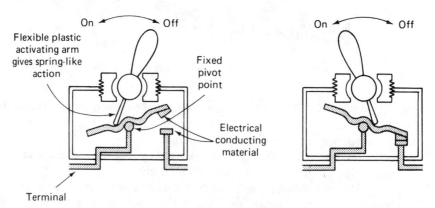

**Figure 3-18** Operation of SPST toggle switch.

closed or open only as long as the pushbutton is depressed (momentary type). Pushbutton switches are especially useful in limited-space applications. They are also easy to activate quickly. Some common uses of pushbutton switches include the selector switches on automobile radios, the dimmer switch on automobile head-lights, the doorbell switch, and safety switches on motors. Pushbutton switches may also be snap-action.

The *rotary switch* is a switch that makes or breaks circuits as it is rotated between positions (Fig. 3-19). A contact attached to a shaft is turned by means connected to the other end of the shaft. The contact moves along a fixed *insulated wafer*, which has strips of conducting material placed along its circumference. As the shaft is rotated from one position to the next, the rotating contact makes a connection to these conducting strips. This closes and opens desired circuits. In some rotary switches, more than one wafer and contact are connected to the shaft. This allows the switch to be used as a multipole switch. A spring-loaded ball bearing aligns itself with indentations in part of the switch, thereby locating the switch positions [Fig. 3-19(b) and (c)].

The rotary switch is used when one switch must be able to be set to many different positions. The channel-selecting dial on a television, the switches on decade resistors, and the function switches on meters and oscilloscopes are all examples of rotary switches.

If the rotary switch is a *shorting-type* switch, this indicates that the rotating contact always makes a connection to the next terminal before breaking contact with the previous one. Such a "shorting" feature provides protection for certain instruments in which this type of switch is used. Conversely, *nonshorting types* of rotary switches always break contact when switching between positions.

*Slide switches* open and close electrical contacts by the translational motion of a level (Fig. 3-20). This level does not need to project out very far from the instrument or appliance on which it is used. Furthermore, slide switches can be marked so that the various positions of the switch are easily seen at a glance.

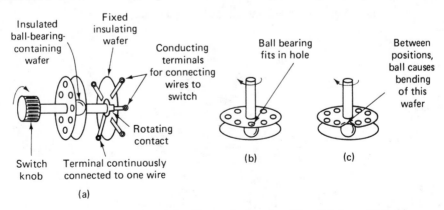

**Figure 3-19**   (a) Rotary switch with one fixed wafer; (b) switch in position; (c) switch between positions.

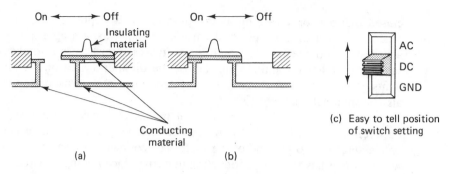

**Figure 3-20**  Slide switch.

Finally, slide switches are very simple, and this makes them attractive for use in inexpensive, low-voltage devices. However, the contacts made by slide switches are not as reliable as toggle or pushbutton contacts. Furthermore, they are subject to arcing, and this is why they are limited to low-voltage applications. Flashlight switches, ON–OFF switches of electric razors, automobile headlight switches, and some control switches on electrical instruments are examples of slide switches.

*Snap-action switches* are designed to be activated by machines rather than by people. They are similar to pushbutton switches, except that the pushbutton is often depressed by a pivoted level (Fig. 3-21). The two most common levers are the straight and the rolling lever.

The *mercury switch* is a position-sensing switch. If the switch is tilted, the contacts remain open because the pool of mercury in the switch remains at the bottom. If the switch is upright, the mercury closes the gap between the two contacts, placing the switch in the ON position. Because the mercury switch has no moving parts, it is a silent switch and is less subject to wear. Common uses of mercury switches include the furnace switch in household thermostats, automobile trunk-lid light switches, and silent lighting switches.

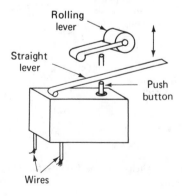

**Figure 3-21**  Snap-action switch.

**Relay Switches**

Relays are widely used in electronic circuits as remotely controlled mechanical switches to turn a sequence of events ON and OFF. Electromagnetic relays are activated by a current that passes through a coil to create a magnetic field (Fig. 3-22). This magnetic field exerts the same attractive force on nearby ferromagnetic materials as would the field of a permanent magnet. If this force is used to attract and move a pivoted piece of metal called an *armature*, and if the motion of this armature is used to open and close electric contacts, the assembly is called an *electromagnetic relay*. (In most relays, when the current is stopped, the spring action of the armature causes the contacts to return to their original positions.)

The nomenclature of relays follows that of switch nomenclature. That is, if the relay has one armature, it is known as a single-pole relay. If relays have two, three, or more armatures, they are called *double-pole, triple-pole, or multipole relays*, respectively. If each armature merely opens or closes one circuit, the relay is called a *single-throw* type. On the other hand, when the armature is double-acting (i.e., when it breaks one circuit while simultaneously making another), it is called a *double-throw relay*. Like switches, relays are thus classified as single-pole, single-throw, double-pole, double-throw, and so on. The contacts of the relay that are open when no current is passing through the relay are called *normally open* (NO) contacts. Those that are closed when no current is passing are called *normally closed* (NC) contacts. Figure 3-23 shows two common relay models.

The current in the relay required for the armature to "pull in" and close NO contacts must exceed a certain specific minimum value. At a lower current value, the armature will "drop out," causing the NO contacts to open. Usually, the switching circuits used to actuate a relay are designed so that they provide several times the minimum relay current, thereby guaranteeing the operation of the relay.

Relays do not open or close instantaneously. Most require several milliseconds after the instant of application of coil power before completing their contact function. In addition, when the contacts make or break, there is always some contact bounce. Although this bounce and its duration can seriously distort the signal being switched, it is not included as a part of the operate and release times. Note that contact bounce is also a characteristic of manual switches.

Relays are used in a tremendous number of industrial and electronic applica-

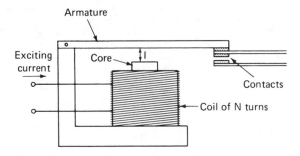

**Figure 3-22**   Electromagnetic relay.

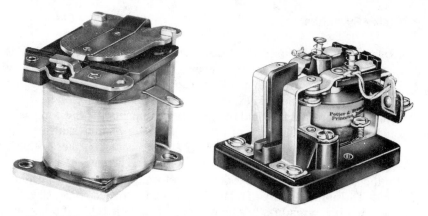

**Figure 3-23**   Electromagnetic relays. (Courtesy of Potter and Brumfield. Division of AMF, Inc.)

tions. They play a key role in the operation of automated machinery. Some common consumer applications of relays include the voltage regulators in automobiles and the release mechanism which opens and closes doors activated by door buzzers. Relays also are responsible for the clicking noises heard when dishwashers, air conditioners, and washing machines are being shifted from one cycle to another during their operation.

Another commonly used type of relay is the *reed relay* shown in Fig. 3-24. In the single-pole, single-throw type shown in this figure, two ferromagnetic reeds are separated by a small distance. These reeds are mounted in a hermetically sealed glass tube, and a coil surrounds the glass tube. When a current is passed through the coil, the magnetic field of the coil magnetizes the two reeds. One reed is thereby made a north pole, and the other becomes a south pole. When enough attractive force is developed, the reeds are drawn together and a connection is established. When the current is halted, spring action due to the stiffness of the reeds forces them apart again. The coil may also be replaced by a permanent magnet. For example, a magnet can be embedded in a door and the reed relay in the door jamb. When the door closes, the magnet closes the reed relay. The closed relay allows an

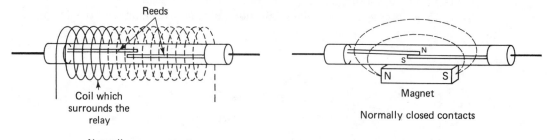

**Figure 3-24**   Reed relay.

electrical signal to activate the lock bolt. Many security systems use this method of actuating a reed relay since both the magnet and relay can be easily concealed.

A normally closed relay can be created by using both a permanent magnet and a coil. The magnet is placed adjacent to the reed relay and the coil encloses both the relay and magnet. The magnet holds the contacts closed until the coil is energized. When the coil is energized, its magnetic field cancels the field of the permanent magnet, allowing the relay contacts to open.

If a large current is flowing through the reed relay when the contacts are opened, the relay can easily be destroyed. That is, when an inductive load (motor) is turned off, the energy stored in the magnetic field of the load is returned to the circuit in the form of a current flow. If the opening of the reed relay contacts is used to turn the load off, the current flows across the open relay contacts. An electric arc provides the path for this current flow. The arc current heats the gas in which the arc occurs, and this will melt the contacts if the gas becomes too hot. Once the arc is extinguished and the contacts close, the molten contacts freeze together, destroying the relay. A contact protection circuit must therefore be used if the maximum life of the reed relay is to be obtained.

## Solid-State Switching Devices

Mechanical switches and electromagnetic relays possess characteristics such as slow response time and contact bounce, which limit their effectiveness in many measurement and control applications. Solid-state devices that overcome these limitations have been developed and have become widely used in many modern industrial and experimental measurement systems. Such switching devices include semiconductor diodes, transistors, silicon-controlled rectifiers (SCRs), and photoconductive and photovoltaic devices. Photoconductive and photovoltaic devices are described in Chapter 14 in the section "Light and Radiation Transducers." Solid-state switching devices controlled by digital logic circuitry are replacing most relay logic systems. The programmable controller and the personal computer are two common systems that can be adapted to generate the control signals that were formerly produced by mechanical relays. The main advantage to this solid-state approach is versatility. Any time a logic change is made to a hard-wired relay system the interconnections must be physically rewired. When a programmable controller is used to generate the control signals, all that is required to change the control logic signals is a software change. The load driving capabilities of solid-state relay systems are now comparable to those of mechanical relays.

The major areas of concern when using solid-state relays are isolation and the failure modes of the solid-state relays. That is, mechanical relays provide excellent isolation because open contacts in a mechanical relay conduct no current, whereas solid-state relays exhibit small, but finite leakage currents in the OFF state. To provide equivalent isolation with solid-state relays, optical or acoustic devices must be used. The failure mode issue is also important because the solid-state relay devices normally fail-shorted (i.e., a short-circuit appears between the input and output terminals of the failed device). This must be carefully considered in designing a solid-state system to control a process or machine. It could be hazardous for a machine to start as the result of the failure of one of the controlling devices.

Solid-state relays can easily control and regulate power to small (e.g. 1 watt) solenoids using a FET or SCR in a T092 package as the output device. On the other hand, a 10,000 ampere source (controlling, for example, the plating of tin on sheet steel) would use water-cooled SCRs for the output device. A detailed discussion of solid-state relays and photoconductive and photovoltaic devices are described in Chapter 14.

## INPUT IMPEDANCE, OUTPUT IMPEDANCE, AND LOADING

The concepts of *input impedance, output impedance*, and *loading* are all commonly used in the description of electrical instruments. The terms are interrelated and are often indications of how effectively a measuring instrument can perform its specified function. These terms can best be explained if we first define the concept of impedance.

*Impedance*, in general terms, is the ratio of the voltage to the current, and it is denoted by the letter $Z$. The units of impedance are ohms ($\Omega$). In dc circuits, the impedance is equal to the ratio of the dc voltage to the dc current. Because resistors are the only effective elements in dc circuits, the impedance is just equal to the resistance of the part of the circuit in which $V$ and $I$ are determined.

$$Z_{dc} = \frac{V}{I} = R \tag{3-3}$$

In ac circuits, impedance is defined as the ratio of the rms voltage to the rms current in the part of the circuit being considered.

$$Z_{ac} = \frac{V_{rms}}{I_{rms}} \tag{3-4}$$

However, in ac circuits the impedance is no longer strictly resistive. Since capacitors and inductors also contribute to the impedance of ac circuits, the impedance contains a reactive as well as a resistive component.

If we have an electrical instrument and if we make a connection to its input or output terminals, the instrument will exhibit some characteristic impedance as seen from these terminals. For the sake of analysis, we can always replace the instrument by this impedance (along with an appropriate voltage source, if the instrument contains active as well as passive elements). If the instrument is a measuring instrument (such as a meter or an oscilloscope), the ratio of the voltage across the input terminals to the current flowing into them is known as the *input impedance* of the instrument.

$$Z_{in} = \frac{V_{in}}{I_{in}} \tag{3-5}$$

This input impedance can be measured if we connect a voltage source across the input terminals and measure the current that flows through the instrument at a particular voltage setting (Fig. 3-25). Note that this ratio (and consequently the input impedance) may be so high in some instruments that it would be very hard to actually measure. If dc signals are used to excite the instrument, $V_{in}$ and $I_{in}$ are dc quantities. If the input signals to an instrument are ac quantities, $V_{in}$ and $I_{in}$ refer to the rms values of the quantities.

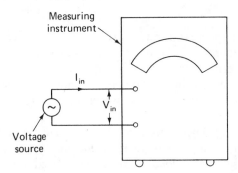

**Figure 3-25** Determining the input impedance of a measuring instrument with the help of a voltage source.

The *output impedance* of a device is defined as

$$Z_{out} = \frac{V_{out}}{I_{out}} \tag{3-6}$$

In most cases we will be interested in the output impedance of devices or instruments that contain active elements and thus serve as signal sources in measurement systems (instruments and devices such as power supplies, oscillators, batteries, amplifiers, and active transducers all fit into this category). For these sources, $V_{out}$ is the voltage appearing across the open-circuited output terminals of the device. $I_{out}$ is the calculated current that would flow if the output terminals were connected by a short circuit. However, the output impedance of sources is not measured by actually short-circuiting the output terminals. (An attempt to measure the value of $I_{out}$ in this manner could result in burning out the source.) As an example of how the output impedance of some devices can be measured, see the discussion in Chapter 12 on the measurement of the internal resistance of a battery. (The internal resistance of a battery corresponds to its output impedance.) Now let us see how the concepts of input and output impedance are related to the concept of *loading*.

Instruments that are used to measure voltage are placed *across* the element (in parallel) or circuit being measured. Ideally, a measuring instrument should not disturb or change the values of the current and voltage in the circuit under test. In the case of voltage-measuring devices, the instruments should draw no current when connected to the two points across which the voltage is being measured. This condition would be satisfied if the voltage measuring device appeared to the test circuit as an open circuit. The input impedance of the voltage measuring device

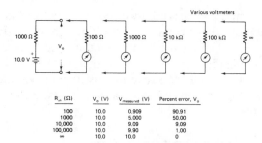

| $R_{in}$ (Ω) | $V_o$ (V) | $V_{measured}$ (V) | Percent error, $V_o$ |
|---|---|---|---|
| 100 | 10.0 | 0.909 | 90.91 |
| 1000 | 10.0 | 5.000 | 50.00 |
| 10,000 | 10.0 | 9.09 | 9.09 |
| 100,000 | 10.0 | 9.90 | 1.00 |
| ∞ | 10.0 | 10.0 | 0 |

**Figure 3-26**   Effect of meter input impedance on measurement errors caused by loading.

describes how it actually appears to the test circuit. Since an open circuit is equivalent to an infinite impedance, the value of the input impedance of a voltage-measuring instrument determines how closely it approaches the open-circuit ideal. However, because voltage-measuring instruments are not ideal, they do draw some current from the circuit being measured. The effect of drawing current is known as *loading*.[3]

If a voltage-measuring device does not have a high input impedance and consequently draws a significant percentage of the current flowing in the test circuit, the measuring device is said to be *loading down* the test circuit. The greater the percentage of the current drawn from the circuit under test, the more a voltage-measuring device disturbs the circuit it is monitoring. Thus the higher the value of input impedance that a voltage measuring instrument possesses, the more accurate a voltage measurement it can make.

For example, if we were to measure the voltage across two points of a circuit with voltmeters of different input impedance values, the loading error would increase as the input impedance of the voltmeter was reduced. As an example, in Fig. 3-26, a voltage $V_o$ of 10 V in series with 1000 Ω is measured with five various voltmeters. The input impedance $R_{in}$ of the voltmeters is varied from 100 Ω to open circuit (∞ Ω). The results of the measurements are tabulated in the figure and they show that the loading error increases from 0 percent (when $R_{in} = ∞$) to 90 percent (when $R_{in} = 100$ Ω). Additional details on the loading effects of voltmeters and oscilloscopes are given in the chapters dealing specifically with these instruments.

If a signal source (e.g., a generator or an oscillator) is providing energy to a circuit, the term *loading* is also used to describe the fact that current is being drawn from the signal source. As the impedance connected across the output terminals of such sources is made smaller, the current output of the source will increase. Hence oscillators or generators are said to be *loaded down* when a low impedance is connected across their output terminals. Additional information relating to the effects of loading on the operation of oscillators is presented in Chapter 13.

---

[3] Loading is also occasionally mentioned in connection with ammeters. However, since an ammeter is connected in *series* with the branch, its resistance reduces the original value of current in the branch, rather than drawing it away into another branch. Thus, for the sake of consistency, we will not refer to the disturbance of a circuit by an ammeter as loading, since it does not satisfy our definition of drawing current.

## POWER TRANSFER AND IMPEDANCE MATCHING

If we have a system in which power is being expended, the part of the system to which power should be delivered is called the *load*. For example, a man on a bicycle may be the system of interest. The man is the power supply, and (on a level path) the mass and friction of the bicycle and the wind resistance comprise the load. However, the man dissipates energy internally (in his heart, muscles, etc.), and thus all the energy he expends cannot be transferred to the load.

In an electrical system, the power source might be a generator, a battery, or an amplifier. The load could be an electric lamp, loudspeaker, or meter. The generator, battery, and amplifier also have internal losses and hence cannot deliver all the power they generate to a connected load. These internal losses become important when we consider the type of load that should be connected to the source in order to transfer a desired amount of power.

There are usually two desired conditions under which power is drawn from a source for delivery to a load. In the first we want maximum *efficiency*; in the second, maximum *amount* of power transferred.

If we are interested in delivering the maximum percentage of the power generated by a source to the load, we want the source to be operating at the *highest efficiency* possible. That is, the ratio of the power generated by the source to the power put into the load should be maximized. But when a generator or other source is operated at maximum efficiency, the source is not transferring the maximum *amount* of power it can generate to the load. At maximum efficiency, the source is run at far below its maximum output capability. This is the price that must be paid for having as little energy dissipated in the source as possible.

For a generator supplying many megawatts of power to a city, a condition of maximum operating efficiency is a necessity. If a sizable fraction of the generator output were dissipated in the generator itself, the resulting heat would quickly melt the entire generating system.

To achieve high efficiency, the output impedance of the generator is kept to only a fraction of an ohm. The loads to which power is supplied have impedances that are larger than the generator output impedances by several orders of magnitude.

In many measurement systems, however, the signal source being measured is very weak and generates only a tiny amount of power. In such cases it is necessary that the maximum *amount* of power that the source can generate be delivered to the measuring instruments. For example, a transducer may be the source and it may generate only a small electrical signal in response to a change in a nonelectrical quantity. We would therefore want as much of the power of this signal to be transferred to the load (i.e., an amplifier or meter) as possible.

When a condition of maximum power transfer from the source to the load is achieved, we say that the load is *matched* to the power source. The procedure used to achieve this condition is called *impedance matching*. To achieve such impedance matching in an electrical system, the impedance of the *load* is usually changed to match the impedance of the source.

Physical principles specify that at the condition of *maximum power transfer* 50 percent of the power being produced by the source is delivered to the load, while 50 percent is being dissipated internally in the source. Of course, the efficiency of this system is only 50 percent, but more power is being delivered by the source than at a point of higher efficiency.

For an electrical system containing only resistance, a condition of maximum power transfer exists if the resistance of the load $R_L$ equals the internal resistance of the source $R_G$.

$$R_L = R_G \qquad (3\text{-}7)$$

A specific example of impedance matching is the connection of an amplifier and a loudspeaker. Usually, the loudspeaker (load) has a low resistance and the amplifier (source) has a high output resistance. To transfer the maximum power from the amplifier to the speaker, a transformer is used. The transformer makes the loudspeaker appear to have a higher resistance in order to match the amplifier's output resistance.

**Example 3-1**

Find the value of $R_L$ for maximum power transfer in the circuit shown in Fig. 3-27.

**Solution.**   The equivalent resistance of the known resistors is

$$R = 6 + \frac{(10)(5)}{10 + 5} = 6 + 3\tfrac{1}{3} = 9\tfrac{1}{3}\,\Omega$$

or
$$R = \frac{V_{\text{Open circuit}}}{I_{\text{Short circuit}}} = \frac{V\dfrac{5}{5 + 10}}{V\dfrac{5\|6}{10 + 5\|6}} = 9\tfrac{1}{3}\,\Omega$$

Then the $R_L = R$ for maximum power to $R_L$ and

$$R_L = 9\tfrac{1}{3}\,\Omega$$

When applied to ac circuits, the maximum-power-transfer theorem states that the impedance of the load should be equal to the complex conjugate of the equivalent impedance of the source, or

$$Z_L = Z_G^* \qquad (3\text{-}8)$$

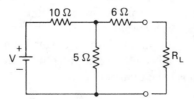

**Figure 3-27**   Circuit for Example 3-1.

Note that in some measurement systems, maximum power transfer is not the important aspect of the data-acquisition situation. The concern may not be to get, for example, a maximum power transfer between a transducer and a meter but to achieve an accurate representation of the transducer output (i.e., in terms of its open-circuit voltage). It is this voltage output that is indicative of the phenomenon activating the transducer. Thus in this case it would be best to use a voltmeter that effectively would not disturb the circuit being measured (i.e., a voltmeter with a very high input impedance that would not load down the circuit).

## PROBLEMS

1. List several factors that can influence the severity of an electric shock in a person.

2. If the resistance of dry skin is 500 k$\Omega$ and accidental contact is made with a conductor that is at a voltage of 120 V, how much current can flow through the human conducting path? If contact with this same conductor is maintained for 10 s and the skin resistance at the point of contact falls to 1 k$\Omega$, what current can then flow through the body to ground?

3. Describe how the third wire of the three-wire power cord acts as a protective mechanism against electric shock hazards.

4. Explain why the following acts lead to hazardous safety conditions when working with electrical equipment.
   (a) Wearing metal rings or bracelets
   (b) Being barefoot
   (c) Working on a damp or concrete floor
   (d) Touching pipes or other grounded conductors while working on electrical equipment
   (e) Working on electrical equipment with sweaty hands

5. If the resistance of the path from the wall outlet end of the neutral wire to ground is 15 $\Omega$, calculate the voltage of that point (relative to ground) if a 75 W bulb is connected across the outlet.

6. The two-wire 120-V supply line from the electric power company to the consumer has a resistance of 0.05 $\Omega$. If a short circuit occurs at the consumer's residence, find the current that flows through the short circuit.

7. Describe (with the help of sketches) the path taken by the currents flowing in circuits whose grounds are the following.
   (a) Earth ground
   (b) Electric circuit in an airplane
   (c) Electric circuit in an instrument whose circuit is floating

8. Describe the difference between fuses and circuit breakers. Where might each type of device find its best use?

9. Explain why it is desirable for the input impedance of a voltmeter to have a very high value. What are the effects on the test circuit if the input impedance of the voltmeter is low?

10. Define the word "loading" and explain in what context it is used when referring to:
    (a) Loading by an oscilloscope

**(b)** Loading by an amplifier
**(c)** Loading of an oscillator

## REFERENCES

1. Dalziel, C. F., et al., "Lethal Electric Currents," *IEEE Spectrum*, Vol. 6, No. 2, 1969, pp. 44–50.
2. Tektronix, *Biophysical Measurements*. Beaverton, Ore.: Tektronix, 1971.
3. Shiers, G., *Design and Construction of Electronic Equipment*. chaps. 6, 7, and 9. Englewood Cliffs, N.J.: Prentice Hall, 1966.
4. Friedlander, G. F., "Electricity in Hospitals; Elimination of Lethal Hazards," *IEEE Spectrum*, Vol. 8, No. 9, 1971, pp. 40–51.

# 4

# Analog DC and AC Meters

There are many different methods and instruments used for measuring current and voltage. Voltage measurements are made with such varied devices as electro-mechanical voltmeters, digital voltmeters, oscilloscopes, and potentiometers. Current-measuring methods use instruments called *ammeters*. Some ammeters operate by actually sensing current, whereas others determine the current indirectly from an associated variable such as voltage, magnetic field, or heat.

Meters that measure voltage and/or current can be grouped into two general classes: *analog meters* and *digital meters*. Those types that use electromechanical movements and pointers to display the quantity being measured along a continuous (i.e., *analog*) scale belong to the *analog* class. In this chapter we discuss such analog meters, together with the basic, conceptual information associated with meter operation. *Digital meters* and their operation are described in Chapter 5.

An ammeter is always connected in *series* with a circuit branch and measures the current flowing in it. An ideal ammeter would be capable of performing the measurement without changing or disturbing the current in the branch. (Such a disturbance-free measurement would be possible if the meter appeared as a short circuit to the current flow.) However, real ammeters always possess some internal resistance, causing the current in the branch to change due to the insertion of the meter.

Conversely, a voltmeter is connected in *parallel* with the elements being measured. It measures the potential difference (voltage) between the points across which it is connected. Like the ideal ammeter, the ideal voltmeter should not change the current and the voltage in the test circuit. Such an ideal voltage measurement can be achieved only if the voltmeter does not draw any current from the test circuit. (It should appear as an *open circuit* between the two points to which it is connected.) However, most actual voltmeters operate by drawing a small but finite current and thereby also disturb the test circuit to some degree. We will discuss the extent of the measurement errors caused by such nonideal aspects of real meters.

## ELECTROMECHANICAL METER MOVEMENTS

### D'Arsonval Galvanometer Movement

The most common sensing mechanism used in electromechanical dc ammeters and voltmeters is a current-sensing device. This mechanism was developed by D'Arsonval in 1881 and is called the *D'Arsonval* or *permanent-magnet–moving-coil movement*. It is also used in some ohmmeters, rectifier ac meters, and impedance bridges. Its wide applicability arises because of its extreme sensitivity and accuracy. Currents of less than 1 μA can be detected by commercially available movements. (Certain special laboratory instruments that use D'Arsonval movements can measure currents as tiny as $1.0 \times 10^{-13}$ A.) The movement[1] detects current by using the force arising from the interaction of a magnetic field and the current flowing through the field. The force is used to generate a mechanical displacement, which is measured on a calibrated scale.

Charges moving perpendicular to the flux of a magnetic field are acted on by a force that is perpendicular to both the flux and the direction of motion of the charges. Since current flowing in a wire is due to a motion of charges, these charges will be subject to the magnetic force if the wire is oriented properly in a magnetic field. The force is transmitted by the charges to the atoms of the wire, and the wire itself is caused to move. As an example, let us place such a wire in a field oriented as shown in Fig. 4-1(a). The direction of the force on the wire carrying the current is easily found by using the right-hand rule. The thumb points in the direction of conventional current, and the middle finger points in the direction of the magnetic field. The vector equation that defines this force is

$$\mathbf{F} = i\, \mathbf{L} \times \mathbf{B}$$

where **F** is the force in newtons on the wire, $i$ is the current in amperes, **L** is the length of the wire in meters, and **B** is the magnetic field strength in webers/square meter. The sine of the angle between **L** and **B** is denoted by **X**. When using the right-hand rule, the angle is 90° and the value of the sine is then 1. If the current flows upward in this wire, the force will cause the wire to move to the right. If we bend the wire into the shape of a rectangular coil and suspend it into the same

[1] The word *movement* is used to denote the sensing devices in electromechanical meters because movements display the electrical quantity being measured by moving a pointer along a calibrated scale. Thus the words "mover" or "movement" describe their action faithfully.

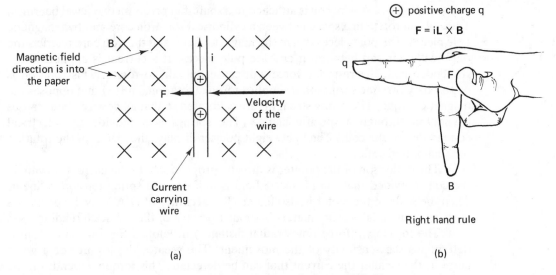

**Figure 4-1**   Moving conductor in a magnetic field.

magnetic field, the resulting force on the wire will now tend to rotate the coil as shown in Fig. 4-2(a).

On some analog meters the scales are nonlinear. This is usually because the magnetic field is not uniform throughout the entire region between the permanent magnet pole pieces [Fig. 4-2(a)]. To allow the meter reading to remain accurate, the scale of the meter must then deviate from linearity [Fig. 4-2(b)] to compensate for this magnetic-field non-uniformity.

The movement that D'Arsonval patented is based on this principle and is

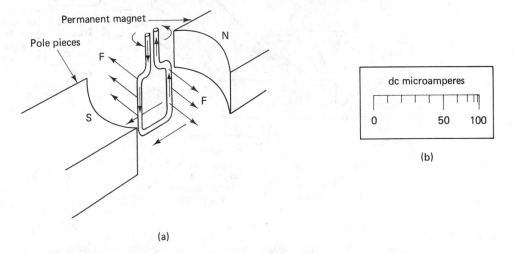

**Figure 4-2**   (a) Loop of current-carrying wire in magnetic field; (b) nonlinear scale.

shown in Fig. 4-3. A wire coil is attached to a shaft that pivots on two jewel bearings. The coil can rotate in a space between a cylindrical soft-iron core and two magnetic pole pieces. The pole pieces create the magnetic field, and the iron core restricts the field to the air gap between it and the pole pieces. If a current is applied to the suspended coil, the resulting force will cause the coil to rotate. This rotation is opposed by two fine springs that supply a torque (rotational force) that opposes the magnetic torque. The spring strengths are calibrated so that a known current causes a rotation through a specific angle. (The springs also provide the electrical connections to the coil.) The lightweight pointer displays the extent of the rotation on a calibrated scale.

The deflection of the pointer is directly proportional to the current flowing in the coil, provided that the magnetic field is uniform and spring tension is linear. Then the scale of the meter is also linear. The accuracy of D'Arsonval movements used in common laboratory meters is about 1 percent of the full-scale reading.

The torque $\tau_D$ (force times radial distance) developed for a given current $i$, determines the sensitivity of the movement. The greater the torque for a given current, the smaller the current that can be detected. This torque depends on the number of turns ($N$), the length ($l$) of the conductor perpendicular to the magnetic

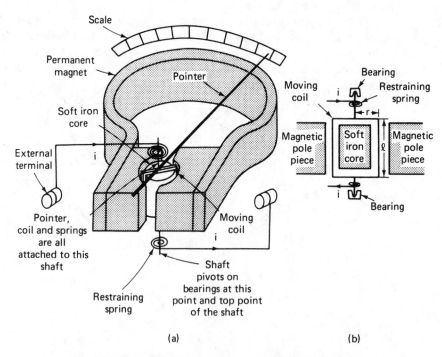

**Figure 4-3**    (a) D'Arsonval movement; (b) cross-sectional view of the moving coil and magnet of the D'Arsonval movement.

field, and the strength ($B$) of the magnetic field. The mathematical expression for the torque is given by[2]

$$\tau_D = f \cdot r = B(2Nl)i \cdot r = 2NBlri \tag{4-1}$$

Because a greater number of turns in the coil also increases the overall length of the wire, this increases the resistance of the movement. Thus this method of increasing the coil sensitivity also makes it less of an ideal movement.

To ensure that a meter can respond to the forces that arise from the quantity it is measuring, any friction that would oppose the rotation of their moving member should be kept as small as possible. The moving member of the meter is usually mounted on a shaft that rotates, and such friction would arise at the points where the shaft is supported. To keep the friction to a minimum and still keep the shaft properly centered, jewel bearings are used.

Another method of supporting a meter movement besides the shaft, jewel bearings, and spring arrangement is the *taut band support*. Here the movement is suspended by two thin metal ribbons or bands (Fig. 4-4). The bands, rather than the springs, provide the electrical connection and restoring torque. An advantage is obtained because there is no friction between moving parts. This method produces highly repeatable measurements. The taut band support is replacing the jewel and pivot bearing in most uses.

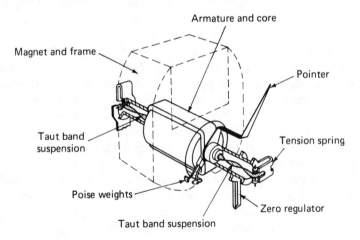

**Figure 4-4** Taut band meter movement.

Two types of scales are generally employed with the D'Arsonval movement: those with a zero at the center of the scale [Fig. 4-5(a)] and those with a zero at the left end of the scale [Fig. 4-5(b)]. (The movements are adjusted to indicate zero on each scale when no current is flowing. The screw shown on both scales is used for

---

[2] The factor of 2 arises because there are two vertical wire sections in each turn of the coil in the field. Typical values of $B = 0.15 - 0.5$ Wb/m$^2$, and $N = 20$ to 100 turns.

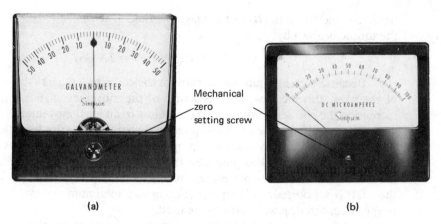

**Figure 4-5** D'Arsonval movement scales: (a) zero-center scale; (b) zero on left end of scale. (Courtesy of Simpson Electric Co.)

making this adjustment.) The scale in Fig. 4-5(a) is used in dc instruments that can detect current flow in either direction or in instruments where an absence of current flow is the desired condition to be detected (such as in the Wheatstone bridge or potentiometer circuits). The scale in Fig. 4-5(b) indicates an upscale reading only when current is passed in one direction through the coil. If the current flows in the opposite direction, a deflection below zero occurs. To obtain a positive reading when this situation develops, one must reverse the connections of the leads to the movement. This reverses the direction of the current flow through the movement. Most meters with scales such as those shown in Fig. 4-5(b) indicate the proper way of connecting the meter into the circuit by polarity markings on the meter terminals.

In many processes, the analog meters are preferred over digital meters because the magnitude of a process variable and its rate and whether it is increasing or decreasing can be determined at a glance by a process operator. This is especially useful when observing process variables that oscillate (such as the quantity of material flowing in a pipe). In another example, a pilot trying to determine the rate of descent by observing a digital meter would be severely handicapped. There are also times when analog meters are used to indicate a process variable that seems unrelated to electric current. For example, the level of liquid oxygen in a tank can be measured by using a variable capacitance probe. The value of capacitance is converted to a current by using an oscillator and other circuitry. The resultant meter is shown in Fig. 4-6.

### Electrodynamometer Movement

The electrodynamometer movement is used in the construction of highly accurate ac voltmeters and ammeters and in wattmeters and power-factor meters. Like the D'Arsonval movement, it also operates as a current-sensitive device. Extremely

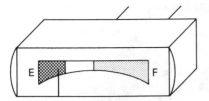

**Figure 4-6** Analog current meter with custom scale. (Courtesy of Electronic Specialists.)

high accuracies can be obtained by using this movement because it uses no magnetic materials (and magnetic materials possess nonlinear properties).

In contrast to the D'Arsonval movement, which uses a permanent magnet to provide a magnetic field, the electrodynamometer creates a magnetic field from the current being measured. This current passes through two field coils and establishes the magnetic field that interacts with the current in the moving coil. The force on the moving coil due to the magnetic fields of the fixed coils causes the moving coil to rotate (Fig. 4-7). The moving coil is attached to a pointer that moves along a scale marked to indicate the value of the quantity being measured. The entire movement assembly is mounted in an iron-lined case to shield it from any external stray magnetic fields. Because the current being measured determines both the strength of the magnetic field and the moving coil's interaction with the field, the resulting deflection of the pointer is proportional to $i^2$. In ac use, the pointer takes up a position proportional to the *average* of current squared. The scale can be calibrated to read to the square root of this quantity (rms). Note that the scale of the electrodynamometer movement shown in Fig. 4-7 is marked in the manner of a typical dynamometer meter.

The electrodynamometer movement produces an extremely accurate reading but is limited by its power requirement. The magnetic field of the stationary coils

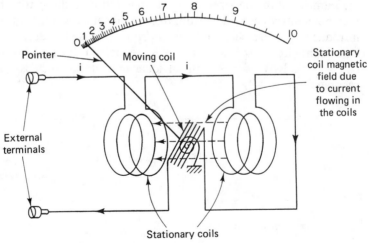

**Figure 4-7** Electrodynamometer movement.

**Figure 4-8** Interior view of an ammeter, with the magnet inside the moving coil. (Courtesy of Weston Instruments, Inc.)

produced by a small current is much weaker than the permanent field of the D'Arsonval movement. (Typical comparison is $6 \times 10^{-3}$ Wb/m$^2$ versus 0.2 Wb/m$^2$.) Thus the sensitivity of the electrodynamometer movement is comparatively poor. When it is used as a voltmeter, the sensitivity is 10 to 30 $\Omega$/V, which, as we will see, is very low.

## ANALOG DC AMMETERS

Laboratory and industrial electromechanical dc ammeters are used to measure currents from 1 $\mu$A ($10^{-6}$ A) to several hundred amperes. Figure 4-8 shows a photograph of the interior of a typical dc ammeter. The D'Arsonval movement is used in most dc ammeters as the current detector. Typical laboratory bench meters of this type have accuracies of about 1 percent of their full-scale readings because of inaccuracies of the meter movement. In addition to this error, the resistance of the meter coil introduces a departure from the ideal ammeter behavior described in the introduction of this chapter. The *model* usually used to describe a real ammeter in equivalent-circuit use is a resistance $R_m$ (equal in value to the resistance of the meter coil and leads) in series with an ideal ammeter (which is assumed to have no resistance) (Fig. 4-9).

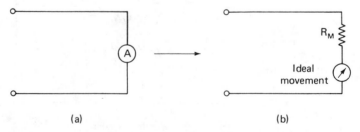

(a)                                          (b)

**Figure 4-9** Ammeter symbol and equivalent circuit model: (a) circuit symbol; (b) equivalent-circuit model.

By using this model, we can calculate the error caused by introducing an ammeter into a circuit, or we can specify the maximum allowable resistance that will cause the ammeter to have a negligible effect. This effect is similar to the effect of voltmeter loading because the additional meter resistance causes less total current to flow in the circuit branch being measured. Table 4-1 shows the internal resistance of typical D'Arsonval movements.

**TABLE 4-1**   INTERNAL RESISTANCES OF
TYPICAL D'ARSONVAL MOVEMENTS

| | |
|---|---|
| 50 μA | 1000–5000 Ω |
| 500 μA | 100–1000 Ω |
| 1 mA | 30–120 Ω |
| 10 mA | 1–4 Ω |

The following example indicates how the error caused by the extra resistance of an ammeter in a circuit can be calculated.

### Example 4-1

We are given a 50 μA ammeter that has an internal resistance of 2.5 K, (K = $10^3$), and we wish to measure the current flowing in a branch that contains a 200 KΩ resistor. Calculate (a) the error introduced by the extra resistance of the ammeter on the circuit and (b) the ammeter reading if 7.2 V is applied across the branch.

**Solution.**   (a) Without the ammeter in the circuit, 7.2 V applied across 200 KΩ will yield a current

$$I = \frac{V}{R_1} = \frac{7.2}{200 \text{ K}} = 36 \text{ μA}$$

When the ammeter is inserted in series with this resistance (Fig. 4-10), the total resistance of the branch is 202.5 KΩ. Thus 7.2 V applied across this resistance will produce a current of

$$I = \frac{V}{R_1 + R_M} = \frac{7.2}{202.5 \text{ K}} = 35.56 \text{ μA}$$

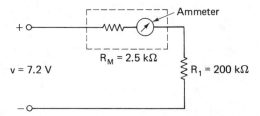

$R_M$ = 2.5 kΩ

v = 7.2 V

$R_1$ = 200 kΩ

**Figure 4-10**

The error in the reading caused by $R_M$ of the ammeter is

$$\text{error} = \frac{(36 \times 10^{-6}) - (35.56 \times 10^{-6})}{36 \times 10^{-6}} \times 100\% = 1.23\%$$

The sensitivity of an ammeter indicates the minimum current necessary for a full-scale deflection. Highly sensitive meters have very small full-scale readings. Commercial meters use movements that have sensitivities as small as 1 μA. However, 50 mA is the upper limit that movement springs can handle with high accuracy. To extend the measuring capabilities of dc ammeters beyond this upper bound, shunts must be used.

A *shunt* is a low-resistance path connected in *parallel* to the meter movement. Figure 4-11(a) shows an ammeter with a shunt. The shunt allows a specific fraction of the current flowing in the circuit branch to bypass the meter movement. If we know exactly how the current is divided, the fraction of the current flowing in the movement can indicate the total current flowing in the branch in which the meter is connected.

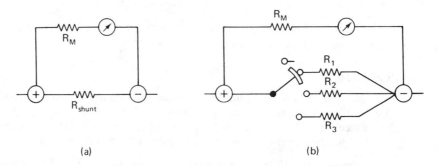

(a)                                         (b)

**Figure 4-11**   (a) Ammeter with shunt; (b) multirange ammeter.

**Example 4-2**

Given a 1-mA meter movement with an internal (coil) resistance of 50 Ω. If we want to convert the movement into an ammeter capable of measuring up to 150 mA, what will be the required shunt resistance?

**Solution.**   If the movement can handle a maximum of 1 mA, the shunt will have to carry the remainder of the current. Thus, for a full-scale deflection

$$I_{shunt} = I_{total} - I_{movement}$$

$$= 150 - 1$$

$$= 149 \text{ mA}$$

Since the voltage drops across the shunt and the movement are equal (by virtue of their being connected in parallel), then

$$V_{shunt} = V_{movement}$$

$$I_{shunt} R_{shunt} = I_M R_M$$

$$R_{shunt} = \frac{I_M R_M}{I_{shunt}} = \frac{(0.001)(50)}{0.149}$$

$$R_{shunt} = 0.32 \ \Omega$$

Many ammeters are multirange instruments. Some of these use several external terminals (binding posts) as a means of changing ranges: others use a rotary switch. If a rotary switch is used to change ranges [as shown in Fig. 4-11(b)], the switch pole must make contact with the adjacent shunt resistance before breaking contact with the shunt resistance in the former branch. By using such a shorting-type rotary switch, insurance that the movement will not be accidentally subjected to the full current in the branch is provided. For meters that can measure up to 50 A, the shunts are usually mounted within the instrument. Higher-range ammeters use external, high-current shunts made of special materials which maintain stability (constant resistance) over a wide temperature range.

Figure 4-12 shows two commonly used external ammeter shunts. Each is rated at a specific current level and voltage drop. For example, a 100-A, 50-mV shunt is designed to drop 50 mV across itself when 100 A is flowing through it. Thus any meter that indicates 50 mV at a full-scale deflection can be used to determine the current in this shunt.

The current is fed through the shunt by heavy current-carrying terminals. These heavy terminals are used to keep the contact resistance of this connection as

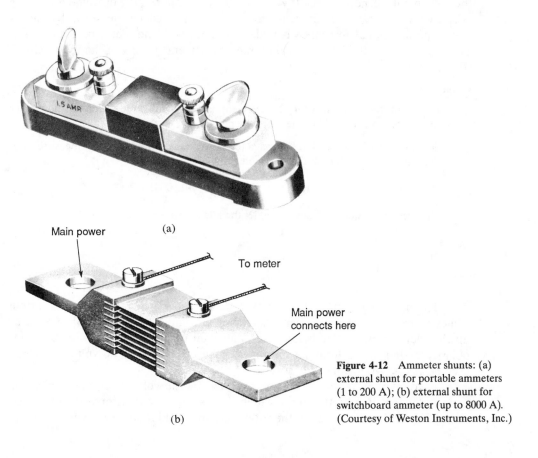

Main power

(a)

To meter

Main power
connects here

**Figure 4-12**  Ammeter shunts: (a) external shunt for portable ammeters (1 to 200 A); (b) external shunt for switchboard ammeter (up to 8000 A). (Courtesy of Weston Instruments, Inc.)

(b)

small as possible. The voltage drop across the shunt is measured by the meter movement which is connected to the two inner "potential" terminals. By measuring voltage drop across the potential terminals, the effect of any contact resistance on the measured value is eliminated.

The two heavy copper blocks that make up the ends of the shunt are welded to sheets of resistive material as shown in Fig. 4-12(b). The resistive material is specially chosen to keep a constant resistance value, even with changes in its temperature. Precision external shunts are built in ranges from 0.1 to 2000 A with accuracies of 0.1 percent.

## ANALOG DC VOLTMETERS

Most dc voltmeters also use D'Arsonval movements. The D'Arsonval movement itself can be considered to be a voltmeter if we note that the current flowing in it, multiplied by its internal resistance, causes a certain voltage drop. For example, a 1-mA full-scale, 50-$\Omega$ movement has a 50-mV drop across it when 1 mA is flowing in the movement. If the scale reads volts rather than amperes, the movement is acting as a 50-mV voltmeter. To increase the voltage that can be measured by such a meter, an additional resistance is added in series with the meter resistance. The extra resistance (called a *multiplier*) limits the current flowing in the meter circuit [Fig. 4-13(a)].

**Example 4-3**

If we want to use a 1-mA, 50-$\Omega$ movement as a 10-V full-scale voltmeter, how much resistance must we add in series with the movement?

**Solution.**   At full scale, 1 mA flows in the movement. If the meter is to measure 10 V, the total resistance required of the meter is

$$R_{total} = \frac{V}{I} = \frac{10 \ V}{0.001 \ A} = 10,000 \ \Omega$$

Since the movement resistance is 50 $\Omega$, the added series resistance must be

$$R_{series} = R_{total} - R_{movement}$$

or

$$R_{series} = 9950 \ \Omega$$

To construct a multirange voltmeter, we can use a switch that connects various values of resistance in series with the meter movement [Fig. 4-13(b)]. To get an upscale deflection, the leads must be connected to the voltmeter with the same polarity as the meter terminal markings. Typical laboratory dc voltmeters have accuracies of ±1 percent of full scale.

A voltmeter's *sensitivity* can be specified by the voltage required for a full-scale deflection. But another more widely used sensitivity criterion is the ohms/volt rating. For each voltage range, the total resistance exhibited by the voltmeter, $R_T$,

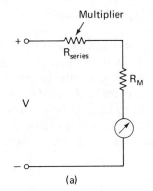

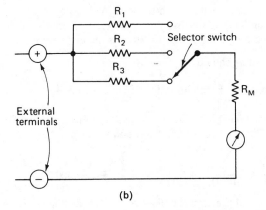

Figure 4-13   (a) Basic dc voltmeter; (b) multirange voltmeter.

divided by the full-scale voltage, yields a quotient, $S$. This quotient is a constant of the voltmeter, and it is called the *ohms/volt* rating. The easiest method for calculating $S$ is to find the reciprocal of the *current* sensitivity of the movement being used in the voltmeter.

**Example 4-4**

What is the ohms/volt rating of a voltmeter with (a) a 1-mA movement and (b) a 50-μA movement?

**Solution.**

(a) $\dfrac{1}{\text{current sensitivity of the meter movement}} = \dfrac{1}{0.001 \text{ A}} = 1000 \ \Omega/\text{V}$

(b) $\dfrac{1}{\text{current sensitivity of the meter movement}} = \dfrac{1}{0.00005 \text{ A}} = 20{,}000 \ \Omega/\text{V}$

The ohms-volt rating is essentially an indication of how well an actual voltmeter approaches the behavior of an ideal voltmeter. An ideal voltmeter would have an infinite ohms/volt ratio and would appear as an infinite resistance (or open circuit) to the circuit it was measuring. Typical basic dc laboratory voltmeters have a 20,000 Ω/V rating.

Because the voltmeter is not ideal, it draws some current from the circuit it is measuring. If a low-sensitivity (small ohm/volt rating) meter is used to measure the voltage across a high resistance, the meter will actually act like a shunt and will reduce the equivalent resistance of the branch. A highly unreliable reading will result. Such a circuit disturbance caused by current being drawn by a voltmeter is called a *loading effect*. Example 4-5 demonstrates a classic situation in which the choice of a voltmeter makes a significant difference in the accuracy of the measured result.

**Example 4-5**

We want to measure the voltage across the 10-kΩ resistor of the circuit shown in Fig. 4-14. We have two voltmeters with which to do it. Voltmeter $A$ has a sensitivity of 1000 Ω/V and voltmeter $B$ has a sensitivity of 20,000 Ω/V. Both use their 50-V scales. (a) Calculate what each meter will read. (b) What is the error from the true reading?

**Solution.**   The true reading should be

$$V_T = \frac{V_s}{R_1 + R_2} R_2 = 100 \times \frac{10\ \text{k}\Omega}{30\ \text{k}\Omega} = 33.3\ \text{V}$$

Now voltmeter $A$ has a 1000-Ω/V sensitivity. It has an equivalent $R$ of 50,000 Ω when using its 50-V scale. The total resistance from point 1 to point 2 with the voltmeter connected in the circuit is found from

$$\frac{1}{R_{12}} = \frac{1}{10\ \text{k}\Omega} + \frac{1}{50\ \text{k}\Omega} \qquad \text{or} \qquad R_{12} \simeq 8300\ \Omega$$

The total resistance, $R_T$, of the circuit is $R_T = R_1 - R_{12} = 28,300$ Ω. Then the voltage between points 1 and 2 of Fig. 4-14, as indicated by the voltmeter, is

$$V_{12} = V_s \frac{R_{12}}{R_T} = 100 \times \frac{8300\ \Omega}{28,300\ \Omega} = 29.0\ \text{V}$$

This is an error of

$$\text{error} = \frac{33.3 - 29.0}{33.3} \times 100\% = 13\%\ \text{low}$$

Voltmeter $B$ has a 20,000-Ω/V sensitivity, and thus its equivalent $R = 50 \times 20,000 = 1\ \text{M}\Omega$. Then

$$\frac{1}{R_{12}} = \frac{1}{10\ \text{k}\Omega} + \frac{1}{1\ \text{M}\Omega} \qquad \text{or} \qquad R_{12} = 9900\ \Omega$$

Therefore, $R_T = 29,900$ Ω, and the voltage indicated by the voltmeter is

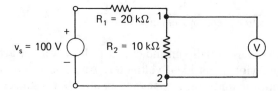

**Figure 4-14**

$$V_{12} = 100 \times \frac{9900\ \Omega}{29,900\ \Omega} = 33.1\ \text{V}$$

The error of this reading is

$$\text{error} = \frac{33.3 - 33.1}{33.3} \times 100\% = 0.6\%\ \text{low}$$

The example shows that the meter with the highest ohms/volt rating will yield the most reliable reading in terms of the possible loading error. We can use the same type of calculation to determine how sensitive a voltmeter must be if we want to reduce the loading error to some maximum percentage of the true reading. We also note that the loading error that may occur when measuring voltages in high-resistance circuits can often be far greater than the error owing to other inherent meter inaccuracies. In some such cases, accurate readings can be obtained only with electronic-type voltmeters that have input resistances of 10 MΩ or more. As a rule of thumb, to reduce the loading error of the voltmeter reading to less than 1 percent, the resistance of the voltmeter should be at least 100 times as large as the resistance of the path across which the voltage is being measured.

### Amplifier-Driven Analog dc Meters

Analog meters driven by electronic amplifiers are often referred to as electronic voltmeters (EVMs). When a meter is driven by an amplifier, significant improvements in performance can be achieved. The amplifier allows the meters to offer more sensitive ranges and higher input resistances than comparable instruments that contain only a simple meter and range-multiplier resistors. For example, analog EVMs typically exhibit input resistances of 20 MΩ or more on dc scales and 1 MΩ on ac scales.

The two types of amplifiers generally used for driving analog dc EVMs are direct-coupled (DC) amplifiers, and chopper amplifiers. The DC amplifiers are used in the less expensive analog dc electronic meters. Their most sensitive scales are usually limited to 0.1 to 1.0 V because of the poor dc stability of DC amplifiers. (The dc stability is a measure of an amplifier's ability to maintain a constant output for a given input. Generally, DC amplifiers change their outputs by several millivolts as their internal temperatures change. Therefore, for ranges of less than 50mV, such changes would create sizable errors.) Related characteristics of DC amplifiers are discussed in more detail in Chapter 15.

The problem of poor stability in DC amplifiers is greatly reduced in higher-priced amplifier-driven dc meters by the use of *chopper amplifiers*. The favorable characteristics of chopper amplifiers allows such meters to be built with ranges down to a few millivolts or microvolts, full scale. The chopper amplifier converts the input dc signal to a proportional ac signal, amplifies it with an ac amplifier, and then converts it back to a dc quantity. Any dc drift at the input of the ac amplifier is not passed to the outputs, thus allowing accurate amplification of very small voltages.

## ANALOG AC AMMETERS AND VOLTMETERS

Electrical signals that change in amplitude and direction periodically with time (Fig. 4-15) are measured with *ac meters*. Such ac meters may respond to the average, peak, or rms value of the applied periodic ac signals. These meters are also calibrated to display their outputs in terms of one of these three ac signal characteristic values. As a result, if accurate measurements of ac signals need to be made, the following guidelines should be followed. First, consider which characteristic value of the wave-form is being sought (average, peak, or rms value). Then, if possible, select a meter that both responds to and is calibrated to display its output in that characteristic. If this is not possible, a correction factor between the actual reading and the desired value of the characteristic needs to be determined. If such is the case, however, it will probably prove easier and more accurate to observe and measure the desired characteristic value from the actual ac waveform with an oscilloscope or spectrum analyzer rather than from the meter at hand. Most manufacturers do not readily indicate how their meters respond. This forces the meter user to test the meter to determine if it is an average, peak, or peak-to-peak responding meter and whether or not the meter scale is calibrated for rms sine waves. Once the meter type is known, tables can be generated for converting the actual meter reading to an rms reading (see references).

### Average-Responding ac Meters

When ac signals whose frequencies are greater than a few cycles per second (Hz) are applied directly to a D'Arsonval meter movement, the inertia and damping of the movement prevents the rapid signal fluctuations from being followed. Instead, the pointer of the meter movement takes up a position in which the average applied torque is balanced by the torque of the shaft springs. The D'Arsonal movement responds to the *average* value of the current through the moving coil. As was pointed out in Chapter 1, the average value of sine waves is zero. Therefore, to achieve a measurable deflection of the D'Arsonval movement pointer when sine waves and

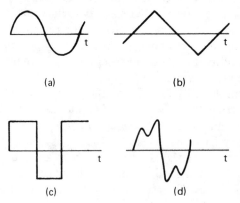

(a)                    (b)

(c)                    (d)          **Figure 4-15**  Examples of ac waveforms.

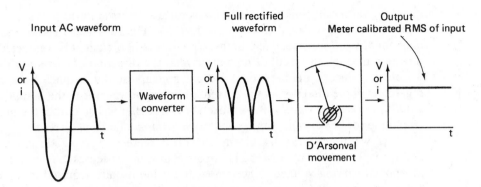

**Figure 4-16**  Block diagram of rectifier-type ac meter that uses a full-wave rectifier for the waveform converter.

other alternating waveforms are measured, some means must be devised to obtain a *unidirectional* torque which does not reverse at each half-cycle.

One method involves rectification of the ac signals by using diode rectifier circuits. The resultant output of the rectifier circuit is a time-varying, unidirectional quantity, which can be used to produce a nonzero deflection of the D'Arsonval meter pointer (Fig. 4-16). The D'Arsonval movement indicates the average value of the quantity that is applied to it. (The scale of the meter is calibrated to display the rms value of the ac waveform that is applied to the meter.) A rectifier is a circuit containing diodes that convert ac quantities to unidirectional (dc) quantities.

The circuit in Fig. 4-17(a) shows the simplest type of rectifier. It contains only a single diode. If an ac sinusoidal signal is applied to this diode, current will be conducted only during the time when the ac current is flowing in the positive direction. During the times that the current has a negative value, the diode prevents the current from flowing. As a result, the output of this rectifier is a unidirectional

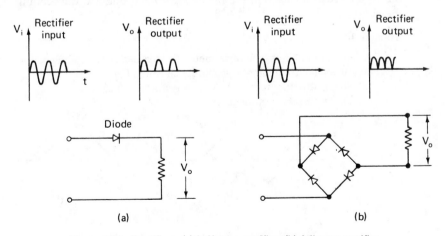

**Figure 4-17**  Rectifiers: (a) half-wave rectifier; (b) full-wave rectifier.

quantity. Since only half of the initial waveform appears at its output, such a simple type of rectifier is called a *half-wave rectifier*. Figure 4-17(b) shows the circuit diagram of a *full-wave rectifier*. In this type of rectifier, current is conducted in one direction through the rectifier no matter what the direction of current in the input waveform. Since a unidirectional current appears at the output of the rectifier regardless of the direction of the current of the input waveform, the output is called a *full-wave rectified waveform*. A third type of rectifier whose output is equal to the amplitude (or peak) value of the input waveform is called a *peak rectifier*.

As we noted, the waveform converter feeds a varying dc current to a D'Arsonval movement. Now the D'Arsonval movement senses the *average* value of an applied varying voltage. Since rms is the value usually required of an ac measurement, the scale must be designed by the manufacturer to read rms rather than average values. This process requires two necessary steps. First, an assumption is made that most of the ac signals to be measured will be pure sine waves. Second, the characteristics of the rectifier circuit are used to calculate the ratio between the average and rms values of the rectified wave. For example, if the waveform converter generates a full-wave rectified waveform to activate the D'Arsonval meter, we can calculate that the ratio of the rms value of a *sinusoid* to the average value of a *full-rectified* sinusoid is 1.11 (see Chapter 1). Then the rms scale of the meter will actually be indicating 1.11 times the actual average current flowing in the D'Arsonval movement. For a *half-wave rectified* sinusoid, this ratio is 2.22.

The first assumption is not always valid, and this can lead to errors in the readings of rectifier meters. If the ac signal is not a pure sine wave [it could be a square wave, Fig. 4-15(c), triangular wave, Fig. 4-15(b), or many sinusoids of different frequencies added together], the meter will still be reading an average value of the rectified waveform. However, the ratio between this average value and the rms value of the rectified signal may no longer be the same as the ratio applicable for sinusoidal signals. A correction factor based on the shape of the ac signal and the type of rectifier used in the meter must be applied to the readings to eliminate the error.

**Example 4-6**

A full-wave rectifier type of ammeter displays a current reading of 4.44 mA rms when measuring the triangular-wave current shown in Fig. 4-18(a).
(a)What is the true rms value of the waveform?
(b)What is the error of the meter?

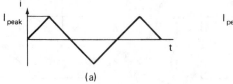
(a)

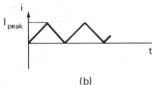
(b)

**Figure 4-18**

[3] Figure 1-7 has the formula for the rms value of this triangular waveform and other common ac waveforms.

**Solution.** The instrument rectifier produces the resulting wave shown in Fig. 4-18(b). The meter displays 1.11 times the average value of the rectified wave. Therefore, the average value of the rectified wave is $I_{av} = 4.44$ mA/1.11 = 4 mA, and the peak value for a triangular wave is $I_{peak} = 2 I_{av} = 8$ mA. The true rms value of the original wave is found in this case from[3]

$$I_{rms} = \sqrt{\frac{I_{peak}^2}{3}} = \sqrt{\frac{(0.008)^2}{3}}$$

$$= 4.6 \text{ mA}$$

Hence, the meter error is

$$\text{meter error} = \frac{(\text{true value}) - (\text{indicated value})}{\text{true value}} \times 100\%$$

$$= \frac{4.60 - 4.44}{4.6} \simeq 4\% \text{ low}$$

Basic rectifier ac meters (i.e., those not augmented with electronic amplifiers) also suffer from variations in their readings with frequency (even though their upper limits are much higher than those of electromechanical ac meters). If accurate readings need to be made by using such rectifier instruments, the manufacturer's specification of the upper frequency limit should be noted. (A rule of thumb that can be applied to rectifier ac meters is that their error can increase by about 0.5 percent for each 1-kHz increase in frequency above their maximum specified frequency.)

Another source of error in basic rectifier ac meters arises from the 0.6 to 0.7 V of forward bias required by the silicon diodes of the rectifier for conduction. As a result of this voltage drop, only that portion of the source voltage in excess of twice the forward diode bias voltage (since two diodes are in series in full-wave rectifiers) is effective in causing a current in the meter. For this reason, rectifier ac meters are provided with nonlinear scale markings for full-scale voltage ranges below 10 V. The diode forward voltage drop also limits the measurement capability of ac rectifier meters to ac signals whose amplitudes are greater than 1 V.

## Analog ac Electronic Voltmeters (Average Responding)

To overcome the drawbacks of not being able to measure ac voltages of less than 1 V with simple rectifier ac meters, amplifiers can be incorporated into the ac meter design. If an ac amplifier is connected in the meter circuit before the rectifier, as shown in Fig. 4-19(a), ac voltages down to the nanovolt range can be measured. In addition, the input impedance of such amplifier-driven rectifier ac meters is typically increased to $10^{10}$ $\Omega$ for ranges less than 3 volts and greater than 10 M$\Omega$ for ranges greater than 30 volts. Modern analog ac meters are designed with an internal microprocessor, and this permits the user to specify whether the output is expressed in average, peak, minimum, or rms form. The accuracy of ac electronic meters ranges from 0.02 to 10 percent of the meter reading depending upon the scale and frequency selected. The specified accuracy is only valid for sinewave inputs. Digital

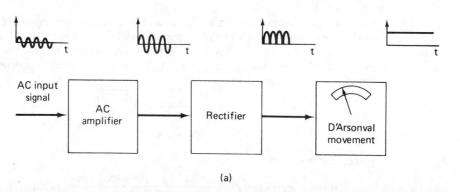

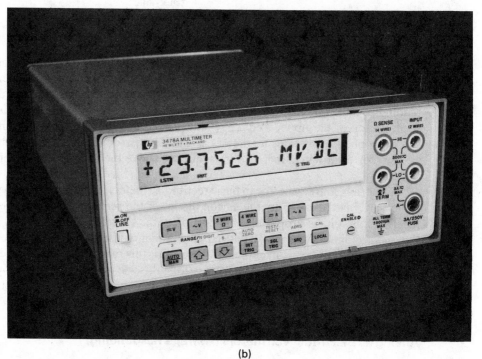

**Figure 4-19** (a) ac meter block diagram; (b) high sensitivity multimeter. (Courtesy of Hewlett-Packard Co.)

meters, however, can be custom calibrated for any waveform since the calibration is performed by making software changes through a port on the rear of the meter case.

### rms-Responding ac Meters

Although most ac meters are calibrated in rms volts or current, very few actually respond to the rms value of the measured quantity. Nevertheless, when an accurate measurement of the rms value of an ac signal is required and the relationship between the average value and the rms value of the signal is not known, a meter that actually responds to the rms value of a signal must be used. Let us discuss three types of meters that actually respond to the rms value of a signal: (1) electro-dynamometer meters, (2) thermocouple meters, and (3) calculating-type meters.

The electrodynamometer movement responds to the square of the applied current and gives a true rms reading. Ac meters built by using this movement can be extremely accurate (especially at power-line frequencies of 60 Hz) but are relatively expensive. They are also limited by their minimum power requirement for activation (1–3 W). Their upper frequency limit is 200 Hz. Above 200 Hz, the inductance of the coils of the movement begins to introduce significant errors. However, for measuring ac signals with frequencies lower than 200 Hz, they are the most accurate instruments available.

*Thermocouple meters* are the second type of ac meter that respond to the rms value of a signal. The thermocouple meter measures ac and dc quantities by connecting the output signal of a thermocouple to a D'Arsonval movement (Fig. 4-20). The major advantage of thermocouple meters over all other ac meters is that frequencies of up to 50 MHz can be measured with up to 1 percent accuracy. The current in the circuit being measured flows through a heating element and causes the temperature of the heating element to increase. This element in turn heats a thermocouple junction and causes a voltage to arise across the junction. The voltage produces a current in the thermocouple wire, and this current activates the D'Arsonval movement. Since the heating effect is proportional to $i^2R$, the resulting current that activates the D'Arsonval movement gives a true rms reading ($i$ is the current flowing in the heating element, and $R$ is the element's resistance). Current in the range 0.5–20 A and voltages up to 500 V can be measured with this instrument. However, the disadvantages of higher cost and sensitivity to burnout under

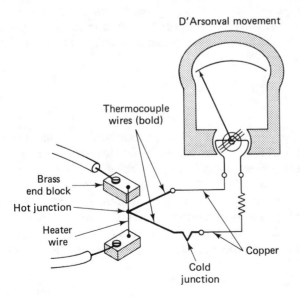

Figure 4-20 Thermocouple meter.

overloads have limited the widespread use of thermocouple meters. The hot and cold junctions must also be in proximity so that they are at the same temperature when no current is flowing through the heater wire. Further discussion on the principles of thermocouples is presented in Chapter 14.

A third type of ac meter that produces a true rms meter is one in which the rms value is calculated by the meter. Single IC chips that perform such calculations are available that also have differential inputs so ground loops are not a problem (cf. Chapter 17). Smart rectifiers are also incorporated. The output is an absolute-value function regardless of the waveform being measured. Two such ICs are the AD736 and AD737.

## Peak Responding ac Meters

If the rectifier used to convert ac to dc voltages in a meter is the *peak rectifier* type, the meter output will indicate the peak value of the input voltage waveform. Peak responding ac meters accomplish this function by using a diode circuit to charge a capacitor to the peak signal voltage. Then the resulting capacitor voltage is measured with a very high input impedance dc voltmeter. The frequency response of such peak responding meters is quite high (often in excess of 1 GHz) and can be made very sensitive when operated together with a differential amplifier. Thus, such peak responding voltmeters find application where both high sensitivity and high-frequency response are required.

Often, the scale of peak responding voltmeters is calibrated in rms volts. Since the scaling factor used for this calibration is also based on the assumption of sinusoidal input waveforms, the rms scale will, in general, be accurate only for sine-wave signals. Some peak responding meters, however, also have peak scales. Since this scale is calibrated in the same characteristic as the one to which the meter responds, it will be accurate for all waveforms within the frequency-response limits. Many instruments use peak responding measurements. One example is a capacitance meter where the output voltage is proportional to the peak voltage across a capacitor as a result of being charged by a constant current source for a predetermined period.

## ANALOG MULTIMETERS

Analog multimeters are very useful and versatile laboratory instruments and are capable of measuring voltage (dc and ac), current, resistance, transistor gain, diode forward-voltage drop, capacitance, and inductance. They are commonly referred to as *volt ohm milliammeters* (or VOMs).

A number of enhancements have recently improved the performance capabilities of such meters. First, the incorporation of microprocessors into VOMs has greatly expanded their capabilities and accuracy. In addition, with the use of FET input amplifiers for dc voltage measurements, their impedances commonly exceed

100 MΩ. Finally, the ohmmeter scale no longer has to be re-zeroed to compensate for internal battery voltage changes or scale changes. Voltage measurements can be made over the range of 0.4 mV to 1000 V with accuracies of 0.1 percent. Current measurements can be made over the range of 0.1μA to 10A with accuracies of 0.2 percent. Resistances as high as 40 MΩ are measured with an accuracy of 1 percent. (It should be noted that while making high resistance measurements, the measuring tip should never be touched with your finger because skin resistance is only several thousand ohms, and this can produce serious measurement errors). Lower resistance measurements are accurate to 0.2 percent.

Digital VOMs have replaced the majority of D'Arsonval movement VOMs for two major reasons: improved accuracy and elimination of reading errors. An analog scale, however, is often still incorporated into the digital scale in order to provide visual indication of time varying inputs (Fig. 4-21). The ability to observe the meter display in analog form is very important when troubleshooting many instrumentation problems. For example, the rate at which a variable changes, as well as its magnitude, can provide valuable insights in many troubleshooting situations.

Analog meters that incorporate D'Arsonval movements, on the other hand,

**Figure 4-21**  Typical volt-ohmmeter.
(Courtesy of John Fluke Mfg. Co., Inc.)

are still used in applications where the displays of many meters must be observed at a glance. For example, most electrical utility substations use analog meters because it is much easier to observe the red mark for maximum current on 30 analog meters than it is to try to remember 30 numbers and their overrange values. Some very inexpensive VOM meters also still incorporate the D'Arsonval movement. The data taken with such meters must be carefully interpreted because their input impedance can be quite low, ranging from 1000 $\Omega$/V (1 mA movement) to 20,000 $\Omega$/V (50 $\mu$A movement). As a result, loading effects can cause significant errors if voltage readings across large resistances are taken. In addition, the ohmmeter in such VOMs must be re-zeroed each time the meter is used and whenever the range is changed. During the re-zeroing process, the probes must be shorted. If the meter will not zero, the internal battery is weak and must be replaced. The state of the battery, however, has no effect on either a voltage or current measurement. As a final word of caution, reverse polarity can easily damage the D'Arsonval movement of such meters as can dropping the instrument.

## SPECIAL-PURPOSE ANALOG METERS

### ac Clamp-On Meters

The *clamp-on ac meter* is an instrument used for measuring ac currents and voltages in a wire without having to break the circuit being measured. The meter makes use of the transformer principle to detect current. That is, the clamp-on device of the meter serves as the core of a transformer. The current-carrying wire is the primary winding of the transformer, and a secondary winding is in the meter. The alternating current in the primary is coupled to the secondary winding by the core, and, after being rectified, the current is sensed by a D'Arsonval movement.

Although the clamp-on ammeter is very convenient for making rapid ac current and voltage measurements, it is limited to relatively high current levels. The meter (Fig. 4-22) can read currents from 1 to 999 amperes for each turn of the current carrying conductor around a jaw. For example, if the jaws are placed around a conductor as shown in Fig. 4-22, the meter will indicate the amperes passing through the conductor. However, if the current carrying conductor is wrapped five times around the jaw, the meter will indicate five times the current passing through the conductor. Wrapping the conductor around the jaw is analogous to increasing the number of turns on a transformer. With five turns wrapped around a jaw, a conductor current of only 400 mA would be displayed as two amperes.

### Electrometers

*Electrometers* are dc voltmeters that have extremely high input impedances (up to $10^{16}$ $\Omega$). Such high input impedances are a necessary property of voltmeters that are used to measure voltages in very-high-impedance circuits. (Recall that a voltmeter

(a)

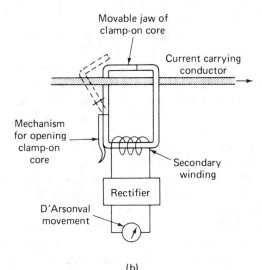

(b)

Figure 4-22 (a) Clamp-on ammeter (Courtesy of Amprobe Instrument); (b) schematic diagram of clamp-on ammeter.

must have an input impedance at least 100 times as great as the impedance of the circuit being measured in order to keep loading errors below 1 percent.) One application in which such high-impedance voltmeters are a necessity is in the measurement of the pH (hydrogen ion concentration) of a chemical solution. In such pH meters, voltages of about 50 mV must be measured across the walls of a glass tube. Such glass walls have resistances of 500 MΩ or more. Therefore, a voltmeter with an

**Figure 4-23**  Electrometer. (Courtesy of Keithley Instruments, Inc.)

input impedance of $5 \times 10^{10}$ Ω or more is needed to measure these voltages accurately. (The standard 10-MΩ input impedance of the multipurpose EVM or DVM would be far too small to make this measurement accurately.) In another application, the meager output current of phototubes is fed through very large valued resistors so that a voltage drop of measurable magnitude can be generated. The voltages across these resistors must be measured by very high input impedance meters. A third application involves the measurement of leakage current in capacitors.

Electrometers are designed to fill these and other applications calling for high-input-impedance instruments. They can also act as current and electric charge detectors of the highest sensitivity. Some electrometers can detect currents of $10^{-16}$ A or less, and measure charge as low as $5 \times 10^{-16}$ C. Figure 4-23 is a photograph of a commercially available electrometer.

The electrometer possesses these unique characteristics because it contains specially designed circuits and materials. There are several types of devices used to build electrometers, depending on the particular application of the electrometer. These devices include MOSFETs (metal-oxide-semiconductor field-effect transistors), electrometer tubes, and vibrating capacitors.

Electrometers do not suffer from drift problems as badly as dc microvoltmeters do because their smallest scales are generally in the millivolt range. A typical electrometer may be specified as drifting less than 2 mV/h. However, the insulation required by electrometers must be extremely good, so that leakage or stray pickup do not contaminate its readings.

## Nanoammeters and Picoammeters

Nanoammeters and picoammeters are specially designed meters for measuring very small current levels. A microvoltmeter measures the voltage drop across a shunt resistor, and the voltmeter reading is calibrated to indicate current rather than voltage. Meters that can measure currents as small as 0.3 pA ($3 \times 10^{-13}$ A) full scale and with accuracies of $\pm 2$ to $\pm 4$ percent (depending on the range) are available.

## HOW TO USE BASIC METERS

**1.** Ammeters are always connected in series with the branch whose current is being measured and never in parallel. The ammeter can be destroyed if connected in parallel by mistake. Its low resistance may allow enough current to flow in the meter to burn it out. *Voltmeters* are always connected in parallel with (across) the portion of the circuit whose voltage is being measured.

**2.** Always make sure that the pointer is set at zero before a meter is connected. If the pointer is not set at zero, adjust the zero-setting screw on the scale face.

**3.** Do not handle meters roughly. The meter shaft and bearings are easily damaged by sharp knocks or jarring.

**4.** To protect the meter movements of multirange meters, start all readings of unknown quantities by setting the meters on their highest scale. Take as the final reading the deflection that is nearest to full-scale. This final reading will be the most accurate value.

**5.** Set portable meters on their backs. This will help keep them from being knocked over and damaged.

**6.** Correct the readings for any loading effect that is caused by the meter's presence in the circuit (see sections on dc voltmeters and ammeters for a discussion of loading).

**7.** To give upscale readings, dc meters must be connected so that the meter terminals are connected to points in the test circuit with matching polarities. Reversed polarity connections can lead to movement damage arising from the banging of the pointer against the reverse stop.

**8.** ac meters—iron-vane, electrodynamometer, and electrostatic ac meters—can be connected independent of polarity considerations.

**9.** Keep meters (especially iron-vane and dynamometer meters) away from conductors carrying large currents. The magnetic fields associated with the currents can interfere with the meter movement's magnetic fields and introduce errors.

**10.** VOM. (a) When not in use, always have the function selector switch set to a high dc-volts scale. This will prevent draining of the battery if an accidental shorting of leads occurs. It also protects the rectifier circuit against accidental connection to a dc source. (b) Check the battery to ensure that it is operating above its minimum allowed voltage. (c) Use each meter function as you would use such a meter alone. (d) If the ohmmeter cannot be zeroed when the leads are shorted, then the battery must be replaced.

**11.** Meters should be calibrated once a year or according to the manufacturer's specifications. Apply a calibration sticker to the meter stating the date calibration was last performed.

## METER ERRORS

**1.** *Scale error.* Inaccurate markings of the scale during calibration or manufacture. Equally probable along the entire scale.

**2.** *Zero error.* Failure to adjust to zero setting before making measurements.

**3.** *Parallax error.* Caused by not having line of sight exactly perpendicular to measuring scale. Somewhat eliminated by mirror under scale.

**4.** *Friction error.* If the bearing is damaged or worn, its friction may prevent the needle from making a true reading. Can be somewhat eliminated by *gently* tapping on meter while making a measurement.

**5.** Temperature effects on magnets, springs, and internal resistances. These errors are proportional to the percent of deflection.

**6.** Error caused by coil-shaft misalignment on bearing; it is reduced by keeping the shaft vertical.

**7.** Bent pointer or pointer rubbing against scale.

**8.** *Poor accuracy.* If a meter is said to be accurate to within some percentage, this usually refers to its full-scale reading. For readings taken at less than full scale, the *actual* percentage error may be much larger. This only applies to analog meters.

**Example 4-7**

If a meter is accurate to 2 percent of its 100-mA scale, any reading will only be guaranteed to within ±2 mA. Thus, if this scale is used to measure 20 mA, the reading may be in error by ±2 mA or 10 percent of its true value. To minimize this error, (1) always use the reading closest to full scale for the final reading and (2) make sure the meter is properly calibrated.

**9.** Loading-effect error owing to introducing a nonideal instrument into a circuit. The disturbance of the circuit by the instrument can be calculated and compensated for in the reading if a more sensitive meter is not available.

**10.** Specific errors associated with a particular meter's operating principles and design. Examples of these are hysteresis effects in moving-iron-vane movements or waveform errors in nontrue-rms-reading ac instruments. The extent of these errors is determined from a knowledge of the meter and its operations.

**11.** *Common-mode noise error.* As was discussed in Chapter 16, common-mode noise can cause serious errors in many electronic measurement systems.

Examples 16-2 and 16-3 examine how common-mode noise appears at the inputs of a voltmeter and how its effects can be reduced through the use of input guarding.

## PROBLEMS

1. Explain, by describing the effects on the meters and on the circuits being measured, why a voltmeter should never be connected in series or an ammeter in parallel with the circuit being measured.

2. If a D'Arsonval movement is rated at 5 mA, 60 $\Omega$, what is the current sensitivity of the movement?

3. What functions do the torque springs perform in the D'Arsonval movement?

4. A D'Arsonval movement has a coil form with 50 turns and with dimensions of $l = 3$ cm and $r = 1$ cm. The permanent magnet of the movement supplies a uniform flux density of 0.4 Wb/m$^2$. If the full-scale current of the movement is 2 mA, find the torque (in pounds) exerted on the coil.

5. If a 10 mA movement has an internal resistance of 4 $\Omega$, calculate the shunt resistance required to increase its range to
   (a) 100 mA
   (b) 200 mA

6. Design a multirange ammeter with ranges of 0–10 mA, 0–100 mA, and 0–1500 mA, using a 1-mA movement whose internal resistance is 50 $\Omega$ (i.e., show a circuit diagram of the meter you have designed).

7. Why is a "make-before-break" switch used in the design of multirange ammeters?

8. If a 1-mA, 50 $\Omega$ movement is inserted into a circuit branch whose resistance is 600 $\Omega$, find the error in the indicated current values caused by the insertion of the meter. If the movement is converted into an ammeter whose full-scale range is 10 mA, find the resulting error.

9. Design a voltmeter that uses a 50-mA, 1500 $\Omega$ movement and has ranges of 0–5, 0–50, 0–500 V. What is the sensitivity of this voltmeter? Use a circuit diagram to show the instrument you have designed.

10. If a voltmeter has the same movement as the meter used in Problem 6, what is the input resistance of the instrument on the following scales?
    (a) 5 V
    (b) 500 V

11. A typical vacuum-tube voltmeter has an input resistance of 11 M$\Omega$. If a D'Arsonval movement were used to build a voltmeter, what sensitivity would the movement have to have if the voltmeter were to have the same input resistance as the VTVM when the voltmeter was used on its 0–10 V scale?

12. If a 0–100 V voltmeter has an accuracy of 1 percent full scale, between what limits may the actual voltage lie when the meter indicated 12 V? What is the range of error at this reading?

13. Two resistors with values of 12,000 $\Omega$ and 6000 $\Omega$ are connected in series across a 100-V voltage source. A voltmeter with a 1000-$\Omega$/V rating is used to measure the voltage across

the 12,000-$\Omega$ resistor. Calculate the error in the measured voltage value caused by the loading effect of the voltmeter on the circuit.

14. What will be the action of a D'Arsonval movement with a zero-center scale if it is used to measure quantities in the following ac circuits?
    (a) Circuit being driven by a 60-Hz sinusoidal voltage source
    (b) With a slowly varying (less than 5 Hz) but nonperiodic source

15. If a dynamometer ammeter responds to $i^2$ and a rectifier ammeter to $I_{av}$, explain how both types of ammeters can be used to indicate the rms values of currents.

16. Predict the readings of (a) a dynamometer ammeter and (b) a rectifier ammeter if the waveform shown in Fig. P4-1(a) is applied to the meter.

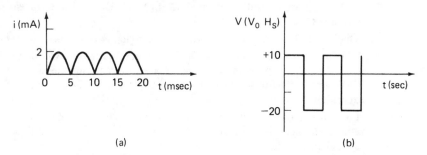

(a)                                   (b)

**Figure P4-1**

17. Repeat Problem 16 for the waveform shown in Fig. P4-1(b).

18. If a thermocouple ammeter reads 10 A at full scale, what will be the current passing through the meter when the meter deflection is one-half of full scale?

19. What is the difference between a taut-band movement and a conventional meter movement?

## REFERENCES

1. Rider J. F., and Prensky, S. D., *How to Use Meters*, 2nd ed. New York: John F. Rider, 1960.
2. General Electric, *Manual of Electric Instruments*, 1958.
3. Stout, M. B., *Basic Electrical Measurements*, 2nd ed. Englewood Cliffs, N.J.: Prentice Hall, 1960.
4. Cooper, W. D., *Electronic Instrumentation and Measurement Techniques*, chaps. 4 and 5. Englewood Cliffs, N.J.: Prentice Hall, 1978.
5. Coombs, C., ed., *Basic Electronic Instrument Handbook*, chaps. 20 to 23, New York: McGraw-Hill, 1972.
6. Kaufman, M., and Seidman, A., *Handbook of Electronics Calculations*, p. 23–28. New York: McGraw-Hill, 1979.

# 5

# *Digital Electronic Meters*

The digital electronic meter (abbreviated DVM, for digital voltmeter, or DMM, for digital multimeter) indicates the quantity being measured by a numerical display rather than by the pointer and scale used in analog meters (Fig. 5-1). The numerical readout gives the DVM the following advantages over analog instruments in many applications:

**1.** The accuracies of DVMs are much higher than those of analog meters. For example, the best accuracy of analog meters is about 0.5 percent, while the accuracy of DVMs can be 0.005 percent or better. Even on simple DVMs and DMMs, the accuracy is at least ±0.1 percent.

**2.** A definite number is provided for each reading made by the DVM. This means that any two observers always see the same value. As a result, such human errors as parallax or misreading of the scale are eliminated.

**3.** The numerical readout increases the speed of reading and makes the task of taking measurements less tedious. In situations where a great number of readings must be made, this might be an important consideration.

**4.** The repeatability (precision) of DVMs becomes greater as the number of digits displayed is increased. The DVM may also contain automatic ranging and polarity features that protect it from overloads or reversed polarity.

**5.** The DVM output can be fed directly to recorders (printers or tape punches) where a permanent record of the readings is made. These recorded data are in a

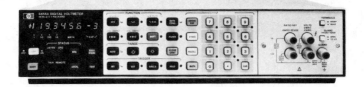

**Figure 5-1**   Digital voltmeter. (Courtesy of Hewlett-Packard Co.)

form suitable for processing by digital computers. With the advent of integrated circuits (ICs), the cost of DVMs has been reduced to the point where some simple models are now competitive in price with conventional analog electronic meters.

At this point it is useful to discuss the "$\frac{1}{2}$" digit in the DVM/DMM specification that refers to $3\frac{1}{2}$ digit, $4\frac{1}{2}$ digit, and so on. What is meant by a "$\frac{1}{2}$" digit? If a DVM displays a "$\frac{1}{2}$" digit, this means that the most significant digit of the DVM/DMM display can only be a "0" or "1," whereas all the other digits can be "0" to "9." The "$\frac{1}{2}$" digit will light up only if the basic range of the instrument is exceeded by less than 100 percent. For example, if a $3\frac{1}{2}$ digit DVM was set on its 1.0-V range, and a voltage of 1.536 V was applied to the meter, the display would show 1.536 V even though the range of the 1.0-V display (0.999 V) would have been exceeded. A three-digit DVM on the 1.0 V range could only read 0.999 V in response to the same signal. (Additional information is contained in Chapter 2.)

The heart of the DVM (and DMM) is the circuitry that converts the measured analog signals into digital form. These conversion circuits are known as *analog-to-digital* (A/D) converters. Therefore, in this chapter, in addition to describing the characteristics of DVMs and DMMs, we also discuss the characteristics of the most widely used A/D converters. We will also cover the circuitry that converts digital signals into analog form, digital-to-analog converters (or DACs). Finally, we shall examine the circuitry and devices that are utilized in digital meters to *display* the measurement information.

To explain the operation of DACs and A/D converters, however, we occasionally have to make reference to devices and circuits which are discussed in other chapters. Therefore, if the reader should want additional information on these devices he should consult the relevant chapters (i.e., for clocks, Chapter 1, and for integrators, Chapter 15).

All A/D converters utilize one or more *comparators* as part of the digitizing process. A comparator is a device whose output signal indicates whether an input voltage $V_{in}$ is larger or smaller than a reference voltage $V_r$ (Fig. 5-2). That is, if $V_{in}$ of Fig. 5-2 is less than $V_r$, the output of the comparator is $V_H$ as shown in Fig. 5-2(b). If $V_{in}$ is greater than $V_r$, the output is $V_L$. Thus the output of a comparator is one of two states, $V_L$ or $V_H$. This means that a digital output is obtained from an analog input. The output depends upon the voltage of the power supply or battery since comparators are operated in the saturated mode. If a typical operational amplifier is used as a comparator, its output will only go to within 1.5 volts of the extremes of the power

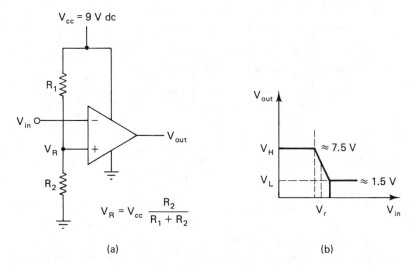

**Figure 5-2**  Comparator.

supply voltage (power supply *rails*). For example, if ground and a 9-volt battery (9 vdc) are used as the power supply of a single operational amplifier comparator, the $V_H$ will be approximately 7.5 vdc and $V_L$ will be approximately 1.5 vdc. Special comparators, however, are available that will allow the output voltage $V_H$ and $V_L$ to swing within millivolts of the power supply rail.

## DIGITAL-TO-ANALOG CONVERTERS

As the name suggests, digital-to-analog converters (DACs) are circuits that convert digital signals into analog electrical quantities directly related to the digitally encoded input number. We will describe the operation of DACs prior to discussing analog-to-digital converters (ADCs) because DACs are key components inside most A/D converters. DACs are also used in many other applications, including cathode-ray tube display systems, voice synthesizers, automatic test systems, and process control actuators. In addition, they are devices that help allow computers to communicate with the analog world. For example, the digital output signal from a process computer cannot be used to directly operate a valve that controls gas flow to a furnace. Most commercial valves are designed to operate on a 4 to 20 mA analog current signal (see Chapter 18). At 4 mA the valve is closed, and at 20 mA the valve is fully open.

DACs can perform their conversions in either parallel or serial form. The decision to use either serial or parallel DACs is based on the end use. For example, military weapon systems and modern instrumentation (such as digital storage oscilloscopes) usually require the high speed of parallel operation. Process control applications (such as filling a bottle with pills) can be performed slowly or serially. A

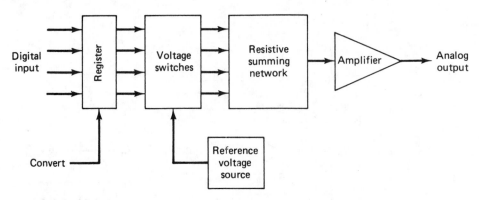

**Figure 5-3**   Block diagram of basic digital-to-analog converter (DAC).

1 millisecond conversion time would typically be considered very slow from a military electronic system standpoint but would be very fast for a typical industrial process. Since the input quantity to the DAC is a digital number, all the conversion techniques convert the number into a corresponding number of units of current (most common), voltage, or charge, and then sum these units with an analog summing circuit.

A basic DAC circuit is shown in block diagram form in Fig. 5-3. The digital input number to be converted is fed into the *input register* of the DAC upon receipt of an external CONVERT command. (The register is designed to accept a digital input only during the duration of the CONVERT command. After acquisition, the register holds constant that digital number until another CONVERT command is received.) The outputs of the register feed the digital input number to voltage switches that provide one of two possible outputs, 0 volts or the value of the precision reference voltage source. Thus these switches are the equivalent to an ordinary SPDT switch (usually in the form of transistor switches) controlled by the binary signals sent from the register. The switches provide access to a resistive summing network (about which more details will be supplied in the subsequent sections) that converts each bit into its weighted current value and then sums them for a total current. The total value is then fed to an amplifier which performs two functions: current-to-voltage conversion and output scaling.

The two most popular types of resistive summing networks are the weighted binary resistance type and the ladder (or $R$-$2R$) type.

## Weighted Binary Resistive Summing Networks

The weighted binary resistive summing networks (Fig. 5-4) are used in the least complex DACs, but they are difficult to build because of nonstandard resistors and the ideal value of the offset compensating resistor $R_c$ is different for each digital word. The magnitude of current proportional to the value of each bit in the digital input word is generated by the reference voltage, $V_{\text{REF}}$, divided by the weighted

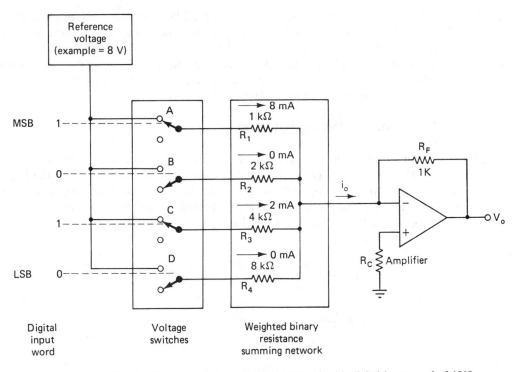

**Figure 5-4** Weighted binary resistance DAC (4-bit DAC with digital input word of 1010 causing $i_0$ to be 10 mA).

precision resistor, $R_n$, or $I_n = V_{REF}/R_n$. The weighting of the resistors is that of straight binary coding (8, 4, 2, 1 for a 4-bit DAC), with the most significant bit (MSB) in a 4-bit DAC having a weighting resistance one-eighth that of the LSB. By way of example, if the 4-bit DAC shown in Fig. 5-4 had a digital input signal of 0000, the switches $ABCD$ would all be open and the output of the DAC would be 0 mA. If the input digital word is 1010 (as shown in Fig. 5-4), the output would be $8 + 0 + 2 + 0 = 10$ mA. Since no current flows into either input terminal of an ideal amplifier, all the current flows through the feedback resistor $R_F$ producing a $-10$ volt output. To simplify the analysis, assume that both input terminals are at the same potential. The plus terminal is at virtual ground since no current flows through $R_c$.

For weighted binary resistance DACs with more bits of resolution, it is seen that larger ranges of resistance values are required as the number of bits increases (i.e., for an 8-bit DAC such as the one shown in Fig. 5-4, the LSB resistor would need to be 128 kΩ, and for a 12-bit DAC, 2.096 MΩ would be required). The large range of resistance values seriously limits the usefulness of weighted binary resistance circuits. This is because the resistance tolerances that must exist in DACs to achieve accurate conversions are difficult to maintain over large resistance value ranges. In addition, the power requirements vary inversely with the resistance value. In a 4-bit DAC circuit, this means that if a $\frac{1}{8}$-W resistor is used for the LSB, a 1-W resistor is needed for the MSB. For the example in Fig. 5-4, assume that a 1 percent resistor is used for $R_1$ (i.e., $R_1$ is 1000 ohms). For the same accuracy in the output, the accuracy of $R_4$ would have to be 0.125 percent. Another difficulty arises from the fact that the resistor values are nonstandard. This would require all the resistors to be laser trimmed in an integrated circuit.

### Ladder Resistive Summing Networks

The ladder resistive summing network largely overcomes the problems of the weighted binary resistive types. The ladder resistive circuit (also known as the $R$-$2R$ ladder) is shown in Fig. 5-5 and Fig. 5-6. Note that only two resistance values are required, making this circuit ideal for integrated circuit manufacturing. The resistance magnitude is set by the input characteristics of the amplifier.

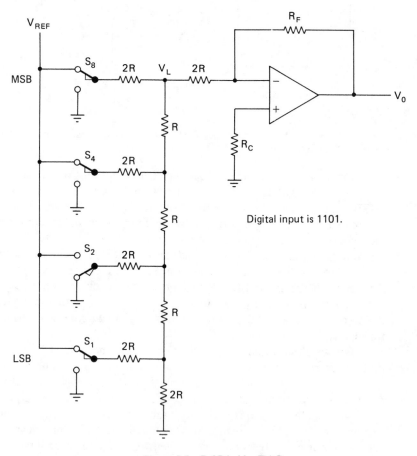

**Figure 5-5**  $R$-$2R$ ladder DAC.

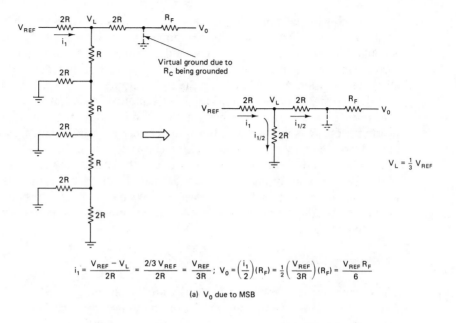

$$i_1 = \frac{V_{REF} - V_L}{2R} = \frac{2/3\, V_{REF}}{2R} = \frac{V_{REF}}{3R}; \quad V_0 = \left(\frac{i_1}{2}\right)(R_F) = \frac{1}{2}\left(\frac{V_{REF}}{3R}\right)(R_F) = \frac{V_{REF}\, R_F}{6}$$

(a) $V_0$ due to MSB

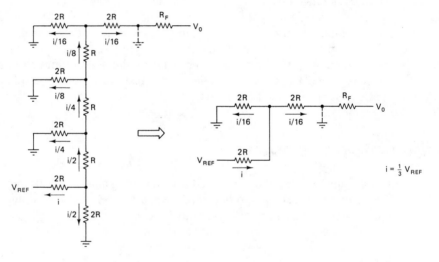

$$V_0 = \left(\frac{i}{16}\right)R_F = \frac{V_{REF}\, R_F}{48}$$

$$\frac{V_{0\,MSB}}{V_{0\,LSB}} = \frac{\dfrac{V_{REF}\, R_F}{6}}{\dfrac{V_{REF}\, R_F}{48}} = 8$$

(b) $V_0$ due to LSB

**Figure 5-6**  Analysis of $R$-$2R$ ladder network.

To explain the operation of this circuit, we analyze it by using the Super-position Theorem. This is done by assuming that an input signal ($V_{\text{REF}}$, see Fig. 5-6) is applied to only one input at a time and all other inputs are at zero. The output current when each of these individual inputs is applied is then determined. Finally, all of these outputs are added to get the total output.

Using this approach, consider a current that flows into any single input. Whenever such a current reaches a node, it divides in half, since an equivalent resistance of $2R$ is exhibited in both branches. As a result, current entering any node divides equally as it leaves a node. This is shown in Fig. 5-6(b). The current flowing through the feedback resistor $R_F$ due to the LSB is $\frac{1}{8}$ of the current due to the MSB. Twelve- and 16-bit DACs can be manufactured by using this technique. Power supply stability and noise, however, becomes a major consideration as the number of bits is increased. The speed of the converter is limited by the output amplifier slew rate.

### Multiplying DACs

The ladder resistive summing networks are specifically used in *multiplying-type DACs*. With these multiplying DACs the reference voltage ($V_{\text{REF}}$) can be varied over the full range of $\pm V_{\text{REF}_{\text{MAX}}}$ and the analog output is the product of the reference voltage and the digital word input. This is one method of constructing a controlled-gain amplifier. A more versatile method uses a memory device such as an EPROM as a look up table that converts a digital input to a digital output that is applied to the $R$-$2R$ converter. The relationship between the digital input and output signals can be any simple or complex mathematical relationship. An example would be the square root for flow measurements.

### ANALOG-TO-DIGITAL CONVERTERS

A large number of methods are used for converting analog signals into digital form. Five of these conversion methods are the most widely used in commercially available analog-to-digital converter circuits (A/D converters).

These conversion methods are as follows:

1. Staircase ramp
2. Successive approximation
3. Dual slope
4. Voltage to frequency
5. Parallel (or flash)

**1.** *Staircase ramp A/D converters.* The simplest A/D converters are staircase ramp converters. A block diagram of their operation is shown in Fig. 5-7. When a

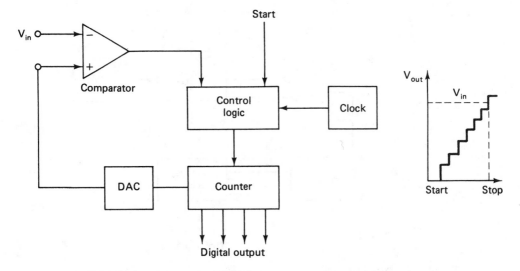

**Figure 5-7** Block diagram of the staircase ramp A/D converter.

*Start* command is issued to the control logic, the analog input voltage is compared to a voltage output of a digital-to-analog converter (DAC). The output of the DAC begins at zero and is increased by one LSB increment with each pulse of the clock, as shown in Fig. 5-7. As long as $V_{in}$ is greater than the output voltage of the DAC, the comparator produces an output signal that continues to allow the clock pulses to be fed to the counter. When the output voltage of the DAC exceeds $V_{in}$, however, the comparator output changes and this action stops the clock pulses from reaching the counter. The counter state at that time represents the value of $V_{in}$ in digital form. The drawback of this converter type is that, in spite of its simplicity, it is quite slow and the conversion time depends on the amplitude of $V_{in}$. For example, if a 5-MHz clock is used and a 10-bit output is required, 0.2 ms is required for a full-scale analog input conversion. The maximum error of the output of the staircase ramp converter corresponds to a counter increment of one LSB.

**2.** *Successive approximation A/D converters.* They are very widely applied because of their combination of high resolution and speed (i.e., they can perform conversions within 1 to 50 μs rather than the milliseconds required by the staircase ramp, dual-slope, and voltage-to-frequency types). However, they are more expensive than these slower types.

A block diagram of the successive approximation converter is shown in Fig. 5-8. The block diagram looks deceptively similar to the staircase ramp diagram, but the difference lies in the special control logic of the converter. Instead of allowing the reference voltage of the comparator (coming from the output of a DAC) to climb from zero through all the steps until $V_{in}$ is reached, the successive approximation converter control logic tries various output codes and feeds them to the DAC and a storage register and compares the result with $V_{in}$ via the comparator.

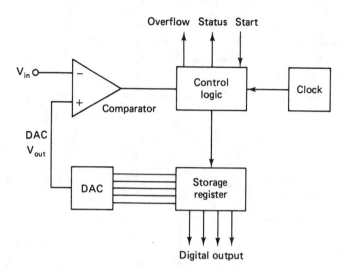

**Figure 5-8**   Block diagram of successive-approximation A/D converter.

   The operation is analogous to weighing an unknown on a laboratory balance scale by using standard weights in a binary sequence such as $1, \frac{1}{2}, \frac{1}{4}, \frac{1}{8}, \ldots, 1/n$ kg. The correct procedure is to begin with the largest standard weight and proceed down in order to the smallest one. The unknown is placed on one pan, and the largest weight is placed on the other; if the scale does not tip, the weight is left on and the next heaviest is added. If the balance does tip, the weight is removed and the next one is added. The same procedure is used for the next largest weight and so on down to the smallest. After the $n$th standard weight has been tried and a decision made, the weighing is finished. The total of the standard weights remaining on the balance is the closest possible approximation to the unknown.
   In the successive approximation converter, the weighing procedure is implemented by a DAC, comparator, storage register, and control logic. The DAC inputs are all initially set to "0." The control logic first turns the MSB of the DAC to "1" and the comparator tests the resulting DAC output against the analog input. A decision is made by the comparator as to whether to store the "1" bit in the storage register or to store a "0." Then the second bit of the DAC is turned to 1 and a second decision is made. After $n$ bits, the storage register will contain all those bits stored as "1" or "0," and the total contents will be a digital approximation of the analog input voltage signal. Figure 5-9 shows the timing diagram of a successive approximation converter as it carries out its search.
   The conversion time of the successive approximation converters is constant and is given by $T_{convert} = n/f$, where $n$ is the number of bits in the converter and $f$ is the clock frequency. Thus a 12-bit converter with a 12-MHz clock can complete a conversion every microsecond. More typical 10-bit and 12-bit converters have conversion times between 4 and 30 μs and are moderately priced.
   One very important requirement of the successive approximation converters,

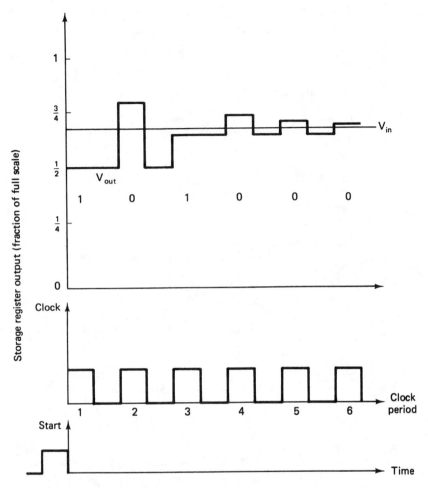

**Figure 5-9** Timing diagrams of a successive-approximation search in a 6-bit converter.

however, is that the input voltage remain constant during the conversion time. If it does not, errors during the test periods may occur, and the output may be highly inaccurate. In Chapter 17, a discussion is undertaken of the use of sample-and-hold circuits to acquire and hold constant the voltage being converted by A/D converter.

**3.** *Dual-slope A/D converters.* The dual-slope A/D converters are widely used in applications where immunity to noise, high accuracy, and economy are of prime concern. Dual-slope converters are able to suppress most input signal noise because they utilize an integrator for performing the conversion. In fact, the noise rejection can be infinite for a specific noise frequency if the first integrating period of the converter ($T_1$ shown in Fig. 5-10) is set equal to the period of the noise. Therefore, to reject the prevalent 60-Hz power-line noise, $T_1$ is required to be 16.667 ms.

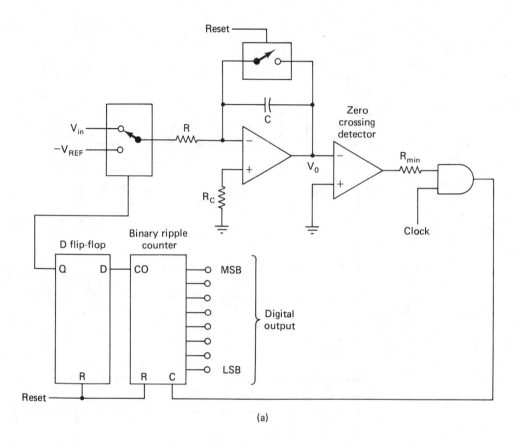

(a)

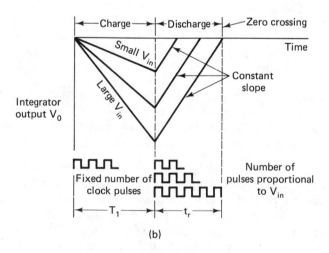

(b)

**Figure 5-10** Dual-slope A/D converter.

However, this advantage also leads to rather long conversion times (typically 10 to 50 ms). The advantages of dual-slope converters, however, do make them well suited for applications in which rapid conversion times are not necessary. Specifically, they find very wide use in such precision instrument applications as digital voltmeters.

The dc voltage to be converted by the dual-slope converter, $V_{in}$, is fed to an integrator, which produces a ramp waveform output. The ramp signal starts at zero and increases for a fixed time interval, $T_1$, equal to the maximum count of the counter multiplied by the clock frequency. An 8-bit counter operating at 1 MHz would thereby cause $T_1$ to be 8 μs. The slope of the ramp is proportional to the magnitude of $V_{in}$. At the end of the interval, $T_1$, the carry-out (CO) bit of the ripple counter causes the switch to move to the $-V_{REF}$ position. In this position, a constant current source $(-V_{REF}/R)$ begins to discharge capacitor $C$. The ripple counter is reset to zero when there is a CO. The count continues until the zero crossing detector switches state as a result of capacitor $C$ being discharged. The counter is stopped by the zero crossing detector, and the resultant count is proportional to the input voltage. In the following derivation it is important to observe that $t_r$ is independent of the value of $R$ and $C$.

$$Q_{CHARGING} = Q_{DISCHARGING}$$

$$\frac{i\,T_1}{C} = \frac{i\,t_r}{C} \tag{5-1}$$

$$\frac{V_{in}}{R}\,T_1 = \frac{V_{REF}}{R}\,t_r \tag{5-2}$$

$$V_{in} = V_{REF}\frac{t_r}{T_1} \tag{5-3}$$

**4.** *Voltage-to-frequency converter.* In these types of A/D converters, the input dc voltage is converted (by a voltage-to-frequency converter) into a set of pulses whose repetition rate (or frequency) is proportional to the magnitude of the input voltage (Fig. 5-11). The pulses are counted by an electronic counter similar to the way the number of wavelengths was counted by the time-interval counter in the

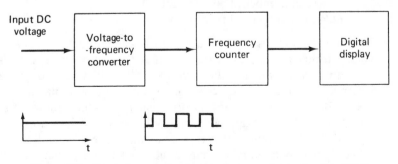

**Figure 5-11**   Block diagram of voltage-to-frequency integrating-type DVM.

ramp-type DVM. Therefore, the count is proportional to the magnitude of input voltage. Since random (normal-mode) noise tends to have an average value of zero, this type of DVM is able to reject ac noise. That is, its displayed value is equal to the average value measured during some specific time interval. This noise-rejection ability is the main advantage of the voltage-to-frequency type of DVM.

The heart of this A/D converter is the circuit that transforms the input dc voltage to a set of pulses. An *integrator* is used to carry out this task. That is, the input dc voltage is integrated and the charge is stored on a capacitor. When the voltage on the capacitor exceeds that of the reference voltage being applied to a comparator, the comparator is used to discharge the integrator capacitor to zero and simultaneously to trigger a pulse generator to emit one constant-width pulse. This pulse is fed to the frequency counter. Since the input voltage to the integrator is still present, the process is reinitiated. The slope of the integration curve depends on the magnitude of the input signal. Therefore, the number of pulses produced per second is directly proportional to the amplitude of the input signal voltage.

Typical V/F output frequencies are in the range 10 kHz to 1 MHz. The popular 10-kHz V/F converter requires a gating interval of 0.025 s for an 8-bit A/D conversion. Since V/F converters are inexpensive, as well as moderately accurate and noise resistant, they are a popular choice for use in three-digit digital panel meters (DPMs).

**5.** *Parallel (or flash) converters.* Parallel converters perform the fastest A/D conversions. In this technique (by way of example, a 3-bit parallel converter, is shown in Fig. 5-12) the input voltage is fed simultaneously to one input of each of $P$ comparators. The other input of each comparator is a reference voltage. As shown in Fig. 5-12, the comparator receives a different reference voltage value, starting at $V_{Rmax}$. By using the voltage-divider principle and equal values of $R$, the voltage reference value $V_{R_p}$ at each comparator will be given by

$$V_{R_p} = V_{R_{max}} \frac{p}{Q} \qquad (5\text{-}4)$$

where

   $p$ = comparator number (1 to $P$)
   $P$ = total number of comparators
   $Q$ = total number of resistors = $P + 1$

The input voltage is thus simultaneously compared to $P$ equally spaced voltage values (from 0 to $V_{Rmax}$). The outputs of all the comparators with $V_{R_p} < V_{in}$ will have one output value, while all those with $V_{R_p} > V_{in}$ will have an output equal to $V_H$. This information is fed to an encoder circuit functionally similar to a Texas Instrument SN54LS147 10-line to 4-line priority encoder, which represents $V_{in}$ in terms of an $n$-bit digital output (and usually on overrange indicator for the case when $V_{in} > V_{R_{max}}$). If the encoder is replaced by a PROM, the output bits can be mathematically manipulated in relation to the input analog signal.

The speed of parallel A/D converters is limited only by the comparator and

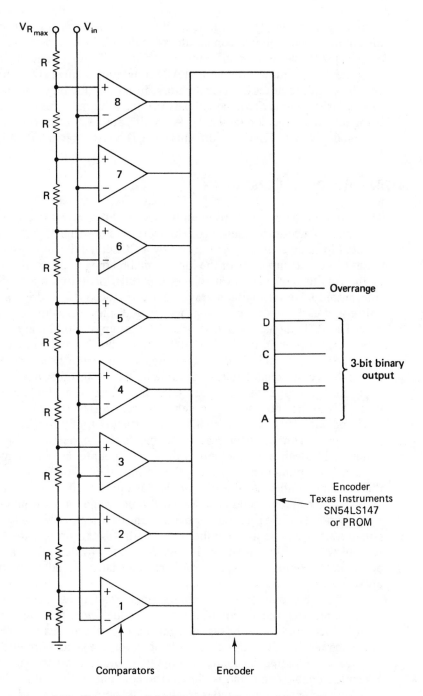

**Figure 5-12** Three-bit parallel A/D converter. The quantizing level for comparator 1 is $V_{R\text{max}}/9$; for comparator 2 it is $2V_{R\text{max}}/9$; etc.

encoder delay times. The resolution is limited by the number of bits used to express the output. The number of comparators required is $2^n - 1$, where $n$ is the number of output bits [because an $n$-bit number can have $(2^n - 1)$] various combinations of "1"'s and "0"'s). Thus a 3-bit parallel A/D converter requires eight comparators, and an 8-bit circuit requires 256 comparators. In spite of the large number of comparators and tremendous circuit complexity needed by parallel A/D converters, they are being manufactured commercially. An 8-bit parallel A/D converter is the Honeywell model HADC 77300. It performs an A/D conversion in 4 ns.

## COUNTING AND DIGITAL ENCODING

The output of the two most commonly used A/D converters in digital meters (dual-slope and voltage-to-frequency A/D converters) is a string of pulses. The pulses are counted for a prescribed time interval by a BCD digital counter, and the resultant count represents the value of the input signal reading in digital form. The contents of the counter are made available as a parallel digital output that can either be transmitted to other digital storage devices or decoded to drive a display mechanism. Let us discuss the counting–decoding–display procedure in further detail.

The BCD digital counter consists of a group of 4-bit binary counters each of which counts only from $0000_2$ to $1001_2$, thus allowing the contents of each 4-bit counter to assume only 10 different states. (Note that a *binary counter* is a digital circuit that counts digital pulses. Its contents are advanced by one count each time a valid pulse appears at its input.) If a pulse is received by a 4-bit binary BCD counter when it is in the $1001_2$ state, the pulse will return (reset) the counter to the $0000_2$ state and will also cause the counter to emit an output pulse. If there is a second 4-bit counter connected in cascade to the first, the output pulse from the first counter will advance the contents of the second counter by one count. The 10 states of each 4-bit counter can represent the decimal digits from 0 to 9. For example, in the decimal number 26, the two would be represented when the contents of the second 4-bit counter would be in the 0010 state, while the six would be represented by the contents of the first counter being in the 0110 state. The contents of each 4-bit counter are available as a parallel 4-bit digital signal through the 8–4–2–1 signal lines of each counter, as shown in Fig. 5-13. To provide a decimal readout of the accumulated counts of the BCD counter, a data latch, decoder, and decimal display are necessary.

When the count that represents a single reading has been completed by the BCD counter, its contents are transferred to a *data latch*. The latch is a digital device designed to "hold" a set of digital signals even if the inputs to it change. When a data latch receives an *enable* command, however, the signals present at its inputs (in this case representing the contents of the BCD counter) are transferred to the latch. When the enable command is halted, the contents of the data latch remain fixed until another enable command is received. The outputs of the latch are then fed into a *decoder*. The decoder converts the BCD representation into signals suitable for

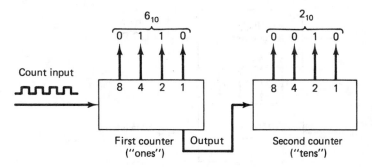

**Figure 5-13**   Two-decimal BCD counter showing a count of 26.

driving the display devices. Most modern display devices are seven-segment light-emitting diodes (LEDs) or liquid-crystal displays (LCDs). The decoder performs the function of converting the BCD signals to signals that light the appropriate segments of the display. When a PROM is used as the decoder, any desired relationship between the input and output can be created. An analog input signal of 1.5 volts could represent temperature, speed, flow or any other process variable. The PROM can change the digital signal from the converter to degrees Fahrenheit, miles per hour, gallons per hour or other desired units based upon the relationship between the analog input signal and a user-friendly output. An example would be the change from metric to English units on an automobile's dash board when only one analog signal is present. In Fig. 5-14, the outputs of the decoder drive the segments of the display to represent the decimal digit 6. (The standard labeling of the segments is *a* to *g*, as also shown in Fig. 5-14.) To display "6," the segments *a,c,d,e,f*, and *g* are lit, while *b* remains dark.

## DISPLAY DEVICES

The most popular devices being used to display digital outputs are the seven-segment displays using light-emitting diodes (LEDs) or liquid crystals (LCDs). Miniature neon and incandescent lamps were widely used prior to the availability of LEDs and LCDs. The Nixie tube (neon), a Burroughs product, and the Numitron tube (incandescent), an RCA product, are two common examples. Many instruments still in use have such tube-type displays.

Light-emitting diodes are diodes fabricated from specially compounded semiconductor materials (gallium-arsenside or gallium-phosphide) that will emit light under conditions of forward bias. LEDs are available in red, green, orange, yellow, and dual red-green, Fig. 5-15(b). The light intensity is linearly dependent on the forward current. Since LEDs are rated at different forward currents, and excess current will greatly reduce the life of an LED, each LED must have a current limiting resistor in series. Typical forward currents range from 10 to 50 mA. There is

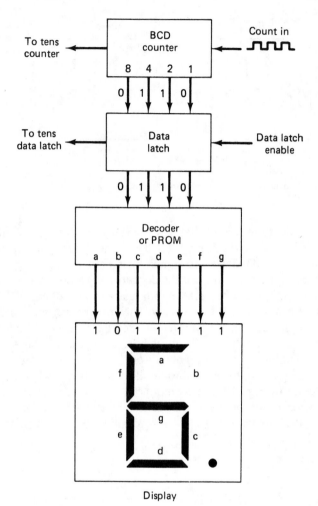

**Figure 5-14** Counter, data latch, decoder, and display for a single decade of counting.

a very wide variety of configurations available to the engineer for creating displays using LEDs. Some of the more popular types are shown in Fig. 5-15.

The most common mistake in using these displays is to disregard the polarity of the LEDs. There are three common configurations: common anode, common cathode, and individual LEDs such as bar graphs. LED displays, however, are usually unsuitable for portable instruments designed for use in bright surroundings (such as outdoors on a sunny day) since the LEDs usually provide insufficient contrast for clearly observing the display.

Liquid-crystal displays (LCDs), on the other hand, produce an output that is easily observable in bright surroundings but is not visible in dark environments. Therefore, in low light conditions, LCDs must be backlighted (an example is the dashboard display of an automobile).

The liquid-crystal material consists of a high concentration of asymmetric

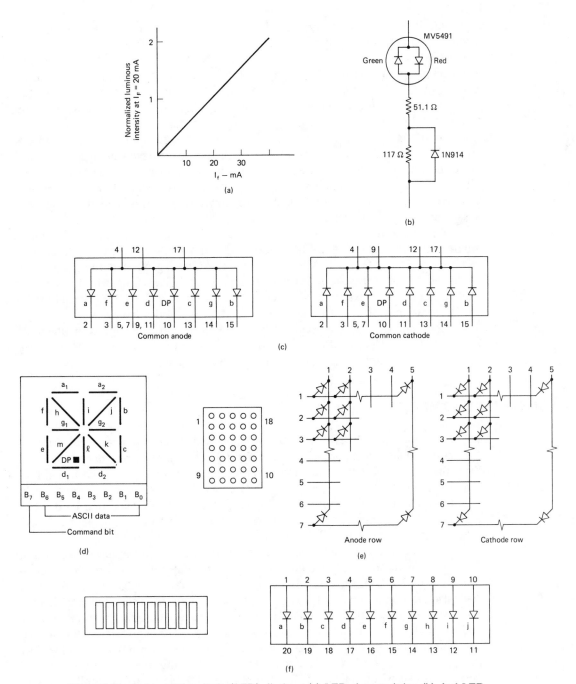

**Figure 5-15** Light emitting diode (LED) displays. (a) LED characteristics; (b) dual LED lamp with matched intensities; (c) seven-segment display typical pin arrangement; (d) 14 segment alpha-numeric display system; (e) 5 × 7 dot matrix alpha-numeric display; (f) 10-segment bar graph array.

molecules in a transparent organic solvent. This combination of chemicals changes both color and transparency under the application of an electric field. The application of a dc voltage will shorten the life of the display. A square wave is applied to the backplane and across a display element. A 5 volt, 100 Hz signal is commonly used; however, manufacturer's recommendations must be followed. When the signal applied to the display element and backplane are out of phase (usually controlled by an exclusive-or-gate) light will not pass through the LCD. Very colorful and informative displays can be created by using colored plastic and back lights. The devices draw current only when the crystal material changes state, and once the number is formed it uses only a small (micro-ampere) leakage current. Thus LCDs find use in applications where low power consumption is important. However, LCDs also respond much more slowly than LEDs, and this may be a disadvantage for some applications.

The Nixie tube is a neon gas discharge device. There are 10 cathodes and a single anode in a single neon-filled tube. The cathodes take the form of thin wires shaped into the digits 0 to 9. When one of the cathodes is grounded, the neon atoms in the vicinity of that cathode surface are ionized and give off light in the shape of that cathode. The Nixie tube also requires a decoder so that the contents of the BCD counter can be used to select the appropriate cathode.

The Numitron device uses a seven-segment array of incandescent filaments similar to the one used with LEDs and LCDs. Their major drawback is that they consume much more power than a comparable LED display.

## DIGITAL VOLTMETERS

Digital voltmeters (DVMs) use A/D converters and BCD counters to convert voltage input signals into binary-coded decimal (BCD) digital words that are used to drive digital display devices. A block diagram of a DVM was shown earlier in Fig. 1-8. Since the analog input signals to A/D converters must be dc (or very slowly varying) quantities, the input voltages measured by DVMs must either be dc voltages or ac voltages changed by an ac converter circuit to an equivalent dc form (average, rms, or peak value).

The simplest and least expensive DVMs typically have the lowest resolution (expressed as the number of digits in the display) and use integrating voltage-to-frequency (V/F) A/D converters to perform their digital conversions. The common $3\frac{1}{2}$-digit digital panel meter (DPM) is a prime example of such a low-cost DVM. Somewhat more sophisticated DVM models are usually designed with integrating dual-slope A/D converters. The dual-slope converters, although relatively slow, can provide excellent resolution, high-noise resistance, and at a moderate price. For most applications the 15 to 30 conversions per second that the dual-slope A/D can perform prove to be adequate. Special-purpose DVMs used for applications in which a larger number of readings per second are required will utilize successive-approximation A/D converters. Up to 1000 readings per second can be achieved with some models.

## Interpreting the Accuracy Specifications of DVMs

The accuracies of DVMs are usually greater than those of analog meters, but the accuracy specifications listed by a manufacturer should be clearly understood. There are three key concepts involved in being able to comprehend the accuracy specifications of DVMs: *resolution, constant error*, and *proportional error*.

The *resolution* of a DVM indicates the number of digits in the display. *Constant errors* are any errors that remain constant over the full range of the instrument. Such errors are expressed in terms of the number of digits or the percentage of full-scale reading (or range). *Proportional errors* are errors that are proportional to the magnitude of the digital indication. Thus proportional errors are expressed in terms of a percentage of reading. Most manufacturers specify the accuracy of a DVM in terms of a combination of constant and proportional errors. For example, the accuracy of a DVM may be expressed by such combinations as "±0.01 percent of reading ±0.01 percent of range"; or as "±0.05 percent of reading ±1 digit." As an example, if 5.000 V is measured with a four-digit DVM whose accuracy is "0.01 percent of reading +1 digit," the maximum error is 0.01 percent of 5 V + 0.001 V, or 0.0015 V total. The resolution of a DVM is important because the resolution should be greater than the accuracy of the meter. For example, it requires an instrument which has a resolution of five digits to enable measurements to be made to 0.01 percent over 90 percent of the total dynamic range of the meter. However, it cannot be automatically assumed that a DVM with a six-digit display has a greater accuracy than a meter with only five digits in the display (even though its resolution is greater). The specifications of both instruments must be examined before the accuracy of either meter is known with certainty.

For measurements of dc voltages, the accuracy of DVMs ranges from 0.1 to 0.001 percent of reading ±1 digit. If the DVM is also capable of measuring ac voltages, resistance, and currents, the accuracy with which the DVM measures each of these quantities is usually different (and less accurate) than the dc voltage accuracy. Specific accuracies of some typical multimeter-type DVMs are presented in the section dealing with these types of DVMs.

A final note should be included on the use of the terms *reference* and *rated* conditions. *Reference* (or short-term) conditions are ideal or laboratory conditions, and specifications at these conditions represent the best accuracy obtainable with the instrument. Specifications listed under *rated* conditions include degradation of accuracy owing to such factors as temperature, component aging, and humidity.

## Additional Features and Specifications of DVMs

**1.** *Input impedance.* DVMs are capable of loading the circuits they measure, just like any other voltmeter. Since the inherent accuracy of a DVM can be made so very high, it is important that such loading effects do not cause an error greater than the uncertainties due to the meter alone. Usually, the input impedances of DVMs are quite high (10 MΩ to 10 GΩ) and should not introduce serious loading. However, the following guideline is presented to enable one to determine if a DVM will cause loading errors that are in excess of the errors caused by its inherent inaccuracies: *The DVM must have an input impedance that exceeds the measured source impedance by at least a factor of $10^n$, where n is the number of digits in the display.* Thus if the number of digits in a DVM's display is 5 and it has an input impedance of $1 \text{ G}\Omega$ ($10^9 \Omega$), the maximum impedance across which the DVM can measure voltage without causing excessively large loading errors is $10 \text{ k}\Omega$.

**2.** *Speed of reading.* In most laboratory applications, a speed of one reading per second is usually satisfactory. However, there are some cases where faster speeds are necessary. Some DVMs are capable of making up to 500 readings per second with $6\frac{1}{2}$-digit resolution and a DC accuracy of 5 ppm and 10 nV sensitivity (if an external recording device is used with the DVM).

**3.** *Range selection.* DVMs can be made to have manual or automatic range selection. Where a large number of measurements having a wide range of random voltages must be made, automatic range selection can be a useful feature.

**4.** *Overranging.* Overranging allows a DVM to measure voltage values above the normal decade transfer points without the necessity of having to change ranges. This allows the meter to keep the same resolution for values near the decade transfer points (1 V, 10 V, 100 V). The extent to which overranging is possible is expressed in terms of the percentage of the range or full scale. Overranging from 5 to 300 percent is available, depending on the model of the DVM.

**5.** *Normal-mode noise rejection.* Normal-mode noise refers to the type of noise that appears superimposed on the high side of the input signal (in the form of ripples or spikes). To eliminate this noise, which prevents the DVM from correctly determining the true dc level, a method of removal or averaging of the noise must be employed. In the integrating types of DVMs, this noise is rejected by averaging. In all other types of DVMs a filter is used to remove as much of the noise as is feasible. Filtering need not degrade the DVMs accuracy, but it does reduce the speed of measurement. The ability of a DVM to reject normal-mode noise is specified by a quantity called the NMR and is expressed in terms of decibels at a specific frequency (e.g., 30 dB at 60 Hz).

**6.** *Common-mode rejection.* Common-mode signals are those that appear at both high and low terminals simultaneously. They arise from ground-loop currents in the circuit to which the DVM is connected. These common-mode signals can be a severe problem in some measurements. So-called *guarding* techniques are used to reject the common-mode signals. These techniques involve completely surrounding the measurement circuitry and the input leads of the DVM with a metal shield that is insulated from the measurement circuitry. This shield is connected to an additional voltage source which provides a voltage whose value is equivalent to that of the voltage at the input lead being measured. Since the guard and the input leads are connected to points of equivalent potential, no potential difference exists between the input terminals and the guard shield. This prevents the ground-loop currents from coupling interference signals into the measuring circuitry; instead, these ground-loop currents flow harmlessly via the guard shield to ground. (See Chapter 16 for further information on ground loops and input guarding.)

The rejection of common-mode signals by a DVM is specified by the quantity called CMR and is also expressed in decibels at a specific frequency. A typical CMR for a DVM might be 120 dB at 60 Hz.

## DIGITAL MULTIMETERS

Although the DVMs we have been describing up to this point are only designed to measure dc voltages, other quantities can also be measured if additional circuitry is included within the meter. Some such multipurpose DVMs are designed to be able to measure all of the following quantities: dc voltage, ac voltage, dc and ac currents, temperature, capacitance, resistance, inductance, $h_{fe}$, conductance, diode forward voltage drop, conductance and accessories for measuring temperature, pressure, and currents greater than 500 amperes, see Figs. 4-21 and 4-22. The upper-frequency limit of such DVMs ranges between about 10 kHz and 1 MHz, depending on the instrument design.

Most digital multimeters (DMM) are built around either a dual-slope or a voltage-to-frequency A/D converter with a fixed range setting (i.e., 1.999 V full scale). To provide the flexibility of allowing voltages to be measured over wider dynamic ranges with sufficient resolution, an input voltage divider is used to scale the input voltage. A block diagram of a complete DMM is shown in Fig. 5-16.

To allow measurement of ac voltages, a rectifier is included in the meter design. Since the accuracies of rectifiers are not as high as the accuracy of the dc voltage-measuring circuitry of the DVM, the overall accuracy of ac voltage measurements is lower then when measuring dc voltages (ac voltage accuracies range from ±0.012 to ±1 percent ±1 digit). Currents are measured by having the DVM determine the voltage drop across an accurately known resistance value. Although the value of a resistor can be very closely specified, there is some additional error owing to the resistance change as a function of the heating effect of the current passing through the resistor. In addition, caution must be exercised when using the current-measuring function. Care must be taken not to allow excessive current to be passed through the resistor. Typical accuracies for dc current measurements range from ±0.03 to ±2 percent of reading ±1 digit, while ac current accuracies are ±0.05 to ±2 percent ±1 digit.

The DVM becomes an ohmmeter when a very accurate current source is included within the meter. This source circulates the current through the resistance being measured, and the remainder of the DVM circuitry monitors the resulting voltage drop across the element. The current source is accurate only for voltages below the full-scale voltage range of the DVM. If the resistance being measured is too large, the test current from the current source will decrease. The accuracies of multipurpose DVMs which are used to measure resistance vary from ±0.002 percent of reading ±1 digit to ±1 percent of reading ±1 digit.

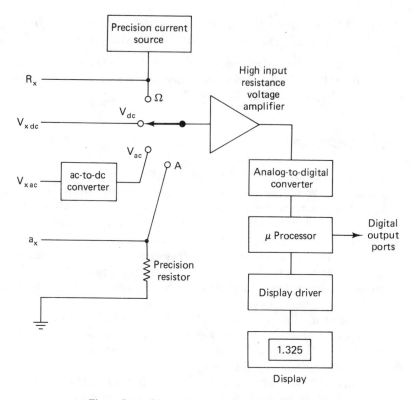

**Figure 5-16**   Block diagram of digital multimeter.

Many DMMs are battery-powered, portable instruments. Some are ruggedly designed to enable them to withstand the rigors of field measurements. Others possess such features as autoranging operation (which means that the meter automatically adjusts its measuring circuits to the correct voltage, current, or resistance range), BCD or IEEE-448 output compatability, and measurement of conductance and even temperature. The prices of DMMs vary from being very low (under $50) to quite high (thousands of dollars). But the least expensive models can now compete economically with their analog DMM counterparts.

## PROBLEMS

1. A 0- to 1.0-V signal is known to possess a $\pm 1$ percent (i.e., 10 mV) error. How many digits of a 0- to 999-mV digital voltmeter will represent valid data about this signal?

2. Each step of an 8-bit D/A converter represents 0.2 V. What will be the output of the D/A converter for the following digital inputs?
   (a) 10101101
   (b) 01000111

3. What is the percent resolution of:
   (a) A 10-bit A/D converter?
   (b) A 15-bit A/D converter?

4. A system is required that represents 0 to 100 V in a 10-bit code. What is the voltage increment that is represented by the LSB of this code?

5. What is the conversion time of a 10-bit successive-approximation A/D converter if its input clock is 6 MHz?

6. What is the function of a *latch*?

7. Draw a block diagram of a decimal counting unit with a latch.

8. Which segments of a seven-segment device must be lit to display the following decimals?
   (a) 1
   (b) 3
   (c) 4

9. What is a "$3\frac{1}{2}$-digit" voltmeter? What is meant by the "$\frac{1}{2}$" digit?

10. Can a $2\frac{1}{2}$-digit voltmeter measure 1 mV if the basic range is 0 to 100V?

11. A DMM measures dc current by passing the current through a low-value fixed resistor, and then measuring the voltage drop across the resistor with a dc voltmeter. Find the current if the voltage drop across a 1-$\Omega$ resistor is 375 mV.

## REFERENCES

1. Sheingold, D. H., ed., *Analog-Digital Conversion Handbook*. Norwood, Mass: Analog Devices, 1977.
2. Carr, J.J., *Digital Interfacing with an Analog World*. Blue Ridge Summit, Pa.: TAB Books, 1987.
3. Hnatek, E.R., *A Users Handbook of D/A and A/D Converters*. New York: John Wiley, 1976.
4. Malmstadt, H.V., Enke, C.G., and Crouch, S. R., *Electronics and Instrumentation for Scientists*. Reading, Mass.: Benjamin/Cummings, 1982.
5. Gothmann, W.H., *Digital Electronics*. Englewood Cliffs, N.J.: Prentice Hall, 1982.

# 6

# The Oscilloscope

When measuring an electrical quantity or a quantity that is converted to an electrical form, the measuring instrument must somehow display the measured result. Two of the more common mechanisms used by instruments to provide the display are found in analog[1] meters and oscillographs. Analog meters (both electromechanical and electronic) use a pointer that is moved along a scale to indicate the value of the measured quantity. The oscillograph and $x$–$y$ recorder employ a moving pen assembly that is deflected along one axis while either the paper or the pen assembly is moved along the other. However, both of these display methods are limited by mechanical inertia when measuring high-frequency signals. The mass of the components that make up the display devices prevents them from rapidly changing direction in response to changes in the applied signal. As a result, meter movements can only follow instantaneous variations up to a few cycles per second (Hz). High speed oscillographs that have no moving parts can record data from dc to 25 kHz with a writing speed of >50,000 inch/second. Paper speeds can be as high as 120 inches/second. As such, oscillographs can be used to monitor industrial process signals. However, many signals of interest in electrical applications have frequencies that are far higher than these limits. If the waveforms of such higher-frequency signals are to be measured or examined, other display mechanisms not hindered by inertia must be found.

---

[1]An instrument that uses a meter movement to display the quantity being measured along a continuous scale is one type of analog instrument.

The display mechanism available in the cathode-ray oscilloscope (CRO) is such a device. Because of this unique display mechanism, CRO instruments are being made which can follow signals with frequencies higher than 1 GHz. In fact, even higher frequencies are being displayed by using the sampling oscilloscope (which is a variation of the basic oscilloscope instrument).

The display device that allows such high-speed variations to be observed is the *cathode-ray tube* (a close relative of the television cathode-ray tube). The tube generates a thin beam of electrons (the cathode ray) within itself. This beam is directed so that it strikes a fluorescent screen which covers one end of the tube. Wherever the beam strikes the screen, a spot of visible light is emitted. As the beam is moved across the screen it "paints" a trace of its path. Since the beam is made of electrons which are electrically charged particles, it can be quickly and accurately deflected by appropriate electric or magnetic fields placed in its path. In addition, because electrons are very light, the beam is hardly hindered by inertia. It can respond almost instantaneously to the rapid variations of high-frequency signals. This capability also allows the cathode-ray tube (CRT) to display virtually any type of waveshape on the oscilloscope screen. The fields which cause the electron beam deflections are created along the path of the beam by *deflection plates*. The strengths of the fields are determined by the voltages applied to the plates, thereby making the amount of deflection directly proportional to the applied voltage. This indicates that the scope[2] display depends upon the voltages impressed upon the CRT plates. It also follows from this conclusion that the oscilloscope is really a voltmeter, that is, a voltmeter with a super-high-speed display mechanism.

Additional components of the oscilloscope extend the capabilities of its CRT so that the image on its screen is not merely a voltage display. Depending upon the mode of operation being used, the pattern displayed on the scope screen is also a *graph* of the voltage variation with time or the graph of the voltage variation of one signal versus another. This may be a useful point to remember if confusing patterns are seen on a scope screen. Furthermore, voltage is not the only quantity that can be measured. By properly interpreting the characteristics of the display, we can use the oscilloscope to indicate current, time, frequency, and phase difference. Finally, by using the oscilloscope to monitor the output of various transducers, we can also measure a large variety of nonelectrical quantities.

Even from this brief introduction, it is easy to see the wide range of measurement applications made possible by an oscilloscope. In fact, the oscilloscope is probably the most versatile and useful single instrument invented for electrical measurement work.

In this chapter we examine the operation of basic oscilloscopes in some detail. The discussion will be aimed at giving the reader a sense of familiarity with the major scope subsystems, as well as the ability to use a scope to measure the more common electrical quantities just described. In line with this objective, some practical suggestions concerning scope operations and limitations will also be covered.

---

[2]The word *scope* is commonly used as an abbreviation for oscilloscope.

## OSCILLOSCOPE SUBSYSTEMS

The oscilloscope is a complex instrument capable of measuring or displaying a wide variety of signals. It possesses such versatility mainly because it actually consists of a group of subsystems, each designed to perform a part of the measurement or display task. The subsystems that usually comprise an oscilloscope follow:

1. Display subsystem (cathode-ray tube)
2. Vertical deflection subsystem
3. Horizontal deflection subsystem
4. Power supplies
5. Probes
6. Calibration circuits

The block diagram of Fig. 6-1 shows how the subsystems interact so that the result is the display of the observed signal. The first part of this section explains the information in this figure more fully. The remainder of the section describes each of the subsystems in detail, including the role that they play in oscilloscope operation.

### How an Oscilloscope Displays a Signal

Figure 6-1 is a block diagram that shows the path of a measured signal as it passes through the various subsystems of an oscilloscope. By following this path, we can see the part that each subsystem performs in causing the signal to be displayed.

The signal is sensed at its source of origin by an oscilloscope probe. The probe either senses a voltage signal or converts the signal from its nonvoltage to voltage form. The signal voltage is then transmitted to the oscilloscope along a cable (usually a coaxial cable) and enters the oscilloscope where the cable is connected to the scope *input terminals*. Often the signal at this point is too small in amplitude to activate the scope *display subsystem* (the cathode-ray tube or CRT). Therefore, it usually needs to be amplified. The function of the *vertical deflection system* is to perform such amplification. After suitable amplification, the input signal is applied to the vertical deflection plates of the scope CRT. Within the CRT an electron beam is created by an electron gun. The electron beam is focused and directed to strike the fluorescent screen, creating a spot of light where impact is made with the screen. The beam is deflected vertically in proportion to the amplitude of the voltage applied to the CRT vertical deflection plates. The amplified input signal is also monitored by the horizontal deflection subsystem. This subsystem has the task of sweeping the electron beam horizontally across the screen at a uniform rate.

The simultaneous deflection of the electron beam in the vertical direction (by the vertical deflection system and vertical deflection plates) and in the horizontal direction (by the time-base circuitry and the horizontal deflection plates) causes the spot of light produced by the electron beam to trace a path across the CRT screen. If

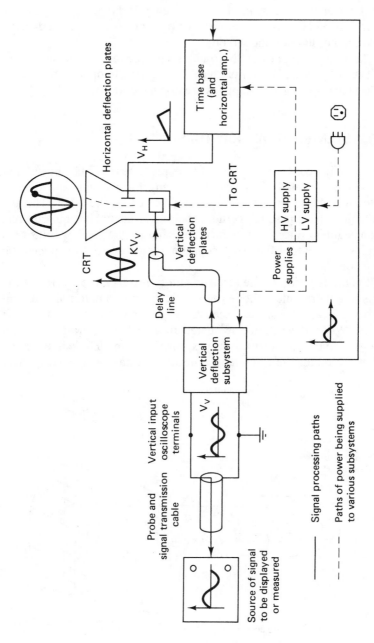

**Figure 6-1** Block diagram of oscilloscope subsystems.

Horizontal deflection plates

Time base
(and
horizontal amp.)

$V_H$

To CRT

CRT

$KV_V$

Vertical
deflection
plates

Delay
line

HV supply

LV supply

Power
supplies

Vertical
deflection
subsystem

Vertical input
oscilloscope
terminals

$V_V$

Probe and
signal transmission
cable

Source of signal
to be displayed
or measured

Signal processing paths

Paths of power being supplied
to various subsystems

the input is periodic and the time-base circuitry properly synchronizes the horizontal sweep with the vertical deflection, the spot of light will trace the same path on the screen over and over again. If the frequency of the periodic signal is high enough (greater than 50 Hz), the repeating trace will appear to be a steady pattern painted by solid lines of light on the screen.

The power supplies of the oscilloscope convert the ac supply power to the correct dc voltages and currents which are required to operate the other subsystems of the oscilloscope. (Power supplies are discussed in Chapter 12.)

## DISPLAY SUBSYSTEM (CATHODE-RAY TUBE)

The heart of the oscilloscope, the cathode-ray tube (CRT) is shown in Fig. 6-2. The tube itself is a sealed glass vessel with an electron gun and deflection system mounted inside one end and a fluorescent screen on the other. The air is removed from the tube, leaving a high vacuum. This high vacuum is required because the fine electron beam produced within the tube would be scattered by collision with any gas molecules in its path. Many such collisions would destroy the pencil-thin character of the beam.

It is the function of the electron gun to produce the electron beam. The gun consists of a thermonic cathode (a cathode made of a material that emits electrons upon being heated), various accelerating electrodes, and controls for focus and intensity. When the cathode is heated to a high temperature, it begins to emit electrons. Some of these electrons pass through a small hole in the *intensity control grid* that surrounds the cathode. If a negative voltage is applied to this grid, only a

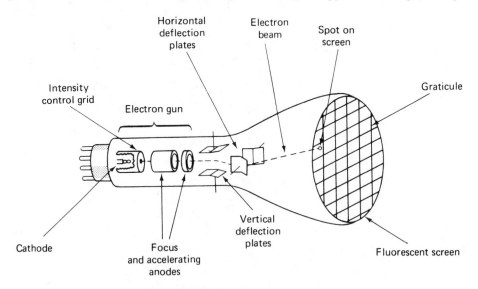

**Figure 6-2**   Oscilloscope cathode-ray tube.

limited number of emitted electrons can pass through the hole. The number can be controlled by varying the magnitude of the voltage. The intensity of the spot of light where the electron beam strikes the fluorescent screen depends on the number of electrons in the beam. There is a nonlinear relationship between intensity and the beam accelerating voltage. Light emitted during phosphor excitation is called fluorescence. The light that is emitted after the beam is turned off is called phosphorescence. The magnitude of the beam current is adjusted by a front-panel control marked INTENSITY.

The emerging electrons are compressed by the focus and accelerating anodes of the gun to form a tight beam. An electrostatic focusing scheme is used in oscilloscope CRTs to achieve this compression. The same electrostatic fields also direct the electrons along the axis of the beam and accelerate them forward toward the fluorescent screen. The difference in potential between the cathode and the accelerating anodes is usually 2–12 kV. The control element that provides adjustment of the voltage on the focusing anode is a front-panel control marked FOCUS.

After leaving the electron gun, the focused and accelerated beam passes between two *deflection plates*. If there is no voltage difference between the plates, the beam continues straight through and strikes the fluorescent screen at its center. If there is a voltage difference between either or both sets of these plates, the beam will be deflected from its straight path. The amount of deflection is determined by the magnitudes of the voltage differences. In typical oscilloscopes, between 10 and 20 V must be applied to a set of plates to deflect the spot 1 cm.

The two sets of deflection plates are placed perpendicular to one another so that they can independently control the beam in both the horizontal and vertical directions. For example, a voltage applied to the *vertical deflection* plates changes the direction of the electron beam only in the vertical direction. The beam can be deflected upward or downward by these plates, depending on the voltage polarity existing between them. If the voltage of the upper plate is made positive with relation to the lower plate, the negatively charged beam will be attracted to the upper plate and be deflected upward. In a similar fashion, a voltage applied to the horizontal plates will deflect the electron beam to the left or right. Figure 6-3 shows how an electron beam is deflected by various dc applied voltages. By adjusting the dc voltages applied to the plates, we can also shift the center of the displayed waveform to any point on the screen. The internal power supplies of the oscilloscope produce such dc voltage levels and their values are adjusted by the POSITION controls available on the scope face.

The *fluorescent screen* of the CRT is coated with a phosphor. At the point where the electron beam strikes the screen, the phosphor emits a spot of visible light. Most phosphors continue to emit light for a short time after the beam has stopped striking it. Thus if the electron beam is repeatedly moved across the screen along the same path, and this retracing is performed rapidly enough, the image "painted" on the scope screen will appear to be a solid line.

The length of time it takes the intensity of the spot to decrease to 10 percent of its original brightness is called the *persistence* of the phosphor. The value of the

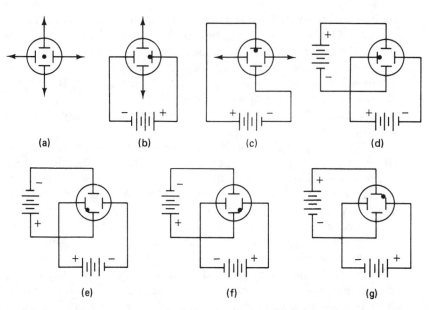

**Figure 6-3**  Deflection of electron beam in CRT: (a) both deflecting plates at zero voltage; (b) positive voltage on right deflecting plate; (c) positive voltage on upper deflecting plate; (d–g) equal positive voltages on adjacent deflecting plates.

persistence varies according to the phosphor type. For laboratory scopes, a green phosphor with medium persistence provides a steady image of a repeated trace. Almost all manufacturers provide their customers with a choice of phosphor materials. Table 6-1 summarizes the characteristics of some of the commonly used phosphors. It should also be mentioned that in addition to standard CRTs, other special CRT types allow a written trace to be stored on the scope screen for up to several hours after the image is first written on the phosphor. Oscilloscopes designed with such storage CRTs are known as *storage oscilloscopes* and are described in more detail in a later section.

Heat as well as light is generated when an electron beam strikes the screen. In fact, 90 percent of the electron beam energy is converted to heat and only 10 percent to visible light. Therefore, care must be taken to prevent the beam from burning holes in the phosphor coating. This is done by keeping the beam intensity (controlled by the INTENSITY knob) set to a low value, especially when the spot is stationary.

The *graticule* is the set of horizontal and vertical lines permanently scribed on the CRT face. These lines allow the waveform display to be visually measured against a set of vertical and horizontal scales. Most oscilloscopes graticule lines are spaced 1 cm apart. The graticule lines are often illuminated by *edge lighting*. The *scale-illumination* control on the front panel allows adjustment of the intensity of the graticule lighting.

The size and aspect ratios of display on the screen vary from 1 : 1 to 1 : 4. The

**TABLE 6-1**  DATA CHART

| Type | Color | | Persistence (to 10% level) | Luminance* | Writing speed* | Intended use |
| | Fluorescent | Phosphorescence | | | | |
| --- | --- | --- | --- | --- | --- | --- |
| P1 | yellow-green | yellow-green | 1–100 ms | 4 | 7 | oscilloscope, radar |
| P2 | yellow-green | yellow-green | 1–100 ms | 2 | 4 | oscilloscopes |
| P4 | white | white | 10 μs–1 ms | 3 | 3 | monochrome television |
| P7 | blue | yellow | 10 μs–1 ms(b) 100 ms–1 s(y) | 4 | 2 | radar, medical |
| P11 | blue | blue | 10 μs–1 ms | 5 | 1 | photographic |
| P15 | ultra-violet | green | <1 μs(uv) 1–10 μs(g) | 9 | 8 | TV (flying spot scanner) |
| P16 | ultra-violet | ultra-violet | <1 μs | 10 | 8 | TV (flying spot scanner), photographics |
| P18 | white | white | 10 μs–100 ms | — | | projection TV |
| P19 | orange | orange | >1 s | 5 | 10 | radar |
| P22 | white (red, blue, green) | white (red, blue, green) | 10 μs–1 ms | — | — | tricolor TV |
| P26 | orange | orange | >1 s | 8 | 10 | radar |
| P28 | yellow-green | yellow-green | 100 ms–1 s | 3 | 5 | radar, medical |
| P31 | green | green | 10 μs–1 s | 1 | 3 | oscilloscopes, bright TV |
| P33 | orange | orange | >1 s | 6 | 9 | radar |
| P39 | yellow-green | yellow-green | 100 ms–1 s | 3 | 6 | radar, computer grahics |
| P40 | blue | yellow-green | 10 μs–1 ms(b) 100 ms–1 s(yg) | — | — | low repetition rate P16 |
| P44 | yellow-green | yellow-green | 1–100 ms | 3 | — | bistable storage |
| P45 | white | white | 1–100 ms | 4 | — | monochrome TV display |

\* 1 = Highest luminance or fastest writing speed

From *Electronic Instruments and Measurement Techniques*; Copyright Cambridge University Press, 1987. All rights reserved.

most usual range is between 4:10 and 8:10. The screen dimensions range from less than 3 in. to more than 7 in.

## VERTICAL DEFLECTION SUBSYSTEM

As mentioned in the preceding section, about 10–20 V must be applied to the CRT deflection plates to deflect the electron beam 1 cm. Therefore, if much weaker signals were to be applied directly to the scope display subsystem, they would cause no measurable deflection of the electron beam. On the other hand, larger-amplitude voltage signals might cause deflections of the electron beam that would be too large to be completely displayed on the CRT screen. Thus the oscilloscope must contain a subsystem that has the capability of amplifying or attenuating input signals, so that a suitable display is produced when the signals of interest are applied to the CRT deflection plates. The *vertical deflection system* is the subsystem of the oscilloscope which performs this function.

The vertical deflection system normally consists of the elements shown in Fig. 6-4.

1. Input coupling selector
2. Input attenuator
3. Preamplifier
4. Main vertical amplifier
5. Delay line

In this section we discuss these elements in more detail.

### Oscilloscope Amplifier Gain and Sensitivity

Our discussion of the vertical deflection subsystem begins with a look at the combined operation of the attenuator, preamplifier, and main vertical amplifier. Together they make up the amplifying portion of the subsystem.

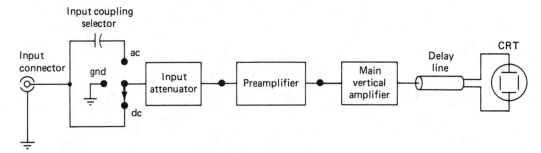

**Figure 6-4**   Vertical deflection subsystem.

To allow the largest range of signal amplitudes to be displayed by a scope, it is desirable for the amplifier subsystem to have a wide range of discrete amplification levels. This is achieved by constructing the vertical deflection subsystem of the oscilloscope as shown in Fig. 6-4.

The main vertical amplifier (together with the preamplifier) is designed to provide a gain, $K$, of fixed value. That is, any signals applied to the inputs of the amplifier are amplified by the same gain factor (typically $K = 1000$–$2000$). The advantage of fixed gain is that the amplifier can more easily be designed to meet and maintain the requirements of stability and bandwidth. (See Chapter 15 for more detailed definitions of these terms.)

The function of the *attenuator* is to reduce the amplitude of the input signals by a selected factor $F$, before such signals are applied to the preamplifier/amplifier section. Thus the voltage, $V_{out}$, presented to the deflection plates of the scope CRT is found from

$$V_{out} = F \times K \times V_{in} \qquad (6\text{-}1)$$

where $V_{in}$ is the voltage applied to the scope inputs (Fig. 6-5). In addition, the frequency response of the attenuator is designed to remain constant at each attenuator setting. Typical attenuator/preamplifier/main amplifier combinations permit general-purpose oscilloscopes to accept a range of signals (also known as the *dynamic range*) of more than $1000 : 1$. As an example of a dynamic range of $1000 : 1$, consider an oscilloscope that has a preamplifier/main amplifier with $K = 1000$. The attenuator can be set to attenuate the input signal by a maximum factor of 1000. Thus at this attenuation factor ($F = \frac{1}{1000}$) the signal that is presented to the CRT deflection plates is, from Eq. (6-1),

$$V_{out} = F \times K \times V_{in} = \tfrac{1}{1000} \times 1000 \times V_{in} = V_{in}$$

On the other hand, the attenuator can be bypassed, and the input signal in this case (i.e., $F = 1$ and thus there is no attenuation) will be amplified by $K$, or

$$V_{out} = F \times K \times V_{in} = 1 \times 1000 \times V_{in} = 1000\, V_{in}$$

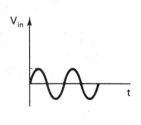
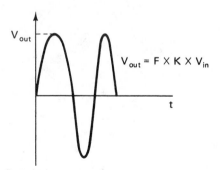

**Figure 6-5**

Almost all amplifiers used in oscilloscopes, however, are calibrated in terms of *sensitivity* rather than gain. That is, each discrete amplification level of the oscilloscope is represented as requiring a specific input signal amplitude per division of deflection (e.g., 1 mV/div, or 0.2 V/div, etc.), rather than having a gain of 1000, or 50, etc. Usually, these sensitivity levels are available in sequences such as 1, 2, 5, 10, ..., or 1, 3, 10, ..., so that there will be some overlap between ranges. The overall range is typically 2 mV/div to 10 V/div. Figure 6-6 shows a vertical sensitivity control switch used to access these various sensitivity levels. It should also be mentioned that the vertical accuracy of the displayed signal is generally between ±3 and ±5 percent. This point is more fully discussed in a later section on oscilloscope measurement.

**Example 6-1**

If the vertical sensitivity of a scope is set to 50 mV/div, how much and in what direction will the following voltages applied to the vertical inputs of the scope, deflect the spot?
(a) 0.2 V
(b) −150 mV

**Solution.**

$$\text{Deflection (div)} = \frac{\text{vertical input signal (mV)}}{\text{vertical sensitivity (mV/div)}}$$

(a) $\text{Deflection} = \dfrac{200 \text{ mV}}{50 \text{ mV/div}} = +4 \text{ div (upward)}$

(b) $\text{Deflection} = \dfrac{-150 \text{ mV}}{50 \text{ mV/div}} = -3 \text{ div (downward)}$

The vertical sensitivity control switch generally also has a *vernier* feature (i.e., a device that makes possible the finer setting of a value or level). This vernier is

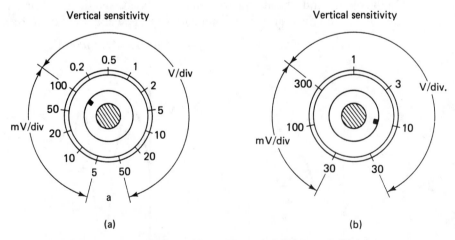

**Figure 6-6** (a) Wide-range amplifier sequence 1–2–5 sequence; (b) lesser range amplifier 1–3–10 sequence.

typically a control knob found in the very center of the vertical sensitivity control switch and allows continuous adjustment of the sensitivity (volts/div) between calibrated positions of the switch.

### Input Coupling Selector

We see in Fig. 6-4 that the first element of the vertical deflection subsystem is the input coupling selector. Its purpose is to allow the oscilloscope more flexibility in the display of certain types of signals. For example, an input signal may be either a dc signal, an ac signal, or a signal consisting of an ac component superimposed on a dc component (ac + dc signal). Sometimes only the ac components of the (ac + dc) type signal are of interest; at other times both the ac and dc components need to be displayed. The input coupling selector allows us to choose which of the signal components will be coupled to the amplifier circuitry for subsequent display.

The input coupling selector is shown in Fig. 6-7 as a three-position switch, with the switch positions being *ac, gnd*, and *dc*. When the *dc* position is selected, the input terminal couples (connects) the entire signal to the subsequent elements of the vertical deflection system. On the other hand, if the *ac* position is chosen, we see that a capacitor is placed in series with the input branch to the amplifier. This capacitor appears as an open circuit to dc components and hence blocks them from entering the subsequent circuitry. In the *ac* position very low frequency ac signals are also blocked (below 2–10 Hz, depending on the oscilloscope design).

Ac coupling is used when a high-frequency ac signal is to be displayed without its dc component. This can be a useful option if the magnitude of the ac component of a signal is much smaller than its dc level. (An example is the ac ripple on the dc output of a power supply.) If the signal with both dc and ac components is displayed, any changes in the larger dc component might drive the entire signal off the screen. By eliminating the dc component, the ac part of such a signal can be more conveniently observed. The dc input coupling is generally appropriate for displaying most other types of signals.

The *gnd* position of the input coupling switch of Fig. 6-7 grounds the internal circuitry of the amplifier. It does *not* convert the input terminal to a ground point for an input signal. The action of selecting the *gnd* position causes any charge that is

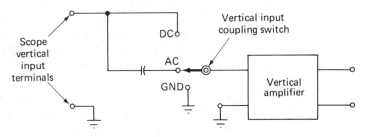

**Figure 6-7**  Functioning of the input coupling selector.

stored in the input attenuator to be removed, and to recenter the electron beam. This capability finds application when it is desired to recenter the beam without removing the input leads to the scope.

## Delay Line

In Fig. 6-1 we see that when the oscilloscope is being operated in the $y$-$t$ mode, a part of the input signal is picked off and fed to the horizontal deflection subsystem. This picked off signal is used to initiate a sweep waveform that is synchronized with the leading edge of the input signal. However, in passing through the various circuits, the picked-off portion of the input signal is delayed by about 80 ns before it reaches the time-base circuitry. Thus the sweep waveform is not started until about 80 ns after the input signal is applied to the vertical deflection plates of the CRT. Since the display does not begin until the sweep waveform starts, the first 80 ns of the input signal cannot be displayed. If the rise time of a fast pulse is to be observed, such an oscilloscope may not be able to be used.

To correct this shortcoming, and thus allow the leading edge of the signal waveform to be observed, the amplified vertical itself must also be delayed by at least 80 ns before it is applied to the vertical deflection plates of the CRT. This is the function of the *delay line*. Although not all oscilloscopes contain a delay line, in the many that do, the delay line introduces delays of 150 ns to 1 ms to the vertical signal.

There are basically two types of delay lines used in oscilloscopes. The first is a special type of coaxial cable designed to produce the desired delay characteristics. The second is the printed-circuit type. It consists of a printed-circuit board with a serpentine pattern on each side. The pattern is defined so that a signal is transmitted along it at a relatively low propagation velocity. Printed-circuit delay lines weigh less than comparable coaxial versions, occupy much less space, and provide good reliability and uniformity.

## Single-Ended and Differential Input Amplifiers

There are two types of inputs through which a signal can be connected to the oscilloscope. They are the *single-ended input* and the *differential input*.

*Single-ended inputs* have only one input terminal (besides the ground terminal) at each amplifier channel, as is shown in Fig. 6-8(a). The single-ended input is the most common form of input on oscilloscopes. Only voltage values relative to ground can be measured with a single-ended input unless the scope is "floated" above earth ground with the use of a three-to-two-wire plug adapter. (Note that the most common input connector to oscilloscopes is the BNC coaxial connector. The external conductor of the BNC connector is the ground terminal of the input.)

A *differential input* signifies that the oscilloscope is equipped with a differential amplifier. The differential input has three terminals (two input terminals besides the ground terminal) at each amplifier channel, as shown on the vertical amplifier channel of the oscilloscope in Fig. 6-8(b). With a differential input, the

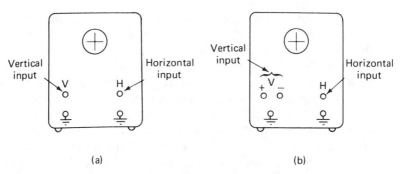

**Figure 6-8**　Oscilloscope inputs: (a) scope with two single-ended inputs; (b) scope with differential input on vertical amplifier and single-ended input on horizontal amplifier.

voltage between two nongrounded points in a circuit can be measured without "floating" the scope. The two nongrounded test points are merely connected to the two nongrounded differential input terminals. The amplifier electronically subtracts the voltage levels applied at the two terminals and displays the difference on the screen. In addition, as described in further detail in Chapters 15 and 16, differential amplifiers are able to reduce unwanted common-mode interference problems. This feature is especially important when it is necessary to measure small signals in the presence of much larger undesired common-mode signals. However, a differential amplifier is more complex than a single-ended amplifier, and hence oscilloscopes equipped with differential inputs are more costly. The differential input is sometimes available as a plug-in unit on those oscilloscopes which have interchangeable plug-in capabilities. A differential input can be obtained by using the *channel 1–channel 2* setting on a dual trace oscilloscope. This mode of operation is sometimes called (*A–B*) or (*1–2*). It is useful when measuring the voltage across an ungrounded component in a grounded circuit. Because an uncompensated differential amplifier is in effect created when this mode is utilized, there is usually a dc offset in the displayed waveform. To observe the amount of offset, connect the same signal to both channel 1 and 2 and set the oscilloscope to 1–2, and then compare the horizontal trace with the gnd trace.

## DUAL-TRACE FEATURE

It is often advantageous to be able to display two signals simultaneously on a scope screen. For example, it can be useful to display and directly compare the input and output waveforms of an amplifier. As a second example, certain troubleshooting procedures can be more rapidly performed by simultaneously examining two signals of interest.

There are oscilloscopes available which can produce such dual displays by utilizing either the dual-trace or the dual-beam technique. *Dual-beam oscilloscopes* are equipped with special CRTs containing two electron guns and two sets of vertical deflection plates. *Dual-trace oscilloscopes*, on the other hand, produce the dual image by means of electronic switching of two separate input signals. Thus the CRTs of dual-trace oscilloscopes require only one electron gun and one set of deflection plates. We discuss the dual-trace feature in this section.

The simplified block diagram for generating a dual-trace is shown in Fig. 6-9. The A and B preamplifier outputs are fed to an electronic switch that alternatively connects the input of the main vertical amplifier to the two signal inputs. When the electronic switch is in the *alternate mode*, the output of one vertical channel is displayed for a full sweep and then the output of the other channel is displayed on the next sweep. The result is that the output of each channel is alternately displayed [Fig. 6-10(b)]. This mode of operation is used for relatively fast sweep rates, and the two images appear as one simultaneous stable display. When two lower-frequency signals are to be observed, the electronic switch is put into the *chopped mode*. The switch is alternately connected to the A and B preamplifier outputs at a fixed rate of about 100 kHz. Thus small segments of the A and B waveforms are successively connected to the main vertical amplifier [Fig. 6-10(a)]. At a rate of 100 kHz, for example, 5-ms segments of each waveform are fed to the CRT for display. The result is that each vertical channel is composed of small chopped segments that appear continuous to the eye.

The selection of either the alternate or chopped mode is performed automatically on many newer dual-trace scope models. The alternate mode is automatically selected when the time/div control is set for $\leq 500\,\text{ms/div}$, the chopped mode when the time/div control $\geq 1\,\text{ms/div}$.

Some dual-trace scopes have the capability of displaying the difference

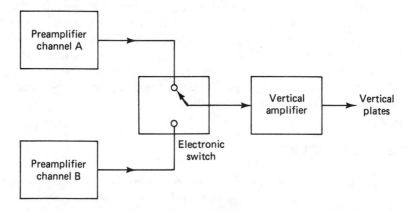

**Figure 6-9**   Block diagram for generating a dual trace utilizing an electronic switch. Alternative mode: Electronic switch will alternate between each amplifier for a sweep period. Chop mode: Electronic switch will alternate between each amplifier for a fixed interval of approximately 5–10 μs independent of the sweep period.

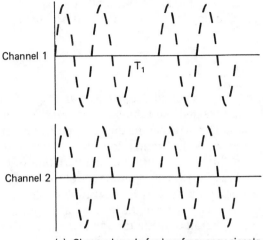

(a) Chopped mode for low frequency signals

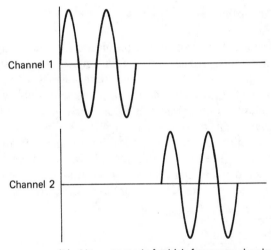

(b) Alternate mode for high frequency signals

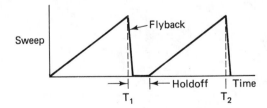

**Figure 6-10** Illustration of the chopped and alternate dual-trace modes. In the chopped mode, switching occurs a number of times during a sweep. In the alternate mode, switching occurs at the end of a sweep.

between the voltages fed to channel *A* and *B*. In this mode (*A–B*), the instrument has the same capability as an oscilloscope equipped with a differential input.

## HORIZONTAL DEFLECTION SUBSYSTEM

The horizontal deflection subsystem of an oscilloscope as shown in Fig. 6-11 consists of the horizontal deflection amplifier and the time-base circuitry. The horizontal amplifier is employed in two ways. The first is in the direct amplification of external input signals (which are then fed to the horizontal deflection plates of the CRT). Since the oscilloscope display in this mode consists of the variation of one signal (displayed in the *Y* or vertical direction) versus another (displayed along the *X* or horizontal direction), the oscilloscope is said to be operating in the *X-Y display* mode (as shown in Fig. 6-12). In the second use of the horizontal amplifier, it is used to amplify the sweep waveforms generated by the time-base circuitry. This type of operation (as shown in Figure 6-1) is known as the *Y* versus *t* mode, because the variation of the input signal (displayed in the *Y* or vertical direction) is observed versus time (displayed along the horizontal direction).

### Oscilloscope Horizontal Amplifiers

In most conventional oscilloscopes the performance requirements (gain/bandwidth) on the horizontal amplifier are not as great as those of the vertical amplifiers. While the vertical amplifier must be able to handle small-amplitude, fast-rise-time signals, the horizontal amplifier is largely required to amplify only sweep signals, with their fairly large amplitudes and relatively slow rise times. Most horizontal amplifiers are capable of being operated at two sensitivity levels (typically 1 V/div and 0.1 V/div). This gives the scope the capability of providing a calibrated magnification to a portion of the displayed waveform (see Fig. 6-19). It also allows one of two horizontal sensitivity levels to be selected when operating in the *X-Y* mode. In some scopes various horizontal plug-in units allow a much wider range of horizontal sensitivities.

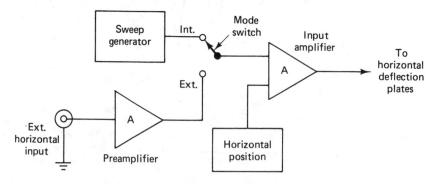

**Figure 6-11**   Block diagram of a basic horizontal amplifier.

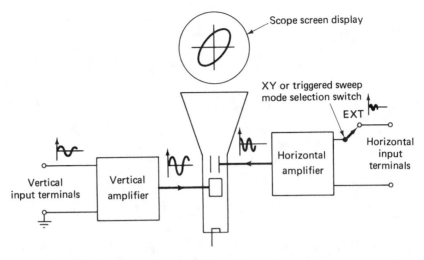

**Figure 6-12** *X-Y* mode operation.

## Time-Base Circuitry

The most common application of an oscilloscope is the display of signal variations versus time (*Y-t* mode). To generate this type of display, a voltage that causes the horizontal position of the beam to be proportional to time must be applied to the horizontal plates of the scope. In addition, this same voltage must be repetitively applied to the horizontal plates so that the beam can retrace the same path rapidly enough to make the moving spot of light appear to be a solid line. Finally, the voltage must be synchronized with the periodic signal being displayed in such a way that the same path actually is retraced and a steady image appears on the scope screen.

The *time-base circuitry* of the oscilloscope performs the task of producing such a repetitive and synchronized voltage signal. To see how it performs this function, let us examine the principles of operation of the time-base circuitry with the following questions in mind. First, what kind of signal must the time base produce to make the horizontal position of the beam be proportional to time? Second, how is this signal repetitively generated? Finally, how is this signal synchronized with the signal to be displayed on the scope screen?

The signal generated by the time-base circuitry is called the *sweep waveform*. It is in the form of a sawtooth, and one cycle is shown in Figure 6-13 (where $V_H$ is the voltage applied to the horizontal plates of the CRT).

If the spot of the electron beam is located at the left edge of the screen when $t = 0$, the increasing voltage of the sweep waveform will cause the beam (and hence the spot) to be pulled horizontally across the screen. At the end of $T_1$ seconds, the spot will have been moved across the full length of the screen. During the time from $T_1$ to $T_2$, $V_H$ will decrease to zero. and the spot will be returned quickly to the left

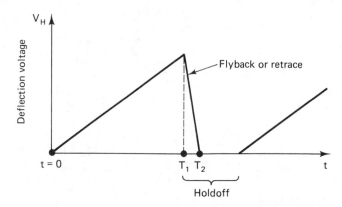

**Figure 6-13**  One cycle of the sweep waveform.

edge of the screen. From $t = 0$ to $t = T_1$, $V_H$ increases linearly with time and thus the position of the spot during this time interval will be proportional to the time elapsed from the beginning of the sweep waveform. The *Time/div.* control on the front panel of the scope determines how much time it takes for the sweep waveform to move the spot across one division of the screen. If no external signal is applied to the vertical plates, a sweep waveform will cause the spot to trace a horizontal line on the scope screen. If there is a vertical input voltage, the sweep waveform will cause a *V* versus *t* plot to be displayed on the scope screen. Figure 6-14 shows how the time variation of an input signal is displayed with the help of the sweep waveform signal.

During the short time the spot is being returned from the right edge of the screen to its starting position, additional circuitry within the scope is used to shut off the beam. This action prevents the beam from leaving a trace during the return trip and is known as *retrace blanking* or *holdoff*.

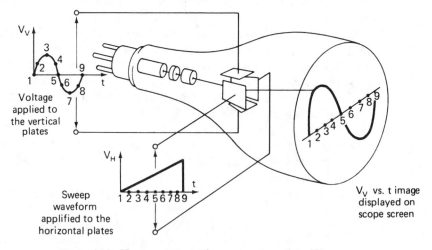

**Figure 6-14**  How a sweep waveform generates a plot of $V_v$ versus $t$.

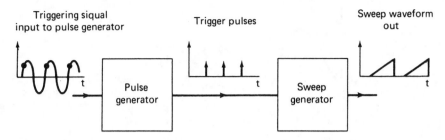

**Figure 6-15**   Triggering signal input to pulse generator.

Let us now look at a block diagram of the time-base circuitry to understand how the sweep waveform is generated (Fig. 6-15). We see that a signal called the *triggering signal* is first fed to the pulse generator of the time base. Every time this triggering signal crosses a preselected slope and voltage level condition, the pulse generator emits a pulse. The emitted pulse triggers the sweep generator to begin producing one cycle of the sweep waveform. Figure 6-16 shows how the triggering signal, emitted pulses, and sweep waveform are related in time. We note that not all pulses from the pulse generator cause the sweep generator to generate a sweep waveform pulse. If the sweep generator receives a pulse during the middle of one sweep cycle, the pulse is ignored. This allows the scope to display more than one period of the input signal without having itself retrigger a new sweep waveform. At the end of each cycle, the sweep generator stops its output and awaits the arrival of the next pulse before producing a new sweep waveform.

The point on the triggering signal at which the pulse generator emits a pulse is controlled by the *Trigger Slope* and *Trigger Level* switches of the scope. The *Trigger Slope* switch allows one to choose whether the *slope* of the triggering signal should be *positive* or *negative* when the pulse generator emits a pulse (Fig. 6-17). Similarly, the *Trigger Level* switch determines the value (sign and magnitude) of the triggering voltage at which a pulse is generated. For example, if the *Trigger Level* switch is set

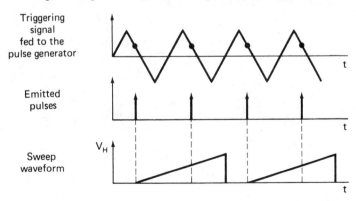

**Figure 6-16**   Relationship of triggering signal, emitted pulses, and sweep waveform in time.

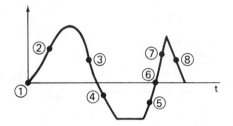

**Figure 6-17** Points 1, 2, 5, 6, and 7 are examples of points that have a *positive* slope. Points 3, 4, and 8 are points of *negative* slope.

to zero and the *Trigger Slope* switch is set to *positive,* a pulse (and hence a sweep waveform) will be triggered when the triggering signal passes through "0" on a positive-going slope (points ① and ⑥ on the curve shown in Fig. 6-17).

We emphasized earlier that the final condition necessary for a stable display of a time-varying signal is that the sweep waveform must be started at the same point of the input signal waveform as the point at which the previous sweep waveform was started (Fig. 6-18).

Since the triggering signal is the stimulus that causes the sweep waveform to be started, the triggering signal and the display on the scope screen must be synchronized to achieve a stable image. This synchronization is easy to achieve if the input signal also acts as its own trigger signal. In such cases the input signal and the trigger signal are always synchronized (being one and the same signal). As a result, the sweep waveform is started by the input signal itself, and the first point of the image on the scope screen will be equal to the point on the input signal where its slope and level triggers the sweep waveform. In an actual oscilloscope, this type of triggering is called *internal triggering* because a part of the input signal is picked off from the vertical amplifier and used as the trigger signal (as shown in Fig. 6-1). The only requirement is that the input signal be large enough to produce a specific deflection (i.e., 0.5 to 1.0 div) at the sensitivity level chosen. (For example, if a 100-mV/div sensitivity level is being used, and a 0.5-div minimum deflection is required for internal triggering to be employed, the input signal must have an

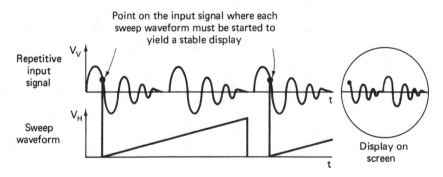

**Figure 6-18** Relationship between consecutive cycles of the sweep waveform and vertical input signal when a stable image is displayed.

amplitude of at least 50 mV to produce a trigger pulse.) Internal triggering is the most commonly used type of triggering.

If the trigger signal is externally applied, the frequency of the external signal $f_e$ must be related to the frequency of the input signal being displayed, $f_s$, by Eq. (6-2) in order for the display to be stable.

$$f_s = Nf_e \qquad\qquad\qquad (6\text{-}2)$$

Here, $N$ must either be an integer or a fraction, such as $\frac{1}{2}, \frac{1}{3}, \frac{1}{4}$, etc. The external triggering signal in this case is introduced through the input terminals marked *External Trigger*, or *Ext. Trig.*

An external triggering signal is useful in the measurement of the phase difference between two sine waves of the same frequency or when the amplitude of the input signal is too small to trigger the pulse generator.

On most scopes, another feature allows the 60-Hz line voltage to be the triggering signal. Such *line triggering* is useful for observing waveforms that bear a fixed relationship to the line frequency. The choice of source of the trigger signal (*Internal, External*, or *Line*) is controlled by the *Trigger Source* switch.

The *Trigger Level* switch often has additional modes of operation. The most common is designated as AUTO. When the Trigger Level switch is set in the AUTO position, the pulse generator will emit a pulse whenever the triggering signal crosses the zero level. The triggering signal in the AUTO mode is also ac coupled to the pulse generator. To be able to trigger the pulse generator, the trigger signal must have enough ac amplitude to cause a deflection of 0.5 div. However, if no triggering signal is received by the pulse generator (or if the signal received is too weak), the AUTO position continues to trigger a sweep waveform automatically. This feature allows the user to display a horizontal trace without feeding in a triggering signal. The AUTO mode is best for most triggered sweep displays. As a result, the Trigger Level switch is usually set to the AUTO position.

The sweep times of the sweep waveforms are selected by using the *Time/div.* control switch on the front panel of the scope. For example, in a 15-MHz scope the range of the sweep times extends from 0.1 μs/div to 500 ms/div in about nine calibrated steps (1-2-5 sequence), accurate to ±3 to ±5 percent.

In addition to the discrete, calibrated switch positions on the *Time/div.* control, there is a feature that allows the sweep rate to be magnified (expanded). By increasing the gain of the horizontal amplifier, the screen display can be expanded to give a more detailed observation of the waveform (Fig. 6-19). This is accomplished by using either the *Expander* or the *Sweep Magnifier* control (the name and its exact functioning depends on the type of scope). In some models, use of the Expander control allows the expansion of the display to be varied continuously up to 10 times (10×) the sweep time on the selected Time/div position. (Of course, the sweep times are not calibrated *if the Expander* vernier control is out of *CAL* detent.) In other scopes, a 5× or 10× sweep magnifier is also available. In these instruments, if the *CAL* vernier of the *Time/div.* switch remains in detent, the magnific..d sweep will still give a calibrated sweep time (at either 5× or 10× the

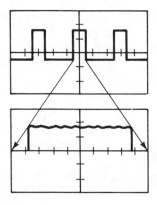

**Figure 6-19** With 10× magnification, a 1-division unmagnified portion of a display can be expanded to cover 10 divisions.

*Time/div.* setting). Accuracy of the sweep time for a time base with 3 percent accuracy will decrease to 5 percent accuracy when such sweep magnifiers are used.

**Example 6-2**

A sine wave of 400 Hz is to be displayed so that four complete cycles are to appear on the scope screen, which has 10 horizontal divisions. To what settings should the *Trigger Source* and *Sweep Time/div.* be set to allow this pattern to be displayed?

**Solution.** The trigger source should be set to INT. The time of one cycle (i.e., period) of the waveform is

$$T = \frac{1}{f} = \frac{1}{400 \text{ Hz}} = 2.5 \text{ ms}$$

The time required to display four periods would be

$$4T = 10 \text{ ms}$$

Thus the Time/div. setting should be 1 ms/div.

## OSCILLOSCOPE PROBES

Oscilloscope probes perform the important task of sensing signals at their source and transferring them to the inputs of the oscilloscope. Ideally, the probes should perform this function without loading or otherwise disturbing the circuits under test. The simplest types of probe would be mere lengths of wire connected to the circuit test points at one end, and to the oscilloscope inputs at the other. Such probes, however, would almost always be unsuitable because they would be prone to picking up and feeding unwanted noise signals to the oscilloscope.

Figure 6-20 is a block diagram showing the elements that are common to most oscilloscope probes. The *probe head* contains the signal-sensing circuitry. This circuitry may be passive (i.e., containing only passive circuit elements, such as resistors and capacitors) or active [containing such active elements as high-input impedance field-effect transistors (FETs)]. A *coaxial cable* is usually used to trans-

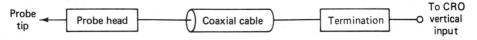

**Figure 6-20**  General block diagram of a CRO probe.

mit the signal from the probe head to the termination circuitry (or directly to the oscilloscope input terminals, if there is no termination circuit). Coaxial cables are capable of transmitting high-frequency signals without distortion and can shield them against external interference pickup. If a *termination circuit* is used, its function is to terminate the coaxial cable in the characteristic impedance of the cable and thus present the cable impedance to the oscilloscope inputs. Properly designed termination circuits reduce the possibility of unwanted reflections along the probe cable, which might otherwise create distortions in the observed signal.

The frequency response of a probe must be matched to the response of the oscilloscope to which it is attached. A probe with a −3 dB frequency of 50 MHz would not be suitable for a 300 MHz oscilloscope. That is, a low frequency signal could be observed without distorting it. The high frequency capabilities of the oscilloscope could not be used. The maximum voltage rating of both the probe and the oscilloscope must also be observed or damage can result. As frequency is increased above 100 kHz, the maximum voltage rating must be reduced in accordance with the manufacturer's recommendations.

## Passive Voltage Probes and Their Compensation

The passive voltage probe is the most commonly used probe for coupling signals of interest to the oscilloscope. Nonattenuating types (1×) are the simplest of such passive probes, but are limited to low-frequency measurement applications. Compensated attenuating passive voltage probes extend the measurement capability of the oscilloscope by increasing the scope input impedance, but such probes attenuate (reduce) the input signal so that the scope CRT beam deflection is less for a given scope amplifier sensitivity setting.

Oscilloscopes are basically voltmeters and thus can load the circuits on which they perform measurements. The input impedance of oscilloscope amplifiers provides a measure of how much the scope will load the test circuit. Typically, the scope input impedance is equivalent to that exhibited by the circuit shown in Fig. 6-21(a). For a typical oscilloscope amplifier, $R$ is about 1 MΩ and $C$ is between 30 and 50 pF.

If a dc voltage is applied to the scope, the input impedance is just 1 MΩ, because the capacitor appears as an open circuit to dc currents. Hence, when measuring dc, such an oscilloscope will begin to seriously load a circuit (by causing a 10 percent or greater indication error) if the equivalent resistance of the test circuit is 100 kΩ or more. (See Chapter 3 for additional discussion of loading by voltmeter-type instruments.) When an ac signal is applied to a scope, the capacitive reactance of the scope amplifier also begins to affect the input impedance. As the frequency of the input signal increases, the capacitance makes the effective input impedance of

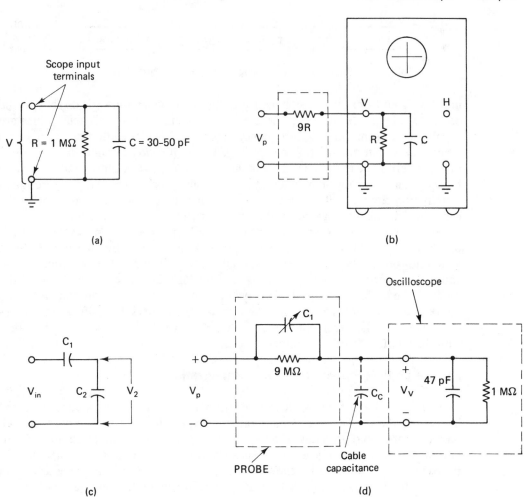

**Figure 6-21**  (a) Input impedance of a typical oscilloscope; (b) input impedance of scope increases by a factor of 10 for dc signals by inserting a resistor of $9R$ in series; (c) two capacitors connected in series $V_2 = [C_1/(C_1 + C_2)]V_{in}$; (d) input impedance of scope plus the 10× probe.

the scope decrease. This happens because the capacitive reactance $X_c$ (in ohms) decreases with increasing frequency according to

$$X_c = \frac{1}{\omega C} \tag{6-3}$$

where $\omega$ is the radian frequency ($\omega = 2\pi f$) and $C$ is the capacitance in farads.

At high frequencies, the magnitude of the capacitive reactance can become much smaller than the magnitude of the input resistance. As a result the input signal current virtually bypasses the higher resistance path and flows through the lower-impedance path offered by the capacitor. At such high frequencies, the input circuit of the scope can be approximated by a capacitor alone.

To counteract the loading effect caused by the capacitance of the coaxial cable (and also the oscilloscope itself), the *compensated attenuating probe* can be used. Such compensated probes accomplish this by increasing the effective input impedance of the oscilloscope. But they also attenuate the input signal by some factor (i.e., by 10×, 50×, or 100×, depending on the probe) and this means that the deflection of the electron beam will be less for a given input and sensitivity setting. However, the probes are designed to attenuate signals of all frequencies over a wide range by an equal amount.

The compensated attenuating probe increases the input impedance of the scope by inserting a circuit (consisting of a resistor in parallel with a variable capacitor) in series with the scope input impedance. As shown in Fig. 6-21(b), the resistor of the probe circuit has a value nine times as large as the input resistance of the oscilloscope. This causes a 10-fold increase in the input impedance to dc signals.

The variable capacitor in the probe circuit attenuates the input voltage and increases the input impedance to ac signals in the following way. If two capacitors are connected in series [Fig. 6-21(c)] the voltage ratio $V_2/V_{in}$ is given by

$$\frac{V_2}{V_{in}} = \frac{C_1}{C_1 + C_2} \tag{6-4}$$

(Note that this ratio is not dependent on frequency.) If the value of $C_1$ is chosen to be $\frac{1}{9}$ of $C_2$, the voltage ratio $V_2/V_{in}$ will be $\frac{1}{10}$. Thus the voltage seen by the inputs of the oscilloscope will again be $V_{in}/10$ (just the same as for the resistive divider created by $9R$ in series with $R$). But with two capacitors in series the value of the total input capacitance will be also decreased by a factor of 10. The total input impedance of the oscilloscope-probe combination as seen by the test circuit is now effectively a resistance of $10R_{in}$ in parallel with a capacitance $C_T/10$ (where $C_T = C_{in} + C_c$).

**Example 6-3**

A 10× probe with a 1.5-m coaxial cable ($C_c = 70$ pF/m) is connected to an oscilloscope with an input impedance of 1 MΩ and 20 pF. What is the input impedance of the probe-scope combination?

**Solution.**  The probe head contains a 9-MΩ resistor in parallel with a variable capacitor. The cable capacitance is $C_c = 70$ pF × 1.5 m = 105 pF. Thus the total capacitance of oscilloscope and cable is

$$C_T = C_c + C_{in}$$
$$= 105 + 20$$
$$= 125 \text{ pF}$$

When the variable probe capacitance $C_p$ is set equal to $C_T/9$, the input capacitance $C_{T_{in}}$ becomes

$$C_{T_{in}} = \frac{C_T}{10} = 12.5 \text{ pF}$$

Thus the input impedance of the probe-scope combination is

$$R_{T_{in}} = 10 \text{ M}\Omega$$
$$C_{T_{in}} = 12.5 \text{ pF}$$

Since each amplifier/cable combination has a unique input impedance value, a different probe capacitance value is required to make $C_p$ exactly equal to $C_T/9$. A variable capacitor is therefore designed into the passive voltage probe circuit to allow adjustment of $C_p$ for the particular amplifier at hand. As a result, each time the probe (and coaxial cable) is connected to another oscilloscope amplifier, the variable capacitor within the probe must be adjusted to *compensate* for the differing scope–coax capacitance values. This assures an optimum frequency response for that particular probe–scope combination. (Since this operation is known as *probe compensation*, the probe is referred to as a *compensated voltage probe*.) Typically, $C_p$ can be adjusted from about 12 pF to 17 pF.

Probe compensation is properly accomplished by observing a square wave with the probe–scope combination. (An appropriate square wave for this purpose is usually internally generated by the oscilloscope.)

The probe capacitance is compensated so that the corners of the calibration square wave observed on the scope screen appear square [Fig. 6-22(c)] with no overshoot or rounding [Fig. 6-22(a) and (c)]. If the probe capacitance is not properly matched to the amplifier, the calibration square wave will appear distorted. That is, too small a value of $C_p$ will cause corner rounding or undercompensation whereas too large a value will cause overshoot or overcompensation.

Figure 6-23 shows a typical 10× passive voltage probe. At one end, there is a BNC connector and at the other is the probe tip and ground lead connection. The capacitance of this probe, $C_p$, is adjusted by rotation of a sleeve on the probe body. Other types of probes may use a recessed screw on the probe head to adjust the probe capacitance.

There are two other types of passive probes, the demodulator probe and the high voltage probe. When aligning rf receivers or tracing rf signals, a demodulator probe is commonly used. It changes the rf signal to a dc voltage level that is proportional to the amplitude of the rf signal. A high voltage probe is used when

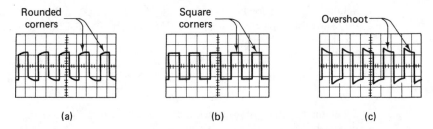

| (a) | (b) | (c) |
|-----|-----|-----|

**Figure 6-22** Square-wave calibrator waveform for adjusting probe compensation: (a) undercompensated, *C* too small; (b) compensated; (c) overcompensated, *C* too large.

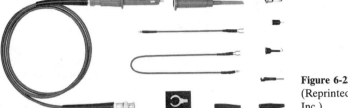

**Figure 6-23** 10× passive voltage probe. (Reprinted with permission of Tektronix, Inc.)

working with electronic equipment that has several thousand volts such as TV sets, laser power supplies, and automotive electronic ignitions. The attenuation of high voltage probes is 1000 : 1.

### Active Voltage Probes

Active voltage probes are so named because they contain active as well as passive circuit elements in their probe circuitry. In general, the active probes have high input impedance (10 MΩ in parallel with 2 to 3.5 pF) and cause less attenuation of the input signal than do passive probes. Because of the electronic circuitry they contain, active probes are bulkier and more expensive than passive probes. However, they considerably extend the measurement capability of the oscilloscope. The power to operate active probes must come from a separate power unit. These probes are useful when long cable runs are required or where loss of signal sensitivity is critical.

Modern active probes have a miniature amplifier with field-effect transistors (FETs) built into the probe lead. Although they have high input impedances, the dynamic range of active probe amplifiers is restricted to the range from 0.5 to 5 volts because of the FET amplifier. To extend this limited voltage range, add-on 10× and 100× attenuators are available. The active probes usually have termination circuits. The coax is terminated in its characteristic impedance ($Z = 50\Omega$) by these termination circuits.

### Current Probes

The use of the *current probe* allows measurements of current without the insertion of a test resistor into the circuit. The current probes are clipped onto a current-carrying conductor, and the voltage produced in the probe is proportional to current in the conductor (Fig. 6–24). The current-sensing element of this type of current probe is a current transformer. The input signal to the probe is the current under test; the output signal is the voltage developed across the probe transformer secondary. Because a transformer requires ac excitation, only ac currents can be measured with this probe. The typical frequency range over which they are effective extends from about 800 Hz to 50 MHz. The sensitivity of these probes is on the order of 10 mA/mV, and they can measure currents from 1 mA to about 1 A.

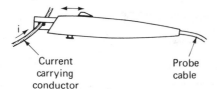

**Figure 6-24** Current probe clamped around current-carrying conductor.

Current carrying conductor

Probe cable

Another useful application of ac current probes is in measurements where minimum loading by the oscilloscope is desired. Ac current probes exhibit input capacitances of 0.5 to 2.0 pF, which is lower than that exhibited by passive voltage probes, and results in less loading at high frequencies. Other types of current probes, based on the combined electrical and magnetic phenomenon known as the *Hall effect*, can measure dc as well as ac currents.

## Calibration Circuits

To ensure that the vertical amplifier of an oscilloscope is accurately amplifying the amplitudes of measured signals, calibration checks need to be performed periodically. That is, a signal that has an accurately known amplitude should be fed to the input terminals of the scope and the display observed. If the display yields a measured value different from the known reference value, it indicates that the vertical amplifier is not properly calibrated. Appropriate adjustments of the instrument are then required to restore adequate accuracy to the display. Similarly, calibration checks also need to be performed periodically to ensure the accuracy of the time base.

Most oscilloscopes contain a provision for supplying a calibration reference signal of the type described. The signal is usually in the form of a square wave (Fig. 6–25) with a specified amplitude and frequency (both guaranteed to within ±1 percent). An output connector on the front panel accesses this reference signal, which can then be fed to input terminals of the vertical amplifiers.

The reference signal just described is also available for adjusting the probe

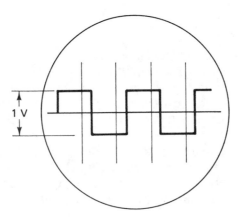

1 V

**Figure 6-25** Example of calibration square wave displayed on scope screen.

compensation. See a more detailed discussion of this important subject in the section "Oscilloscope Probes."

## OSCILLOSCOPE CONTROLS

The control knobs and switches on the front panel (Fig. 6-26) of an oscilloscope can appear to be a bewildering array to a user who is unfamiliar with the instrument. This confusion might best be relieved by explaining the function of each control and how it is connected to the inner subsystems of the scope. The operating manual that accompanies each scope is designed to do just that, and the reader is strongly advised to study such manuals carefully before operating an instrument with which the reader is unfamiliar. Most of the control settings are also displayed on the screen so that operator errors are minimized

Unfortunately, a user may find the instruction manual of the scope unclear or incomplete, or the manual may not be available. Therefore, this section will present a list of common oscilloscope controls and their functions. Readers using most other

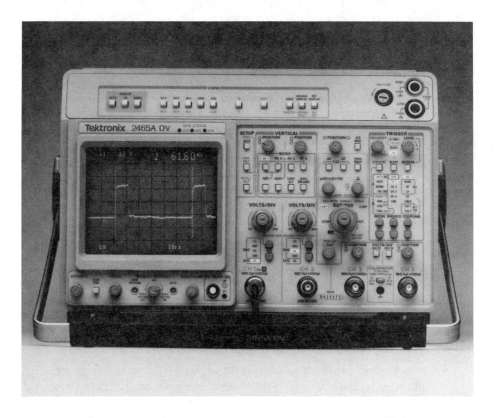

**Figure 6-26**   Oscilloscope controls for model 2465A. (Courtesy of Tektronix, Inc.)

conventional oscilloscopes can also scan this list of controls and very probably find a description of the particular control function about which they may have a question. Of course, very high frequency and other special-purpose oscilloscopes (e.g., storage scopes, spectrum analyzers, curve tracers, or vectorscopes) are not likely to have many of their controls listed herein.

Although the following list describes the most commonly used scope controls, their names may be slightly different on a given scope model. When more than one name is apt to be used, an attempt is made to also list the alternative names.

- *Power (or line)*. Turns the oscilloscope on and off (after it has been plugged in).

- *Intensity*. Controls the brightness of the scope trace. This knob provides a connection to the intensity control grid of the electron gun in the CRT. When the knob is turned clockwise, the grid repelling voltage is decreased and more electrons can emerge from the hole in the cathode grid to form the beam. A larger number of electrons in the beam causes a brighter spot to appear on the screen. *Caution*: Care must be taken to prevent the electron beam from burning spots on the screen. A stationary spot should be kept on very low intensity. If the intensity is kept high, the spot must be kept moving. If a "halo" appears around the spot, the intensity is *too high*. Before turning the scope on, turn down the intensity.

- *Focus*. The focus control is connected to the electron gun anode which compresses the emerging electron beam into a fine point. When this control is adjusted. the trace on the scope screen is made as sharp and well defined as possible.

- *Beam finder*. Returns display to viewing area of the CRT regardless of other control settings by reducing the vertical and horizontal deflection voltages. By noting the quadrant in which the beam appears when the beam finder is activated, you will know which directions to turn the horizontal and vertical position controls to reposition the trace on the screen once the scope is returned to operation.

- *Position*. The position knobs are used to shift the trace or the center of the displayed image around the screen. The position knobs provide this control by adjusting the dc voltages applied to the deflection plates of the CRT.

1. *Vertical position*. Controls the vertical centering of the trace. Use this control with the input *coupling* control set to dc to locate or set the chassis ground trace.

2. *Horizontal position*. Controls the horizontal centering of the trace.

- *Scale illumination*. Provides illumination for the graticule. The etched lines of the graticule are brightened by light applied from the edge of the screen, thereby producing no glare to interfere with the displayed image.

- *Vertical sensitivity or V/div or V/cm*. Determines the necessary value of voltage that must be applied to the vertical inputs in order to deflect the beam 1 division (or cm). This control connects a stepped *attenuator* to the scope amplifier and allows the vertical sensitivity to be controlled, by discrete steps. Typical range is from 10 mV/cm to 10 V/cm. There are many measurements that are more convenient if the vertical sensitivity is not in the calibrate position. For example, when determining the −3 dB frequency, set the trace peak to 1 division. Then increase the frequency until the peak decreases to 0.707 divisions. In this case you are finding a voltage ratio rather than an absolute value.

- *Variable V/div* (*usually a red dial VAR*). Allows a smooth (rather than a stepped) variation of the vertical sensitivity. This knob *must* be set to the calibrated position (usually fully clockwise past the click stop) for the vertical sensitivity of the scope to be equal to the value marked by the Vertical Sensitivity switch. When measuring sine wave amplitude, set the control to the largest possible time such that the peaks appear as a line. This makes reading amplitude much easier.

- *Sweep time or time/div.* Controls the time it takes for the spot to move horizontally across one division of the screen when the triggered-sweep mode is used. A very small Time/div. setting indicates a very short sweep time. Typical sweep times from 1 μs/cm to 5 s/cm.

- *Variable time/div* (*usually a red dial VAR*). This vernier control allows a continuous but noncalibrated Time/div. sweep rate to be chosen. Some very inexpensive scopes only have a continuously variable Time/div. control.

- *Source or trigger source.* Selects the source of the triggering signal. By using this control, one chooses the type of signal being used to synchronize the horizontal sweep waveform with the vertical input signal. The possible selections usually include

1. *Internal.* The output of the vertical amplifier is used to trigger the sweep. This choice makes the input signal control the triggering. For most applications, this type of triggering is appropriate.
2. *Line.* This position selects the 60-Hz line voltage as the triggering signal. Line triggering is useful when the frequency of the vertical input signal is related to the line frequency.
3. *Ext.* When this position is used, an external signal must be applied to trigger the sweep waveform. This signal must be connected to the *External Trig.* input. The external trigger signal must have a compatible frequency with vertical input signal in order to get a stable display.

- *Sweep magnifier* (*×10 mag*). This control allows one to decrease the time per division of a sweep waveform. However, the reduction is accomplished by magnifying a portion of the sweep waveform rather than by changing the time constant of the internal circuits that generate it.

- *Trigger slope.* This switch determines whether the pulse circuit in the time base will respond to a triggering signal of positive or negative slope. This topic was discussed in more detail in the time-base section (see Fig. 6-17).

- *Trigger level*. Selects amplitude point on trigger signal that causes the sweep to start.

- *Coupling*. Selects capacitive (ac) or direct (dc) coupling of the input signal to scope amplifier.

- *Probe adj*. (*probe comp*. or calibrator). Provides a square wave (usually 1 or 2 kHz and 0.5 or 1.0 V) for probe compensation.

- *Triggering mode*

1. AUTO permits normal triggering and provides a baseline in the absence of a triggering signal. Requires 0.5-div deflection to cause triggering.
2. NORM permits normal triggering, but sweep is off during absence of adequate triggering signal.
3. TV provides triggering on TV field or TV line.

- *Vertical presentation* (*or display*). In dual-trace scopes, selects the type of display that scope will exhibit.

1. *Channel A* (*or 1*) selects channel *A* for display.
3. *Channal B* (*or 2*) selects channel *B* for display.
3. *Dual-trace* (or *channel A and B*). Both channels displayed.
4. *A + B*. Summation of signals from channels *A* and *B* is displayed.
5. *B INV*. Inverts polarity of channel *B* signal. Allows display of the difference between signals of channels *A* and *B* when, the (*A + B*) mode is simultaneously engaged.
6. *Differential input* (*A − B*). Displays difference between signals of channels 1 and 2.

## HOW TO OPERATE AN OSCILLOSCOPE

The previous sections of this chapter introduced the functions of the oscilloscope subsystems and the controls on the scope face. They also discussed the two modes in which an oscilloscope is used to display the quantities being measured. This information can now be used to learn how to operate an oscilloscope. Proper operation includes making correct connections to the instrument, knowing how to turn it on, and how to display an accurate trace of the signal being measured. It also involves knowing how to measure such quantities as voltage, current, time, and frequency.

### Making Connections to an Oscilloscope

A voltage measurement always involves placing the voltmeter across the two points being measured. Therefore, when making a measurement with an oscilloscope, at least two leads must be connected from the circuit under test to the scope inputs.

The number of leads and the type of connection depends upon the type of amplifier input and whether the voltage is being measured relative to ground or some other nongrounded level. The following rules should spell out how connections are made in the proper manner.

**1.** If some point in the circuit being measured is earth grounded, connect the scope ground to this circuit ground with a separate lead.

**2.** If the voltage of the point in question is being measured relative to ground, then one additional connection to the scope must be made. Depending on whether the scope input is single-ended or differential, this connection is made in the following way:

(a) *Single-ended input.* Connect the probe to the point in question of the test circuit and to the nonground input terminal of the oscilloscope (i.e., the center conductor of the BNC connector).

(b) *Differential input.* Connect the probe to the point of interest in the circuit and to the *plus* input terminal of the differential input. Ground the *minus* terminal of the input by setting its input coupling switch to the GROUND position or by ensuring that the grounding strap to that terminal is connected. (The grounding procedure depends on the design of the particular input involved.)

**3.** If the voltage being measured by a scope is a voltage between two non-grounded points in a circuit, the method of connection also depends on the type of input available on the scope being used.

(a) *Single-ended inputs.* With a single-ended input, the measurement of un-grounded voltages involves a method which is quite hazardous (and may sometimes lead to erroneous results). This method involves the use of the three-to-two-wire adapter on the scope power-line plug (Fig. 6-27). This

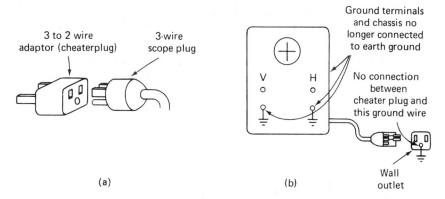

**Figure 6-27** (a) Three-to-two wire adaptor. (b) Scope with cheater plug attached to its three-wire plug. This disconnects its ground terminals and chassis from earth ground.

adapter (also called a *cheater plug*) disconnects the third wire of the three-wire cord from the building ground and thereby also disconnects the scope ground terminals and chassis from the building ground. (See Chapter 3 for a review of grounds and grounding.) Since the ground terminals of the scope are no longer connected to earth ground, a nongrounded voltage can be connected between the two vertical input terminals, and the scope will display the voltage difference between them.

The possible hazard that can arise when using this method involves the fact that the chassis of the scope assumes the same voltage level above ground as the voltage connected to any of its ground terminals. If the ground terminal is connected to a point in the circuit which is at 115 V above ground, *the entire chassis of the scope* is like an exposed 115-V power-line wire sitting on the workbench. If a user somehow touches the scope and an earth ground point, a potentially lethal shock can occur. Therefore, the use of cheater plugs is mentioned in this section primarily to warn the user of the danger associated with their use and *not* as a recommendation.

On some dual-trace scopes with single-ended inputs, another method for measuring nongrounded voltages (which is not hazardous) is sometimes possible. The scope must have a feature that allows it to display the difference of the voltage levels fed to its two input channels. Then the voltage of one nongrounded point in the test circuit can be fed to channel *A* and the other to channel *B*. With the *Presentation* control set to the subtracting position, the scope will display the voltage difference between the two points without having to resort to a cheater plug. This technique will have offset errors since the amplifiers for each channel are not identical. Correcting for the offset error is discussed earlier in the chapter.

It is important to note that if the $(A - B)$ mode of operation is used to reject a high common voltage that is present on both channels, (e.g., a power-line alternating voltage of 115 vac), the input amplifiers of the oscilloscope can be damaged. Consequently, the manufacturer's maximum voltage rating of the oscilloscope input amplifier should never be exceeded. Whenever possible, use a battery powered oscilloscope when making differential voltage measurements in the middle of a circuit.

(b) *Differential input.* Connect one point being measured to the plus input terminal of the differential input and the other to the minus input. The scope will electronically substract the one voltage from the other and display the difference of the voltages between the two points.

## Turning on the Oscilloscope

1. *Read the instruction manual*, if available.
2. Before plugging in the power cord, do the following:
    (a) Be sure that the Power switch is off and that the Intensity control is set to its lowest position.

  (b) Set the Vertical and Horizontal Position controls to their approximate midpoint positions.

  (c) Set Time/div. controls to 1 ms/div and Trigger Source to INT.

  (d) Make sure that the Triggering Mode switch is set to AUTO.

**3.** Plug in the power cord. Turn on the Power switch.

**4.** Wait about a minute, then slowly turn up Intensity until a trace appears.

**5.** Using the Vertical Position control, position the trace on the center horizontal graticule line.

**6.** Use the FOCUS control to obtain the sharpest trace possible.

**7.** The scope is now ready for use.

## Voltage Measurements

When connecting a scope to a circuit to measure voltage, follow the rules listed in the section "Making Connections to an Oscilloscope." Voltage can be measured from the resulting display as described next.

The oscilloscope is primarily a voltmeter. If it is used in the triggered-sweep mode, it displays the time variation of the voltage being applied to its vertical input. The height of the vertical deflection of the displayed trace combined with the setting of the V/div. switch yields the *peak-to-peak* voltage of the input signal. For example, the waveform shown on the screen as in Fig. 6-28 has a vertical excursion of four divisions. If the V/div. switch is set to 0.1 V/div. [and the Variable (red) V/div. switch to the *calibrate* position], the peak-to-peak voltage being displayed is 0.4 V.

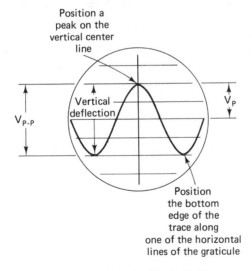

$$V = V_p \sin (\omega t) = V_p \sin (2\pi ft)$$

$$= V_p \sin \frac{2\pi}{T} t$$

**Figure 6-28** Measuring voltage from a scope display.

To get an accurate reading from the display, position the trace so that its bottom edge is aligned with one of the lines of the graticule. Also position one of the peaks near to the vertical centerline.

If the waveform being examined is a *sinusoid*, one can convert the peak-to-peak reading to get an rms reading by using the relation

$$V_{\text{rms}} = \frac{V_{\text{peak-to-peak}}}{2} \left(\frac{\sqrt{2}}{2}\right) \cong 0.3536\, V_{p-p} \tag{6-5}$$

For the above example

$$V_p = \frac{0.4}{2} = 0.2\ V \qquad V_{rms} = 0.1414$$

$$V = V_p \sin \omega t = 0.2 \sin \omega t$$

The relations between peak and rms values of some other waveforms are given in Fig. 1-6.

### Current Measurements Using a Test Resistor

Although the oscilloscope really measures voltage, indirect measurements of current can also be made. One way to do this is to pass the current through a known test resistor and measure the resulting voltage drop. Figure 6-29 shows how such a connection can be made. $R$ is often chosen to be a 1-$\Omega$, noninductive resistor. Since $v = iR$, a 1-$\Omega$ resistor eliminates calculations and yields the current immediately. When this connection is used, the power rating of the test resistor should be large enough to handle the anticipated power resulting from the current flow. Note that, unless a differential input is available or unless the scope ground is being floated, this method requires that one side of the resistor be grounded. Therefore, current cannot always be measured using this method. The use of current probes, as described in an earlier section, is an alternative approach.

### Measurements of Time

When used in the triggered-sweep mode, the time-base circuit of a scope is used to provide sweep waveforms with various values of sweep times (s/Div.). If a signal is

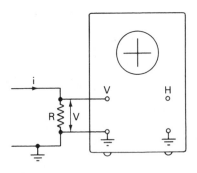

**Figure 6-29**  Measuring current by using a test resistor.

displayed when the scope is set to specific sweep time per division, the number of horizontal divisions between two points along the signal waveform is a measure of the time elapsed. The following relation can be used to calculate the time from such a reading:

$$\text{time} = \begin{pmatrix} \text{horizontal distance} \\ \text{between points of} \\ \text{display} \end{pmatrix} \times \begin{pmatrix} \text{horizontal sweep} \\ \text{setting} \end{pmatrix} \qquad (6\text{-}6)$$

$$= d \times \text{s/div}$$

### Example 6-4

The horizontal distance between points 1 and 2 of the signal waveform shown in Fig. 6-30 is 5 div. The horizontal sweep is set to 0.5 ms/div. What is the time duration between points 1 and 2?

**Solution.**   Using Eq. (6-6), we obtain

$$t = 5 \times 0.5 \text{ ms/div} = 0.0025 \text{ s} = 2.5 \text{ ms}$$

For accurate measurements, position at least one point of the waveform of interest on the horizontal centerline of the screen.

### Frequency Measurements (Triggered-Sweep Method)

Measuring the frequency, $f$, of periodic waveforms using the triggered-sweep mode is essentially the same technique that is used for measuring time. However, one additional calculation must be made for determining $f$. The frequency of a waveform is the number of cycles per second. Therefore,

$$f = \frac{1}{T} \qquad (6\text{-}7)$$

where $T$ is the time of one cycle, or the period. To find $f$ we measure the time of one period and use Eq. (6-7).

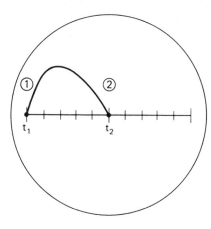

**Figure 6-30**   Measuring time from the scope display.

**Example 6-5**

If a periodic function displayed on the scope screen has a distance of 4 cm between the beginning and end of a cycle, and if the Time/div. control is set to 1 ms/div, what is the frequency of the waveform?

**Solution.**    First find the time duration of one waveform:

$$t = \text{horizontal distance} \times \text{horizontal sweep setting}$$
$$= 4 \text{ div} \times 0.001 \text{ s/div}$$
$$= 0.004 \text{ s}$$

Since in this case $t = T$, use

$$f = \frac{1}{T} = \frac{1}{0.004} = 250 \text{ Hz}$$

then if $V_p = 0.2$ volts

$$V = V_p \sin \omega t = 0.2 \sin (2\pi f)t = 0.2 \sin (1570t)$$

## Phase Measurements (Triggered-Sweep Method)

The phase difference between two waveforms of the same frequency can be found by using the triggered-sweep method and the Lissajous figures method. In this section the triggered-sweep method is discussed.

The triggered-sweep method for determining phase difference compares the phase of two signals by using one signal as the reference. The shift in the position of the second signal compared to the reference signal can be used to calculate the phase difference between the signals.

To make the measurement, the phase of one signal is chosen as zero and the scope display is calibrated to indicate this choice. The calibration procedure involves setting the scope to External Trigger, the Level to zero, and the Slope to plus so that the sweep triggers when a trigger signal crosses zero with a plus slope. The first signal, $A$, is connected then to both the vertical inputs and the external trigger terminals. The waveform displayed by the scope is like the waveform shown in Fig. 6-31. Next, the vertical input signal is changed from signal $A$ to signal $B$. Signal $A$

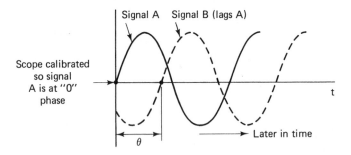

**Figure 6-31**    Phase-difference measurement using the triggered-sweep method.

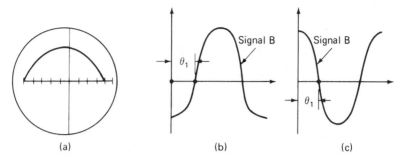

**Figure 6-32** How to determine the phase angle from the triggered sweep display: (a) calibration of horizontal axis so that 180° equals nine divisions; (b) phase of signal $B$ is equal to $-\theta$ for this type of position shift; (c) phase of signal $B$ is $\theta = 180° - \theta$ for this type of position shift.

remains connected as the external triggering signal. Thus, if signal $A$ triggers a sweep when signal $B$ is not at the same level and slope, the display of signal $B$ will be shifted in position along the horizontal (time) axis. To calibrate the time axis so that it corresponds to 20°/div, use the *Variable Sweep Time* control to adjust the display of the waveform so that half a cycle of signal $A$ corresponds to nine divisions [Fig. 6-32(a)]. Then the phase shift can be found by measuring the distance to the first zero-crossing of signal $B$ [Fig. 6-32(b) and (c)].

**Example 6-6**

Let $V_p = 3.4$ volts and $\omega = 23$ radians/second for signal $B$ in Fig 6-31. If $\theta = 0.78$, write the equation for signal $B$ if signal $A$ is used as the reference.

$$V = V_p \sin(\omega t + \theta) = 3.4 \sin(23t - 0.78)$$

*Note:* $\theta$ is lagging (or minus) since it arrives later in time.

## Lissajous Figures

If two sine waves are simultaneously fed to an oscilloscope (one to the vertical inputs and the other to the horizontal inputs) and the scope is set to operate in the *X-Y* mode, the resulting display on the scope screen is referred to as a *Lissajous pattern*. If the two sine waves are of the same frequency and phase, the Lissajous pattern will be a diagonal line. If the sine waves are of the same frequency but out of phase by 90°, the pattern will be an ellipse (if the amplitudes are also equal, the ellipse will instead be a circle). Figure 6-33 shows how Lissajous patterns result from the input of two sine waves.

The numbered dots on these figures trace the position of the electron beam as time and the magnitudes of the applied sine waves change. If the two signals are not of equal frequencies, the pattern will not be a diagonal, ellipse, or circle, but it will be some other unusual gyrating pattern. Thus, if the frequency of one signal is known, the other can be found by varying the known frequency source until a steady Lissajous pattern is displayed.

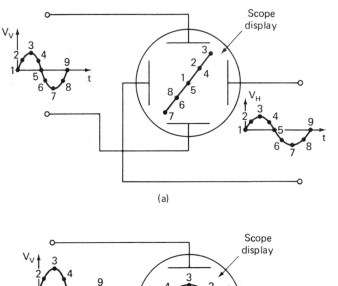

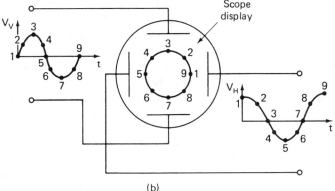

**Figure 6-33** How Lissajous patterns are generated: (a) sine waves of equal frequency and phase applied to both horizontal and vertical plates; (b) sine waves of equal frequency and amplitude but with a phase difference of 90° applied to both the horizontal and vertical plates.

In addition to Lissajous patterns for measuring frequency, there other methods such as the modulated ring pattern, broken ring pattern, and broken line pattern. All of these are obtained by using a procedure very similar to the Lissajous pattern method and equipment interconnection. None of these methods, however, are used in modern laboratories where a high degree of accuracy and speed is required. Instead, digital frequency counters and phase meters are used. A 9-digit meter can resolve a period, frequency, or phase with a resolution of 1 LSD. Determining frequency or phase shift is only one application of the *x-y* mode. For any time two variables that are interdependent but not time dependent, the *x-y* setting will produce a display of the relationship. Some instrumentation examples would be the simultaneous display of the pressure and volume of a liquid, the speed and torque of a motor, and the deflection and force on a structural beam.

## Frequency Measurements Using the X-Y Mode

Since the sweep times of the sweep waveform are usually calibrated to within 5 percent of their stated values, frequency measurements using the triggered-sweep method can be in error up to this amount. However, if an accurate, adjustable frequency source is applied to the horizontal input of a scope, an unknown frequency can be determined much more accurately by comparison (Fig. 6-34). This is done by varying the frequency of the accurate frequency source until either a Lissajous pattern of a circle or an ellipse appears on the screen. The appearance of the steady Lissajous pattern indicates that the frequencies of both applied signals are equal.

If it is not possible to adjust the frequency of the source to obtain a circle or ellipse, the known frequency should be adjusted until a stationary Lissajous pattern with a number of loops is reached. The ratio of the number of horizontal to vertical loops of the stationary pattern yields the unknown frequency (Fig. 6-35).

## Phase Measurement Using Lissajous Patterns

To measure phase difference between two sine waves, they must by definition be of the same frequency. (The phase difference between two sine waves of different frequencies is meaningless.) Therefore, if two equal-frequency sine waves are fed to the vertical and horizontal inputs, respectively, the display on the scope screen will be a stable Lissajous pattern. The characteristics of the shape of the pattern allow the phase difference between the two signals to be determined. If the equations of the two waves are

$$X = C \sin \omega t \tag{6-8}$$

and

$$Y = B \sin (\omega t + \theta) \tag{6-9}$$

the phase difference $\theta$ is found from the Lissajous pattern by the equation

$$\frac{A}{B} = \sin \theta \tag{6-10}$$

where $A$ is the point where the ellipse crosses the $Y$ axis (Fig. 6–36).

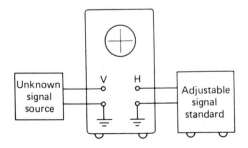

**Figure 6-34** Connections for measuring an unknown frequency by comparison to a known frequency.

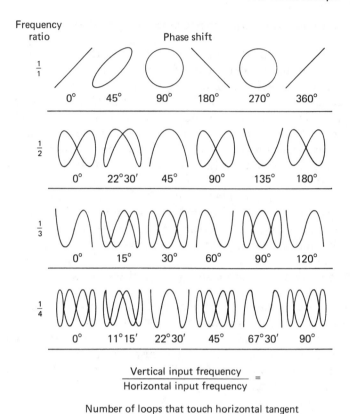

$$\frac{\text{Vertical input frequency}}{\text{Horizontal input frequency}} =$$

$$\frac{\text{Number of loops that touch horizontal tangent}}{\text{Number of loops that touch vertical tangent}}$$

**Figure 6-35**   Lissajous patterns obtained when measuring frequency and phase.

## Voltage Versus Current Display of Two-Terminal Devices

The determination of voltage versus current characteristics (*V* versus *I*) of two- and three-terminal devices is usually a preliminary step towards the useful application of the devices as circuit elements. In the case of nonlinear devices, such as diodes and transistors, a graphical display of the *V–I* characteristic is usually the most efficient means of displaying the *V* versus *I* data related to device operation. In this section we see how the oscilloscope can be used to display the *V–I* characteristics of two-terminal devices by using semiconductor diodes as the demonstration vehicle.

In a later section of this chapter dealing with curve tracers, we shall see how the *V–I* characteristics of three-terminal devices (such as bipolar and FET transistors) can be displayed.

Actual diodes are two terminal devices that have nonlinear *V–I* characteristics. The current, $I_D$, flowing in semiconductor diodes is approximately found from the equation

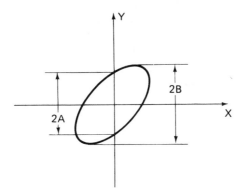

**Figure 6-36** Determining phase angle from Lissajous patterns.

$$I_D = I_o(e^{qV/kT} - 1) \qquad (6\text{-}11)$$

The constant, $I_o$, is the reverse-saturation current of the diode (typically very small, $\approx 10^{-12}$A); $V$ is the voltage applied across the diode; $q$ is the electronic charge, $1.6 \times 10^{-19}$ C; $k$ is Boltzmann's constant, $1.38 \times 10^{-23}$ J/°K; and $T$ is the temperature in °K. Therefore, the quantity $q/kT$ is equal to 0.026 V at room temperature ($T = 300$°K $= 26$°C). The graphical form of the voltage versus current characteristic of semiconductor diodes as described by Eq. (6-11) is shown in Fig. 6-37(a). In this figure we see that when the voltage across the diode, $V$, is positive (forward-bias condition), and is several times the value of $q/kT$ (i.e., $V \gg q/kT$), the current increases rapidly with increasing voltage. When the applied voltage is negative (reverse-bias condition) Eq. (6-11) predicts that $I_D \approx -I_o$. Therefore, the reverse current of the diode is constant and independent of the applied reverse-bias voltage. Most commercially available diodes exhibit an essentially constant value of $I_o$ for negative values of $V$. However, some diodes possess a pronounced (and possible unacceptable) increase in reverse current with increasing reverse voltage. In addition, at some value of reverse-bias voltage, real diodes exhibit an abrupt departure from Eq. (6-11). At this critical voltage (called the reverse-breakdown voltage), a large reverse current flows and the diode is said to be operating in the breakdown region.

An oscilloscope can be used to graphically display the $V$–$I$ relationship of the diode. The circuit shown in Fig. 6-37 demonstrates how this can be done. (Note that the same circuit can also be used to display the $V$–$I$ characteristics of virtually all other two terminal devices as well.) The oscilloscope (one with a single-ended input is shown in Fig. 6-37) is used in its $X$-$Y$ mode of operation. The sine-wave oscillator applies a sinusoidal voltage (60Hz is a convenient frequency) across the diode. The voltage appearing across the diode is also applied to the horizontal input of the scope. The current through the diode, $I_D$, is displayed as a vertical deflection, since $I_D$ is proportional to the voltage appearing across the current sampling resistor, $R_1$, of the circuit. If $R_1 = 1$ k$\Omega$, as in Fig. 6-37, the vertical voltage sensitivity of the scope display (in V/div) is automatically converted to mA/div. The resistor, $R_1$, also performs the function of limiting the maximum power dissipation in the diode. For

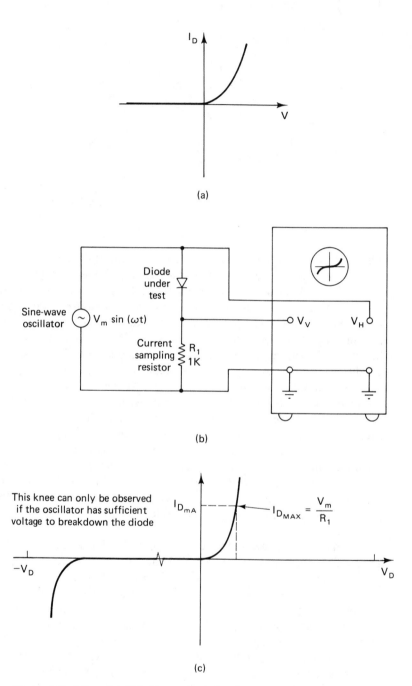

**Figure 6-37** Measuring *V-I* characteristics of diodes using a single-ended oscilloscope in the *X-Y* mode: (a) *V-I* characteristic as predicted by Eq. (6-11); (b) test set up; (c) oscilloscope display.

example, if the amplitude of applied sine wave is $V_m$, the maximum current in the diode is limited to $V_m/R_1$. The $V$–$I$ characteristic as displayed by this circuit on the scope screen is shown in Fig. 6-37(c). Note that since the oscilloscope in Fig. 6-37(b) is equipped with a single-ended input, the "minus" terminal of both the $X$ and $Y$ scope inputs have to be connected to the same ground point. Therefore, the connection as shown in Fig. 6-37(b) must be used.

An additional important point needs to be mentioned. For the diode $V$–$I$ display to appear as shown in Fig. 6-37(c), the phase difference between the two signals applied to the deflection plates of the scope must be zero. If a *nonzero* phase difference is created between the two signals because the vertical and horizontal amplifiers of the scope are not identical, a $V$–$I$ characteristic that is a closed curve rather than a single line will be displayed. This can be observed by increasing the oscillator frequency above 10kHz. To determine whether a loop in the display is caused by a phase shift due to nonidentical amplifiers, a carbon-composition resistor can be substituted for the diode in the test circuit. If the resultant $V$–$I$ display is still an ellipse with the resistor in place (rather than a straight line), it is indicative that the scope amplifiers are the cause of the phase shift.

A useful circuit and the resultant oscilloscope curves for testing any two terminal device is shown in Fig. 6-38. The curves will vary depending upon the component value and the frequency used. The basic principles of this circuit have been incorporated into commercial testers that are very versatile. For digital circuits

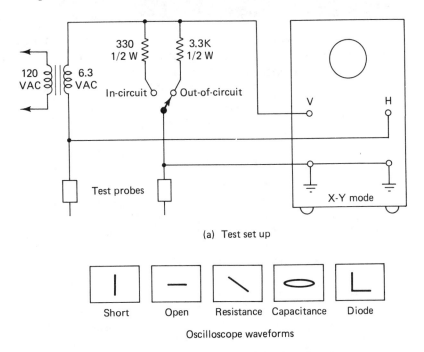

(a) Test set up

Short    Open    Resistance    Capacitance    Diode

Oscilloscope waveforms

**Figure 6-38**  Test set up for in-circuit and out-of-circuit testing of components.

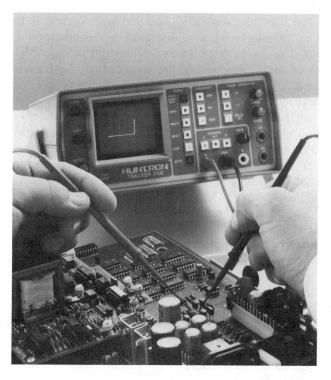

**Figure 6-39**   In-circuit component tester. (Courtesy of Huntron Inc.)

the technique of applying a known signal to a circuit and then observing the output at every terminal in the circuit is known as signature analysis. The same technique is also applied to analog and digital circuits but in an analog fashion. By knowing the response of a component to a given set of signals we can determine if a component is, within specifications, defective or acceptable. Figure 6-39 shows a tester being used to test components in a circuit board without removing the components.

Figure 6-40 contains patterns displayed on a Huntron tester for a diode and capacitor being tested. When the pattern is not known, a switcher can be used to compare the pattern of a known good device with the pattern of an unknown device (Fig. 6-41). Acceptable limits on the mismatch of patterns can be established, for example, for use in an inspection department. When the known patterns are well documented, troubles in an instrumentation circuit can be located and repaired considerably faster than by using an oscilloscope and a multimeter.

### Oscilloscope Errors

**1.** *Reading error.* It is usually difficult to read the position of a scope trace to better than $\frac{1}{10}$ of a major division. Hence the reading error can be $\pm\frac{1}{20}$ of a division at best, and more if the observer is careless. This error, as a percentage of the reading, becomes larger if the deflection involves fewer divisions. To minimize the error,

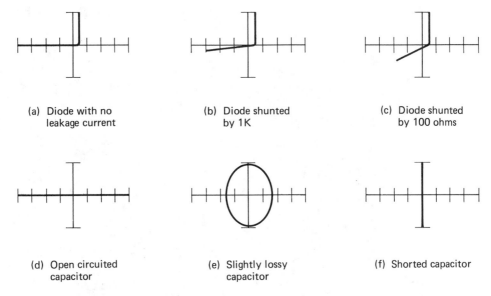

(a) Diode with no
    leakage current

(b) Diode shunted
    by 1K

(c) Diode shunted
    by 100 ohms

(d) Open circuited
    capacitor

(e) Slightly lossy
    capacitor

(f) Shorted capacitor

**Figure 6-40**  Typical displays on Huntron in-circuit component tester. (*Courtesy of Huntron Inc.*)

always use the V/div. setting which yields the largest deflection while still displaying entire vertical excursion of the signal on the screen.

**Example 6-7**

If the deflection of a signal displayed on a scope screen is five divisions, what is the minimum reading error? If the deflection is only one division, how large is this error?

**Solution.**

(a) If the deflection is five divisions and the error is $\pm\frac{1}{20}$ or 0.05 div, the possible error is

$$\text{error} = \frac{\pm 0.05}{5} \times 100\% = 1\%$$

(b) If the deflection is 1 div, the possible error is still $\pm 0.05$ div or

$$\text{error} = \frac{0.05}{1} \times 100\% = 5\%$$

**2.** *Parallax error.* In some scopes, the graticule is positioned a short distance in front of the screen (Fig. 6-42). If the observer does not look at the screen properly, the spot will appear to be at a slightly different position on the graticule than it really is. The error due to this effect is called *parallax error.*

There are other scopes that are made with the graticule etched directly on the CRT screen. In these scopes parallax error is eliminated. Check the operation manual of the particular scope being used to see if this feature exists.

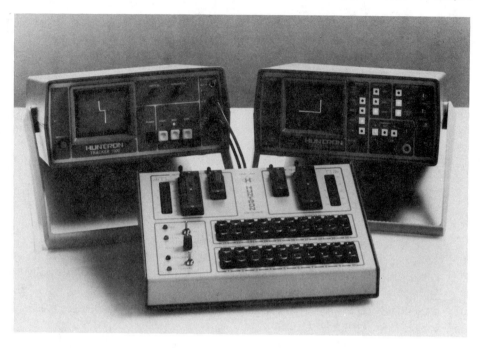

**Figure 6-41** Huntron tester used to compare component signatures or response patterns. (Courtesy of Huntron Inc.)

**3.** *Scope calibration error.* The sensitivity of a scope may be set to various levels. When the *calibration position* is employed, the deflection per division is set equal to the sensitivity step selected. But even when the scope is properly calibrated, there is still an error inherent in the calibration circuits. This calibration error varies with the instrument, but common values are from 1 to 3 percent. Thus a scope with a 3 percent calibration error, which is set to 1 V/div sensitivity, may actually be displaying 0.97 to 1.03 V/div. Check the scope manual for the value of this error.

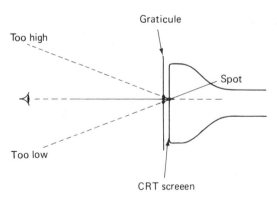

**Figure 6-42** Parallax error in oscilloscopes.

**4.** *Loading error.* The general concept of loading was introduced in Chapter 3. The details of how loading causes errors in the readings of a scope were discussed in the section "Oscilloscope Probes," together with information as to how probes are used to reduce these errors.

**5.** *Hum and noise pickup.* The oscilloscope is designed to be capable of amplifying and displaying small input signals. This capability also makes it susceptible to amplifying small unwanted signals and noise, especially when set to its most sensitive V/div capability. To prevent the scope from picking up such interference signals, special precautions must be taken.

The most serious interfering signals in a scope come from the electric and magnetic fields generated by the power lines that carry 60-Hz electric power to the home and laboratory. Motors, fluorescent lights, and other electrical equipment being run from the power lines are also sources of such time-varying fields. These stray fields are always present in the lab and cause 60-Hz signals to be induced in the circuits being measured by the scope and in the scope input leads. This 60-Hz signal is referred to as *hum*.[3] The magnitude of hum picked up by an oscilloscope depends upon the impedance of the circuit being measured, the length of the scope leads, and the type of shielding used by these leads. Hum signals are largest when long, unshielded leads are connected to high-impedance circuits. The longer the lead lengths, the greater is the stray capacitance between the leads and nearby power-line cords. Since

$$i = C \frac{dv}{dt} \tag{6-12}$$

a larger unwanted signal will be induced in a test lead when a greater capacitive effect exists between it and some other conductor at a different potential. This equation also predicts that higher-frequency signals (larger value of $dv/dt$) will also cause a greater pickup current to flow in the leads. Thus any wires carrying higher-frequency signals (from nearby oscillators, radios, etc.) can also be sources of interfering signals.

To verify the fact that a scope picks up and displays interfering signals, connect one end of an unshielded cable (or a shielded cable with an *ungrounded* shield) to the scope inputs. Leave the other end unconnected and lying on the workbench. The vertical amplifier should be set to the position of greatest sensitivity and the trigger level to the AUTO triggering position. The sweep time should be set to about 5 ms/div. A 60-Hz sine wave should be displayed on the scope screen. If the resulting display is a distorted 60-Hz sine wave, the distortion may be caused by high-frequency harmonics or spikes arising from fluorescent lights, etc.

The best way to reduce the level of capacitive coupled interference signals is to use short cables which have a shield that can be connected to ground. Most scope

---

[3]See Chapter 16 for a discussion as to why 60-Hz noise is called "hum" and for a more detailed description of externally generated noise.

probes are provided with a shield that can be connected to the scope ground. (Typically, the outer conductor of the coaxial cable of the probe serves as this shield.) When the shield is grounded, it remains at approximately zero potential and thus protects the signal-carrying cable from being influenced by the external time-varying electric fields. This fact can also be verified by grounding the shield of the input cables in the experiment mentioned above. In practice, shielded cables are about 90 percent effective in protecting the leads from external interfering signals.

In addition to the use of shielded cables, obvious noise sources such as fluorescent lights, oscillators, and motors should be turned off or kept far from the scope during low-level measurements. Differential inputs may need to be used to reduce common-mode noise pickup in low-level measurement applications.

**6.** *Bandwidth/rise time errors.* If the frequency of the signals applied to the scope is greater than the frequency-response capabilities of the scope amplifiers, the displayed images will not be faithful replicas of these input signals. The frequency-response specifications of amplifiers (*bandwidth* and *rise time*) are discussed in Chapter 15, together with the errors introduced when these specifications are exceeded. The results of the information presented in that section can be directly applied to estimating bandwidth/rise time errors in oscilloscopes. Thus readers are directed to Chapter 15 for information concerning such errors.

## OSCILLOSCOPE PHOTOGRAPHY

To obtain a permanent record of the signals displayed by an oscilloscope or transistor curve tracer, a hand trace or photograph can be made.

For hand tracing, special stick-on, transparent papers with inscribed graticules are available. This recording method is not as accurate as a photograph, but it is cheaper. When making hand traces, care must be taken to avoid scratching the screen or getting the intensity of the scope display too high, thus causing the screen phosphor to be burned.

For making photographs of scope traces, special cameras employing Polaroid films are used (Fig. 6-43). These special cameras are designed to mount directly onto the scope face. The Polaroid film allows photographs to be made quickly, and it is unnecesary to expose the whole roll before developing the pictures. Some cameras, including the model shown in Fig. 6-42, have a viewing hood so that the operator can observe the waveform on the scope screen before taking the picture.

Usually high-speed film (ASA rating of 3000 or 10,000) is used so that short exposure times are possible. After Polaroid film is developed, it must be carefully coated to protect the photograph from deteriorating. Suitable coating material is usually supplied with the film.

The *photographic writing speed* of an oscilloscope is the maximum speed at which a spot of light can move across the face of CRT and still be photographed. High-quality oscilloscopes have writing speeds of 5000 cm/μs. The writing speed

**Figure 6-43** Typical oscilloscope camera. (Reprinted by permission of Tektronix, Inc.)

determines the highest-frequency waveform or fastest-rising pulse that can be photographed from the screen.

Postfogging is a technique that allows photographs of oscilloscope traces to be made at two to three times the writing speed. This is achieved by exposing the film to a light source after it has been exposed to the CRT trace. Postfogging is performed by graticule illumination, storage flood gun, or UV light source, depending on the specific instrument design.

When attempting to photograph a single transient event, it is better practice to externally trigger the scope with the same circuit that initiates the transient rather than use internal triggering. Noise and other spurious signals may accidentally trigger the scope if *Internal* triggering is used. An appropriate external trigger signal provides a predictable, noise-free command to both the oscilloscope and the test setup to produce the transient. Better yet, use a storage scope (if available) and photograph the stored image.

### Guidelines for Oscilloscope Photography

**1.** Consult the instruction manual of the oscilloscope camera being used, for details on loading the film, focusing the camera, and adjusting shutter speed.

**2.** When estimating exposure conditions, adjust the scope trace for normal brightness (i.e., a brightness level that seems proper for normal visual observation). Then set the camera aperture opening ($f$-stop) at 8 and exposure at 0.1 s.

**3.** Keep graticule illumination low if a control for adjustment of this lighting is available.

**4.** Take the photograph.

**5.** If the photograph is not satisfactory when the $f$-stop and exposure conditions noted above are used, change the exposure time by a factor of 4 or 8. A change by a factor of 2 will have only a mild effect on the outcome of the print because most films have a wide exposure latitude.

**6.** When photographing multiple traces by making repetitive shots on one photograph, turn off the graticule illumination after the first shot.

## SPECIAL-PURPOSE OSCILLOSCOPES

In addition to the basic oscilloscope described in this chapter, there are other special-purpose oscilloscopes and features which extend some of the basic scope capabilities. The most important types will be briefly described here, and a few facts about their special features and functions will be mentioned. However, the operating details and applications of such scopes are too specialized (and complex) to be fully explored by this text.

### Delayed Time Base (Delayed Sweep)

In the section describing the time-base circuitry of oscilloscopes, we examined the sweep magnifier and its function. We noted that the sweep magnifier could expand the horizontal scale by some factor (5× or 10×), but that a loss of accuracy in the time scale would occur when this was done. In addition, it was pointed out that under high magnification it might be difficult to distinguish the expanded waveform segments of interest. By using a delayed-time-base feature instead of a sweep magnifier, the desired segments of a waveform can still be expanded and the problems mentioned above overcome. In addition, use of a delayed time base can have other advantages in certain measurement applications (including time-interval measurements of about 1 percent accuracy).

The delayed sweep feature is created by using two times bases in the oscilloscope; a main time base $(A)$ and a secondary time base $(B,$ referred to as the delayed sweep). Time base $B$ is typically 10 or 100 times faster than time base $A$. Time base $A$ generates the same type of sweep waveform as in the basic oscilloscope [Fig. 6-44(a)]. On oscilloscopes with the delayed sweep feature, the sweep control is set to *Sweep A* when only time base $A$ is used. Time base $B$ also generates a sweep waveform, but its starting time occurs when the amplitude of the sweep waveform $A$ reaches a selected level (corresponding to desired delayed time, $t_d$) which has been set by the *Delay Time Multiplier* control. At this time, the delay sweep generator is activated, and its function is to cause the selected portion of the display to be intensified [Fig. 6-44(b)]. (The horizontal display switch is set to *A intensified* when this mode of operation is being employed.) Next the horizontal display switch is set to the *B Sweep* position. The original display is now replaced with the portion of the display that was formerly intensified [Fig. 6-44(c)]. In this position the intensified portion can be expanded on the screen [Fig. 6-44(d)]. Since time base $B$ is 10–100 times faster than time base $A$, the expansion of the display can be achieved up to such magnification.

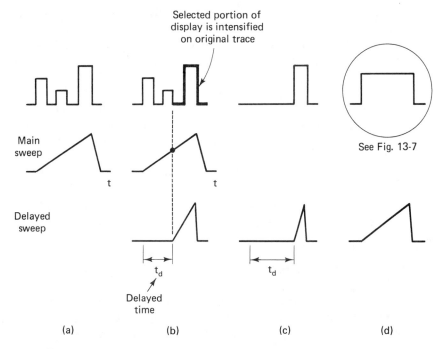

**Figure 6-44** (a) Original trace on scope screen; (b) selected portion of trace intensified by activation of delayed sweep waveform; (c) main sweep no longer causes a signal to be displayed, only delayed sweep waveform yields display; (d) selected portion of original trace in expanded display.

The delay time between the start of the main and delayed sweeps is related to the main time base Time/div. setting. This is why the delay control is labeled *Delay Time Multiplier*. The delay control is usually a precision 10-turn potentiometer with an accompanying scale from 0.00 to 10.0 (uncertainty of 0.1 percent of full scale). A comparator compares the precision dc voltage (obtained from the output of the delay potentiometer) to the rising voltage of the main sweep ramp. When the two voltages are equal, the comparator emits a pulse that starts the delayed sweep. Time interval measurements with an accuracy of about 1 percent are possible by using the following procedure. The center of the scope screen is used to mark the start and stop points of the delayed time interval, and the potentiometer readings of these two points are determined. Next the readings are subtracted (stop-point potentiometer reading minus start-point potentiometer reading) and the time interval between the two points is obtained. The accuracy of the measurement is limited by the uncertainty in the reading of the 10-turn potentiometer reading (0.1 of full-scale) plus the uncertainty of the reading due to the sweep generator (0.5 percent of reading), or less than 1 percent total error.

Some delayed time bases can also be operated in a *mixed-sweep mode*. In this mode, the normal display as caused by the main time base (MTB) is displayed.

However, the delayed sweep time base (DTB) is also activated as before. When the amplitude of the DTB sawtooth waveform exceeds the amplitude of the MTB sawtooth, the display is electronically switched to show the expanded image. Thus a mixed display consisting of the nonexpanded portion and the expanded portion is simultaneously produced. In some DTBs, the nonexpanded and magnified traces can be alternately displayed and separated to give a dual-trace type of display. The mixed sweep finds its best use in some digital applications. An entire pulse train can be minutely observed and counted by expanding each pulse using the DTB. When used in the mixed-sweep mode each pulse in the train can be observed merely by varying the delay-time controls.

Use of the delayed-sweep-time base also has the following advantages:

1. Jitter and drift in the observed waveforms can be eliminated.
2. Greater accuracy of time measurements can be accomplished than is possible with a conventional time base.

### Sampling Oscilloscopes

To display recurrent signals whose frequencies are higher than the limits of the ultrawideband (high-frequency) scopes, *sampling* techniques must be used. With these techniques, the signal is reconstructed for display from sequential samples of its waveform.

A sampling scope is analogous to a stroboscope, a device that permits visual observation of rapidly rotating machinery by momentarily lighting it at slightly advanced positions on successive revolutions. A sampling system measures the amplitude of a small portion of the waveform and displays this instantaneous amplitude on the CRT in the form of a dot. After displaying the dot, the beam is shut off and moved horizontally over a short distance. During this time interval, the scope measures another sample of the input waveform, but during a subsequent cycle of the waveform, and at a slightly later position in the cycle. Now the scope displays the new instantaneous amplitude as another dot, positioned horizontally a very short distance from the first dot. In this way, the oscilloscope plots the input waveform point by point, using as many as 1000 samples to reconstruct the original waveform (Fig. 6-45). The sample frequency may be as low as one-hundredth of the input signal frequency. In such a case, if the input signal has a frequency of 1 GHz (1000 MHz), the required bandwidth of the oscilloscope amplifier need be only 10 MHz. Sampling scopes are being built with capabilities of displaying recurring signals with frequencies of over 10 GHz, and pulses having risetimes of less than 30 ps.

There are certain limitations possessed by the sampling oscilloscope besides the requirement of recurrent waveforms for observation. First, there is a maximum input voltage, which must never be exceeded (typically 1–2 V). The sampling diodes can be destroyed by larger input voltages. However, attenuators attached to the probe tips allow larger signal voltages to be attenuated to a safe level and fed to the

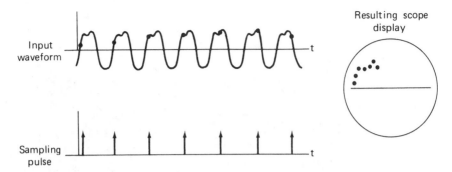

**Figure 6-45**  Relationship between input waveform, sampling pulses, and scope display in sampling oscilloscopes.

sampling inputs. In addition, the input impedance of most sampling units is 50-$\Omega$, and normally they are used to view pulses from a 50-$\Omega$ source. Thus special active probes must be used to measure signals coming from higher-impedance signal sources.

## Storage Oscilloscopes

In conventional oscilloscopes, the persistence of the phosphor typically ranges from milliseconds to seconds (refer to Table 6-1). This means that a display that represents a nonrepetitive waveform or event will disappear in a short time. *Storage oscilloscopes* have been invented to overcome this drawback of image impermanence. They have the capability of storing an image of the display on the scope screen for a much longer time (typically from seconds to hours). The stored image can be erased at any time to allow a new one to be displayed. Also, for measurements that do not require the storage capacity, the storage scope may be operated in a conventional nonstorage mode.

Storage scopes are especially useful in a number of applications. Besides being able to record nonrepeating phenomena (which would flash across an ordinary CRT too quickly to be evaluated), storage scopes can effectively display the waveforms of phenomena that change very slowly with time. Similarly, recurring signals with very low repetition rates can also be displayed without having the image flicker. Storage can also reduce the time required to photograph scope traces. That is, unwanted displays can be erased as many times as necessary before the photograph is taken. On the other hand, if an event cannot be run repetitively, storage will protect against the possibility of a photographic mishap.

The storage (or *retention*) feature is made possible because of the invention of special storage cathode-ray tubes. There are several types of these tubes: the *bistable phosphor* tube, the *variable-persistence* tube (sometimes called the halftone tube), and the *fast transfer* tube. (Digital storage is also available, and this is discussed in the section on digital oscilloscopes.)

The *bistable phosphor* tube (Fig. 6-46) employs a special phosphor that will

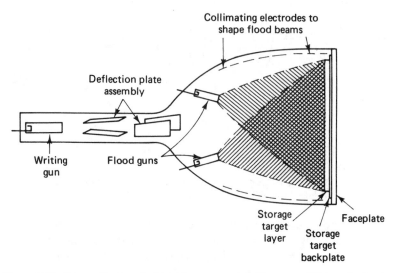

**Figure 6-46**   Schematic view of a bistable storage tube. (Courtesy of Tektronix, Inc.)

either store or not store an image and will therefore produce only one level of image brightness (no halftones are possible). The stored display can be retained for up to several hours. Bistable storage tubes were the first commercially available storage tubes. They are the easiest to use and are the least expensive. A *split-screen* option on some bistable storage tubes divides the viewing area of the screen into two halves. Each half can be used independently for storage displays. (Furthermore, one half can be used in the storage mode, to retain, say, a reference waveform. The other half can be operated in the nonstorage mode to monitor, say, an external output.)

The second storage tube, the *variable-persistence CRT*, allows a gradation in the display brightness (typically five brightness levels) and also permits selection of the time a stored image will be retained. Variable-persistence tubes are best used for real-time displays, spectrum analysis, time-domain reflectometry, sampling, and other measurements that require slow sweep speeds. This type is the most widely used of the storage tubes.

The *fast-transfer tube* is a storage tube especially designed to be able to capture rapidly the waveform to be stored. A special intermediate mesh target captures the waveform and transfers it to a slower, longer-storing electrode. The slower electrode can offer either the bistable or variable-persistence mode of operation.

The storage capability of all three types of storage tubes arises from the phenomenon of secondary electron emission. (Secondary electrons are those that are separated from the phosphor surface when struck by the electrons of the electron beam. Beam electrons are called primary electrons.) The secondary electron emission phenomenon is exploited to build up and store charges on the surface of an insulated target (the phosphor-coated screen of the CRT).

The storage tube contains two electron guns: a *flood gun* and *writing gun*. The writing gun is identical to the gun in a conventional CRT, and serves the same function of providing a pencil-thin electron beam. The flood gun, on the other hand, floods the screen with electrons in the way that a floodlight illuminates a wall. These low-velocity electrons strike the phosphor but do not have enough energy to dislodge many secondary electrons, or cause the phosphor to glow.

As the primary electrons of the writing beam (from the writing gun) strike the screen in a localized area, they dislodge many secondary electrons. This occurs because the primary (beam) electrons are focused and of much higher energy than those of the flood gun. The areas of the screen that have been struck by the primary beam lose electrons and therefore become positively charged. The flood-gun electrons are thus attracted by these positive regions and are accelerated toward them. If they strike the positive regions with enough energy, they will keep the phosphor glowing and cause it to emit enough secondary electrons to maintain the positive charge on these regions. The displayed image thus constantly renews itself as the flood gun electrons illuminate the screen. As a result, it can be seen that the storage feature is not due to any extraordinary phosphor persistence. It is instead caused by the continuing flood-gun electron bombardment of the phosphor along the pattern that was traced previously by the writing beam. (Note that because of the insulating properties of the target, the lit phosphor areas do not spread.)

As the sweep rate of the writing beam is increased, a maximum limit is reached. This characteristic is known as the *stored writing speed* and is the most significant specification of storage CRTs. Typical values for each of the tube types are 0.2 to 5 div/ms for bistable CRTs, 0.2 to 12 div/ms for variable-persistence crts, and up to 3500 div/ms for fast-transfer CRTs. With the latter storage CRTs, single-shot pulses with 3 ns rise time can be captured.

### Digital Storage Oscilloscopes

The digital storage oscilloscope differs from its analog storage counterpart in that it "digitizes" or converts the analog input waveform into a digital signal that is stored in a semiconductor memory and then converted back into analog form for display on a conventional CRT. A typical digitizing-scope architecture applies extensive processing power to waveform acquisition, measurement, and display (Fig. 6-47). A wide variety of plug-in modules are available for special timing functions and additional channels. The data are displayed most frequently in the form of individual dots that collectively make up the CRT trace. The vertical screen position of each dot is given by the binary number stored in each memory location, and the horizontal screen position is derived from the binary address of that memory location. The number of dots displayed depends on three factors: the frequency of the input signal with respect to the digitizing rate, the memory size, and the rate at which the memory contents are read out. The greater the frequency of the input signal with respect to the digitizing rate, the fewer the date points captured in the oscilloscopes memory in a single pass and the fewer the dots available in the reconstructed waveform.

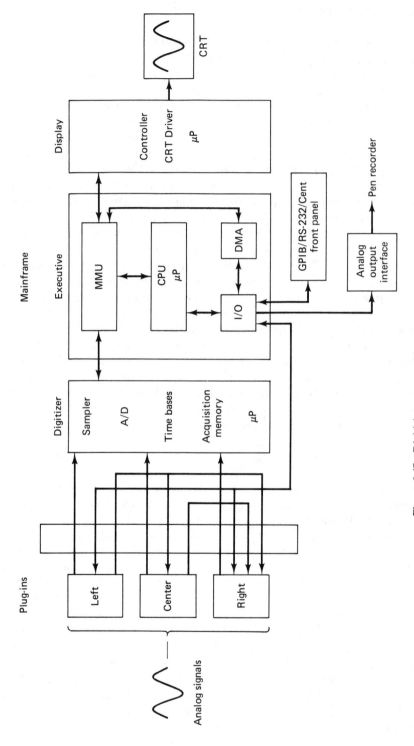

**Figure 6-47** Digitizing-scope architecture and block diagram.

Digital storage scopes and analog scopes each have distinct advantages, but digital scopes have created the most recent excitement because they have undergone some dramatic improvements in their performance. System bandwidths to 1 GHz are available with a time interval accuracy of 100 ps and 1 giga-samples/second (Gsa/s). Because of the modularity of many of the oscilloscopes, additional channels and features are available by just plugging in the appropriate accessory to the mainframe. In addition to merely capturing and displaying waveforms, the digital scopes can also perform the following functions: the indefinite storage of waveform data for comparison and the transferring stored data to other digital instruments. Furthermore, pulse-parameter analysis can be computed and made available for presentation in decimal form on the scope screen. Digital storage scopes are also well suited for a myriad of special tasks such as diagnosing "intermittent" problems. By setting the scope to single-sweep mode, data can be automatically captured and stored on the occurrence of the trigger event. Thus constant monitoring of a test setup until the "intermittent" shows up can be eliminated. The data once captured not only includes the data that occurred *after* the trigger signal but data *before*. This is very important when trying to determine the cause or source of a false signal.

The analog-to-digital converter of the scope determines some of its most important operating characteristics. The voltage resolution is dictated by the bit resolution of the A/D converter, and the storage speed by the maximum speed of the converter. For example, a 10-V converter using 8, 10, or 12 bits is able to resolve 0.0391 V, 0.0098 V, and 0.0025 V, respectively. The time resolution is selectable, in that the user can define how much memory space is needed for each waveform stored. The output of digital scopes is available in other forms than the trace on the CRT. Analog output is provided for driving pen recorders. Digital output formats in RS-232C and IEEE-488 are available as options. (See Chapter 18 for a discussion of these terms.)

Most of the front-panel controls of digital scopes are similar to the controls of conventional analog scopes, and they are operated just as conventional scopes. However, the additional features possessed by digital scopes are controlled by additional dials and switches on the scope's face. A photograph of a digital storage scope is shown in Fig. 6-48.

## GRAPHICAL DISPLAY OF THREE-TERMINAL DEVICES/THE CURVE TRACER

Three-terminal devices are electronic components that have three input terminals (i.e., as compared to such two-terminal devices as resistors, capacitors, inductors, or diodes). The most significant examples of three-terminal devices are components used to amplify signals: bipolar transistors, junction field-effect transistors (JFETs) and metal-oxide-semiconductor field-effect transistors (MOSFETs). The current-voltage relationships for these devices must be known in order for them to be properly incorporated in electronic circuits. Such characterization of three-terminal

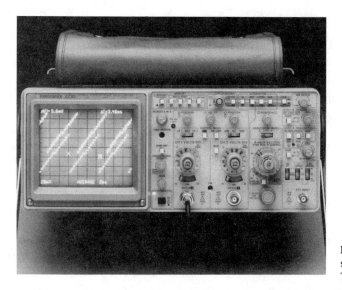

**Figure 6-48**   Photograph of a digital storage oscilloscope. (Courtesy of Tektronix, Inc.)

devices, however, is more complex then the characterization of two-terminal components. (We saw earlier in this chapter that even for such non-linear two-terminal components, a single *V–I* curve completely characterizes the current vs. voltage relationship of a two-terminal device.) However, for three-terminal devices there must be two *V–I* driving point characteristics (Fig. 6-49), since there are two independent driving points (or *ports*). In addition, instead of a single *V–I* curve at each driving point, there will be a family of parametric curves, controlled by the voltage or current at the third terminal.

A special oscilloscope-type instrument, the *curve-tracer* (Fig. 6-50), is designed to allow a graphical display of the *V–I* characteristics of three-terminal devices. We shall describe the operation of the curve tracer in this section and use the information to show how the *V–I* characteristics of bipolar and field-effect transistors can be displayed using this instrument. Three-terminal devices require two independent *V–I* characteristics since there are two driving points. In Fig. 6-49, terminal ③ is common to both driving points, and thus the two *V–I* characteristics

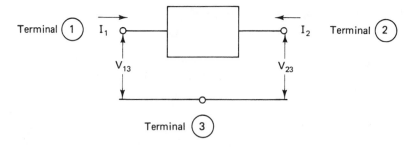

**Figure 6-49**   Three-terminal device.

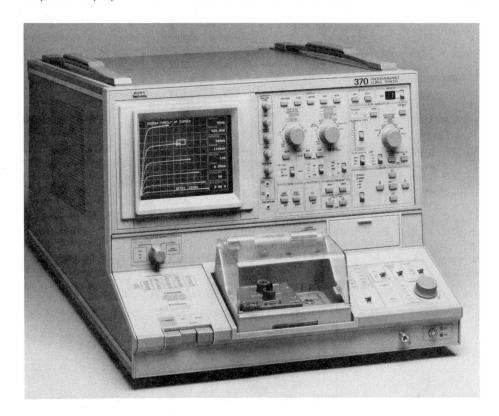

**Figure 6-50**  Semiconductor curve tracer used to display the characteristic curves
of semiconductor diodes, transistors, and FETs. (Courtesy of Tektronix, Inc.)

would be ($V_{13}$ versus $I_1$), and ($V_{23}$ versus $I_2$), as the driving points would be between
(a) terminals ① and ③ or (b) terminals ② and ③.

Many device parameters required by electronic circuit designers are obtainable from these *V–I* characteristics. We shall briefly discuss the procedures for extracting such information from these graphically displayed curves.

A simplified block diagram of the curve tracer is shown in Fig. 6-51. We see that a waveform generator called a *collector sweep source*, internal to the curve tracer, generates a full-wave rectified sine wave. It does this by rectifying the 60-Hz line voltage. The full-rectified waveform is applied to the device under test (which is connected across terminals *C* and *E* of the curve-tracer test panel terminal shown in Fig. 6-51) and to the horizontal amplifier of the scope. As a result, the electron beam of the curve-tracer CRT is deflected back and forth in a horizontal direction and the deflection is proportional to the voltage across the device under test.

Assume that a bipolar transistor (whose three terminals are respectively labeled collector, emitter, and base) is connected to the curve-tracer test panel terminals with its collector attached to terminal *C*, its emitter to terminal *E*, and its

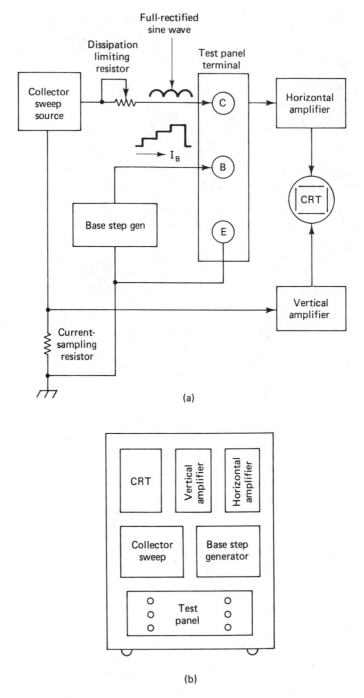

**Figure 6-51** (a) Curve tracer block diagram; (b) curve tracer control panel layout.

base to terminal $B$. Now the rectified voltage waveform from the collector sweep source is applied across the collector and emitter terminals ($V_{CE}$) of the bipolar transistor. As the collector voltage rises and the beam sweeps horizontally across the CRT screen, the collector current may also rise. This current is measured by sensing the voltage across a 1-kΩ current-sampling resistor (which is also internal to the curve tracer). By applying the current-dependent signal to the vertical plates of the scope, the current through the collector of the transistor, $I_C$, is displayed in the vertical direction on the CRT, and the voltage being applied across collector and emitters terminals ($V_{CE}$) is displayed along the horizontal axis. Thus, the V–I characteristic of $I_C$ versus $V_{CE}$ of the transistor is displayed. (Note that if a diode or any other two-terminal device were connected across terminals $C$ and $E$ of the curve-tracer test panel, the display would be either the forward or reverse portion of the V–I diode characteristic, depending on the relationship between the polarities of the applied rectified sweep waveform and the terminals of the connected diode.)

If a current of known value is fed into the base terminal of the transistor, the current–voltage characteristic can be used to determine the gain (or β) of the bipolar transistor (where $β = I_C/I_B$). The curve tracer internally contains a current source that can supply such a known current. This source (called the base-step generator) also has the capability of feeding a number of selected current values into the base of the transistor. It does this by generating the current in the form of a staircase waveform (see Fig. 6-52). The staircase current level is raised one step after each sweep of the rectified voltage between collector and emitter. In Fig. 6-52 three

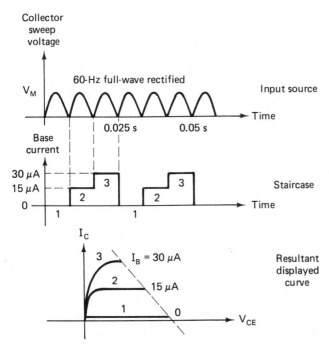

**Figure 6-52** How a family of characteristic curves is generated in a curve tracer.

steps are shown (0,15, and 30μA). One curve of the family of the *I–V* characteristic curves is generated for each step, and then the process repeats. Since the full-rectified waveform is derived from the 60-Hz line voltage, 120 sweeps are generated per second. The repetition rate of each single curve is thus 120 divided by the number of curves in the display. On a typical curve tracer, as many as 10 curves may be displayed or as few as 1.

The family of curves representing $I_C$ versus $V_{CE}$ of the bipolar transistor is one of 12 possible characteristics of the bipolar transistor when it is connected so that the emitter is common to both driving points. [Note that since there are four terminal variables as shown in Fig. 6-53(a), theoretically six characteristic curves are possible ($I_C$ versus $V_{CE}$, $I_C$ versus $V_{BE}$, $I_C$ versus $I_B$, $I_B$ versus $V_{CE}$, $I_B$ versus $V_{BE}$, and $V_{CE}$ versus $V_{CB}$.] In addition, each of these six can have two choices of parametric variation, bringing the possible total to 12.) To require all 12 families of curves, however, would involve overspecification, since in general two sets of parametric curves are sufficient to describe a three-terminal device. However, in the case of most bipolar transistor applications, the $I_C$ versus $V_{CE}$ curves with $I_B$ as the parameter (frequently called the *output curves*) provide the most useful data about the device operation.

Let us see how the gain (β) of the bipolar transistor, as well as three other of its important parameters—collector leakage current ($I_{CEO}$), collector-emitter breakdown voltage ($BV_{CEO}$), and collector-base breakdown voltage ($BV_{CBO}$)—can be determined by using the curve tracer to display $I_C$ versus $V_{CE}$.

The gain (β) of the bipolar transistor is found from the equation

$$\beta = \frac{I_C}{I_B}\bigg|_{V_{CE}} \qquad (6\text{-}13)$$

When a bipolar transistor is connected to the curve-tracer test panel with the collector to terminal *C*, the base to terminal *B*, and the emitter to terminal *E*, a single curve of $I_C$ versus $V_{CE}$ will be displayed on the CRT for each applied value of $I_B$ [Fig. 6-54(a)]. The value of β for a given $V_{CE}$ will be found as shown in Fig. 6-54(a).

The collector-to-emitter leakage current ($I_{CEO}$) with $I_B = 0$ (base open) is found

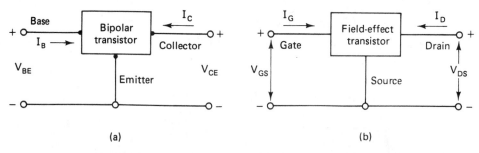

(a)                                                                (b)

**Figure 6-53**   (a) Bipolar transistor connected in the common-emitter configuration; (b) field-effect transistor connected in the common-source configuration.

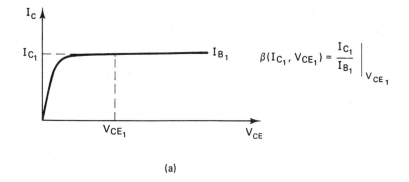

(a)

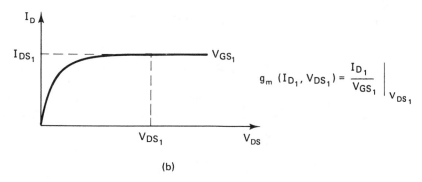

(b)

**Figure 6-54**   (a) Determination of the common-emitter current gain, $\beta$, of a bipolar transistor from the display of $I_c$ versus $V_{ce}$; (b) determination of the common-source transconductance, $g_m$, of field-effect transistors from the display of $I_D$ versus $V_{DS}$.

by connecting the bipolar transistor to the curve tracer in the same manner as for the determination of $\beta$. The input base current, $I_B$, is set to zero and $I_{CEO}$ can be read for the desired $V_{CE}$ directly from the display. The most sensitive collector current switch position will probably need to be used since $I_{CEO}$ is generally quite small.

The collector–emitter and collector–base breakdown voltages are found in this manner: To find $BV_{CEO}$, connect the transistor as before. Set $I_B = 0$, and increase the amplitude of $V_{CE}$ until the I–V curve rises steeply. Limit the current, $I_C$, to some reasonable small value (i.e., 1 mA) to prevent device damage.

To find $BV_{CBO}$, connect the collector lead of the transistor to terminal $C$, the base lead to terminal $E$, and leave the emitter terminal unconnected. Slowly increase the amplitude of the horizontal voltage (if at 0.6 V the current starts to increase steeply, reverse the polarity of the applied horizontal voltage, as this indicates that the collector–base *pn* junction is forward biased). If the proper polarity (reverse-bias) voltage has been applied, continue to increase the horizontal voltage until breakdown (signified by a sharp increase in $I_C$) is reached. Limit the maximum collector current at breakdown to about 1 mA to prevent device damage.

Field-effect transistors (both JFETs and MOSFETs) have their terminals labeled *drain, source,* and *gate.* They are most often used in the common-source configuration [Fig. 6-54(b)]. The figure of merit of most importance, the *transconductance,* or $g_m = (I_D/V_G)|_{V_{DS}}$, can be found from the drain current ($I_D$) versus drain–source voltage ($V_{DS}$) characteristic, with $V_{GS}$ as the parametric variable. Figure 6-54(b) shows how $g_m$ is determined from the graphical display of $I_D$ versus $V_{DS}$ at a given value of $V_{GS}$.

Curve tracer capabilities have been expanded to include two-, three- and four-terminal devices such as optoelectronic devices, SCRs, triacs, and ICs. The testers are now programmable and have printer interfaces for documenting not only the device curves but also the test parameters. As shown in Fig. 6-50, the screen displays the major test parameter settings so that operator errors will be minimized.

### Measurement of Transistor Forward Current Transfer Ratio ($h_{fe}$)[4]

The following procedure is applicable to the measurement of $h_{fe}$ of a small signal transistor such as the NPN 2N3904 or 2N2222. The emitter will be the common terminal [i.e., that which is connected to both the base step generator and the collector supply (Fig. 6-55)]. The voltage drop across resistor $R_s$ is proportional to

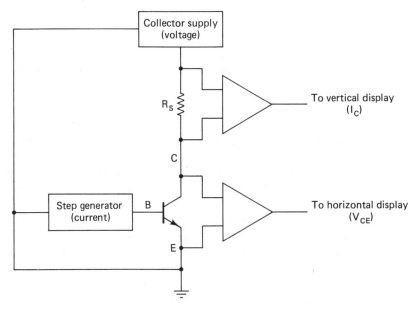

**Figure 6-55** Curve tracer configuration for $h_{fe}$ measurement. (Courtesy of Tektronix, Inc.)

[4]Adapted from Tektronix application note 48W6756 (Courtesy of Tektronix, Inc.)

$I_c$. This voltage is applied to the vertical display of the curve tracer. The voltage drop $V_{ce}$ is applied to the horizontal display. Do not install the device being tested until initial settings on the curve tracer are completed.

**1.** Turn on power, and allow the curve tracer to warm up. Place the curve tracer in the standby mode (i.e., do not apply power to the transistor socket).

### Create a Collector Circuit.

**2.** Set the correct collector supply polarity, that is, plus for NPN transistors and minus for PNP transistors.

**3.** Set the collector voltage *range* to a value based upon the manufacturer's specifications for the transistor. Always start at the lowest possible setting, and increase the setting as required by the test. Use the variable *Vernier* control to set the exact voltage.

**4.** Select the peak power (in watts) that the device can dissipate. Do not exceed 50 percent of the manufacturer's recommended maximum power. If the tester permits you to select a series resistor, then select a 1K resistor for a 2N2222.

### Create a Base Drive Circuit.

**5.** Set the polarity of the base *step generator*. The polarity of the collector and base drive circuits should be the same when measuring $h_{fe}$ of a (NPN) transistor (e.g., the 2N2222).

**6.** Set the size of the current steps (not voltage) to be applied to the base. Start at the lowest setting.

**7.** Set the number of steps to be displayed in the family of curves (e.g., five steps).

**8.** Set the offset to zero.

### Create the Horizontal and Vertical Display.

The horizontal and vertical display controls do not affect the collector and step generator settings. They are display controls only.

**9.** Set the vertical display to approximately 1 mA/div.

**10.** Set the horizontal display to approximately 1 volt/div.

**11.** Use the dot or trace position controls to set the beginning (origin) of the family of curves.

### Create a Family of Curves for the Transistor.

**12.** Insert the transistor in the curve tracer socket. Then place the curve tracer in the active mode (i.e., apply power to the transistor).

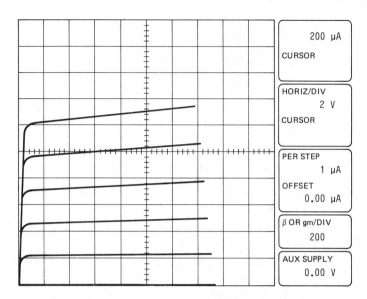

| 200 µA |
| CURSOR |

| HORIZ/DIV |
| 2 V |
| CURSOR |

| PER STEP |
| 1 µA |
| OFFSET |
| 0.00 µA |

| β OR gm/DIV |
| 200 |

| AUX SUPPLY |
| 0.00 V |

**Figure 6-56**    $H_{fe}$ curves on curve tracer. (Courtesy of Tektronix, Inc.)

**13.** Increase the collector voltage until a family of curves is seen on the display (see step 3).

**14.** Increase the base step generator setting until the desired number of curves are displayed. The spacing and length of the base curves can be adjusted by using the *vertical* and *horizontal* controls.

A typical family of curves is shown in Fig. 6-56. The digital windows display the selected settings of a device being tested on a curve tracer of the type shown in Fig. 6-50. Calculate the $h_{fe}$ by using the method shown in Fig. 6-54. The curves can be either copied by hand, printed, or photographed (Fig. 6-43) for a permanent record.

### Measurement of Power MOSFET Forward Transconductance $(g_m)$[5]

The following procedure is applicable to a MOSFETs. The source will be common (connected to) to both the base step generator and the collector supply (Fig. 6-57). The collector supply is connected to the MOSFET drain. The voltage drop across resistor $R_s$ is proportional to $I_{DS}$. This voltage is applied to the vertical display of the curve tracer. The voltage drop $V_{DS}$ is applied to the horizontal display. Do not install the MOSFET until initial settings are completed.

**1.** Turn on power, and allow the curve tracer to warm up. Place the curve tracer in the standby mode (i.e., do not apply power to the MOSFET socket).

[5]Adapted from Tektronix application note 48W6757 (Courtesy of Tektronix, Inc.)

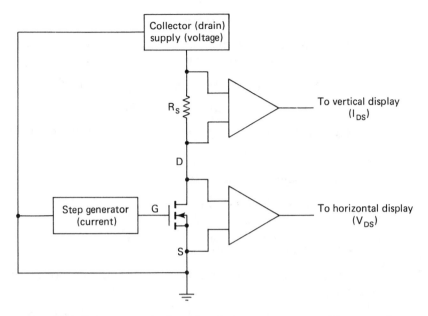

**Figure 6-57** Curve tracer configuration for $g_m$ measurement. (Courtesy of Tektronix, Inc.)

### Create a Drain Circuit.

**2.** Set the collector (drain) supply polarity, that is, plus for N channel MOSFET and minus for P channel.

**3.** Set the collector voltage *range* to a value based upon the manufacturer's specifications for the MOSFET. Always start at the lowest possible setting, and increase the setting as required by the test. Use the variable *Vernier* control to set the exact voltage.

**4.** Select the peak power or watts that the device can dissipate. Do not exceed 50 percent of the manufacturer's recommended maximum power.

### Create a Gate Drive Circuit.

**5.** Set the polarity of the *step generator*. The polarity of the drain and gate drive circuits should be the same when measuring $g_m$ for a MOSFET. The voltage should be opposite for a JFET or a depletion mode MOSFET.

**6.** Set the size of the voltage steps (you can use current if you connect a 1K resistor between the gate and source) to be applied by the base step generator to the gate of the MOSFET. Start at the lowest setting.

**7.** Set the number of steps to be displayed in the family of curves (e.g., five steps).

**8.** Set the offset to zero.

*Create the Horizontal and Vertical Display.*

The horizontal and vertical display controls do not affect the collector and step generator settings. They are display controls only.

**9.** Set the vertical display to approximately 1 mA/div.

**10.** Set the horizontal display to aproximately 1 volt/div.

**11.** Use the dot or trace position controls to set the beginning (origin) of the family of curves.

*Create a Family of Curves for the MOSFET.*

**12.** Insert the MOSFET in the curve tracer socket. The drain terminal should be inserted into the collector socket, the gate into the base socket, and the source into the emitter socket. Then place the curve tracer in the active mode (i.e., apply power to the MOSFET).

**13.** Increase the collector voltage until a family of curves is seen on the display (see step 3).

**14.** Increase the base step generator setting until the desired number of curves are displayed. The spacing and length of the base curves can be adjusted by using the *vertical* and *horizontal* controls.

A typical family of curves is shown in Fig. 6-58. The digital windows display the selected settings of a device being tested on a curve tracer of the type shown in

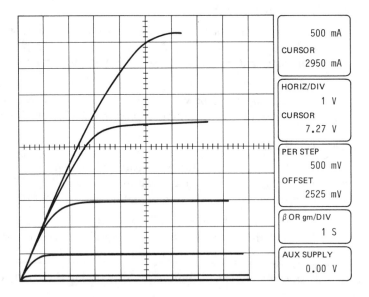

| 500 mA |
| CURSOR |
| 2950 mA |
| HORIZ/DIV |
| 1 V |
| CURSOR |
| 7.27 V |
| PER STEP |
| 500 mV |
| OFFSET |
| 2525 mV |
| $\beta$ OR gm/DIV |
| 1 S |
| AUX SUPPLY |
| 0.00 V |

**Figure 6-58**   $g_m$ curves on curve tracer. (Courtesy of Tektronix, Inc.)

Fig. 6-50. Calculate the $g_m$ by using the method shown in Fig. 6-54. The curves can be either copied by hand, printed, or photographed (Fig. 6-43) for a permanent record.

## PROBLEMS

1. What are some of the advantages that oscilloscopes possess over other types of electronic-measuring instruments?

2. Describe how the small spot of light seen on the oscilloscope screen is created by the CRT. Explain why permanent damage may be caused to the CRT if the spot is allowed to remain stationary on the screen.

3. Referring to Example 6-1, if the CRT shown in Fig. 6-2 is set to vertical sensitivity of 2 V/cm and a horizontal sensitivity of 5 V/cm, where will the spot appear on the scope screen if 0.2 V were applied to the vertical input and −150mV were applied to the horizontal input (assuming it would appear at the center of the screen if no voltages were applied to the plates)?

4. What would be the position of the deflected spot on the scope screen if the polarity of the battery attached to the horizontal plates was reversed and the horizontal sensitivity was also 1 V/cm?

5. In Fig. P6-1 the horizontal and vertical sensitivities of the scope are set to 2 V/cm. With the circuit connected to the scope as shown, determine the position of the deflected spot.

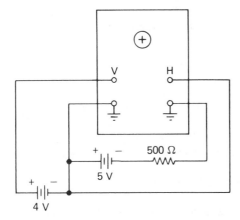

Figure P6-1

6. Explain the purpose and operation of the input coupling switch of the oscilloscope. What would be the waveform displayed on the scope screen if the waveform shown in Fig. P6-2 were applied to the scope and the input coupling switch was set to:
   (a) *dc* position?
   (b) *ac* position?

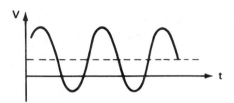

**Figure P6-2**

**7.** What is the purpose of the *ground* position of the input Coupling switch?

**8.** Describe the function of each of the following oscilloscope controls.

(a) Focus                          (b) Astigmatism

(c) Vertical position            (d) External horizontal input

(e) Sweep vernier

**9.** List one application in which each of the following connections that trigger the time base are used.

(a) Internal

(b) External

(c) Line

**10.** What will be the peak-to-peak distance of the display shown on an oscilloscope of a sine wave whose amplitude is 0.15 V if the vertical sensitivity switch of an oscilloscope is set to the following?

(a) 50 mV/div

(b) 1 V/div

(c) On which setting could you read the input voltage more accurately?

**11.** Describe the difference in the capabilities of measurement between an oscilloscope that has a single-ended input and one that has a differential input.

## REFERENCES

1. Van Erk, R., *Oscilloscopes*. New York: McGraw-Hill, 1978.

2. Coombs, C., *Basic Electronic Instrument Handbook*, Chap 24. New York: McGraw-Hill, 1972.

3. Roth, C.H., *Use of the Oscilloscope: A Programmed Text*. Englewood Cliffs, N.J.: Prentice Hall, 1971.

4. Lenk, J. D., *Handbook of Oscilloscopes*. Englewood Cliffs, N.J.: Prentice Hall, 1982.

5. Shackil, A. F., "*Digital Storage Oscilloscopes*," *IEEE Spectrum*, Vol. 17, pp. 22–25, July 1980.

6. Sessions, K. W., and Fisher, W. A., *Understanding Oscilloscopes and Display Waveforms*. New York: John Wiley, 1978.

7. Mazda, F. F., *Electronic Instruments & Measurement Techniques*. New York: Cambridge University Press, 1987.

# Potentiometers
# and Recorders

The purpose of this chapter is to discuss the operation and use of potentiometers and recorders. *Potentiometers* may be used to make very accurate determinations of unknown voltages. There are many measurement applications where such accuracy is a necessity. For example, the calibration of voltmeters must be undertaken by comparing their readings to an accurately known *voltage standard*. Because the accuracy of the potentiometer is so high, it can be used in the laboratory in conjunction with such a standard to calibrate voltmeters and other instruments.

The accuracy of potentiometers arises from the fact that they measure voltage through the use of a comparison technique. The unknown voltage is compared to an accurately known but adjustable voltage within the potentiometer. When the known voltage is set equal to the value of the unknown voltage, the unknown voltage becomes identified. The existence of the equality between the two voltages is indicated as a null reading on a sensitive meter. The value of the unknown voltage is read from the dial setting of the potentiometer at this *null* or *balance* point. (Note that a similar comparison measuring technique is used in the Wheatstone bridge circuit for measuring resistance.)

There are two basic classes of potentiometers, the *manually operated* and the *self-balancing* types. The manually operated models require an observer to adjust the dials until the equality between the known and unknown voltages is located. The self-balancing models are automatic devices which seek out the condition of equality themselves and do not need a human manipulator. Also, the self-balancing models are usually equipped with a marking device and a moving chart system. This allows, their indications to be permanently and automatically recorded on a graph.

*Recorders* are devices that provide a permanent graphical record of the quantity being measured, as well as carrying out the measurement itself. We note that these functions are also performed by the self-balancing potentiometer which was just described. This coincidence is not accidental. Rather, it is because self-balancing potentiometers actually belong to one class of recorders. Since this overlapping of functions between some recorders and potentiometers exists, it is logical to undertake the discussion of the other types of recorders in this chapter, too. Consequently, the latter part of the chapter is devoted to an examination of *X-Y* recorders, galvanometer recorders, and other forms of null-balancing recorders.

The recorders just described give an output that requires human interpretation. In some cases, however, the output data of a measurement system must be recorded in a form that can be "read" or interpreted by a machine. Machine-interpretable outputs include information stored on magnetic tape, magnetic disks, punched paper tape, punched cards, and electrical signals. If data are recorded on magnetic tape in analog form, the signals stored on the tape can be used to drive various analog output display devices such as the CRT, strip-chart recorded, or analog meter. Data recorded in digital form on magnetic tape (or magnetic disk or drum), punched paper, or punched cards, can activate digital display devices such as printers, digital computers, or digital display units. The latter part of this chapter provides a brief introduction to the recording devices used to record and store measurement data in machine-interpretable form.

## POTENTIOMETERS

### Manual Potentiometers

Manual and self-balancing potentiometers both achieve a null or balance condition by the use of the same basic measuring principle. However, the manual potentiometer is the simpler of the two instruments. Since it is easier to understand the general principles of potentiometer measurements from an examination of a simple instrument, we will undertake a discussion of the manual potentiometer first.

The operation of the basic manual potentiometer can best be understood with the help of a block diagram which shows its major elements (Fig. 7-1). The four most important of these elements are the precision slide-wire resistance, the sensitive current detector, the working voltage source, and the highly accurate reference voltage source.

The precision slide-wire resistance element is really the heart of the entire device. It basically consists of a length of wire whose total resistance is very accurately known (i.e., to within $\pm 0.02$ percent or better). In addition, the cross section of the wire is kept extremely uniform so that any fraction of its total length will contain the same fraction of the total resistance. For example, if the wire has a total

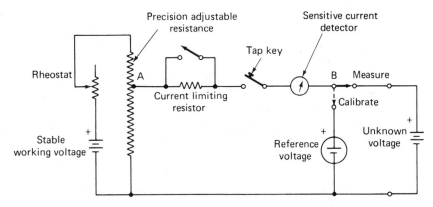

**Figure 7-1**  Block diagram of a manual potentiometer.

resistance of 150 Ω and is 150 cm in length, a portion of the wire whose length is 53.65 cm will have a resistance of 53.65 Ω (±0.01 Ω). If a current of a 10.0 mA is caused to flow such as in a 150-Ω resistance element, a voltage drop of 1.50 V will exist across its entire length.

A moving slider is attached to the precision resistor in such a way that an accurately known fraction of the total voltage across the resistor can be "picked off" by setting the slider to that fractional length of the resistor. The position of this slider along the length of the resistance element is indicated in voltage units by the dials on the potentiometer.

The purpose of the working voltage and rheostat is to provide an accurate amount of current flow in the precision resistor element. The value of this current is chosen so that it provides convenient numbers for calculating the voltages being measured by the instrument. To ensure that the working voltage and rheostat are actually providing the desired value of current to the precision resistor, a very accurately known voltage source is used as a calibrating device. (If a secondary standard cell is the calibrator, its voltage is known to be 1.019 V.)

In the process of calibrating the potentiometer, the slider on the precision resistance is set to a position which will ensure that some convenient current value (e.g., 10 mA) is flowing in the wire when it is properly calibrated. For example, if the 150-Ω, 150-cm resistor is being used, the slider is set to the 101.9-cm point on the wire. The resistance of the length of the wire from that point to the potentiometer circuit ground is 101.9 Ω. If a current of 10 mA were flowing in the wire, a 1.019-V drop would exist between that point and the ground point. To calibrate the potentiometer (i.e., to make sure that 10 mA is actually flowing in the wire resistor), the positive terminal of the standard cell (which is also at +1.019 V) is connected to the same point on the wire as the slider. A current detector (in the form of a very sensitive D'Arsonval galvanometer) is connected between the slider on the precision resistor (point *A* in the circuit of Fig. 7-1) and the positive terminal of the standard cell (point *B* of the same figure). If the voltage drop across the precision resistor is actually equal to 1.019 V, no current will flow in the circuit branch containing the detector because there will be no voltage drop between points $A$ and $B$. Since the value of the resistance of the potentiometer wire between point $A$ and ground is known to be 101.9 $\Omega$, we know that exactly 10.0 mA is flowing in the wire.

If, on the other hand, 10.0 mA is not flowing in the resistor, point $A$ will not be at a potential of 1.019 V. In that case, there will be a potential difference between point $A$ and $B$, and a current will flow in the branch between them. This current flow will be sensed by the galvanometer detector. The rheostat connected in series with the working voltage battery and the precision resistance must then be adjusted so that the current flowing becomes 10 mA. At that point, the galvanometer detector will indicate that no current is flowing between points $A$ and $B$ of the circuit.

Once the magnitude of the current flowing in the precision resistance is calibrated in this way, the potentiometer is ready for making voltage measurements. From that time on, the rheostat (which was adjusted to allow this exact current flow) should no longer be touched. When an unknown voltage is connected to the UNKNOWN VOLTAGE terminals of the instrument, the slider of the precision resistor is moved until a position is reached where the galvanometer detector again shows no indication of current flow. At that point, the unknown voltage is known to be equal to the voltage drop across that fraction of the precision resistance that exists between the slider tap and ground. In the example of the potentiometer (which was calibrated to have 10.0 mA flowing in it), a position equal to 94.72 cm at no-current (or *null*) would indicate that the unknown voltage was equal to 0.947 V.

A few additional features of the potentiometer and its components should also be mentioned at this point. First, when a null condition is reached, the unknown voltage is being measured without any current being drawn by the potentiometer. This means that the impedance of the potentiometer at balance is essentially infinite. An infinite impedance corresponds to the condition existing in an *ideal voltmeter*. Second, the mechanical aspects of the galvanometer operation (such as bearing friction and component nonlinearities) are bypassed because there is no deflection required of the detector at the time the final measurement is made. As a result of these features, two major sources of error which exist in other voltage measurement procedures are eliminated (the loading effect and the mechanical uncertainties).

## Commercial Manual Potentiometers

Commercially available manual potentiometers are built with full-scale voltage ranges of 0.01 mV to 1.5 kV. However, the most commonly used instruments generally have ranges from 150 mV to 1.5 V. The accuracy of portable units (Fig. 7-2) is typically 0.05 percent of the *reading*.

The precision variable-resistance element of most commercial potentiometers consists of a group of precision decade resistors (for adjusting the coarser resistance values), plus a slide wire for making fine resistance adjustments. The value of the

**Figure 7-2**  Manual potentiometer.
(Courtesy of Leeds and Northrup Co.)

measured voltage is then found from the dial settings by adding the values of the decade resistors to the value of the slide-wire resistance.

Some manually operated potentiometers are also equipped for measuring the voltage outputs of thermocouples. These models ordinarily have a control marked *Reference Junction*. When making ordinary voltage measurements, this control should be set to zero.

### Self-Balancing Potentiometers

When manual potentiometers were the only types of potentiometers available, potentiometric methods for determining voltages were limited to highly precise laboratory measurements. The major limitation that prevented potentiometers from being more widely used was that they required an operator to perform the careful manipulations necessary to bring them to a condition of balance. With the development of automatic or *self-balancing* methods for achieving the balance conditions, the advantages of potentiometric measurement techniques were applied to a much wider range of measurements.

In addition to merely providing a measurement indication, self-balancing potentiometers are also capable of *permanently recording* the values of the voltages as they are being measured. This is done by putting a marking device on the pointer and using a motor-driven chart as the recording surface.

The automatic balancing feature of the self-balancing potentiometer relies on a servomotor[1] working together with several other pieces of electronic equipment. Figure 7-3 is a block diagram of a self-balancing potentiometer which shows how these elements perform the balancing act.

In this figure we see that the voltage signal to be measured ($V_i$) is applied to one input terminal of a device called an *error detector*. Another voltage ($V_s$), ob-

---

[1]A *servomotor* is defined to be a motor that responds to a command. This response is in the form of a motion that corrects any difference between the actual and desired state of the system to which the motor is connected.

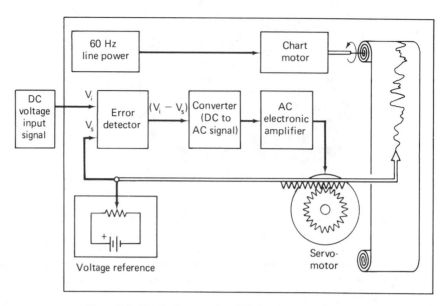

**Figure 7-3**   Block diagram of a self-balancing potentiometer.

tained from an adjustable voltage reference source, is fed to the second input terminal of the *error detector*. The error detector electronically subtracts $V_i$ from $V_s$ and utilizes the result as its output signal. The output signal from the error detector is called the *error signal*, and it is used to drive the servomotor. However, before the error signal is applied to the servomotor, it must be amplified by an electronic amplifier. This amplification endows the signal with a magnitude that is sufficiently large to activate the servomotor.[2] The amplified error signal applied to the servomotor causes the shaft of the motor to rotate.

The servomotor shaft has two mechanical connections attached to it. The first couples the shaft to a voltage-divider circuit that is part of the variable dc reference source that provided $V_s$. As the shaft turns, it moves the slider of this voltage divider. The direction of the motion is such that $V_s$ changes in value so that it approaches the value of $V_i$. When the values $V_s$ and $V_i$ are equalized, the error signal from the error detector becomes zero ($V_i - V_s = 0$) and the servomotor halts. (Note that the movement of the slider by the servomotor replaces the action that the observer performs in the manually operated models.) The second mechanical connection attached to the servomotor shaft is coupled to a pointer which indicates the degree of rotation of the shaft. When the servomotor comes to a halt, the reference voltage, $V_s$, is equal to the input voltage, $V_i$. Since the value of the reference voltage is known very accurately, the position of the pointer (as controlled by the motor shaft position) can be calibrated to indicate the value of the input voltage directly. If

---

[2]To avoid certain problems involved with dc amplifiers, the error signal is generally converted to an ac form before being fed to the amplifier.

the pointer has a pen or other type of marking device connected to it, and if a separate motor is used to drive a ruled paper chart under the marker, a permanent record of the voltage measured versus time is produced.


## RECORDERS

A *recording instrument* is a device whose function is to record the value of a quantity as it is being measured. Such instruments may include graphic recording devices, computer printers, tape recorders, computer disks (magnetic and optical), EPROMS, and cathode-ray tubes (CRTs). With the exception of the CRT (which was discussed in Chapter 6), the operation of these instruments is covered in the remainder of this chapter.

The major types of recorders used by industry for recording process variables are the galvanometric, oscillographic, potentiometric, and linear array recorders. These devices contain signal conditioning circuitry so that incoming signals can be scaled or otherwise manipulated before being recorded. For example, thermocouple inputs can be linearized and scaled so that temperature is recorded, not the original thermocouple output voltage signal (whose value is in mV). As a second example, the input from the differential pressure across an orifice plate can be converted so that flow is recorded. In some cases a recorder with an inexpensive preamplifier is used because it is less expensive, and chart paper is used for scaling and linearizing the output. There are numerous standard chart scales that can be used for most process variables such as flow, pressure, and temperature.

In the self-balancing and galvanometer recorders, a pen assembly is moved over a paper chart that is simultaneously being moved in an orthogonal direction by a separate motor. (These types of recorders are often called *oscillographs* as well.) In the *X-Y* recorders, the pen is simultaneously moved in two perpendicular directions while the paper chart is kept stationary.

The charts of the self-balancing and galvanometer recorders consist of either *round* or *strip-type* graphs (Fig. 7-4). The round graphs are circular pieces of paper that have concentric circles ruled on them to form their scales. In addition, there are printed arcs extending from the center of the chart to the paper's edge. As the pen of the recorder is moved, it swings along these arcs. The arcs are known as *time arcs*. (Since the chart is rotated at a uniform rate, the angular position between arcs indicates elapsed time.) The radial position of the pen at any time indicates the instantaneous value of the quantity being measured. The *strip-chart recorders* have graphs of paper in the form of a long roll. As the graph is unrolled, it also moves along under the pen or marker at a uniform rate of speed. The lines on the paper are ruled parallel to the direction of motion and form the scales of the graph. The lines that are ruled perpendicular to the direction of the motion are the time lines. (It is also common to use this axis for other parameters which can be made proportional to time.) The pen assembly moves in a direction perpendicular to the motion of the paper. Thus the value of the measured variable versus time is traced on the chart as

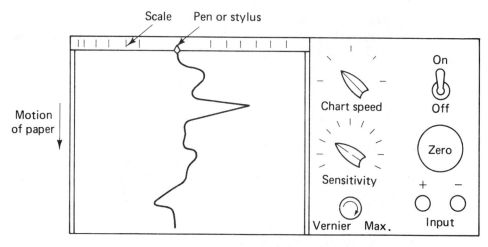

**Figure 7-4**    Recorder: Front control panel of a typical strip-chart recorder.

the paper rolls by the pen. The instantaneous value of the measured quantity is given by the position of the pen. Typical chart papers are shown in Fig. 7-5.

There are several common marking devices used in recorders. The stylus or writing system may be direct pressure, gravity ink pen, pressurized ink pen, thermal stylus, thermal or optical heating arrays, or felt-tip pens. The choice depends upon the application specifications. For slow recordings (such as a furnace's temperature), a felt pen may be the best choice. For high speed events, such as transient velocity changes on a paper mill, a pressurized ink pen would be appropriate. When real time observation of an event is necessary, a thermal pen would not be appropriate because there is a time lag before the trace is visible since the thermal writing head covers a significant portion of the chart paper and is usually inside the recorder. Ink pens are simple, but they require constant cleaning and filling of the ink reservoir. Ink spills are also a frequent problem. When the ink dries in the carrier tube, the entire pen assembly usually has to be replaced. This problem has resulted in the widespread use of disposable pens (Fig. 7-5).

As a consequence of the disadvantages of pen-and-ink assemblies, several other making devices have been developed. These inkless methods are particularly useful in those recorders that must remain unattended for long recording periods. One type of inkless marking device utilizes a heated pen on a heat-sensitive paper. In this type, an electric current is passed through the tip of the movable stylus. The current heats the stylus, and the heat causes a thin, clear line to appear on the special heat-sensitive paper. Other inkless graphing methods include light-sensitive and pressure-sensitive chart papers. These methods must also be used in conjunction with special styluses. The biggest disadvantages of inkless methods are that they are generally more complex than pen-and-ink assemblies, and they also require more expensive paper for their operation. The current to the thermal stylus must be

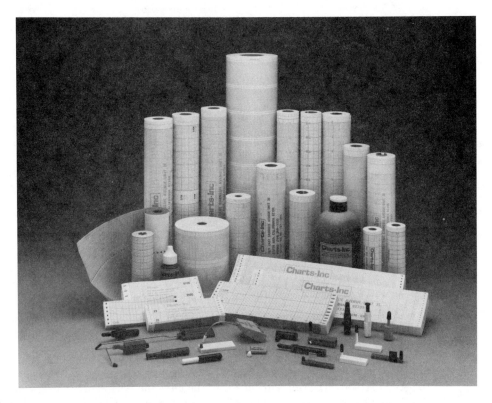

**Figure 7-5**   Typical charts and disposable pens. (Courtesy of Charts, Inc.)

adjusted to the chart speed. A slowing of the chart speed can result in burning the chart if the current is not readjusted. An increase in speed can result in poor contrast.

The pen assembly always possesses some inertia because of its mass. The inertia prevents the pen from moving until some certain minimum signal is applied. The range of signal amplitudes that are too small to actuate the movement of the pen assembly is called the *deadband* of the recorder (Fig. 7-6). Typical deadband signal amplitudes run between 0.05 and 0.1 percent of full scale. Deadband can cause distotion of the recorded trace, especially on low-amplitude signals comparable to the deadband range. The frequency response of the recorder is also limited by the inertia of the pen and the galvanometer coil. High inertia pen recorders have an upper frequency limitation just above 100 Hz. Inertia also results in *overshoot* and *undershoot* if the pen system is not properly damped. Most recorders have a *damping* control. A square wave is applied to the input and the *damping* control is adjusted until the the best (*critically damped*) square wave is recorded (Fig. 7-6(b).)

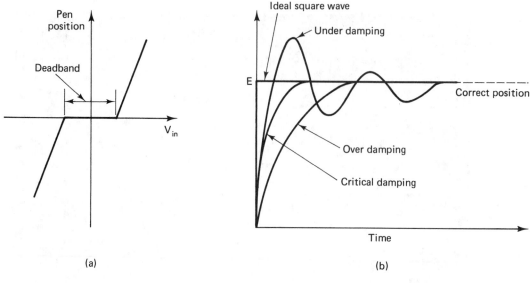

**Figure 7-6** Deadband and damping effect in recorders: (a) deadband; (b) damping.

## Galvanometer Recorders

Galvanometer recorders use a pen assembly mounted on the end of the pointer of a rugged D'Arsonval movement similar to the movement used in basic dc meters (Fig. 7-7). The movement is also known as a permanent magnet moving coil (PMMC) movement. The recorders are also called direct writing galvanometers. By connecting the writing stylus directly to the coil a curvilinear recording results. Mechanical linkages are used between the coil land the stylus when rectilinear recording is required. When a quantity is being measured by such galvanometer recorders, the restraining springs of the movements (rather than a self-balancing signal) provide a counterforce which balances the force created by the quantity being measured. Strip charts are usually used with this type of recorder.

Although galvanometer recorders are not as sensitive as self-balancing models, they have some other advantages that self-balancing recorders do not possess. First, their frequency response can be made much greater than that of self-balancing types. Second, the D'Arsonval movement which directs their marking devices can be made much more compact than the pen-motor assemblies of the self-balancing recorders. Thus galvanometer recorders with multichannel outputs are feasible (Fig. 7-8). As a result, commercially available recorders are built which contain up to 36 output channels. This feature makes galvanometer recorders attractive for simultaneously monitoring and recording a large number of slowly varying quantities. For example, a multichannel recorder can be used to record the outputs of transducers which simultaneously monitor several physiological functions of a hospital patient (such as body temperature, blood pressure, and respiration rate).

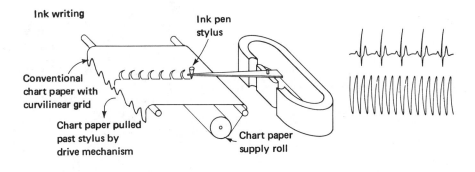

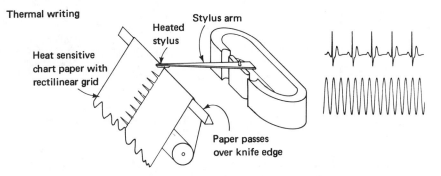

**Figure 7-7**   Sketch of galvanometer recorder mechanisms. (Reprinted with permission of Tektronix, Inc.)

Another application involves the simultaneous monitoring of various conditions that exist aboard a satellite in space.

The maximum frequency of ordinary galvanometer recorders is about 100 Hz (for small pen excursions), while their maximum sensitivity is on the order of 25 mV/in. The input impedances of these instruments are typically 100 kΩ or more, and corresponding accuracies are ±1.0 to ±2.0 percent, full-scale.

Oscillographic recorders are similar to galvanometer recorders except that additional electronic and mechanical subsystems increase the sensitivity the frequency response by several orders of magnitude. A typical oscillographic recorder (Fig. 7-9) has a frequency response of 5 kHz. To obtain the high frequency response, the writing system is either thermal or light; in such cases the paper must also be either temperature or light sensitive. A significant increase in frequency response is obtained by using a linear array recorder (Fig. 7-10).

These are capable of responding to a 25 kHz signal but can just as easily be used to respond to a dc signal. Two major advantages of the linear array recorders are that they can use the full width of the recording paper for all channels, and they can print alphanumeric characters and chart grids simultaneously. Data skew, which results from chart wander or lateral weave in a galvanometric movement, is elim-

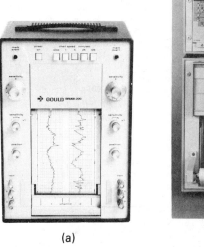

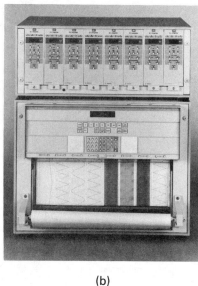

(a)                                           (b)

**Figure 7-8** Galvanometer recorders: (a) 2-channel; (b) 8-channel. (Courtesy of Gould, Inc., Instrument Systems Division.)

inated since the grid is automatically aligned with the recorded data. An eight-channel recorder can be switched from eight separate recordings to eight overlapping recordings with the touch of a button. The thermal printing head of the array recorder is an integrated circuit array of closely spaced resistors. The print heads are a byproduct of the facsimile (FAX) machine industry. The resistors are strobed at about 1 kHz. Since the strobe is a digital signal, an array recorder can

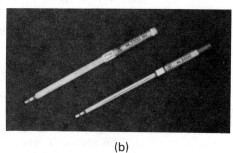

(a)                                           (b)

**Figure 7-9** (a) Oscillographic recorder. (Courtesy of Honeywell, Inc.); (b) miniature galvanometers used in oscillograph.

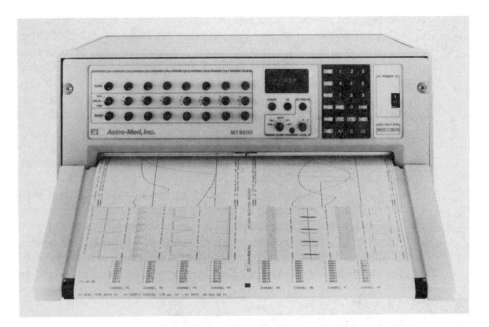

**Figure 7-10**   8-channel linear array recorder. (Courtesy of Astro-Med, Inc.)

receive digital data directly from an IEEE-488 bus and can also print analog data without an analog to digital converter, or it can print numerical data in a similar manner as a data logger. By reducing the number of moving mechanical parts, the recorder is more tolerant to dust and other contaminants. Printed reports can also be generated by the recorders. By transmitting digital signals instead of analog signals to the recorder, most noise sources are completely eliminated. Many transducers have signal conditioning as part of the transducer in the form of a hybrid circuit that converts the basic signal to digital signal and then transmits the signal over an optical fiber to a time multiplexed recorder. A few high volume transducers are manufactured in the form of a VLSI circuit with transducer, signal conditioning, and line drivers on a single chip.

The higher the frequency response, the higher the required chart speed. For example, if we wanted to record a 25-kHz signal, such that one cycle was 0.5 cm long, the chart speed would have to be 1,250,000 mm/s. This, therefore, represents the major limitation of chart recorders. As a result, such recorders are not intended to permit recording a 25-kHz signal. Instead, they are used to capture high-speed transients that occur over very short time periods. A 200 μs spike can easily be recorded. The maximum chart speed on most recorders is only 200 mm/s.

## X-Y Recorders

*X-Y* recorders are instruments that have the special capability of displaying two separately varying quantities on the *X* and *Y* axes of Cartesian coordinates

(Fig. 7-11). This means that one of the variables can be applied to the $X$ input of the recorder and the other to the $Y$ input, and the recorder will plot their variations against one another. This feature exists because the $X$-$Y$ recorder can simultaneously move a pen in both the $X$ and $Y$ directions across a fixed paper in response to electrical signals applied to its two input terminals. Since most $X$-$Y$ recorders also contain a time base, they can be used to portray the variation of one variable versus time as well. This allows them to perform some measurements in the same manner as strip-chart recorders.

Another feature of $X$-$Y$ recorders is that they are rather inexpensive to operate. Because most use a pen-and-ink marking system, ordinary low-cost paper can be used for the chart. In addition, they are rather easy to use.

On the negative side, $X$-$Y$ recorders are slower than strip recorders and cannot be used for continuous monitoring applications. They are also considerably more expensive to buy than many simple strip recorders.

There are a wide variety of applications in which $X$-$Y$ recorders are used. Some of these applications include plotting current versus voltage ($I$–$V$) curves of diodes and transistors, plotting $B$–$H$ curves of magnetic materials, reproducing

**Figure 7-11** Measurement plotting system capable of recording, analyzing, and annotating data. (Courtesy of Hewlett-Packard Co.)

analog and digital (indirectly) computer readouts, and graphing voltage vs frequency plots from sweep-frequency oscillators.

The mechanism that locates the position of the pen along both the $X$ and $Y$ axes of the recorder is a closed-loop servosystem (very similar to the servomechanism used in the self-balancing potentiometer). The rotation of the servomotor moves the marking pen to its position by using a string-and-pully arrangement. Because this mechanism is electromechanical, the frequency responses and speeds of $X$-$Y$ recorders are inherently limited. Their typical frequency response is about 5 Hz for a 1-in. peak-to-peak signal. Similarly, their *slewing speed* (defined as the maximum velocity at which the pen assembly can be moved after its acceleration has stopped) is about 20 in/s.

The other specifications of $X$-$Y$ recorders include input and output characteristics, paper sizes, maximum sensitivities, and their accuracies. Typical values for these quantities are as follows. Input impedance, 100 k$\Omega$ to 1 M$\Omega$ on the less-sensitive ranges and 50 M$\Omega$ or more on the most sensitive range; paper sizes, $8\frac{1}{2} \times 11$ in. or $11 \times 17$ in.; maximum sensitivity, 100 mV/in. to 1 mV/in, depending on the model; accuracy, $\pm 0.1$ percent, full scale, number of pens and channels.

## MACHINE-INTERPRETABLE RECORDERS

### Analog Magnetic Tape Recorders

Magnetic tape recorders can be used to record measurement data in either analog or digital form. When data are recorded in analog form, electrical signals fed into the magnetic tape recorder are stored on the tape and the same signals are available for reproduction from the recorder when needed. Thus the analog magnetic tape recorder permits data to be reproduced at some later time in the *original electrical format*. Analog magnetic recorders are the only convenient devices possessing this capability. They also have the advantage of being able to record data at one speed, and then to be able to reproduce the intact data at a different speed. This feature allows time expansion or time compression of data. For example, by recording at a slow speed and reproducing at a faster speed, many hours of information may be compressed and reproduced in a few minutes.

In general, an *instrumentation magnetic recorder* is the type of analog magnetic recorder used in measurement applications (rather than the magnetic recorders used for voice, music, or video recording). Three recording methods are available for use in various measurement requirements: (1) direct recording, (2) frequency-modulation recording, and (3) pulse-modulation recording. Direct and frequency-modulation recording meet the requirements of most measurement applications. Pulse-modulation recording is used in only a limited number of special situations. *Direct recording* requires the least expensive supporting electronics, provides the greatest available bandwidth, and depends only on the intensity of the magnetization for recording the instantaneous amplitude of the input signal. Its drawbacks, however, include the problem that during playback, frequencies below 50 to 100 Hz are not reproduced. In addition, amplitude-variation errors occur, for a number of reasons. Thus the technique of direct recording finds best use in applications where the amplitude-variation errors are not critical (such as audio recording, or in recordings where the signal frequency, but not amplitude, are of primary importance). *Frequency-modulation (FM) recording* overcomes the problem of the signal-amplitude instability exhibited by the direct recording technique, but at the expense of reduced high-frequency response and more complex electronics. However, since the method can reproduce signals with frequencies down to dc, FM recording is used when the dc component of the input signal must be preserved and when errors in the amplitude of the recorded data cannot be tolerated.

A block diagram of a magnetic recorder is shown in Fig. 7-12(a). Magnetic tape of $\frac{1}{4}$-, $\frac{1}{2}$-, or 1-in. widths is moved past the recording (and playback) heads at various speeds (ranging from $1\frac{7}{8}$ in./s to 120 in./s). A number of separate heads may be positioned across the width of the tape, allowing individual data to be recorded along each of these positions or *tracks* (and thereby increasing the amount of data that can be recorded on a single tape). A $\frac{1}{2}$-in. tape width allows 7 tracks, and 1-in. width permits 14 tracks.

Recording of the data on the tape occurs at the point where the *recording head* contacts the magnetic tape. The head is similar to a torroidal transformer with a single winding as shown in Fig. 7-12(b). The core of the transformer has a short nonmagnetic gap in it. A single current fed to the transformer causes the magnetic flux in the core to detour around the core gap and through the magnetic material on the tape with which the head is in contact. When the magnetic tape is moved past the recording head, the magnetic tape material is thus subjected to a magnetic flux pattern proportional to the current in the recording head winding. As it moves away from the gap, each tiny magnetic particle retains the state of magnetization to which it was subjected by the detouring magnetic flux. The tape transport mechanism moves the tape past the heads at a constant, selected speed. The amplifier conditions and applies the input data to the recording head.

### Digital Machine-Interpretable Recorders

The increasingly widespread use of digital computers to store and process experimental data has created a need for instruments that can provide the data in a digital format suitable for direct input into the computer. In contrast to analog data (in which the signal amplitudes or frequencies represent data values directly), digital data must be formatted in a digital code compatible with the computer or digital display device being serviced. The digital code selected is usually one of those introduced in Chapter 1; that is, binary code, binary-coded decimal (BCD), or ASCII code. Devices that can record digital data in a form suitable for digital machine interpretation include digital magnetic tape recorders and the memory of the digital computer itself (which can store the digital data on various digital storage

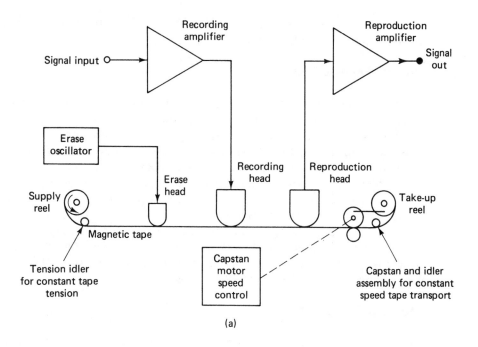

(a)

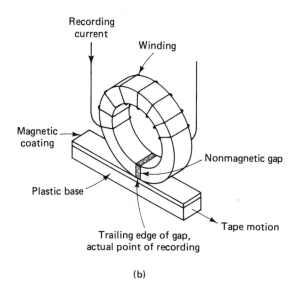

(b)

**Figure 7-12** (a) Block diagram of a magnetic tape recorder; (b) simplified diagram of a magnetic recording head.

systems, such as magnetic tape, magnetic disks, magnetic drums, magnetic-core memory, or semiconductor memory).

### Digital Magnetic Tape Recorders

Digital magnetic tape recorders usually use $\frac{1}{2}$-in.-wide tape with either seven or nine tracks. The combination of bits recorded on the several tracks across the tape comprises the code for each character (Fig. 7-13). The recording density along the axis of the tape ranges from 100 to 1600 characters per inch. Thus with tape speeds of up to 200 in./s, 320,000 characters per second may be recorded.

There are two types of digital magnetic tape recorders, *incremental* and *synchronous* (continuous motion). The *incremental digital recorder* will record one character at a time and at any recording rate, from zero to some maximum (typically 600 characters per second). They find best use when the data rates are low or discontinuous. For higher data rates, the *synchronous recorder* must be used. In the synchronous recorder, the tape moves at a constant speed (i.e., 100 in./s), and a large number of data characters are recorded.

### Magnetic Disk Storage

Digital data can also be stored on magnetic disks. The access time of disk storage is much faster than tape storage, and a system with removable disks can have unlimited storage space. A magnetic disk system may contain several disks on the same drive (Fig. 7-14), or just a single disk with either one or two surfaces. (The latter device is known as a *floppy disk*.)

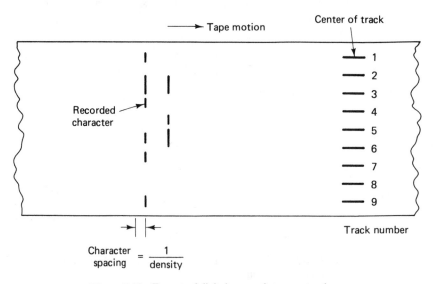

**Figure 7-13**   Format of digital magnetic tape; a track.

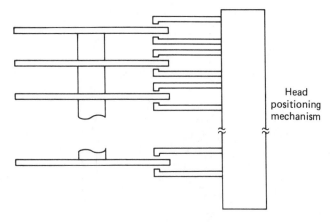

**Figure 7-14**   Disk storage.

The systems with several tracks on one drive have a separate head for each disk surface. The heads float over the surface on a cushion of air, reading and writing onto the disk. The heads are all driven simultaneously by the head-positioning mechanism toward or away from the center.

Digital information is stored on the disk surfaces along concentric circles called *tracks*. If the heads are positioned over a given track, say track 29, the heads are addressing track 29 of all the tracks of all the disk surfaces. An example of a disk system is one that contains 10 disks (20 surfaces), with 200 tracks per surface, and with each track capable of storing over 7000 characters. Thus this system has a storage capacity of 29.5 million characters. Since the disk drive operates at 2400 rpm, an average access time of 87.5 ms is provided.

Floppy disks are $\frac{1}{16}$ in. thick with one or two magnetic recording surfaces on a flexible plastic substrate permanently housed in a protective jacket. An 8-in. version is shown in Fig. 7-15. The disk is spun by the disk drive at 360 rpm and the read/write head contacts the tracks through the access slot. Data can be recorded on 77 concentric tracks, each of which contains 41.7 kilo bits of storage. This allows up to 3.2 megabits of storage per recording surface. Since the data are formatted in byte form, this means that 256 kilobytes of data can be stored per surface. Access times average from 25 to 120 ms. Because of the low cost and relatively fast access times (compared to magnetic tape cassettes), floppy disks are becoming very popular storage devices in many systems, including home personal computers.

The $3\frac{1}{2}$-in. floppy disks used on most personal computers are now able to store 1.44 M bytes of formatted data at a density of 17,000 bpi (bits per inch). This is very near the theoretical limit for longitudinal recording on cobalt/ferric oxide media. With the disk rotating at approximately 200 rpm, a bit cell is only 2 μs wide and the data window only extends ±500 ns from the center of the cell. Changes in the recording media now permit recording densities near 51,000 bpi. However, to read this bit density, the head must be positioned by closed-loop servo system and a

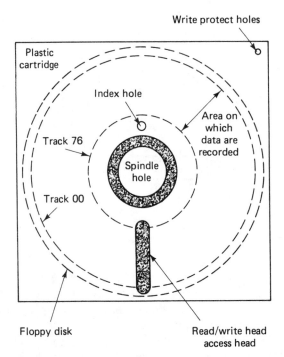

**Figure 7-15**  Eight-inch floppy disk.

Bernoulli Box. The Bernoulli effect is the creation of a partial vacuum between the head and a high-speed disk. Manufacturers are operating their disks at 1800 and 3600 rpm to take advantage of the Bernoulli effect. A vacuum is caused by the high-speed rotation, and it is strong enough to pull the head close to the disk while allowing a sufficient air cushion to prevent disk damage.

Personal computer disk drives use stepper motors to position the head. The head moves approximately 0.025 inches between tracks or 40 tpi (tracks per inch). There are three ways in which the storage density can be inexpensively increased: (1) The recording media can be changed to decrease the size of the magnetized area and thus increase the bpi, (2) a closed loop servo for head positioning can be used (this will increase the tpi by an order of magnitude to 480 tpi), and (3) an order of magnitude in the recording density can be achieved by vertical plane recording. By using a closed-loop servo a $3\frac{1}{2}$-in disk can store 6M-bytes of data. High performance electronics in conjunction with closed-loop servo drives have achieved 20M-bytes on a $3\frac{1}{2}$-in. disk by using vertical plane recording techniques.

Many manufacturers must maintain manufacturing records of every item produced. This requires vast amounts of storage for even medium-sized companies. These records are maintained for reasons such as historical trend analysis, taxes, defending product liability suits, and military requirements. Many companies are changing from microfiche to optical disks, which are connected directly to manufacturing process computers. Optical disks offer an economical solution for storing large amounts of data. The disk is constructed of a light-weight plastic or metal that is coated with a photosensitive emulsion. This emulsion is then coated with a thin protective layer of plastic or glass. A laser beam striking the emulsion causes it to change its ability to reflect light, thus enabling data to be read from the disk. Unlike the floppy disk, once the writing is completed the written data cannot be changed. Data can be recorded at a rate of 5M bits per second, and a disk can store 2 G bytes of data. This is equivalent to approximately 400,000 typewritten pages.

## PROBLEMS

1. Explain why measurements that involve the detection of a null to indicate the quantity being measured can be made more accurately than those involving the reading of a pointer deflection made along a scale.

2. What is the difference between the definitions of the term *potentiometer* as introduced in Chapter 7 and as used in Chapter 10?

3. Describe the differences between *manual* and *self-balancing* potentiometers.

4. Explain why the loading effects of potentiometers and conventional voltmeters differ greatly from one another in magnitude.

5. A manual potentiometer contains the following components: a working battery of 5.0 V and negligible internal resistance; a standard cell whose voltage is 1.0191 V and whose internal resistance is 200 $\Omega$; a 200-cm adjustable precision resistance; a galvanometer with an internal resistance of 50 $\Omega$. The rheostat of the instrument is set so that the potentiometer is calibrated at the 101.91-cm mark of the precision adjustable resistance.
   (a) Calculate the current flowing in the rheostat as well as its resistance value.
   (b) If a protective resistance is placed in series with the galvanometer, calculate the value of the resistance necessary to limit the current in the galvanometer to 10 $\mu$A.

6. List some of the factors which determine the accuracy of potentiometers.

7. When using the potentiometer it is recommended that the key should only be depressed momentarily to avoid damage to either the galvanometer or the standard cell. Explain in more detail why this recommendation should be followed.

8. What would be the effects of a sizable temperature change on the accuracy of a potentiometer?

9. Describe the functions of the following components of a self-balancing potentiometer.
   (a) Error detector                    (b) Amplifier
   (c) Servomotor                        (d) Reference voltage source

10. Draw the charts of both the round and strip-chart recorders. Indicate how both time and the magnitudes of the quantities being measured are exhibited on these charts.

11. Explain the advantages that each of the following types of recorders have over the other types:
    (a) Self-balancing recorders
    (b) Galvanometer recorders
    (c) *X–Y* recorders

12. What causes the frequency response of recorders to be limited?

13. Describe two types of inkless recording methods.

# REFERENCES

1. Stout, M. B., *Basic Electrical Measurements*, 2nd ed. Chap 7. Englewood Cliffs, N.J.: Prentice Hall, 1960.
2. Harris, F. K., *Electrical Measurements*. Chap 9. New York: John Wiley, 1952.
3. *Leeds & Northrup Catalog, 1970*. North Wales, Pa.
4. Coombs, C., ed., *Basic Electronic Instrument Handbook*. Chaps. 9 and 36. New York: McGraw-Hill, 1972.
5. *Measurements and Control*, Issue 129, June 1988.

# 8

# *Time and Frequency Measurements*

The accurate measurement of time is an important task in all fields of scientific investigation. Since time is one of the fundamental units from which all other units are derived, it is necessary that time measurements be very accurately made. Although time intervals are measured by many different techniques and devices, the most accurate measurements involve electronic measuring devices (i.e., for accuracies of up to 1 part in $10^{10}$). We will discuss such accurate time-interval measuring instruments as well as some simpler electrical timing devices.

Frequency is defined as the number of recurring events taking place in a unit interval of time. When we refer to electrical signals, frequency ($f$) is usually meant to be the number of cycles of a periodic waveform that occur per second. The length of time (in seconds) of one complete cycle of such a waveform is called the period ($T$) of the waveform. As pointed out earlier in the text, the frequency and the period of a periodic waveform are related by

$$f = \frac{1}{T} \tag{8-1}$$

The unit of frequency is the hertz (Hz, formerly referred to as cycles per second). Equation (8-1) indicates that if we can measure either time or frequency, we can determine the other of these two quantities.

As a final topic in this chapter, we discuss the instruments used for harmonic analysis of waveforms. Harmonic analysis involves the determination of the various frequency components of which signals are composed. Harmonic-analysis instruments are also used to detect the extent of distortion of various waveforms.

### Time Standards

The earliest time standards were based on the rotation of the earth about its axis. One *day* is the time it takes for the earth to make one complete revolution about its axis. When clocks were first developed, the *second* was chosen to be approximately the time of one complete oscillation of a pendulum. Eventually, the second was made into the unit which was 1/86,400 of a day (24 hours × 60 minutes × 60 seconds).

As more precise astronomical measurements of the earth's rotation were made, however, it was learned that slight variations occur in the time of the axial rotation of the earth. Thus it was realized that a time standard which was tied to the earth's rotation would never be completely constant. Since modern measurements require very accurately known time intervals, various standards were developed that are not dependent on astronomical observations.

The modern standard of time is defined in terms of an *atomic standard*. In this standard, the resonant frequency of cesium atoms is measured and converted to time. Since the frequency is dependent only on the internal structure of the cesium atoms (as long as the atoms are not disturbed by external conditions such as magnetic fields), the accuracy of this standard is guaranteed to within approximately 1 part in $10^{10}$. One second is identified as 9,192,631,770 periods of oscillation of such cesium atoms.

An atomic clock based on this principle of operation (and whose precision exceeds 1 μs per day) is now operated by the National Bureau of Standards as the primary time standard. The time signals from this clock are broadcast on WWVB (60 kHz) and are accurate to 5 parts in $10^{10}$.

In the near future, it is likely that the time standard will be based on the number of wavelengths in a beam of laser light. The accuracy of such a standard promises to be several orders of magnitude higher than the atomic standard.

## TIME MEASUREMENTS

Time can be measured by various mechanical, electrical, and astronomical methods. Our discussion will be concerned with only the electrical and electronic methods. The complexity and expense of time-measuring equipment increases as the length of the time interval being measured decreases (and as the accuracy with which the measurement must be made increases). We will discuss time-measuring instruments in an ascending order of complexity and accuracy. The most accurate time-measuring instruments (i.e., digital timers) are discussed in the section on universal timer-counters.

### Electronic Timers

One method for measuring time intervals utilizes *electronic* timers. Such electronic timers can measure time intervals whose durations are as small as one microsecond.

The most common principle used in electronic timers involves the charging of a capacitor in an *RC* circuit. Electronic timers make use of the fact that it takes a finite amount of time to charge a capacitor when a dc voltage source is suddenly applied across the series resistor–capacitor connection. The variation of the voltage versus time across the capacitor will have a curve whose shape is shown in Fig. 8-1. The rapidity with which the capacitor charges depends on the values of the resistance and capacitance in the connection. Specifically, a quantity known as the time constant ($\tau$) of the resistance–capacitance combination is defined as the time it takes for the voltage across the capacitor to reach 63 percent of its full-charged value. The time constant is equal to

$$\tau \,(\text{seconds}) = R\,(\text{ohms}) \times C\,(\text{farads}) \tag{8-2}$$

For example, if we had a 500-pF capacitor and a 4-M$\Omega$ resistor in series, it would take

$$\tau = R \times C = (4 \times 10^6) \times (5 \times 10^{-10})$$

$$2 \times 10^{-3} = 2 \text{ ms}$$

to charge the capacitor to 63 percent of the value of any dc voltage applied across the series connection.

If we apply a known voltage across the series *RC* connection at time $t = 0$ and then interrupt the voltage at the end of the measured time interval, the voltage across the partially charged capacitor will be an indication of the time interval.

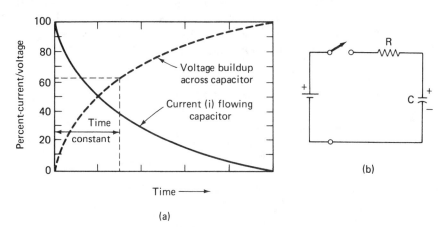

**Figure 8-1** (a) Current and voltage of a capacitor when a dc voltage is applied to it at $t = 0$; (b) *RC* circuit of the electronic timer.

## FREQUENCY MEASUREMENT

Various types of instruments are used to measure frequency, depending on the frequency range and accuracy desired. The following is a list of some of the instruments commonly used for measuring various frequency ranges.

1. Oscilloscopes
2. Wien bridge frequency meters (audio-frequency range)
3. Zero-beat frequency meters (radio-frequency ranges)
4. Digital frequency counters (wide frequency range, very accurate)

The oscilloscope can measure frequencies over a wide range by using the triggered-sweep and Lissajous pattern methods described in Chapter 6. However, the accuracy of the frequency measurements made with an oscilloscope is somewhat limited. Thus, for more accurate measurements, other frequency-measuring instruments are more likely to be used.

### Wien Bridge Frequency Meters

In Chapter 13 we will see that the Wien bridge oscillator is a device designed to produce accurately known audio-frequency sine-wave signals. The bridge portion of the oscillator circuit can also be used as an instrument for measuring frequencies in the audio-frequency range. In this type of meter, the signal of interest is applied to the bridge (Fig. 8-2). The arms of the bridge contain passive adjustable electrical components. For such combination of values of the components, there is a specific frequency at which the bridge is balanced (i.e., the value of the voltage difference between points $A$ and $B$ is zero). A set of headphones, a CRT, or an electronic voltmeter can be used to detect this balance condition. In most Wien bridges, the values of the components are chosen such that $R_1/R_2 = 2$, $R_4 = R_3$, and $C_4 = C_3$. Then, at balance, the unknown frequency is found from

$$f = \frac{1}{2\pi R_3 C_3} \tag{8-3}$$

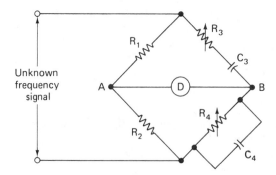

**Figure 8-2** Basic circuit of the Wien bridge used to measure audio frequencies.

### Zero-Beat Frequency Meter

Zero-beat frequency meters are used to make accurate measurements of the frequencies of radio-frequency (RF) signals. The operation of such meters is based on the zero-beat (or heterodyne) principle. This principle says that if two signals of different frequencies are combined in a nonlinear circuit[1] (in this case a *mixer* circuit), the output of the circuit will contain a signal whose frequency is equal to the difference between the two original signals. This frequency is called the *difference heterodyne*.

In the zero-beat frequency meter, the signal whose frequency is being measured is one of the two signals applied to the mixer (Fig. 8-3). The other signal is generated within the meter from a variable-frequency oscillator (VFO). As the frequency of the VFO is adjusted so that it approaches the value of the unknown frequency, the difference between the two frequencies gets smaller. If a pair of headphones is used to detect the difference heterodyne signal, a sound will be produced in the headphones when the frequency difference becomes smaller than 15,000 Hz. As the frequency of the VFO continues to be varied so that it gets closer and closer to the unknown frequency, the pitch of the sound produced by the headphones will become lower and lower. When the two signals are of equal frequencies, the sound will disappear. If the frequency of the VFO is raised above that of the unknown frequency, the sound in the headphones will again reappear because the difference in the two frequencies will produce a nonzero difference signal. Thus if the frequency of the VFO is accurately known at the point when the headphone sound disappears, the unknown frequency is identified. A recorder can be used instead of headphones to determine the point at which the difference heterodyne is zero.

The accuracy of the VFO can be checked by means of a crystal oscillator built into the instrument. The crystal oscillator produces signals whose specific output frequencies are known with a very high accuracy. By comparing the frequency of

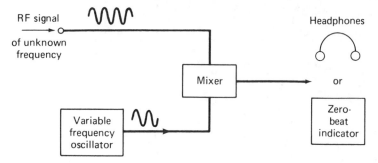

**Figure 8-3** Block diagram of a zero-beat frequency meter.

[1]In a nonlinear circuit there are elements, such as vacuum tubes or transistors, whose current versus voltage characteristics are not straight lines. The fact that these characteristics are curved in some fashion, and not straight, implies that their operation is nonlinear. The heterodyning takes place in the region represented by the curved portion of these characteristics.

the VFO to the frequency of the crystal oscillator, we can calibrate the VFO so that the zero-beat frequency meter will yield values of the measured frequencies to within 0.01 percent of their true values.

The range of the zero-beat frequency meter can be extended beyond the maximum frequency range of the VFO by virtue of the nonlinear characteristics of the mixer. In other words, because it is a nonlinear circuit, the mixer also acts on each individual signal applied to it and produces components whose frequencies are double, triple, and so on, those of each input signal. For example, if a 100-Hz signal is applied to the mixer, the mixer will produce an output signal with components whose frequencies are 100 Hz, 200 Hz, 300 Hz, and so on. These higher-frequency components can also be used to produce a difference heterodyne signal with the unknown frequency if the unknown frequency is higher than the maximum frequency of the VFO.

### Digital Frequency Counters

Digital frequency counters are the most accurate and flexible instruments available for measuring unknown frequencies (Fig. 8-4). The highest accuracy digital frequency counters can approach the accuracy of the atomic time standards described earlier. Frequencies from dc to the gigahertz range can be measured with digital frequency counters. In addition, since most events can be converted into an electrical signal consisting of a train of electrical pulses, digital frequency counters can also be used as counters of almost all types of quantities. Thus events such as heartbeats, the passing of radioactive particles, motor of shaft revolutions, light flashes, and meteorites can all be counted with digital counters. However, digital frequency counters are usually considerably more expensive than the other methods

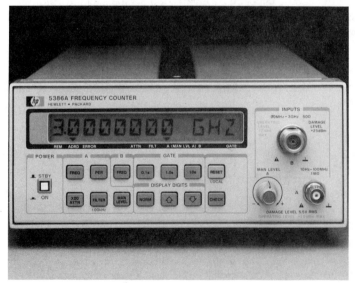

**Figure 8-4**   Digital frequency counter: HP Model 5386A. (Courtesy of Hewlett-Packard Co.)

used for measuring frequency. If the very high accuracy of a digital counter is not required by a frequency measurement, another device may be an adequate choice.

The major components which make up digital frequency meters are shown in Fig. 8-5. They are the digital counting and display assembly, the time-base generator, the pulse-forming circuit (usually a Schmitt trigger type of circuit), and the gate circuit. (If a quantity other than the frequency of an electrical signal is to be counted, a transducer must also be used to convert the events into electrical pulses.)

In Fig. 8-5 we see that the signal of unknown frequency is fed into the counter and enters a pulse former. This circuit creates a pulse for every cycle of the input signal. These pulses are then applied to a gate. If the gate is open, the pulses can pass through it and be counted and displayed by a digital counting and display assembly. (Note that the pulse-counting circuits are reset to zero before the gate is opened.) If the gate is open only for a fixed and accurately known time period, the number of pulses counted during the time period will yield the frequency (i.e., cycles per second) of the unknown signal.

The method by which the gate is allowed to be open for an accurately known interval is shown in Fig. 8-6. A crystal oscillator produces a signal at 1 MHz or 100 kHz (depending on the particular instrument design). To keep the output of this oscillator at exactly the desired frequency, the crystal is enclosed in a temperature-controlled oven. The oscillator output signal is then fed to another pulse-forming circuit. If the oscillator has a frequency output of 1 MHz, the pulse-forming circuit will produce a train of narrow pulses at 1-μs intervals. These pulses are applied to a number of decade frequency dividers designed to increase the time between pulses

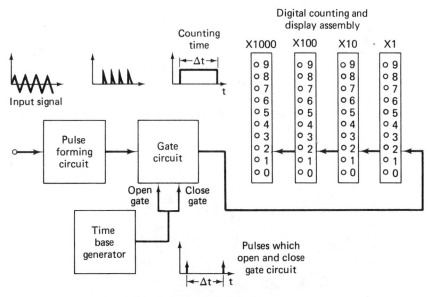

**Figure 8-5** Block diagram of a digital frequency meter.

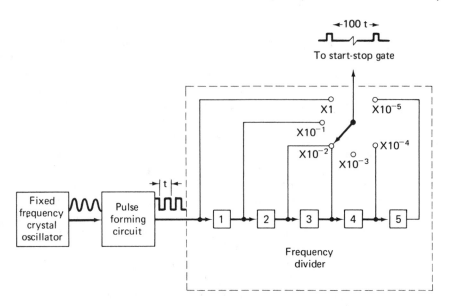

**Figure 8-6**  Block diagram of the time-base generator of a digital frequency counter.

by factors of 10. As a result, we can choose to have pulses separated in time from 1 μs to 1 s. Any two of these pulses will act as the pulses that open and close the gate circuit of the counter. The accuracy to which the time interval between pulses produced by the time-base generator is known will determine the accuracy to which the frequency can be measured. Typically, this accuracy is 1 part in $10^8$. After a short time, the counter is reset and another two pulses from the time-base generator again open and close the gate circuit. The frequency of this reset cycle is called the *sampling rate* of the counter. The microwave counter (Fig. 8-7) has an acquisition time of less than 200 ms and a resolution of 0.1 Hz from 10 Hz to 26.5 GHz. The instrument can accurately measure a carrier frequency even with a frequency modulation of 30 MHz. Many satellite communication systems operate in the upper range of this instrument.

**Figure 8-7**  Microwave counter: Model 2442. (Courtesy of Marconi Instruments.)

If intervals longer than one second are required for counting quantities other than frequency, the opening and closing of the gate circuit can be achieved by means of an external time-base generator. Even higher-accuracy time intervals can be achieved by applying external, high-accuracy time-base generators.

### Universal Time-Counters

Time intervals can also be determined with an instrument that operates very much like the digital frequency counter. In fact, since the two instruments are so similar, *universal timer-counters* are built that perform the functions of measuring both frequency and time intervals. The time-interval counter is also used instead of the digital frequency counter to accurately measure low frequencies. This is done because measuring the period of a low-frequency signal allows more counts to accumulate during one period. Thus the resolution and accuracy of the measurement are both improved.

The time-interval counters employ an oscillator whose output frequency (typically 1 MHz) is constant and very accurately known (Fig. 8-8). They count the number of cycles emitted by this oscillator during the time interval of interest. Therefore, since the time of each cycle is known, the time interval can be calculated, The gate that controls the starting and stopping of the counter is activated by pulses that signify the beginning and end of the time interval being measured.

The front panel controls of a typical universal timer-counter are shown in Fig. 8-9. The *signal input* is through a connector (marked Ⓐ in the figure). Most counters have an input impedance of 1 MΩ shunted by 25 pF, with an input sensitivity of 100 mV or less. The *visual digital display* Ⓑ may be a light-emitting-diode readout, a liquid-crystal readout, or an in-line gas-tube readout. Most counters have an attenuator Ⓒ between the signal input connectors and the input amplifier. This allows large input signal amplitudes to be measured without overloading the input amplifier of the counter. The *level* control Ⓓ allows the operator to adjust the counter to count positive or negative pulses. The *function* switch Ⓔ

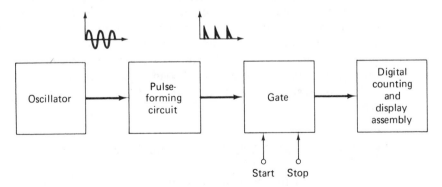

**Figure 8-8**  Block diagram of time-interval counter.

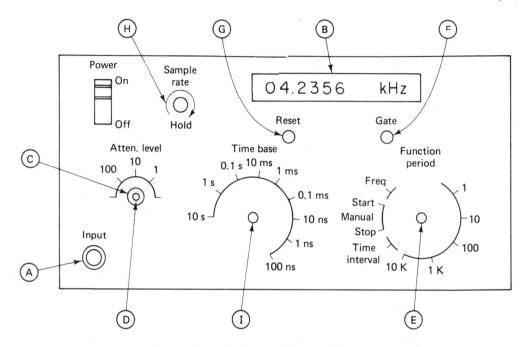

**Figure 8-9** Front panel controls of a typical universal timer-counter.

selects the operating mode of the counter. Available operating modes of the counter modes usually include the frequency, period, manual start-stop, and time interval. Other special modes may be offered as well. A *gate indicator light* Ⓕ is turned on when the "gate" opens to start a count and goes off at the end of the count. A *reset switch* Ⓖ allows the operator to reset the counting circuit to zero manually, thus making the counter ready to start a new measurement. A *sample rate* control Ⓗ allows the time between the end of one measurement and beginning of the next to be adjusted. Thus time between measurements can be varied from 100 μs to 5–10 s, depending on the counter. A HOLD position on this control puts the counter into a mode where it will take only one measurement and hold it indefinitely every time the reset switch is activated. The *time base* Ⓘ control selects the length of time the gate remains open during frequency measurements or determines the counted frequency if time-interval or period measurements are being made. Rear-panel controls often include power-input connectors, BCD or IEEE-488 connectors, 115/230-V change-over switches, various auxiliary outputs, and power-line and dc fuses.

## HARMONIC ANALYSIS AND SPECTRUM ANALYZERS

In the course of this book we have encountered waveforms with many different waveshapes. Some of them were periodic waveforms (i.e., waveforms that repeat their shapes at regular intervals). These periodic waveforms included sine waves, square waves, triangular waves, half-rectified sine waves, and full-rectified sine waves. There are also many other irregularly shaped, but still-periodic, waveforms. Although each of these waveforms can be plotted on a graph versus time as one signal waveform (or displayed as such on the oscilloscope screen), the same signal waveform can be decomposed into a number of sinusoidal waveform components. Conversely, any periodic waveform can also be constructed by adding together the waveforms of a particular group of sine-wave components. (These components are known as the *harmonics* of the waveform.) Of course, the amplitudes and frequencies of each of the components must be of the proper value for the group of components to mesh together to form the waveform that is to be reconstructed. The generalized mathematical procedure for determining the amplitudes and frequencies of these harmonics was developed by the French mathematician, Fourier. We will not present the mathematics of this procedure (called *Fourier analysis*), but we will content ourselves with a description of its qualitative ideas.

As an example that shows that an actual waveform can be decomposed, or analyzed, into a number of harmonic components, let us consider the square wave shown in Fig. 8-10. The first four harmonic components of the waveform are shown in the column on the left. The column on the right indicates how the shape of the *fundamental* component changes as additional harmonic components of the appropriate amplitudes and frequencies are added to it. We see that the composite waveform begins to look more and more like a square wave as each higher harmonic is added (the sides of waveform become steeper and the top and bottom, flatter). To completely reconstruct the square wave, we would have to use an infinite number of harmonics. However, we can see that the principle of adding together sine waves to get a square wave is not at all far fetched. The equation of the harmonics of a square wave having an amplitude of +1 is given by

$$y = \frac{4}{\pi}(\sin \omega t + \tfrac{1}{3}\sin 3\omega t + \tfrac{1}{5}\sin 5\omega t + \tfrac{1}{7}\sin 7\omega t + \cdots) \qquad (8\text{-}4)$$

The same procedure can be applied to any periodic wave, and the sum of the waveforms can be expressed in an equation such as Eq. (14-4). This equation, made up of an infinite sum (or series) of harmonics, is called the *Fourier series* of the waveform.

The frequency composition of a signal as expressed by the Fourier series is called the *frequency spectrum* of a signal. Such a frequency spectrum of a waveform can be plotted on a graph, with the frequencies of the harmonics on the abscissa ($X$ axis) and their amplitudes on the ordinate ($Y$ axis). The *frequency spectrum* of a square wave is shown in Fig. 8-11.

Although the mathematical calculation for a regularly shaped waveform (such as a square wave) can be performed rather easily, actual waveforms encountered in systems rarely have such mathematical simplicity. The calculations of the frequency spectrums of such irregularly shaped waveforms become prohibitively complex and thus are rarely made. Instead, either a *spectrum analyzer*, *Fourier analyzer*, or *wave analyzer* are used to determine the frequency spectrum of a waveform.

The way the *spectrum analyzer* (Fig. 8-12) performs this analysis is by using a

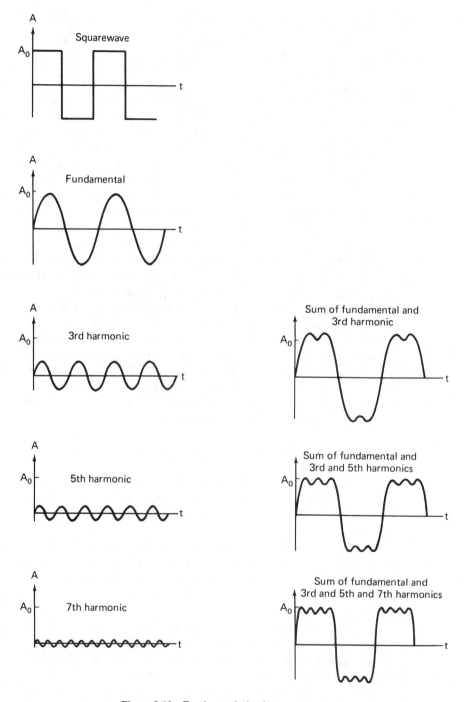

**Figure 8-10**   Fourier analysis of a square wave.

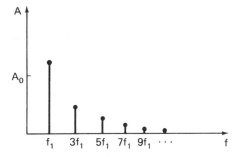

**Figure 8-11** Frequency spectrum of a square wave.

filter that rejects all but a very narrow band of frequencies. The center frequency of the narrow band of the filter is swept over the range of interest by the spectrum analyzer. The frequency components of the signal being analyzed are passed through to the display only at the time when their frequency matches that of the swept-filter frequency. A CRT is used as the display device to show the amplitude of each harmonic of the signal versus frequency over the swept frequency range [Fig. 8-13(b)]. Spectrum analyzers are built to display signals over the range from 5 Hz to 40 GHz.

The *Fourier analyzer* is another form of spectral analyzer, in which the signal is simultaneously presented to a large number (up to 2048) of parallel filters [Fig. 8-13(c)]. These filters are actually very specialized digital filters so that precise, repeatable results can be obtained. With the analysis being performed on the signal of interest in a simultaneous, parallel manner, the frequency spectrum of a signal can be displayed very rapidly. In fact, since the complete display is generated in the same time that it takes a conventional spectrum analyzer to analyze the lowest-frequency component of the signal, Fourier analyzers are also known as "real-time" analyzers. A CRT is again used to present the display of the amplitude of the signal components versus frequency. Fourier analyzers presently cover the range of dc to 100 kHz.

The *wave analyzer* [Fig. 8-13(d)] uses a tunable filter much like a conventional spectrum analyzer. However, the frequencies of interest are manually selected and adjusted. Thus the frequency window of interest can be precisely set to specific

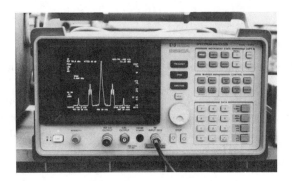

**Figure 8-12** Spectrum analyzer. (Courtesy of Hewlett-Packard Co.)

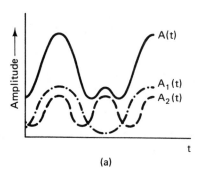

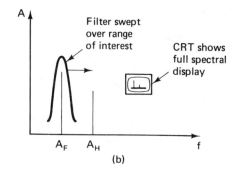

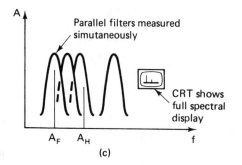

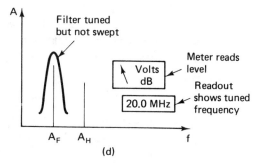

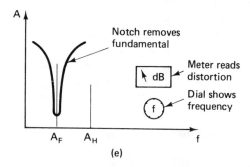

**Figure 8-13** (a) Waveform; (b) spectrum analyzer; (c) Fourier analyzer; (d) wave analyzer; (e) distortion analyzer.

frequencies so that the amplitudes of various signal harmonics can be accurately compared. An ac voltmeter is used rather than a CRT to display the amplitude of the harmonic components of interest. Spectrum analysis using wave analyzers is practical from 15 Hz to above 32 MHz.

### Distortion Analyzers

The harmonic distortion caused by a circuit or electronic device is defined to be the ratio of the total portion of the output signal produced by the harmonics to the portion of the output signal at the fundamental frequency:

$$\text{harmonic distortion } (\%) = \frac{\left[\sum_{n=2}^{\infty} A_n^2\right]^{1/2}}{A_1} \qquad (8\text{-}5)$$

In this expression, $A_1$ is the amplitude of the *fundamental*, and $A_n$ (from $n = 2$ to $n = \infty$) are the amplitudes of the harmonics. As an example, if the rms value of the signal due to the harmonics is 3 mV while the amplitude of the output due to the fundamental was 100 mV, the harmonic distortion would be 3 percent.

The instrument used to measure the total harmonic distortion caused by amplifiers or other electronic equipment is called a *distortion analyzer* [Fig. 8-13(e)]. It consists of an amplifier that suppresses the signal at the fundamental frequency and amplifies all the others. A Wien bridge circuit is used in the instrument as the *rejection filter*; that is, the Wien bridge circuit allows all the harmonics to be passed and amplified and fed to a voltmeter. The voltmeter indicates the rms value of the total signal due to the harmonics.

Distortion analyzers are used for fast, quantitative determinations of the *total distortion* in a waveform. Wave analyzers provide detailed information concerning each harmonic component of a test waveform.

## PROBLEMS

1. Find the period of the periodic waveform whose frequencies are
   (a) 60 Hz
   (b) 1 Hz
   (c) 550 Hz

2. Find the frequency of a repeating waveform with a period of
   (a) 40 s
   (b) 60 μs

3. Find the angular frequency of a sinewave whose frequencies are
   (a) 60 Hz
   (b) 10 MHz

4. Find the frequency of a sine wave whose angular frequency is 628 rad/s.

5. List the applications and advantages of each of the following time-measuring devices.
   (a) Electric timers
   (b) Electronic timers
   (c) Universal timer-counters

6. An electronic timer consists of a resistor whose value is 10 k$\Omega$, a capacitor of 30 pF, and a 10-V source which is applied across the *RC* connection at time $t = 0$. How much time will have elapsed if the voltage across the capacitor is 8.6 V at the instant the applied voltage is interrupted?

7. Describe the role of the mixer in the zero-beat frequency meter.

8. In what type of circumstances might a grid-dip meter be used to measure frequency?

9. Describe the three most commonly used display methods used in digital frequency meters.

10. Explain the operation of a simple timer-counter with the help of a block diagram. Explain the function of each block in the diagram.

11. Calculate or look up the Fourier series of a full-wave rectified sine wave. Draw the fundamental and the first four harmonics on a graph by using the same scale on each graph. Finally, graphically add the waveforms of each of the harmonics together.

12. Draw the frequency spectrum of the first five harmonic components of the (full-wave) rectified sine wave.

13. Define the following terms:
    (a) Distortion analyser
    (b) Rejection filter

14. Under what circumstances are harmonic analysers and distortion analysers used, respectively.

# REFERENCES

1. Millman, J., and Taub, H., *Pulse, Digital, and Switching Waveforms*. New York: McGraw-Hill, 1965.

2. Coombs, C., ed., *Basic Electronic Instrument Handbook*. Chaps. 33 and 34. New York: McGraw-Hill, 1972.

3. Malmstadt, H. V., Enke, C., and Crouch, S., *Electronics and Instrumentation for Scientists*. Menlo Park, Calif.: W. A. Benjamin, 1981.

4. "Spectrum Analysis," *Electronics Test*, Vol. II, pp. 34–38, June 1979.

# Power and Energy Measurements

In this chapter we discuss the measurement of both dc and ac power. When measuring power, various types of instruments and methods are used. Their design depends on the frequency range, the power dissipation level, and the type of load being supplied.

Power is an indication of the amount of work done in a specified amount of time; that is, it is the rate of doing work. In equation form

$$\text{power } (p) = \frac{\text{work}}{\text{time}} = \frac{w}{t} \tag{9-1}$$

where the electrical unit of measurement of power is the *watt*.

The power delivered to electrical devices can be expressed in terms of current and voltage. To express Eq. (9-1) in terms of current and voltage, we use Eqs. (1-2) and (1-3):

$$w = qv \tag{9-2}$$

$$i = \frac{q}{t} \tag{9-3}$$

where $q$ is the charge in coulombs, $v$ the potential difference in volts, and $i$ the current in amperes. By substituting Eqs. (9-2) and (9-3) into Eq. (9-1), we get

$$p = \frac{qv}{q/i} = vi \tag{9-4}$$

For dc circuits, we can use Ohm's law to write an expression for power in two other forms as well:

$$P = I^2R \qquad \text{(9-5a)}$$

and

$$P = \frac{V^2}{R} \qquad \text{(9-5b)}$$

From Eqs. (9-4), (9-5a), and (9-5b), we see that in dc circuits, power can be determined from measurements of any two of the three parameters $I$, $V$, and $R$. Therefore, power measurements can be made with dc ammeters and voltmeters. In fact, use of dc meters is encouraged for measuring power in dc circuits, since their employment will usually yield more accurate results than will the use of wattmeters. Here, dc meters are more sensitive than wattmeters and thus draw less power from the circuit under test. The usual connection for measuring dc power in a load by using a dc voltmeter and ammeter is shown in Fig. 9-1.

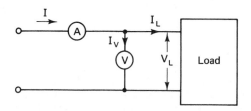

**Figure 9-1** Meter connections for measuring dc power.

Since the voltmeter and ammeter consume some power themselves, their presence in the circuit will introduce some error into the value of the readings. To find the true power being consumed by the load in Fig. 9-1, we need to subtract the power needed to operate the voltmeter from the product of $V_L$ and $I$.

**Example 9-1**

We are given a 50-V dc voltmeter with a 1000-$\Omega$/V rating and a 100-mA dc ammeter. They are connected to measure the power in a load as shown in Fig. 9-1. The voltmeter reads 40 V and the ammeter reads 50 mA. How much power is being dissipated by the load?

**Solution.**    We find the resistance of the voltmeter from its ohms/volt rating.

$$R_V = 50 \text{ V} \times 1000 \text{ } \Omega/\text{V}$$

$$= 50 \text{ k}\Omega$$

From Eq. (9-4) we find that the power being dissipated by the load is

$$P = 2.00 - \frac{1600}{5 \times 10^4} = 2.00 - 0.03$$

$$= 1.97 \text{ W}$$

## POWER IN AC CIRCUITS

In all circuits (ac or dc), the instantaneous power, $p$, being delivered to a load can be found by taking the product of $v$ and $i$ in the load. However, in ac circuits, this instantaneous product varies from moment to moment. Hence, a more useful quantity than the instantaneous power is the *average power*, $\bar{P}$, dissipated by a load. Because the ac quantities of voltage and current are not only sinusoidal but also tend to differ in phase, the calculation of this average power is more complex than for dc loads. These factors tend to complicate the *measurement* of ac power as well. For example, the procedure of measuring the values of $V_{rms}$ and $I_{rms}$ with ac meters, and then taking the product of the readings, is usually not a valid technique for measuring ac power. (Only if $v$ and $i$ are in phase in a load is the average power correctly found by using this method.)

To see how the average power in an ac load may be calculated, let us examine the relevant voltage and current waveforms. For a typical load, these waveforms would look like those shown in Fig. 9-2(a). Note first that the phase difference $\theta$ between $v$ and $i$ can be determined from their waveforms. Now the power at any instant is found by multiplying the magnitudes of $v$ and $i$ at the time of interest, and the resulting power waveform of such products is plotted as in Fig. 9-2(b). We see from this figure that the power being delivered to a load at any instant can be positive or negative. Positive power (the plus portions of the curve) indicates that the load is *absorbing* power from the generator or source. Negative power (the negative portions of the curve) means that the load is *returning* power to the source and is helping to run it (as in the charging of a battery or the supplying of power to turn the shaft of a generator).

The average power dissipated by the load is calculated by finding the average value of the power waveform of Fig. 9-2(b). It can be shown mathamatically that this average value is given by the expression

$$\bar{P} = V_{rms} I_{rms} \cos \theta \qquad (9\text{-}6)$$

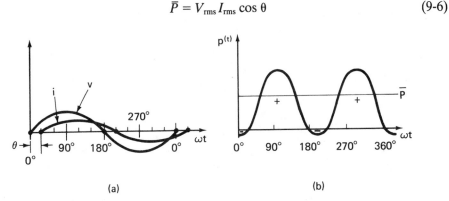

(a)                                    (b)

**Figure 9-2** (a) Voltage and current waveforms in an ac circuit where $\theta = 30°$; (b) power waveform in the same circuit.

where $\theta$ is the phase angle that exists between $v$ and $i$. We see from Eq. (9-6) why the product of the rms voltage and current readings would not ordinarily yield a correct value for $\bar{P}$. Their product would not take into account the factor owing to the phase angle $\theta$. This factor, the $\cos \theta$ term in Eq. (9-6), is known as the *power factor* (pf).

If the load is purely resistive, the waveforms of $v$ and $i$ are in phase, and $\theta = 0$. Since $\cos (0) = 1$, $\bar{P}$ in this special case is equal to

$$\bar{P}_{\theta = 0} = V_{\text{rms}} I_{\text{rms}}$$

As noted in the introduction, if the resistance $R$ of such a load is known, $\bar{P}$ can also be found from

$$\bar{P} = I_{\text{rms}}^2 R$$

or

$$\bar{P} = \frac{V_{\text{rms}}^2}{R}$$

The instruments designed to sense the effect of the phase difference and yield correct values for the average power are called *wattmeters*. The name is derived from the unit used to describe power.

The energy that flows in an ac circuit but is not dissipated in the load is known as the reactive power, $Q$ (var) Fig. 9-3. The two devices that do not dissipate power are the ideal capacitor and inductor. An ideal capacitor and inductor can only store and release energy. In practice, neither device is ideal insofar as they also exhibit resistive behavior. The losses arising from such nonideal resistive behavior cannot be ignored in many circuits. This reactive power $Q$ is described by the units vars (volt-amperes-reactive) and is calculated from the expression

$$Q = V_{\text{rms}} I_{\text{rms}} \sin \theta \qquad (9\text{-}7)$$

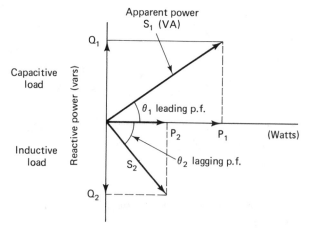

**Figure 9-3**   Power triangle.

where $\theta$ again is the phase angle between the $v$ and $i$ waveforms. $Q$ alternately flows into and out of the load. Even though this energy is not consumed by the load, there is some loss associated with its transmission back and forth in the circuit because transmission lines are not perfect conductors. Hence it is important to minimize this quantity, especially when large current flows are involved. $Q$ is measured with instruments known as *varmeters*.

The combination of $\bar{P}$ and $Q$ present in the load is called the *apparent power*, $S$, of the load. $S$ is found from $\bar{P}$ and $Q$ by the expression

$$S = \sqrt{\bar{P}^2 + Q^2} \tag{9-8}$$

In addition, $S$ is also equal to the product of $V_{rms}$ and $I_{rms}$ in the load.

$$S = V_{rms} I_{rms} \tag{9-9}$$

Therefore, the product of the two ac meter readings (connected to a load as the dc meters were connected for dc power measurements) generally yields $S$ and not $\bar{P}$. The units of $S$ are the volt-ampere (VA), but they are also often expressed in thousands of volt-amperes (kVA). For example, 1500 kVA transformer.

Finally, we see from Eqs. (9-6) and (9-9) that the ratio of $\bar{P}$ to $S$ will also yield the power factor

$$\text{pf} = \cos\theta = \frac{\bar{P}}{S} \tag{9-10}$$

**Example 9-2**

The voltage and current being applied to a load are sinusoidal waveforms whose amplitudes are 100 V and 5 A, respectively. The phase angle between them is 30°. Calculate the power, the reactive power, and the apparent power of the load.

**Solution**

(a) The rms values of any sinusoidal waveforms with amplitudes $V_0$ and $I_0$ are

$$V_{rms} = \frac{V_o}{\sqrt{2}} = \frac{100}{1.412} = 70.7 \text{ V}$$

$$I_{rms} = \frac{I_o}{\sqrt{2}} = \frac{5}{1.41} = 3.53 \text{ I}$$

(b) The phase angle between $v$ and $i$ is 30°, so

$$\cos\theta = 0.866$$

$$\sin\theta = 0.50$$

(c) $\bar{P} = V_{rms} I_{rms} \cos\theta = 216 \text{ W}$
(d) $Q = V_{rms} I_{rms} \sin\theta = 124 \text{ var}$
(e) $S = V_{rms} I_{rms} = 70.7 \times 3.53 = 249 \text{ VA}$
      or    $S = \sqrt{\bar{P}^2 + Q^2} = 249 \text{ VA}$

## Power Ratings of ac Equipment

Here, ac electrical equipment (machines, transformers, etc.) is rated in terms of apparent power $S$ as well as average power $\bar{P}$. Thus an ac motor will have both a wattage and kVA rating. The dual ratings must be used because of the effect of the power factor. If the power factor of the circuit in which the equipment is connected is small, excessively high current through the equipment could result. (Since voltage from the power station remains relatively constant, only $I$ can increase when $S$ increases.) High currents could burn out the insulation or other parts of the equipment. Such an occurrence would be possible even if the average power dissipated by the actual operation of the equipment remained below its wattage rating. By specifying and observing a maximum kVA rating, such high current damage is avoided. All power transformers are rated in VA or kVA.

### Example 9-3

A plant has a motor load of 1000 kVA at a pf of 0.85 lagging and lighting load of 500 kW. What is the current input to the plant, the watt and var load and the pf as seen by the utility company. The plant operates at 480 vac.

$$\cos^{-1}(0.85) = 31.788°$$

total power in watts $= P + S \cos \theta = 500 + 1000 (0.85) = 1350$ watts; reactive power in vars $= S \sin \theta = 1000 \sin (31.788°) = 526.7$ vars

$$S = \sqrt{P^2 + Q^2} \quad \underline{/-31.78°} \text{ (lagging)} = 1449 \quad \underline{/-31.78°}$$

$$\cos \theta = \frac{W}{S} = \frac{1350}{1449} = 0.9317 \text{ as seen by the utility company}$$

$$I = \frac{S}{V} = \frac{1449}{480} = 3014 \text{ amperes}$$

## SINGLE-PHASE POWER MEASUREMENTS

### Wattmeters

The electrodynamometer movement is utilized as the sensing mechanism for the vast majority of low-frequency (below 400 Hz) power instruments. Dynamometer-type instruments can be built to measure the average power dissipated in a load, the power factor, or the reactive power in a circuit. They can determine these quantities even if the waveforms being measured are not sinusoidal. This allows them to be used to measure power in dc circuits as well as in other applications where the alternating waveform has a nonsinusoidal shape.[1]

In Chapter 4 we saw how coils of the dynamometer movement were connected so that it could be used to measure ac currents and voltages. A series connection was employed, and the same current flowed in both the stationary and the rotating coils.

---

[1] However, as we noted earlier, dc power measurements can be made more accurately by using dc meters.

In the dynamometer *wattmeter*, the stationary coils and the rotating coil of the movement are connected differently (Fig. 9-4). The current coming from the power source, $i_c$, is made to pass through the stationary coils, thus connecting them in series with the load. (These coils are also called the *current coils* or *field coils*.) The movable coil (rotating coil) has a large resistance $R$ connected in series with it. This coil and resistance branch is connected *across* the load. The movable coil branch is also called the *voltage branch* and carries a small current $i_p$ (usually 10–50 mA).

The current in the stationary coils sets up a magnetic field in the space between them, which is proportional to $i_c$. The current in the moving coil is proportional to the voltage across the load ($i_p \approx V_L/R$). Since the moving coil is located in the space between the field coils, $i_p$ interacts with their magnetic field and causes the moving coil to rotate. The pointer attached to the rotating coil displays this rotation on a scale. Since the power at any instant is defined as

$$p = v \times i \tag{9-4}$$

the torque developed in the moving coil is proportional to the instantaneous power.

At frequencies above a few hertz, the inertia of the pointer is too great for it to follow the variations of $p$. Instead, the pointer takes up a position proportional to the average of $v \times i$. This position is calibrated to indicate the average power $\bar{P}$.

In the dynamometer ammeter, the current $i$ was equal in both coils, so the deflection was proportional to the average of $i^2$. In the dynamometer wattmeter, the currents in the two coils are different, and the deflection is proportional to the average of $v \times i$. Figure 9-5 is a photograph of a typical dynamometer wattmeter.

The dynamometer wattmeter has four external terminals to which connections must be made in order to measure power. Two of them are designated as the *voltage terminals* and the other two as the *current terminals*. The current terminals provide connections to the stationary coils, while the voltage terminals provide connections to the rotating-coil branch. One terminal of each type is marked $\pm$. It is necessary to connect the $\pm$ current terminal and the $\pm$ voltage terminal to the same wire of the incoming power line. In that way the stationary coils and the movable coil will be at

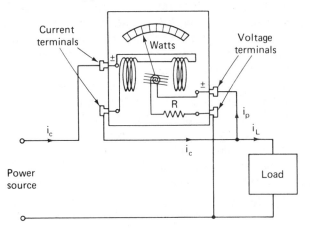

**Figure 9-4** Dynamometer wattmeter.

KILOWATTS

**Figure 9-5** Dynamometer wattmeter. (Courtesy of Weston Instruments, Inc.)

about the same potential. (Because the value of the resistance of the series resistor *R* in the voltage branch is much greater than the resistance of the voltage coil, most of the voltage across the voltage branch is dropped by *R*.) Then, no electric field will exist between the stationary and the moving coils. An electric field would arise between the voltage and current coils if they were at different potentials. The force of attraction due to the field could slightly restrict the rotation of the movable coil and produce an erroneous reading.

Dynamometer wattmeters are built with current ratings up to 20 A and voltage ranges up to 300 V. It is good practice, however, to limit the input current to the wattmeter to a maximum of 5 A. This can be done by using a current transformer to step down the value of the input current. When this practice is followed, the large magnetic fields associated with leads carrying heavy current are reduced. Such magnetic fields could sizably alter the relatively weak magnetic fields of the instrument coils. If the voltage being applied to the load exceeds 300 V, it is also wise to step it down to 115 to 125 V with a voltage transformer (commonly called a potential transformer). In this way, damage to the voltage circuit of a wattmeter is avoided.

The overall errors in commercially manufactured dynamometer instruments run between ±0.1 and ±0.5 percent when operated between their specified frequencies. The highest-accuracy meters are used as laboratory standards of power.

The wattmeter is rated in terms of its maximum current, voltage, and power. Each of these ratings must be observed to prevent damage to the wattmeter. Excess current could harm the current coils and their insulation. Excess voltage could cause the voltage-coil branch to suffer in a similar fashion. In low-power-factor circuits, either of these limits could be exceeded without exceeding the wattage rating.

### Errors in Dynamometer Wattmeters

Even if the connections to a dynamometer wattmeter are made properly, an error is still present in its readings. This error is caused by the power needed to maintain the magnetic field of the stationary coils and the power consumed by the voltage drop across the voltage branch. At 5 A, the power loss through the current coils is about 0.8 W. At 115 V, the power loss through the voltage coil circuit is about 2.9 W. For large power measurements this error is small, but it can become appreciable if the power levels being measured are small (5 W or less).

A wattmeter can be connected so that the voltage branch is placed either before or after the current coils. When the wattmeter is connected in the manner shown in Fig. 9-6(a), the movable coil branch is connected at point *A*. In this connection, the voltage branch senses the true voltage across the load. However, the magnetic field of the current coils is too large because some of the current causing their field goes to the voltage branch of the movable coil instead of going to the load. The result is an average power reading that is higher than the actual power dissipated by the load. On the other hand, if the movable-coil branch is connected to point *B*, the current in the current coils is now correct. But the voltage across the voltage branch is too large, that is, the voltage branch is sensing the voltage drop of the load and the stationary coils in series. Thus a high reading again results.

There are two methods that can be used to reduce these errors. The first is to make use of the connection that will yield the lowest error and apply a correction factor to the resultant readings. When this method is used, the following guidelines may be helpful. The connection of Fig. 9-6(a) will be better for high-current, low-voltage loads, while connection (b) of the same figure is better for high-voltage, low-current loads.

The second solution is to use a *compensated wattmeter* (Fig. 9-7). The compensated wattmeter is constructed by winding the wire carrying the movable-coil cur-

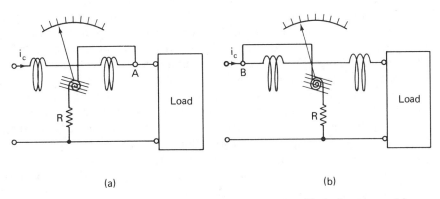

(a)                                           (b)

**Figure 9-6**  Connecting a wattmeter to measure power with the least error: (a) connection *A*, good for high-current, low-voltage loads; (b) connection *B*, good for high-voltage, low-current loads.

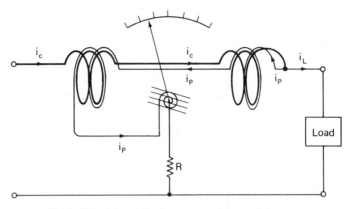

**Figure 9-7**    Schematic of a compensated wattmeter.

rent $i_p$ back into the current coils that originally carried $i_c = i_L + i_p$. The extra winding is carried out in the opposite direction to the windings of the current coils. Then the magnetic flux caused by the current in the extra winding will cancel that part of the magnetic flux due to $i_p$ flowing in the original current coils. As a result, the compensated wattmeter eliminates the above errors and indicates low-power readings much more accurately.

## Measuring $\bar{P}$ and S Simultaneously

Because the power dissipated in a load does not reveal all the power relationships of an ac circuit, the ac currents and voltages in the load are also sometimes monitored simultaneously. This allows calculations of $S$ and the power factor to be made. Unfortunately, when three meters are connected into a circuit together, their total effect may markedly disturb the true currents and voltages. This is particularly true of ac meters, which are usually less sensitive than comparable dc meters. In general, the same effects arise as those described in the discussion on errors in wattmeters and in the discussion on the errors present in dc power measurements. If it is suspected that the errors caused by the three meters are large enough to be significant the choice of connections that will minimize the errors should be used (along with a computation that corrects the resultant readings).

For example, if we are using a connection as shown in Fig. 9-8 we can calculate the true average power, the true voltage, and the true current in the load by making the following corrections. First we note that the voltmeter is reading the correct voltage across the load, and its value does not need correction. However, the reading of the wattmeter is too high because it is indicating the power of the load, the voltmeter, and its own voltage-sensing circuit. Thus the true value of $\bar{P}$ is found from

$$\bar{P} = W - \frac{V^2}{R_v} - \frac{V^2}{R_w} \qquad (9\text{-}11)$$

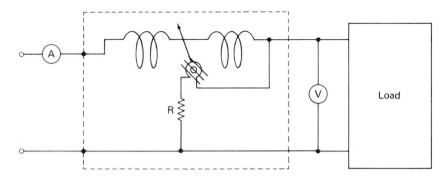

**Figure 9-8**   Simultaneous measurement with $\bar{P}$ and $S$.

where $W$ is the wattmeter reading in watts and $R_w$ and $R_v$ are the resistances (in ohms) of the wattmeter voltage branch and the voltmeter, respectively. Similarly, the true current in the load $I_L$ is found from

$$I_L = \sqrt{I^2 + V^2\left(\frac{1}{R_w} + \frac{1}{R_v}\right)^2 - 2W\left(\frac{1}{R_w} + \frac{1}{R_v}\right)} \qquad (9\text{-}12)$$

where $I$ is the reading taken from the ammeter and $V$ is the voltage reading of the voltmeter. By applying the foregoing corrections to readings obtained from this connection, more accurate data are obtained.

### Power Measurements Using an Oscilloscope

The power dissipated by a load in an ac circuit can also be measured with an oscilloscope. The voltage across the load is displayed directly by the scope. The current through the load is measured indirectly as discussed in Chapter 6 (using either current probes or the voltage dropped across a small-valued resistance placed in series with the load). For sinusoidal ac waveforms, the rms values of the voltage and current are then calculated according to Eq. (1-13). The phase difference between the voltage and current waveforms is then measured (note that a dual-trace oscilloscope is particularly convenient for this measurement since both the voltage and the current waveforms can be displayed simultaneously) and the power dissipated is calculated from Eq. (9-6).

## POLYPHASE POWER AND MEASUREMENTS

Most electric power for commercial use is transmitted in the form of three-phase ac power (Fig. 9-9). At the load end of a three-phase ($3\phi$) system, a three-phase load connection is used. The two possible $3\phi$ load connections are shown in Fig. 9-10. They are called the *wye* (Y) and *delta* ($\Delta$) loads, respectively. In the Y load, a fourth

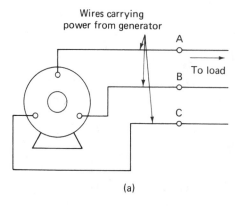

(a)

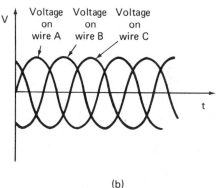

(b)

**Figure 9-9** Three-phase power generator: (a) three phase generator; (b)
phase relation between the waveforms of
the voltages on wires *A, B,* and *C* of a
three-phase generator.

wire connected to point *N* (called the neutral connection) may also be used.
Usually, *N* is connected to ground.

In most urban centers, the common load connection is the Y load. The circuit
shown in Fig. 9-11 is used for domestic power delivery. In this circuit, the voltage
between the phases of the power mains is 4160 V. The voltage between any of the
phase wires and the neutral point is 2400 V. For each household, a single-phase
transformer is used to step down this 2400 V to 240 V. The single-phase transformer
is center-tapped so that there are two 115-V, single-phase circuits available for each
house. (A 5-V drop is allowed for the connection between the transformer and the
house wall outlets.) Single-phase power is all that is required to run most common
appliances and lighting.

For schools, factories, and farms, however, 3φ power is sometimes required to
drive motors and other 3φ electrical equipment. The circuit used to deliver this 3φ
power is a little different than the single-phase circuit. That is, as shown in Fig. 9-12,
a 10:1, 3φ transformer is used to step down the 2080 V between phases to 208 V
between phases. This 208 V, 3φ voltage can then be applied to 3φ equipment. The
voltage between the phase wires and the neutral wire is still 120 V, and this connection can be used to obtain the same single-phase power as in the previous circuit.

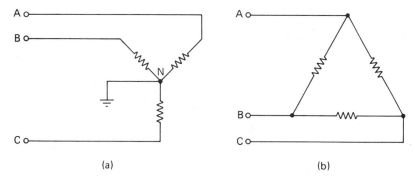

**Figure 9-10** Three-phase loads: (a) wye-load connection; (b) delta-load connection.

The line-to-line voltage on a three phase Y system is always $\sqrt{3}$ larger than the line-to-neutral because of the 120° phase difference between the line voltages. For example, 120 V$\sqrt{3}$ = 208 V, and 277 V$\sqrt{3}$ = 480 V are two very common voltage systems.

## Polyphase Measurements

The power delivered to any one of the elements of the 3ϕ load can be measured by a wattmeter placed across the element of interest as shown in Fig. 9-13. However, if it is desired to measure the total power $P_T$ being delivered to the entire Y or Δ load, the measurement is not as straightforward. The power delivered to a three-phase

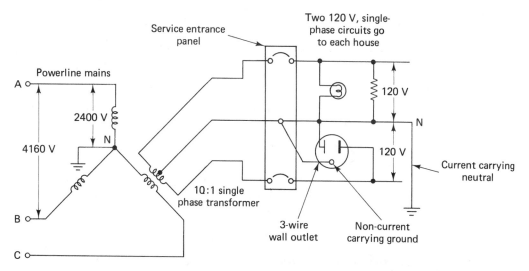

**Figure 9-11** Common urban household power connection.

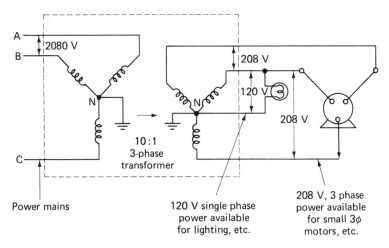

**Figure 9-12**  Common power connection for schools, light industry, etc.

load, connected in either Y or Δ configuration, must be measured by means of a polyphase wattmeter or by a proper connection of two wattmeters.

One possible two-wattmeter connection is shown in Fig. 9-14. The same side of the voltage branch of each wattmeter is connected to the phase wire that does not have a wattmeter connected in series (in this case, wire *B*). We find the power dissipated by the entire load by taking the *algebraic sum* of the two wattmeter readings. (However, in this method, neither wattmeter is reading power by itself

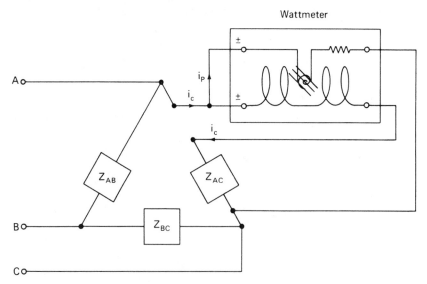

**Figure 9-13**  Using a wattmeter to measure power in one arm of a three-phase load.

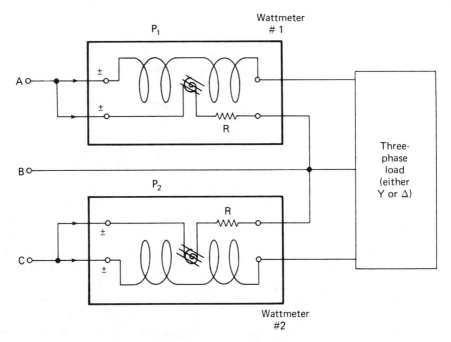

**Figure 9-14**   One possible connection used to measure power in a three-phase load using two wattmeters.

because there is no measurement being made of both $v$ and $i$ in a single arm of the load.) In making the initial connections, it is important to ensure that the connections to the phases are symmetrical (as shown in Fig. 9-14). A convenient guideline for ensuring symmetrical connections is to connect the wattmeters so that, as current flows from the source, it always enters the wattmeter through the $\pm$ terminals. This arrangement will establish a reference for determining the sign of the wattmeter readings. If the pf of the load is $>0.5$, both wattmeters will read positive and $\bar{P}_T = P_1 + P_2$. If the pf of the load is $0.5 <$, one of the meters (e.g., meter 2) will give a negative reading. In this case, the current leads of the negative-reading meter should be reversed, and the resultant reading should then be subtracted from the positive wattmeter reading to yield the power dissipated in the load ($\bar{P}_T = P_1 - P_2$).

If only one wattmeter is available, this method can still be used if two separate readings are made. As long as the proper connection rules are followed, a correct result is obtained.

## Polyphase Wattmeters

Polyphase wattmeters are made by attaching two electrodynamometer movements onto one shaft (Fig. 9-15). When the proper connections are made to the meter (as

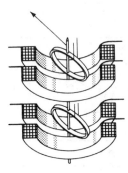

**Figure 9-15** Polyphase wattmeter movement. (Courtesy of Weston Instruments, Inc.)

in the two-meter method), the torque acting on the assembly will be the sum of the two torques acting on each of the moving coils separately. The total power is then summed automatically and indicated directly on one scale. The use of the polyphase wattmeter eliminates calculations and conserves space (since only one instrument performs the job of two). The instrument can also be used to measure single-phase power. In this case, connections are made to only one of the movements.

Do not attempt to install a three-phase power monitoring system in a large industrial plant unless adequate training in the proper calibration and installation techniques has been received. There are too many areas in which the neophyte can make a mistake. For example, just because the disk of the watt-hour meter (Fig. 9-16) is rotating, it does not mean that the installation is indicating properly. As shown in Example 9-2, even a small plant can draw several thousand amperes from the electric utility company. This necessitates current transformers that reduce the current before it is applied to the wattmeter. Potential transformers are also used to reduce the voltage. If any of the transformers leads are reversed, the output will be reversed. There are eight phase-sensitive connections that have to be made on a typical industrial wattmeter. In addition, the wattmeter calibration constants are dependent on the type of current and potential transformers used.

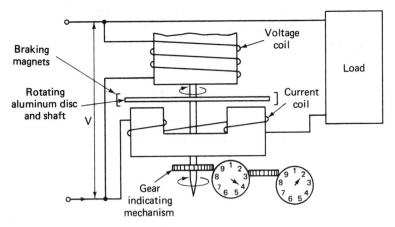

**Figure 9-16** Watt-hour meter.

## How to Use the Dynamometer Movement Meters to Measure Power Quantities

**1.** If the current exceeds 5 A, use an instrument transformer to step it down below 5 A before applying it to the wattmeter or other meters.

**2.** Try to avoid placing the wattmeter into regions where strong external magnetic fields are present.

**3.** Connect both ± terminals of the wattmeter to the same side of the load and line. This keeps the potential of the current and the voltage coils approximately equal.

**4.** Do not exceed the voltage, current, or power rating of the meter. Remember that the current or voltage ratings can be exceeded even if the power reading is not indicating a full-scale deflection.

**5.** When using a wattmeter, voltmeter, and ammeter together, corrections caused by the meter disturbances of the circuit should be made (especially if it is suspected that the reading errors are significantly large).

**6.** When a high-current, low-voltage reading is taken, connect the voltage coil to position *A* of Fig. 9-6.

**7.** When a high-voltage, low-current reading is made, connect the voltage coil to position *B* as shown in Fig. 9-6.

**8.** If a negative reading occurs in one of the wattmeters during a two-wattmeter measurement of polyphase power, reverse the connections of the current leads of this wattmeter and treat the resultant reading as a negative number.

## ELECTRICAL ENERGY MEASUREMENTS

When the power being dissipated in a load is calculated in terms of time, the amount of energy consumed by the load can be found. If one watt is delivered for 1 second, the energy consumed in that time is equal to 1 *joule*. The joule is therefore also called the watt-second. In electric power calculations, the watt-hour, or the kilowatt-hour are also used, since these are often more convenient units than the watt-second. One kilowatt-hour represents 1000 watts delivered for 1 hour.

Earlier in this chapter we saw how electric energy is generated and distributed to various consumers of electric power. At the distribution points there must be measuring devices to record the amount of energy used by each household or industrial consumer. In this way the supplier can bill the consumer for energy delivered during a given time span. The unit of energy that is sold by electric power companies is the kilowatt-hour (kWh). The approximate cost per kWh is about 5 cents in the United States. Table 9-1 gives a list of common electrical appliances and their typical wattage ratings.

**TABLE 9-1**   TYPICAL WATTAGE RATINGS OF
ELECTRICAL EQUIPMENT

| Equipment | Wattage rating |
|---|---|
| Air conditioner | 2800 |
| Clothes washer | 400 |
| Dishwasher | 1400 |
| Electric heater | 1650 |
| High-fidelity equipment | 230 |
| Electric iron | 1000 |
| Radio | 30 |
| Electric oven | 10,000 |
| Refrigerator | 320 |
| Color television | 420 |
| Oscilloscope | 50–150 |

The most common energy-measuring device is the watt-hour meter. The type of watt-hour meter that is most widely used today was developed by Schallenberger in 1888. This meter is an inexpensive and accurate instrument that can operate properly for long periods with little maintenance. Furthermore, it is not significantly affected by large changes in the load, power factor, or environmental conditions. It records the energy consumption of a load by counting the turns of a spinning aluminum disk. The spinning of the disk is caused by the power that passes through the meter.

A sketch of the important components of the meter is shown in Fig. 9-16. Its operation is similar to that of an induction-type motor. The current coil and the metal core on which it is wound set up a magnetic field. The voltage coil and its metallic core set up an additional magnetic field. In the aluminum disk (which is positioned to be influenced by both fields), eddy currents arise from the variation of the magnetic field of the current coil. These eddy currents interact with the magnetic field of the voltage coil, and a torque is exerted on the disk. Since there are no restraining springs, the disk continues to turn as long as power is fed through the meter (Fig. 9-17). The torque on the disk is proportional to the product of $v \times i$. Thus the greater the power passing through the meter, the faster the disk turns. The number of turns is a measure of the energy consumed by the load. The shaft on which the disk is mounted is geared to a group of indicators with clocklike faces. By reading the values on their faces at different times, one can determine how much energy passed through the meter during the interval between readings.

In order that the speed of rotation remains proportional to the power in the load (i.e., so that the disk does not continue to pick up speed as long as a torque is being applied to it), a retarding or braking torque must be applied to the rotating disk as well. Permanent magnets placed at the rim of the disk are designed to produce this retarding torque and hence are known as "braking magnets." As the disk rotates through the permanent magnetic fields, eddy currents in the disk are caused to arise. These currents themselves are proportional to the angular velocity

**Figure 9-17**   Watt-hour meter for industrial and domestic application. (Courtesy of Westinghouse Electric Corp.)

of the disk and cause their own magnetic fields. The permanent and disk-current magnetic fields interact to cause a retarding torque on the disk, proportional to the speed of rotation. Hence, a disk speed of rotation proportional to the power in the load is maintained.

## POWER MEASUREMENTS AT HIGHER FREQUENCIES

To measure power at frequencies above about 400 Hz, other types of measurement techniques and instruments other than the electrodynamometer meters must be used. The three general high-frequency power measurement techniques follow:

1. Measurement of voltage across a resistor, $R_L$, which is a substitute for the actual load
2. Measurement of the heat absorbed by a load, which again is a substitute of the actual load
3. Measurement of the transmitted power (does not require substitution of the actual load)

Techniques 1 and 2 both involve the use of so-called *absorption-type meters*, whose operation relies on being able to absorb all the power which would otherwise be destined for the load. Technique 3, on the other hand, is based on the use of transmission-type meters, which draw upon only a small fraction of the power being delivered to the load.

As pointed out, the first technique for high-frequency power measurements involves the measurement of voltage across a load resistor, $R_L$, and the calculating of power from $P = V^2/R_L$. For frequencies in the audio range (200 Hz to 20 kHz), power is usually measured by connecting the output of the power source (i.e., an audio amplifier) to a standard-value resistor (4 Ω and 8 Ω are common values). An ac meter then measures the voltage across this load resistor and the power is calculated from the meter reading. For high frequencies (RF frequencies up to 500 MHz), the voltmeter and load resistor are usually combined into a single absorption-

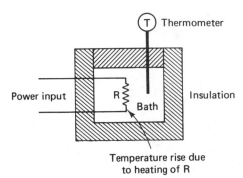

Figure 9-18  Basic calorimeter power meter.

type power meter instrument. The load resistor is designed to maintain a constant resistance value over the entire range of frequencies of interest, and the voltmeter is capable of acculately responding to such high-frequency signals. The scale of the meter is also calibrated to display its readings in units of power. Technique 2 for measuring power at high frequencies utilizes sensing of the heating effects caused by the signal of interest. Two meters (also, in fact, *absorption-type* power meters) that perform such measurements are known as the *calorimeter* and the *bolometer*.

The *calorimeter* is an instrument which contains a resistor that is completely surrounded by a well-insulated calorimetric body. The high-frequency (RF) input signal is fed to the calorimeter, and heat is dissipated by the resistor within the meter. The calorimetric body that surrounds the resistor is in turn heated by the warmed resistor and the rate of temperature rise is proportional to the power dissipated in the load. A temperature-measuring device, calibrated to display its output in terms of power, serves to indicate the power absorbed (Fig. 9-18). Calorimeters are used to measure RF power in the range 10 mW to 10W.

The *bolometer* is basically a bridge circuit in which one branch contains a temperature-sensitive resistor. This resistor is placed in the field of the RF power and its temperature rises as it absorbs power from the field. The change in resistance is measured by the bridge circuit. Bolometers are much more sensitive than calorimeters.

The two most common types of temperature-sensitive resistors used in bolometers are the *thermistor* and the *barretter*. The thermistor is a resistor that has a large temperature coefficient, usually negative. Principles of thermistors operation are discussed in more detail in Chapter 14. The *barretter* consists of a short length of fine wire or a thin film encapsulated so that it can be mounted in a bolometer mount. Barretters also indicate power absorption by changing resistance in response to heating, but have a positive temperature coefficient. Bolometers are customarily used to measure RF power in the power range 0.1 to 10mW. Figure 9-19 shows a photograph of a bolometer.

The third technique for measuring high-frequency power does not require that the power be delivered to a substitute (or "dummy") load. A transmission-type meter is inserted in series in the path of the power being delivered to an arbitrary

**Figure 9-19** Power meter capable of measuring power from 100 kHz to 50 GHz and from −70 to +44 dBm. Also shown are 50 GHz, 2.4mm connector, and two waveguide connectors. (Courtesy of Hewlett- Packard Co.)

load. A device known as *directional coupler* is built into the meter, which allows a small, fixed fraction of the input power to be sampled by the meter. This sampled fraction is then used to activate an absorption-type sensor to indicate the power being delivered to the load. The absorption element can be either the voltage-sensing or bolometer type as described in techniques 1 and 2. The directional coupler samples a fixed fraction of both the input power to the load and the reflected power returning from the load. The power being dissipated in the load, $P_{\text{delivered}}$, is then the difference between the input power, $P_{\text{into}}$, and the reflected power, $P_{\text{refl}}$.

$$P_{\text{delivered}} = P_{\text{into}} - P_{\text{refl}} \qquad (9\text{-}13)$$

The error to be expected with transmission-type meters is the loading error caused by the absorption-type element plus the errors contributed by the directional couplers.

## PROBLEMS

1. What is the power output of a 9-V battery when it is generating a current of 0.5 A?
2. A power supply can deliver a dc output of 200 mA at 115 V. What is the power output of this supply?
3. What is the power dissipated by a 4.7-Ω resistor if 4 mA of current is flowing through it?
4. What is the resistance of a resistor that dissipates 320 W when a voltage drop of 35 V exists across it?
5. A sine wave whose amplitude is 30 V and whose frequency is 60 Hz is applied across a 15-Ω resistor. Find the power dissipated by the resistor.
6. A 250-V voltmeter with a 20,000-Ω/V rating and a 5-A ammeter are used to measure the power being dissipated in a dc circuit. The voltmeter reads 150 V and the ammeter 4.3 A. What is the actual power being dissipated by the circuit?
7. The voltage and current being applied to a single-phase load are in the form of sinusoidal

quantities whose amplitudes are 250 V and 3.4 A, respectively. The power factor of the load is 0.41. Calculate $\bar{P}$, $S$, and $Q$ of the load.

8. Explain why one pair of terminals in a single-phase wattmeter are called the current terminals and the other pair of terminals are called the voltage terminals.

9. Explain why it is necessary that the ($\pm$) terminal of the current terminals and the ($\pm$) terminal of the voltage terminals of the dynamometer wattmeter must be connected to the same wire of the input power line.

10. Describe why connection $A$ of Fig. 9-6 is better for measuring loads that have high current and low voltage levels.

11. If the cost of electric power is 12 cents per kilowatt-hour, what is the cost of using each of the following?
    (a) A 1200-W toaster for 30 min
    (b) Twelve 50-W light bulbs for 4 h
    (c) A 205-W television for 3 h
    (d) A 5600-W electric clothes dryer for 20 min

12. Solve Example 9-3 for the percent change in plant current if the pf is increased to 1.0 by adding capacitors (var) in parallel with the motors. Does the current increase or decrease?


# REFERENCES

1. Hayt, W., and Kemmerly, J., *Engineering Circuit Analysis*. Chap. 12. New York: McGraw-Hill, 1978.
2. Stout, M. B., *Basic Electrical Measurements*, 2nd ed. Chap. 17. Englewood Cliffs, N.J.: Prentice Hall, 1960.
3. Harris, F. K., *Electrical Measurements*. Chaps. 11 and 12. New York: John Wiley, 1952.
4. Coombs, C., ed., *Basic Electronic Instrument Handbook*. Chap. 25. New York: McGraw-Hill, 1972.

# 10

# *Resistors and the Measurement of Resistance*

## RESISTANCE AND RESISTORS

Resistance, roughly speaking, describes the tendency of a material to impede the flow of electric charges through it. The unit of measurement of resistance, $R$, is the ohm ($\Omega$). If a circuit or device requires the effect that a specific amount of resistance produces (such as limiting the current flowing through it, or dissipating energy), an element that increases the overall resistance of the circuit is used. Such an element is called a *resistor*. Resistors are made of materials that conduct electricity but possess a large resistance compared to the resistance of the wires and the contacts. The instantaneous voltage across a resistor is directly proportional to the current flowing through it. The equation that describes this relation was discovered by George Ohm in his work with dc circuits in 1836. It is given by

$$v = Ri \tag{10-1}$$

and is known as *Ohm's law*.

If we wish to express how well an element conducts rather than impedes electricity, Ohm's law can instead be written in the form

$$i = Gv \tag{10-2}$$

where $G = 1/R$ is called *conductance* and its SI units are siemens (S). To say that a circuit element has a low conductance implies that it conducts electricity poorly and has a high resistance. For example, a conductance of $10^{-6}$ S (very low conductance) is equivalent to a resistance of 1 M$\Omega$.

## RESISTOR TYPES

Resistors are used for many purposes such as electric heaters, voltage and current dividing elements, and current-limiting devices. As such, their resistance values and tolerances vary widely. Resistors of 0.1 $\Omega$ to many megohms are manufactured. Acceptable tolerances may range from $\pm 20$ percent (resistors serving as heating elements) to $\pm 0.001$ percent (precision resistors in sensitive measuring instruments). Since no single resistor material or type can be made to encompass all the required ranges and tolerances, many different designs must be employed. The most common of these are discussed in this section. Table 10-1 summarizes the properties of the most common commercially available resistors.

The most common of the resistor types is the *carbon composition resistor* [Fig. 10-1(a)]. It is made of hot-pressed carbon granules mixed with varying amounts of filler to achieve a large range of resistance values. Such resistors have the advantages of being inexpensive, reliable, and remarkably free of stray capacitance and inductance. However, their tolerances of 5–20 percent compare unfavorably with most other resistor types, and their temperature coefficients (i.e., the percentage change in resistance value per degree of temperature change) are relatively high.

*Wirewound resistors* [Fig. 10-1(b)] are made primarily for three applications: high precision, low resistance, and high power dissipation. They consist of lengths of wire wound about an insulating cylindrical core. Typical tolerances range from 0.01 to 1 percent and when they are made from low temperature coefficient alloy wire, very precise and stable resistors are the result. They can be fabricated to dissipate up to about 200 W of power.

Very thin metal and carbon films can be deposited on insulating materials to provide very high resistance paths. *Metal-film* and *carbon-film resistors* [Fig. 10-1(c)] made from such processes can range in value up to 10,000 M$\Omega$. The accuracy and stability of such resistors can be comparable to those of wirewound resistors. In addition, these resistors possess the attributes of low noise and low inductance.

Some resistors are also made to have an adjustable or variable resistance. Such *variable resistors* usually have three leads: two fixed and one movable. If

**TABLE 10-1** CHARACTERISTICS OF VARIOUS TYPES OF RESISTORS

| Type | Available range | Tolerance (%) | Temperature coefficient (%/°C) | Maximum power |
|------|-----------------|---------------|--------------------------------|---------------|
| Carbon composition | 1 $\Omega$ to 22 M$\Omega$ | 5 to 20 | 0.1 | 2 W |
| Wirewound | 1 $\Omega$ to 100 k$\Omega$ | 0.0005 up | 0.0005 | 200 W |
| Metal film | 0.1 to $10^{10}$ $\Omega$ | 0.005 up | 0.0001 | 1 W |
| Carbon film | 10 $\Omega$ to 100 M$\Omega$ | 0.5 up | $-0.015$ to 0.05 | 2 W |
| Steel | 0.1 to 1 $\Omega$ | 20 | | 250 kW |
| Liquid ($H_2O$ + $CaCO_3$) | 0.01 to 1 $\Omega$ | 20 | | >250 kW |

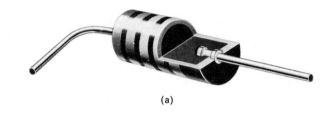

(a)

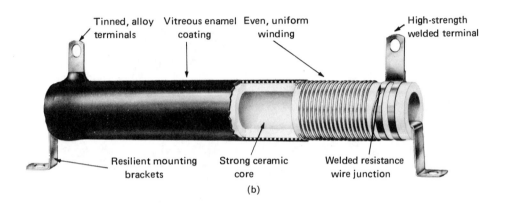

Tinned, alloy    Vitreous enamel   Even, uniform                                High-strength
terminals            coating            winding                               welded terminal

Resilient mounting       Strong ceramic        Welded resistance
brackets                   core                 wire junction

(b)

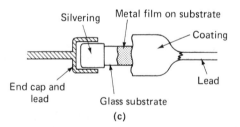

Silvering    Metal film on substrate

Coating

End cap and
lead                                            Lead

Glass substrate

(c)

**Figure 10-1**   Resistor types: (a) cutaway view of a carbon composition resistor
(Courtesy of Allen-Bradley Co.); (b) cutaway picture of a wirewound resistor
(Courtesy of Ohmite Manufacturing Co.); (c) construction of thin-film resistor.

contacts are made to only two leads of the resistor, the variable resistor is being
employed as a *rheostat* [Fig. 10-2(a)]. If all three contacts are used in a circuit [Fig.
10-2(b)], it is termed a potentiometer or "pot." Pots are often used as variable
voltage dividers in circuits. The value of the overall resistance and the power ratings
of variable resistors are usually stamped on their cases.

   If the resistance of a device is a known and reproducible function of temperature, light, strain, voltage, magnetic field (or other physical or chemical parameter),

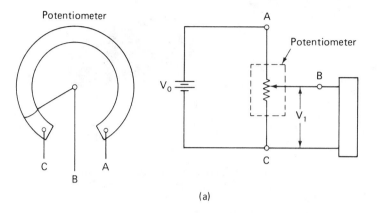

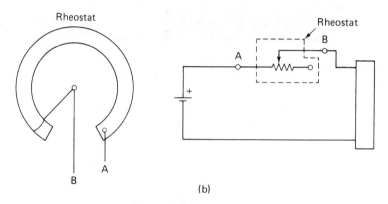

**Figure 10-2**   (a) Potentiometer; (b) rheostat.

the device can be used as *resistive transducer*. The resistance can be measured and the magnitude of the physical or chemical parameter can be inferred from the resistance value. A variety of such resistive transducers and their operation are discussed in Chapter 14. For example, a string of carbon resistors in series can be used to measure the level of liquid oxygen in a flask. Each resistor will have a large change in resistance when immersed in liquid oxygen. As the liquid level drops, more of the resistors are exposed to higher temperatures, and the total resistance value increases.

## COLOR CODING OF RESISTORS

Most larger resistors have their resistance values and tolerances stamped on their bodies. However, carbon composition resistors and some wirewounds are too small to use this method of identification. To be able to visually identify the resistance and

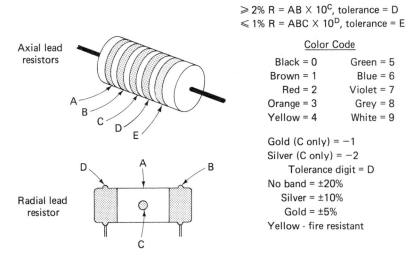

$\geqslant 2\% \ R = AB \times 10^{C}$, tolerance = D
$\leqslant 1\% \ R = ABC \times 10^{D}$, tolerance = E

Color Code

| | |
|---|---|
| Black = 0 | Green = 5 |
| Brown = 1 | Blue = 6 |
| Red = 2 | Violet = 7 |
| Orange = 3 | Grey = 8 |
| Yellow = 4 | White = 9 |

Gold (C only) = −1
Silver (C only) = −2
Tolerance digit = D
No band = ±20%
Silver = ±10%
Gold = ±5%
Yellow - fire resistant

Axial lead resistors

Radial lead resistor

**Figure 10-3**  Resistor color code.

tolerance of carbon resistors without actually having to measure them, a color code is utilized. Resistors that have a tolerance of 5 percent or greater have four or more color bands, and those with 1 percent or less tolerance have five or more color bands (Fig. 10-3). In small resistors ($<\frac{1}{3}$ watt), the bands are so close that it is impossible to determine which end of the resistor has the first significant digit. In this case, an ohmmeter should always be used to verify the resistance value.

Resistors are available in standard resistance values as shown in Fig. 10-4 and Table 10-2. Resistors used in military or critical instrumentation applications are specified under MIL-R-55182. These resistors have specific temperature coefficients, failure rates, and are manufactured under specific quality control requirements. They are coded as shown in Fig. 10-5. Resistors are also available as surface mount chips. For both surface mount chips and military usage, the physical size and shape of resistors and capacitors are the same. In many cases, it is impossible to tell the difference except by testing the component. Even diodes look like resistors except for the small polarity band on one end.

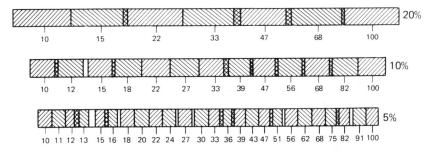

**Figure 10-4**  Nominal values and tolerance ranges for resistors (logarithmic scale).

**TABLE 10-2**  STANDARD RESISTOR VALUES FOR 10 TO 100 OHMS. DECADE MULTIPLES AND SUBMULTIPLES ARE ALSO AVAILABLE

Resistance tolerance (±%)

| 0.1% 0.25% 0.5% | 1% | 2% 5% 10% | 0.1% 0.25% 0.5% | 1% | 2% 5% 10% | 0.1% 0.25% 0.5% | 1% | 2% 5% 10% | 0.1% 0.25% 0.5% | 1% | 2% 5% 10% | 0.1% 0.25% 0.5% | 1% | 2% 5% 10% | 0.1% 0.25% 0.5% | 1% | 2% 5% 10% |
|---|---|---|---|---|---|---|---|---|---|---|---|---|---|---|---|---|---|
| 10.0 | 10.0 | 10 | 14.7 | 14.7 | — | 21.5 | 21.5 | — | 31.6 | 31.6 | — | 46.4 | 46.4 | 47 | 68.1 | 68.1 | 68 |
| 10.1 | — | — | 14.9 | — | — | 21.8 | — | — | 32.0 | — | — | 47.0 | — | — | 69.0 | — | — |
| 10.2 | 10.2 | — | 15.0 | 15.0 | 15 | 22.1 | 22.1 | 22 | 32.4 | 32.4 | 33 | 47.5 | 47.5 | — | 69.8 | 69.8 | — |
| 10.4 | — | — | 15.2 | — | — | 22.3 | — | — | 32.8 | — | — | 48.1 | — | — | 70.6 | — | — |
| 10.5 | 10.5 | — | 15.4 | 15.4 | — | 22.6 | 22.6 | — | 33.2 | 33.2 | — | 48.7 | 48.7 | — | 71.5 | 71.5 | — |
| 10.6 | — | — | 15.6 | — | — | 22.9 | — | — | 33.6 | — | — | 49.3 | — | — | 72.3 | — | — |
| 10.7 | 10.7 | — | 15.8 | 15.8 | — | 23.2 | 23.2 | — | 34.0 | 34.0 | — | 49.9 | 49.9 | — | 73.2 | 73.2 | — |
| 10.9 | — | — | 16.0 | — | 16 | 23.4 | — | — | 34.4 | — | — | 50.5 | — | — | 74.1 | — | — |
| 11.0 | 11.0 | 11 | 16.2 | 16.2 | — | 23.7 | 23.7 | 24 | 34.8 | 34.8 | 36 | 51.1 | 51.1 | 51 | 75.0 | 75.0 | 75 |
| 11.1 | — | — | 16.4 | — | — | 24.0 | — | — | 35.2 | — | — | 51.7 | — | — | 75.9 | — | — |
| 11.3 | 11.3 | — | 16.5 | 16.5 | — | 24.3 | 24.3 | — | 35.7 | 35.7 | — | 52.3 | 52.3 | — | 76.8 | 76.8 | — |
| 11.4 | — | — | 16.7 | — | — | 24.6 | — | — | 36.1 | — | — | 53.0 | — | — | 77.7 | — | — |
| 11.5 | 11.5 | — | 16.9 | 16.9 | — | 24.9 | 24.9 | — | 36.5 | 36.5 | — | 53.6 | 53.6 | — | 78.7 | 78.7 | — |
| 11.7 | — | — | 17.2 | — | — | 25.2 | — | — | 37.0 | — | — | 54.2 | — | — | 79.6 | — | — |
| 11.8 | 11.8 | — | 17.4 | 17.4 | — | 25.5 | 25.5 | — | 37.4 | 37.4 | — | 54.9 | 54.9 | — | 80.6 | 80.6 | — |
| 12.0 | — | 12 | 17.6 | — | — | 25.8 | — | — | 37.9 | — | — | 55.6 | — | — | 81.6 | — | — |
| 12.1 | 12.1 | — | 17.8 | 17.8 | 18 | 26.1 | 26.1 | — | 38.3 | 38.3 | 39 | 56.2 | 56.2 | 56 | 82.5 | 82.5 | 82 |
| 12.3 | — | — | 18.0 | — | — | 26.4 | — | — | 38.8 | — | — | 56.9 | — | — | 83.5 | — | — |
| 12.4 | 12.4 | — | 18.2 | 18.2 | — | 26.7 | 26.7 | 27 | 39.2 | 39.2 | — | 57.6 | 57.6 | — | 84.5 | 84.5 | — |
| 12.6 | — | — | 18.4 | — | — | 27.1 | — | — | 39.7 | — | — | 58.3 | — | — | 85.6 | — | — |
| 12.7 | 12.7 | — | 18.7 | 18.7 | — | 27.4 | 27.4 | — | 40.2 | 40.2 | — | 59.0 | 59.0 | — | 86.6 | 86.6 | — |
| 12.9 | — | — | 18.9 | — | — | 27.7 | — | — | 40.7 | — | — | 59.7 | — | — | 87.6 | — | — |
| 13.0 | 13.0 | 13 | 19.1 | 19.1 | — | 28.0 | 28.0 | — | 41.2 | 41.2 | — | 60.4 | 60.4 | — | 88.7 | 88.7 | — |
| 13.2 | — | — | 19.3 | — | — | 28.4 | — | — | 41.7 | — | — | 61.2 | — | — | 89.8 | — | — |
| 13.3 | 13.3 | — | 19.6 | 19.6 | — | 28.7 | 28.7 | — | 42.2 | 42.2 | 43 | 61.9 | 61.9 | 62 | 90.9 | 90.8 | 91 |
| 13.5 | — | — | 19.8 | — | — | 29.1 | — | — | 42.7 | — | — | 62.6 | — | — | 92.0 | — | — |
| 13.7 | 13.7 | — | 20.0 | 20.0 | 20 | 29.4 | 29.4 | 30 | 43.2 | 43.2 | — | 63.4 | 63.4 | — | 93.1 | 93.1 | — |
| 13.8 | — | — | 20.3 | — | — | 29.8 | — | — | 43.7 | — | — | 64.2 | — | — | 94.2 | — | — |
| 14.0 | 14.0 | — | 20.5 | 20.5 | — | 30.1 | 30.1 | — | 44.2 | 44.2 | — | 64.9 | 64.9 | — | 95.3 | 95.3 | — |
| 14.2 | — | — | 20.8 | — | — | 30.5 | — | — | 44.8 | — | — | 65.7 | — | — | 96.5 | — | — |
| 14.3 | 14.3 | — | 21.0 | 21.0 | — | 30.9 | 30.9 | — | 45.3 | 45.3 | — | 66.5 | 66.5 | — | 97.6 | 97.6 | — |
| 14.5 | — | — | 21.3 | — | — | 31.2 | — | — | 45.9 | — | — | 67.3 | — | — | 98.8 | — | — |

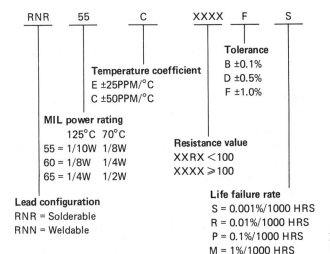

RNR    55    C    XXXX    F    S

**Tolerance**
B ±0.1%
D ±0.5%
F ±1.0%

**Temperature coefficient**
E ±25PPM/°C
C ±50PPM/°C

**MIL power rating**
       125°C 70°C
55 = 1/10W 1/8W
60 = 1/8W  1/4W
65 = 1/4W  1/2W

**Resistance value**
XXRX <100
XXXX ≥100

**Lead configuration**
RNR = Solderable
RNN = Weldable

**Life failure rate**
S = 0.001%/1000 HRS
R = 0.01%/1000 HRS
P = 0.1%/1000 HRS
M = 1%/1000 HRS

**Figure 10-5**  Resistor code for MIL style RNR and RNN.

Resistors are also available in numerous network configurations. When selecting a resistor it may also be important to consider such second and third order error sources as high frequency equivalent circuit, Johnson thermal noise, current noise energy, insulation resistance, insulation voltage limit, maximum change in resistance due to applied voltage, thermal change in resistance, the high frequency equivalent circuit, and failure rate.

**Example 10-1**

Given a resistor with bands

$$A = \text{blue} \qquad C = \text{orange}$$
$$B = \text{gray} \qquad D = \text{silver}$$

Find the resistance value and tolerance.

**Solution.**

$$R = 68 \times 10^3, \qquad \pm 10\%$$
$$= 68,000 \ \Omega, \qquad \pm 68000 \ \Omega$$

The resistance value as indicated by the color bands of carbon composition resistors is called the *nominal* value of resistance. Carbon resistors are manufactured only in a specific set of nominal values. These values are determined according to a formula which states that each nominal value is approximately $(1 + 2N)$ times the value of the preceding nominal value (where $N$ is tolerance of the resistor). By using this formula, the resistance of every resistor manufactured is within the tolerance range of each nominal value. Figure 10-4 shows the nominal values for the 5, 10, and 20 percent tolerance resistors.

## MEASUREMENT OF RESISTANCE

Resistance measurements are usually encountered in the tasks of testing and trouble-shooting circuits, in measuring component resistance values, and in determining the varying resistance values of transducers. We present a discussion of resistance measurement techniques with these applications in mind. At first, we cover the basic methods of resistance measurements, in order of increasing accuracy: the voltmeter–ammeter method, the ohmmeter, and the Wheatstone bridge. We shall see that the ohmmeter and the voltmeter–ammeter methods are popularly used for testing and rough resistance-value determination. Then it will be shown that the Wheatstone bridge becomes the instrument of choice when a very accurate measurement of resistance values or the monitoring of the output signals from resistive transducers is desired. Finally, we examine additional techniques that must be employed for measuring very low and very high resistance values. The resistance-measurement devices utilizing these techniques, including the Kelvin bridge, the megohmmeter, and the megohm bridge are to be introduced in this discussion.

### Voltmeter–Ammeter Method

The voltmeter–ammeter method is a rapid, simple, and moderately accurate technique for measuring resistance when one has only voltmeters and ammeters at hand. As shown in Fig. 10-6 the ratio of $V/I$ is the dc resistance of the device, $R_x$. The accuracy of this method depends on the calibration and stability of the two meters and the loading effect of the voltmeter. Use of a high-input-impedance digital voltmeter and a digital ammeter is a convenient way to minimize the error in the measured resistance values. This is the only method that can be used for very large wattage resistors. Resistors used to limit the inrush current in dc motors and large wound-secondary induction motors are rated as high as 1000 amperes. The resistance value is usually between 0.1 and 1 ohm, and this value must be known within a tolerance of 20 percent. To determine the value, an ammeter is inserted in the power cable supplying the motor armature. A digital voltmeter is used to determine the voltage across the resistor. The value of $R$ is then found by using $R = V/I$. This same technique is used to find the resistance to ground of an industrial electrical installation. The resistance of a grounding electrode cannot exceed 25 ohms as specified by the National Electric Code. The value of resistance is determined as shown in Fig. 10-7.

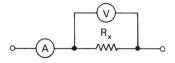

**Figure 10-6**   Meter connections for measuring resistance using voltmeter-ammeter method.

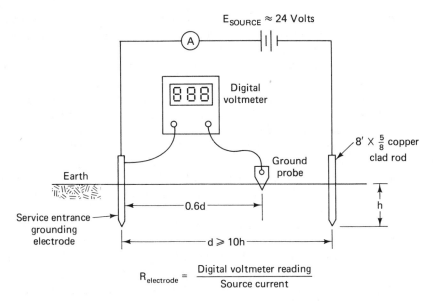

$E_{SOURCE} \approx 24$ Volts

Digital voltmeter

$8' \times \frac{5}{8}$ copper clad rod

Ground probe

Earth

Service entrance grounding electrode

0.6d

h

$d \geqslant 10h$

$$R_{electrode} = \frac{\text{Digital voltmeter reading}}{\text{Source current}}$$

**Figure 10-7**  Schematic diagram for measuring grounding electrode resistance.

## Ohmmeters

The ohmmeter[1] is a simple instrument that applies the fixed voltage of a battery across two resistors in series. One is a resistor of known value and the other is the one being measured. The voltage across the known resistor is measured by a dc voltmeter. The measured voltage causes an indication on the dc voltmeter that is calibrated to display the unknown resistance value directly.

Ohmmeters are useful for making quick measurements of resistance values under many common conditions and ranges. They are used particularly often in servicing communication equipment. Resistance values that can be measured with the ohmmeter vary from milliohms to 50 MΩ. However, there are some limitations on their use. Because their best accuracy is about ±1 percent, they are generally not suitable for highly accurate measurements. Also, certain special precautions must be followed in using ohmmeters to measure circuits with high inductance or capacitance. Finally, because they contain batteries, ohmmeters should be used only on passive circuits or on circuits that will not be damaged by them. Circuits possessing active sources may contribute currents that would change the voltage/current ratio and might injure the D'Arsonval movement of the ohmmeter. Circuits

[1]Note that the ohmmeter described in this section is the *series-type* ohmmeter. Such series-type ohmmeters are used extensively in portable meters for general-purpose measurement tasks. *Shunt-type* ohmmeters, on the other hand, are better suited for low-valued resistance measurements and are used primarily in low-resistance measurement applications. We will not discuss shunt-type ohmmeters in further detail.

that contain sensitive devices (such as some semiconductors and fuses) might be burned out by the passage of even the small amount of current put out by the ohmmeter battery.

The circuit of a simple series-type ohmmeter is shown in Fig. 10-8. The unknown resistor, $R_x$, is connected to the meter probes at points $A$ and $B$. If the probes are short-circuited ($R_x = O$), maximum current flows in the circuit and part of this current flows in the meter movement and part in resistor $R_2$. For this condition, $R_2$ is adjusted so that the position of the meter pointer is set to the "O $\Omega$" mark on the scale. The current flowing in the meter movement under this condition is called the full-scale current, $I_{fsd}$. When the probes are open-circuited ($R_x = \infty$), no current flows and the pointer resets on the "$\infty$" mark of the meter scale. Values of $R_x$ between "O" and "$\infty$" cause the pointer to move to some point in between these two extremes.

As the battery of the meter ages, its voltage output gradually decreases. Thus the current flowing in the meter will drop, and the meter will no longer read "0 $\Omega$" when the probes are shorted. The resistor $R_2$, however, is adjustable, and, by varying the value of $R_2$, the effect of the battery voltage change can be counteracted. The battery must be replaced when $R_2$ can no longer zero the meter.

When using the series-type ohmmeter, the most accurate results will be obtained when the unknown resistance $R_x$ causes a half-scale deflection of the meter. If it is desired to design the meter so that a given value of resistance will cause such a deflection, it is useful to define this resistance value as $R_h$. Then, given the internal resistance of the meter movement $R_M$ and the battery voltage $V_b$, resistors $R_1$ and $R_2$ can be chosen to allow a half-scale deflection.

The total battery current required to provide full-scale deflection $I_t$ when $R_h$ is connected (i.e., $R_h = R_x$ in this case) can be shown to be

$$I_t = \frac{V_b}{R_h} \tag{10-3}$$

The shunt current $I_2$ through resistor $R_2$ in this case is the difference between the total battery current and the full-scale current flowing in the meter movement:

$$I_2 = I_t - I_{fsd} \tag{10-4}$$

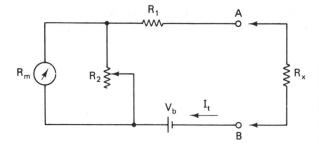

Figure 10-8  Series-type ohmmeter: $R_1$ = current-limiting resistor; $R_2$ = zero-adjust resistor; $V_b$ = internal battery; $R_m$ = internal resistance of the d'Arsonval movement; $R_x$ = unknown resistor.

The voltage across $R_2$ is $V_2$ and is also equal to the voltage across the meter movement $V_m$:

$$V_m = V_2 \tag{10-5}$$

or

$$I_{\text{fsd}} R_m = I_2 R_2 \tag{10-6}$$

or

$$R_2 = \frac{I_{\text{fsd}} R_m}{I_2} \tag{10-7}$$

Substituting Eqs. (10-4) and (10-3) into Eq. (10-7), we find $R_2$ to be

$$R_2 = \frac{I_{\text{fsd}} R_m}{I_t - I_{\text{fsd}}} = \frac{I_{\text{fsd}} R_m R_h}{V_b - I_{\text{fsd}} R_h} \tag{10-8}$$

and

$$R_1 = R_h - \frac{R_2 R_m}{R_2 + R_m} \tag{10-9}$$

$$R_1 = R_h - \frac{I_{\text{fsd}} R_m R_h}{V_b} \tag{10-10}$$

An ohmmeter's range Fig. 10-9 can be changed by varying its meter sensitivity. This is done by means of a switch that can connect various resistors of different values to replace resistor $R_1$. Since a half-scale deflection gives the most accurate reading with an ohmmeter, various scales (from the highest on down) should be tried until such a half-scale deflection is approximately reached.

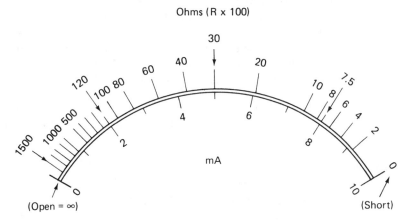

**Figure 10-9**  Nonlinear ohms used on ohmmeters.

## Using the Ohmmeter

**1.** If the resistance to be measured is known approximately, switch the ohmmeter to the scale that will indicate that value most accurately. If the resistance is not known at all, set the ohmmeter to the highest resistance scale. (Note that this step or step 2 are not applicable to autoranging digital ohmmeters.)

**2.** Before connecting the probes of an analog ohmmeter across the unknown resistor, touch them together to close the battery circuit. The adjustment knob should be turned until the resulting pointer position indicates an exact full-scale deflection. (This corresponds to a zero-ohms reading.)

**3.** Disconnect the probes from each other and put them across the resistance to be measured (first making sure all other electric power to the resistor being measured is turned off). Switch the ohmmeter scale settings until approximately a half-scale deflection is achieved (making sure to readjust the zero of the final scale before recording the resistance value). This scale will yield the most accurate result. The resistance can be read directly from the pointer deflection on the proper scale.

**4.** Shut the ohmmeter off to keep the battery from draining.

**5.** When making *in-circuit resistance measurements* with an ohmmeter, the following precautions should be observed: Even when the power supply of a circuit is turned off, capacitors in the circuit and in the power supply itself may remain charged. During measurement of an in-circuit resistance value, these capacitors may slowly discharge and cause erroneous readings or damage to the ohmmeter. Thus prior to measuring in-circuit resistance values in even apparently nonpowered circuits, all capacitors should be discharged.

**6.** Since the ohmmeter itself supplies a dc voltage whose polarity depends on the instrument design, some care needs to be exercised before using the ohmmeter to make resistance measurements on devices that might be damaged by even moderate currents (i.e., meter movements, fuses, semiconductor devices, as well as circuits containing these elements). This warning applies most stringently when the ohmmeter is to be used at its highest ranges. On these ranges, 30–40 V may be supplied by the ohmmeter in the course of making the measurement.

**7.** When making in-circuit resistance measurements, it should also be remembered that the ohmmeter will indicate the parallel combination of the resistor and all other dc conducting paths shunting it.

### Example 10-2

The ohmmeter of Fig. 10-8 uses a 50-$\Omega$ basic movement requiring a full-scale current of 1 mA. The internal battery voltage is 3 V. The desired scale marking for half-scale deflection is 2000 $\Omega$. Calculate (a) the values of $R_1$ and $R_2$, (b) the maximum value of $R_2$ to compensate for a 10 percent drop in battery voltage, and (c) the scale error at the half-scale mark (2000 $\Omega$) when $R_2$ is set as in part (b).

**Solution.**

(a) The total battery current at full-scale deflection is

$$I_t = \frac{V_b}{R_h} = \frac{3 \text{ V}}{2000 \ \Omega} = 1.5 \text{ mA}$$

The current through the zero-adjust resistor $R_2$ then is

$$I_2 = I_t - I_{fsd} = 1.5 \text{ mA} - 1 \text{ mA} = 0.5 \text{ mA}$$

The value of the zero-adjust resistor $R_2$ is

$$R_2 = \frac{I_{fsd} R_m}{I_2} = \frac{1 \text{ mA} \times 50 \ \Omega}{0.5 \text{ mA}} = 100 \ \Omega$$

The parallel resistance of the movement and the shunt ($R_p$) is

$$R_p = \frac{R_2 R_m}{R_2 + R_m} = \frac{50 \times 100}{150} = 33.3 \ \Omega$$

The value of the current-limiting resistor $R_1$ is

$$R_1 = R_h - R_p = 2{,}000 - 33.3 = 1966.7 \ \Omega$$

(b) At a 10 percent drop in battery voltage,

$$V_b = 3 \text{ V} - 0.3 \text{ V} = 2.7 \text{ V}$$

The total battery current $I_t$ then becomes

$$I_t = \frac{V_b}{R_h} = \frac{2.7 \text{ V}}{2000 \ \Omega} = 1.35 \text{ mA}$$

The shunt current $I_2$ is

$$I_2 = I_t - I_{fsd} = 1.35 \text{ mA} - 1 \text{ mA} = 0.35 \text{ mA}$$

and the zero-adjust resistor $R_2$ equals

$$R_2 = \frac{I_{fsd} R_m}{I_2} = \frac{1 \text{ mA} \times 150 \ \Omega}{0.35 \text{ mA}} = 143 \ \Omega$$

(c) The parallel resistance of the meter movement and the new value of $R_2$ becomes

$$R_p = \frac{R_2 R_m}{R_2 + R_m} = \frac{50 \times 143}{193} = 37 \ \Omega$$

Since the half-scale resistance $R_h$ is equal to the total internal circuit resistance, $R_h$ will increase to

$$R_h = R_1 + R_p = 1966.7 \ \Omega + 37 \ \Omega = 2003.7 \ \Omega$$

Therefore, the true value of the half-scale mark on the meter is 2003.7 $\Omega$, whereas the actual scale mark is 2000 $\Omega$. The percentage error is then

$$\% \text{ error} = \frac{2000 - 2003.7}{2003.7} \times 10\% = -0.185\%$$

## Digital Ohmmeters

The limitations owing to the inaccuracies of analog ohmmeters are overcome to some extent in digital ohmmeters. Precision-regulated power supplies increase the accuracy of the internal voltage sources of digital ohmmeters, and the high-resolution digital displays reduce the reading errors associated with analog meter scales. Since digital readouts cannot, however, be made nonlinear, the digital ohmmeter internally generates a constant current that is applied to the unknown resistance. The voltage across the unknown is then measured. The measured voltage thereby becomes proportional to the resistance. The basic circuit of the digital ohmmeter also usually employs the "four-terminal" technique (see the section "Low-Valued Resistance Measurements" later in this chapter for more details on "four-terminal" measurements). Note that the same precautions must be followed when using digital ohmmeters for resistance measurements as when using their analog counterparts.

## WHEATSTONE BRIDGES

A *bridge* is the name used to denote a special class of measuring circuits. They are most often used for making measurements of resistance, capacitance, and inductance. Bridges are used for resistance measurements when a very accurate determination of a particular resistance value is required. The most well known and widely used resistance bridge is the Wheatstone bridge. It was invented by Samuel Christie but improved to the point of being a commercial product by Charles Wheatstone. It is used for accurately measuring resistance values from milliohms to megohms.

Most commercial Wheatstone bridges are accurate to approximately 0.1 percent. Thus the values of resistance obtained from the bridge are far more accurate than the values obtained from the ohmmeter or the voltmeter-ammeter method.

The circuit of dc Wheatstone bridge is shown in Fig. 10-10, where $R_x$ is the resistance to be measured. The bridge works on the principle that no current will flow through the very sensitive D'Arsonval galvanometer connecting points $b$ and $c$ of the bridge circuit if there is no potential difference between them. When no current flows, the bridge is said to be *balanced*. The balanced condition is achieved if the voltage $V_o$ is divided in path $abd$ by resistors $R_1$ and $R_2$ in the same ratio as in path $acd$ by resistors $R_3$ and $R_x$. Then points $b$ and $c$ will be at the same potential. Thus the conduction of no current flow through the galvanometer implies that

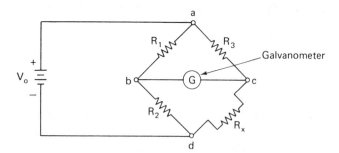

**Figure 10-10**   Wheatstone bridge circuit.

$$\frac{R_x}{R_3} = \frac{R_2}{R_1} \tag{10-11}$$

Now if $R_x$ is unknown and $R_1$, $R_2$, and $R_3$ are known, we can find $R_x$ from

$$R_x = R_3 \frac{R_2}{R_1} \tag{10-12}$$

In practical bridges, the ratio of $R_2$ to $R_1$ is controlled by a switch that changes this ratio by decades (i.e., factors of 10). Thus the ratio $R_2/R_1$ can be set to $10^{-3}$, $10^{-2}$, $10^{-1}$, 1, 10, $10^2$, and $10^3$. $R_3$ is a continuously adjustable variable resistor. When a null is achieved, the resistance can be read directly off the dials because these dial settings correspond to the variables of Eq. (10-12). Figure 10-11 is a photograph of a Wheatstone bridge.

Since the resistance value of a resistor is known to vary with frequency, resistors used in high-frequency applications should be measured at the frequency of use. When such measurements are performed, an ac source rather than a battery is used. A number of detectors, including the oscilloscope or even earphones, are available to determine a null or balanced condition.

### Slightly Unbalanced Bridge

If a Wheatstone bridge is slightly unbalanced (i.e., if there is a change in the value of $R_x$ from that at balance and this change, $\Delta R$ is less than about 10 percent of the balanced value of $R_x$), we can obtain the Thévenin equivalent of the bridge as seen from terminals $b$ and $c$ in Fig. 10-10. We can use this Thévenin equivalent to determine quickly the unbalance current flowing in the branch containing the meter.

**Figure 10-11** Wheatstone bridge. (Courtesy of Beckman Instruments, Inc., Cedar Grove Operations.)

In general, for a bridge circuit (Fig. 10-10) in slight unbalance,

$$R_{TH} = (R_1 \| R_2) + (R_3 \| R_x) \tag{10-13}$$

and

$$V_{TH} = V_o \frac{R_3 \Delta R}{2R_3 R_x + R_3^2 + R_x^2} \tag{10-14}$$

if all four resistances at balance are equal, as is often the case,

$$R_{TH} = R \tag{10-15}$$

$$V_{TH} = V_o \frac{\Delta R}{4R} \tag{10-16}$$

If the value of the internal resistance of the meter, $R_m$, is known, the current flowing through the meter at slight unbalance, $I_m$, is found from

$$I_m = \frac{V_{TH}}{R_{TH} + R_m} \tag{10-17}$$

### Example 10-3

Given a Wheatstone bridge as shown in Fig. 10-10 with $R_1 = 100\ \Omega$, $R_2 = 40\ \Omega$, $R_3 = 500\ \Omega$, $V_o = 10$ V, and $R_m$ (the galvanometer resistance) $= 600\ \Omega$. Find (a) the value of $R_x$ when the bridge is balanced and (b) if the value of $R_x$ changes by $+2\ \Omega$ from its value at balance, find the current $I_m$ that will flow through the meter.

### Solution.

(a) Using Eq. (10-12), which describes a bridge at balance,

$$R_x = R_3 \frac{R_2}{R_1} = 500 \left(\frac{40}{100}\right) = 200\ \Omega$$

(b) If the value of $R_x$ changes to $202\ \Omega$, the value of $I_m$ is found by using Eq. (10-17). However, $V_{TH}$ and $R_{TH}$ must be found first. These are found using Eqs. (10-13) and (10-14).

$$V_{TH} = V_o \frac{R_3 \Delta R}{2R_3 R_x + R_3^2 + R_x^2}$$

$$= 10 \left[ \frac{500 \times 2}{(2 \times 500 \times 202) + (500)^2 + (202)^2} \right]$$

$$= 0.02\ V$$

and

$$R_{TH} = (R_1 \| R_2) + (R_3 \| R_x)$$
$$= (100 \| 40) + (500 \| 202)$$
$$= 171\ \Omega$$

Then $I_m$ is found from Eq. (10-17):

$$I_m = \frac{V_{TH}}{R_{TH} + R_m} = \frac{0.02}{171 + 600} = 26\ \mu A$$

### Bridge-Circuit Applications Involving Resistive Transducers

As was mentioned earlier, another important application of the Wheatstone bridge involves the monitoring of resistance values of resistive transducers. The advantages of the bridge in these situations are twofold: (1) The balance of the bridge is not affected by changes in the excitation voltage, and (2) "out-of-balance" signals from the bridge can be kept sufficiently linear to provide convenient measurement signals for many applications. Let us examine the importance of these advantages in more detail.

**1.** Given a resistive transducer $R_x$ in a circuit as shown in Fig. 10-12 (a) (a potentiometric rather than a bridge-circuit configuration). The excitation voltage $V_A$ is 10.000 V, and $R_1$ is a resistor of constant value. If $R_1 = R_x$, the output voltage $V_o = 5.000$ V. If the physical parameter being monitored changes value and causes $R_x$ to change, $V_o$ will also change. In the circuit of Fig. 10-12(a), if $R_x$ changes by 0.1 percent, $V_o$ will change by 5 mV. An accurate digital voltmeter monitoring $V_o$ would read 5.005 V, and the positive deviation, +0.005 V, would be observed. Unfortunately, if at the same time, either the excitation voltage were to change by +0.02 percent or if the meter error was 0.02 percent, the displayed value $V_o$ would also be changed by 1 mV. Thus the measurand would be in error by 20 percent.

On the other hand, if a bridge circuit were used, as shown in Fig. 10-12(b) (with $R_1 = R_x$ at balance, and $R_2 = R_3$), for the small unbalance as described above ($R_x$ changes by 0.1 percent) the meter would read the difference $V_{AB}$ as 0.005 V. Any small changes occurring in the power supply voltage would be canceled because the voltages at points $A$ and $B$ would both be changed by essentially the same amount. The error caused by meter inaccuracy could also be improved by using a meter with a lower full-scale voltage range or by amplifying the difference voltage with an instrumentation amplifier (discussed in Chapter 15). Thus the use of a bridge circuit rather than the simpler circuit of Fig. 10-12(a), would result in a much more accurate measurement setup.

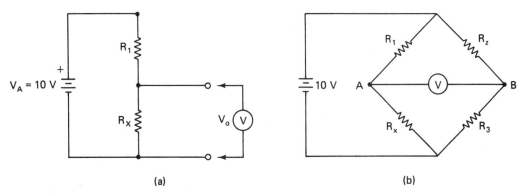

(a)                 (b)

**Figure 10-12** (a) Potentiometric circuit configuration to measure changes in resistive transducer $R_x$; (b) bridge circuit configuration used to measure changes in resistive transducer $R_x$.

**2.** In the example just described the resistive transducer in the bridge circuit changed its nominal resistance value by 0.1 percent, and it was implied that the bridge output voltage would change linearly as $R_x$ changed. Such out-of-balance signals in Wheatstone bridges will, in fact, vary quite linearly with changes in $R_x$, provided that the percentage change in $R_x$ remains less than ±10 percent. For example, if all the resistances in the bridge circuit in Fig. 10-12 had equal values of 100 Ω at balance, and $R_x$ were a resistive transducer whose value varied from 90 to 110 Ω, the voltages and currents developed across the voltmeter in the circuit would be those shown plotted in Fig. 10-13 From the data of Fig. 10-13 we can see that the assumption of a linear relationship between changes in $R_x$ and changes in $V_{AB}$ or $I_m$ (the current in the branch between $AB$) will yield data with an error smaller than 3 percent, provided that $R_x$ changes in value by less than 10 percent of its balanced value. As a result of this fact, the out-of-balance signal of a bridge

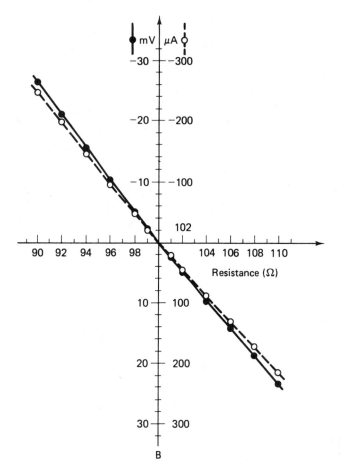

**Figure 10-13**  Out-of-balance resistance versus current and voltage.

circuit can be usefully monitored to provide an accurate measure of the change in $R_x$ even though the bridge is not in balance. This technique is widely used to measure the output of such transducers as resistance temperature devices (RTDs), strain gauges, and thermistors. Linearization techniques, as discussed in References 4 and 5, can reduce even further the error in the data obtained from unbalanced bridge signals.

## Commercial Resistance Bridges

Several resistance bridges are available commercially. An example of one is shown in Fig. 10-14. These instruments are basically similar to the Wheatstone bridge, but some also have the capability of measuring capacitance and inductance values. To measure resistance, the dc source and dc detector are used. Several scales are provided for a wide range.

## Substitution Method

The technique for making an extremely accurate determination of an unknown resistance involves comparing its resistance to the resistance of a high-precision resistor. The unknown resistor is first used to balance a resistance bridge. The unknown is then replaced by an *adjustable* precision resistor. The precision resistor is varied until the null is found again. The value of the unknown equals the value of the precision resistor at the null point. By this method, the errors of the bridge are bypassed, and the error depends on the accuracy of the adjustable resistor only.

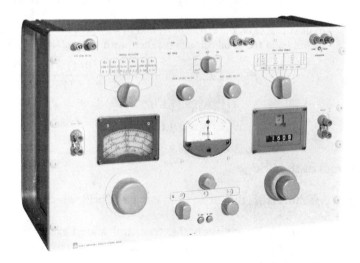

**Figure 10-14**   Commercial resistance bridge. (Courtesy of General Radio Corp.)

## MAKING "BALANCED" WHEATSTONE BRIDGE MEASUREMENTS

1. Connect the unknown resistor $R_x$ to the terminals of the bridge with good, tight contacts. This will minimize contact resistance.

2. Set the scale of the galvanometer to the least sensitive setting. (A variable shunt resistor is connected across the galvanometer to allow variation of its sensitivity.) This will prevent damage to the D'Arsonval movement if the bridge is severely unbalanced.

3. Adjust the variable resistor dials until a null is reached (zero deflection of the galvanometer needle).

4. Move to a more sensitive scale setting, and null again.

5. Continue until the most sensitive scale setting is reached.

6. Calculate the resistance from

$$R_x = R_3 \left(\frac{R_2}{R_1}\right)$$

or read from the dial settings.

### Example 10-4

A resistor is measured by using a Wheatstone bridge, and a null is reached for $R_2 = 100 \ \Omega$, $R_1 = 1000 \ \Omega$, and $R_3 = 120.3 \ \Omega$. The fixed resistors are known to be accurate within $\pm0.02$ percent and the variable resistor within $\pm0.04$ percent. What is the unknown resistance?

**Solution.**

$$R_x = R_3 \times \frac{R_2}{R_1} = (120.3) \times (1 \pm 0.0004) \times \frac{(100) \times (1 \pm 0.0002)}{(1000) \times (1 \pm 0.0002)}$$

In the worst case, the errors of $R_3$ and $R_2$ will be in the plus direction, and the errors of $R_1$ will be in the minus direction. Then the errors will add to give

$$R_x = (120.3) \times (1 + 0.0004) \times \frac{(0.1) \times (1 + 0.0002)}{(1 - 0.0002)}$$

$$= (12.03) \times (1 \pm 0.0008)$$

$$= 12.03 \pm 0.01 \ \Omega$$

### Errors of the Bridge

The possible errors that arise from using the bridge include:

1. Discrepancies between the true and stated resistance values in the three known branches of the bridge circuit. This error can be estimated from the resistor tolerances.

2. Changes in the known resistance values due to self-heating effects.

**3.** Thermal voltages in the bridge or galvanometer circuits caused by different materials in contact and at slightly different temperatures.

**4.** Balance-point error caused by lack of galvanometer sensitivity.

**5.** Lead and contact resistances introduced when making low-resistance measurements.

## LOW-VALUED RESISTANCE MEASUREMENTS

When attempting to determine the value of relatively small resistances (i.e., below 1 Ω), the connecting leads of the measuring instruments and the actual contacts between the probes and the unknown resistance may cause significant errors. The Kelvin bridge of Fig. 10-15 solves this problem and allows accurate measurements in the range from 1 Ω down to 0.0001 Ω.

The right-hand voltage divider contains extremely low resistance ($R_3$ and the unknown, $R_x$), and the wiring resistance $R_w$ (between $R_3$ and $R_x$) would cause an error of 10 percent if the galvanometer were connected directly to the bottom of $R_3$. Connecting the meter to the top of $R_x$ would not correct the problem because $R_w$ would then cause a 5 percent change in $R_3$.

The solution is to split the voltage drop that occurs across $R_w$ in the same ratio

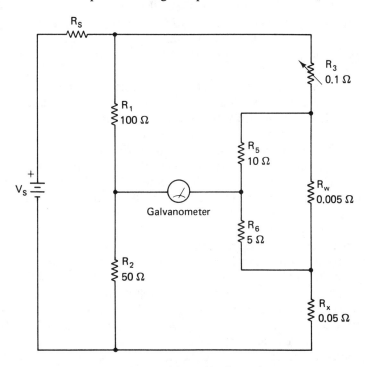

**Figure 10-15**   Kelvin bridge used for making low-resistance measurements.

as $R_1$ over $R_2$. This is most easily accomplished by adding two resistors to the circuit, $R_5$ and $R_6$, as shown in Fig. 10-15 with

$$\frac{R_5}{R_6} = \frac{R_1}{R_2} \tag{10-18}$$

## Milliohmmeters

The *milliohmmeter* can also be used for measuring low-valued resistors. An amplifier is typically employed to increase the sensitivity of the instrument voltmeter. A low-frequency ac current is sometimes provided by an internal source so that an ac amplifier can be used to avoid the drift that is inherent in dc amplifiers.

## Megohmmeter

Megohmmeters are a special class of ohmmeters used to measure very high resistances. The act of using this instrument is called *hipotting*. The resistance range that can be measured is from 0.01 M$\Omega$ to 10,000 M$\Omega$. To enable the measurements to be made, these instruments provide from 50 V to 15,000 V. As a result, even though these instruments are current limited, they must be used with caution. That is, if the wrong instrument is used, the insulation of a system under test can be damaged by applying too high a voltage. For example, if 15,000 volts was applied to a 300 volt instrumentation cable, this would certainly damage the cable. An electric arc would probably be produced, and the carbon paths created by such an arc cannot be removed.

Megohmmeters are used to determine if low resistance paths exist in areas such as the motor winding to ground, between wires or cables, from apparatus to ground, and from electrical apparatus to mechanical apparatus. All underground cable installations should be meggered before power is applied.

The megohmmeter shown in Fig. 10-16 has an output of 500 V and is capable of measuring to 100 M$\Omega$. It is also capable of measuring ac voltage and low values of resistance from 0 to 100 $\Omega$. It has a built-in human operator safety feature. That is, one hand must be used to push the button in order to apply the 500 volts to the probes.

IEEE Standard No. 43 suggests the minimum resistance values that should exist in equipment designs to operate at standard voltage ratings (Table 10-3). These ratings are based on an insulation temperature of 40°C. At such a temperature, correction must be applied. For every 10°C above 40°C, the values given in Table 10-3 must be doubled. For every 10°C below 40°C, the reading should be half the reading shown in the table. For example, a resistance reading of 15 M$\Omega$ taken at an insulation temperature of 20°C is equivalent to a resistance of $(15 \text{ M}\Omega)(\frac{1}{2})(\frac{1}{2}) = 3.75$ M$\Omega$ at 40°C. There is an additional correction factor based on the year and type of insulation being tested. Consult either the original manufacturer or the IEEE Standard for these factors.

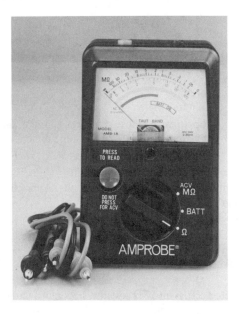

**Figure 10-16**  Megohmmeter. (Courtesy of Amprobe Instrument.)

The following guidelines should be observed whenever making measurements with a megohmmeter. First, the resistance reading should be taken after the voltage has been applied for a constant time period (normally 60 seconds). Second, the voltage applied should be neither higher nor lower than the recommendation. Third, always use the same time and voltage when comparing readings. Look for trends in resistance changes, and do not expect the same reading at different places in equipment. A megohmmeter test will not detect holes in insulation, however, which over a period of time can result in very high voltages stresses, corona discharge, and eventual failure of the insulator. This is why documentation of resistance trends is very important.

**TABLE 10-3**  RECOMMENDED RESISTANCE READINGS FOR STANDARD INDUSTRIAL EQUIPMENT

| Equipment voltage rating | Recommended resistance ($M\Omega$) | Recommended megohmmeter test voltage | Recommended voltage for destructive insulation test |
|---|---|---|---|
| 115 | 1.1 | 500 V | — |
| 230 | 1.23 | 500 | 2480 V dc |
| 460 | 1.46 | 500 or 1000 | 3300 V dc |
| 2300 | 3.3 | 1000 or 2500 | 9480 V dc |
| 4000 | 5.0 | 1000 to 5000 | 16,330 |
| 6600 | 7.6 | 1000 to 5000 | 24,100 |

Source: IEEE Standard No. 43.

## PROBLEMS

1. How is the resistance of a wire changed by
   (a) Tripling?
   (b) Tripling its cross-sectional area?
   (c) Replacing copper with aluminum?

2. Calculate the resistance of a copper wire whose diameter is 3 mm and whose length is 5 m. Repeat the calculation for a aluminum wire of the same dimensions.

3. An aluminum wire whose length is 50m has a resistance of 2.0 $\Omega$. Find its diameter in inches.

4. Convert the following quantities from inches to circular mils. Approximately what gauge numbers would wires with these diameters have?
   (a) 0.054 in.
   (b) 0.29 in.
   (c) 0.76 in.

5. What is the area, in circular mils, of a conductor whose cross-sectional dimensions are 0.3 in. by 0.25 in?

6. If a 115 V source is connected across an 57 $\Omega$ resistor, how much current will flow in the resistor?

7. A resistor whose value is 0.71 M$\Omega$ has a current of 0.33 mA flowing through it. What is the voltage across the resistor?

8. If the armature of an automobile starting motor has a resistance of 0.028 $\Omega$, and the automobile uses a 12-V battery, find the current drawn by the starting motor when it is connected to the battery. What is the conductance of the starting motor?

9. What is the unit used to describe the temperature coefficient of a resistor? Explain the meaning of the unit.

10. If a resistor wound of Nichrome wire has a resistance of 75 $\Omega$ at 42°C, find the resistance of the resistor at 70°C. (The temperature coefficient of Nichrome is +0.0004.)

11. Give the color code that would be used to identify the following composition resistors.
    (a) 4700 $\Omega$, ±20%        (b) 330 $\Omega$, ±5%        (e) 16.2 k 1%
    (c) 5.1 $\Omega$, ±10%        (d) 1000 $\Omega$, ±20%        (f) 845 k 1%

12. A 70-W, 120-V tungsten incandescent lamp has a resistance of 10.5 $\Omega$ when no current flows through it. What is the resistance of the lamp when it is connected to a 120-V source?

13. Two resistors with unmarked values are received. However, it is known that one is a 250-$\Omega$, 5-W resistor and the other is a 50-$\Omega$, 25-W resistor. How can the 50-$\Omega$, 25-W resistor be identified by inspection?

14. If a 4700-$\Omega$ resistor carries a current of 20 mA, how much power does it dissipate? What should be its power rating for safe design?

15. Design a voltage divider whose possible voltage outputs are 1.0 V, 3.0 V, 7.5 V, and 15 V. Assume that a 15-V battery is the source of voltage and that no current is drawn from the output terminals. What is the power dissipated?

16. Explain the difference between a *rheostat* and a *potentiometer*.

17. List some of the factors that limit the accuracy of the Wheatstone bridge.

**18.** Find the value of the unknown resistance in the balanced Wheatstone bridge shown in Fig. P10-1.

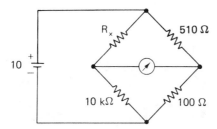

**Figure P10-1**

**19.** The Wheatstone bridge of Fig. P10-2 has $R_1 = 250 \, \Omega$, $R_2 = 750 \, \Omega$, and $R_3$ is a decade box with steps from 100 $\Omega$ to 0.1 $\Omega$. $R_1$ and $R_2$ are known to within $\pm 0.02$ percent and the resistors of the decade box are known to within $\pm 0.05$ percent. If $R_3$ is set to 153.7 $\Omega$ when a balance is found, determine

**(a)** $R_x$

**(b)** The percentage error that exists in the calculated value of $R_x$

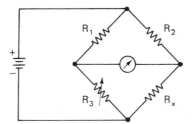

**Figure P10-2**

## REFERENCES

1. Stout, M. B., *Basic Electrical Measurements*, 2nd ed. Chaps. 4 and 5. Englewood Cliffs, N.J.: Prentice Hall, 1960.
2. Coombs, C., ed., *Basic Electronic Instrument Handbook*. Chap. 27. New York: McGraw-Hill, 1972.
3. Oliver, B. M., and Cage, J. M., eds., *Electronic Measurements and Instrumentation*. Chap. 27. New York: McGraw-Hill, 1972.
4. Sheingold, D. H., *Transducer Interfacing Handbook*. Norwood, Mass.: Analog Devices, Inc., 1981.
5. Sheingold, D. H., ed., *Non-Linear Circuits Handbook*. Norwood, Mass.: Analog Devices, Inc., 1976.
6. Kaufman, M., and Seidman, A. H., eds., *Handbook for Electronic Engineering Technicians*. Chap. 1. New York: McGraw-Hill, 1985.

# Measurement of Capacitance, Inductance, and Impedance

In this chapter we concern ourselves with the concepts of capacitance, inductance, and impedance. The first half of the chapter describes the properties of capacitors, inductors, and transformers as electrical components, in addition to the roles they play in the behavior of electrical circuits. In the latter half of the chapter, we discuss various techniques for measuring capacitance, inductance, and impedance.

## CAPACITANCE AND CAPACITORS

Material bodies that possess opposite electric charges will be attracted to one another by a force whose strength is found from Coulomb's law. To help represent this force, an electric field and a voltage between these bodies can be calculated. It has been observed that, for each particular configuration of two charged bodies in which the shape of the bodies and their separation remains fixed, the ratio of the charge to the voltage existing between them is a constant. This observation is expressed mathematically as

$$\frac{q}{v} = C \tag{11-1}$$

The constant $C$ is known as the *capacitance* of the particular geometrical configuration. To put it another way, *capacitance* refers to the amount of charge the

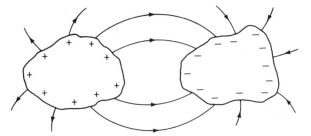

**Figure 11-1** Two bodies that are separated by a fixed distance (and are not connected by any conducting path) will store a constant amount of charge for each volt of potential difference between them.

configuration can store for each volt of potential difference that exists between the two bodies (Fig. 11-1).

If a circuit element is built so that it deliberately possesses a particular capacitance value, this element is called a *capacitor*. The unit of capacitance is the farad (F), and it is expressed as

$$1 \text{ farad} = \frac{1 \text{ coulomb of charge stored}}{1 \text{ volt}} \tag{11-2}$$

One coulomb is a very large amount of charge, and the quantity of charge stored for each volt in most real capacitors is much smaller than a coulomb. This makes the farad too unwieldy to describe the capacitance of actual capacitors. As a result, it is more common to see the capacitance of particular configurations and capacitors expressed in picofarads ($1 \text{ pF} = 10^{-12} \text{ F}$) or microfarads ($1 \text{ }\mu\text{F} = 10^{-6} \text{ F}$). For example, the large capacitors used in power supply filters have capacitance values of 10 to 1000 $\mu$F. The small-valued capacitors used in radio communication instruments have capacitance values of 25–500 pF.

The circuit symbol used for the capacitor is either ⊣⊢ or ⊣⊢ or ⊣⊢ (the last represents variable capacitors). The special configuration of two closely spaced, parallel metal plates is used to construct almost all circuit elements that are used as capacitors. Such capacitors are called *parallel-plate capacitors*, and an example of their form is shown in Fig. 11-2. The capacitance value of parallel-plate structures is found from the expression

$$C = \frac{K\epsilon_0 A}{d} \tag{11-3}$$

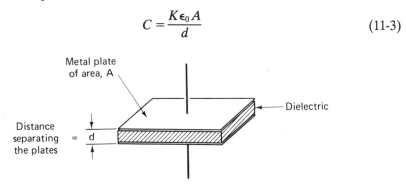

Metal plate of area, A

Dielectric

Distance separating the plates $= d$

**Figure 11-2** Parallel-plate capacitor.

where $K$ is the relative dielectric constant, $\epsilon_0$ is the permittivity of free space (and is constant of value $\epsilon_0 = 8.85 \times 10^{-12}$ F/m), $A$ is the area of the plates (in square meters), and $d$ is the distance between the plates (in meters). From Eq. (11-3) we can see that to increase the value of the capacitance of a parallel-plate structure, we can either increase its plate area or dielectric constant value, or decrease the distance between the plates.

Capacitance may also exist between conductors of other shapes and non-parallel separations (e.g., Fig. 11-2). When irregularly shaped configurations give rise to capacitance, more elaborate computational techniques than Eq. (11-3) must be used to calculate their resulting capacitance values. Such calculations are not covered by this text.

## Dielectrics

A dielectric is an insulating material placed between the plates of a capacitor to increase the capacitance value. Many different capacitance values can be obtained from two parallel plates of the same size and separation through the use of various dielectrics.

The relative dielectric constant $K$ introduced in Eq. (11-3) is the parameter that indicates how much a particular dielectric inserted between a capacitor's plates can increase the capacitance relative to vacuum.

## Capacitive Reactance

If the voltage across a capacitor varies with time, we see from Eq. (11-1) that the changing voltage causes a change in the charge stored on the capacitor. This charge storage must be accompanied by a current flow to, or from, the capacitor. Since current is the rate of charge flow, the current required is directly proportional to the rate of change of the capacitor voltage. The voltage across a capacitor, however, cannot change instantly, because a discontinuous instantaneous voltage change would require an infinite current. The capacitor thus reacts against changes in voltage across it; that is, it exhibits an impedance or *capacitive reactance*, $X_c$. Capacitive reactance has units of ohms and is dependent of frequency according to

$$X_c = \frac{1}{2\pi f C} = \frac{1}{\omega C} \tag{11-4}$$

where $f$ is the frequency of the applied signal. The higher the frequency, the less reactance exhibited by the capacitance to the flow of charge. Equation (11-4) also shows that a capacitor appears as an open circuit (infinite impedance) to a dc voltage ($X_c \to \infty$ as $f \to 0$).

## Capacitor Safety

A charged capacitor stores energy. If the capacitor has a large capacitance value and is charged to a high voltage, the quantity of energy stored can become quite large. During discharge, the energy is released by the current flowing in the connection between the plates. If this discharge accidentally takes place through a human conducting path, the resulting electric shock can be painfully nasty or sometimes even fatal. Since a charged capacitor looks no different from an uncharged one, a charged capacitor represents a disguised safety hazard. This means if a capacitor is ever charged during use, it must be discharged before being handled or put back on the shelf.

The discharge of the capacitor should always be made through an appropriate resistor. By simply short-circuiting its leads we can easily damage a sound capacitor. Therefore, such methods as connecting the capacitor terminals of a charged capacitor together with a screwdriver blade are not acceptable ways of discharging capacitors.

### Stray Capacitance

As was noted earlier, capacitance can and does exist between conductors that are at different potentials, regardless of their shape. Various configurations of circuit elements and leads that are at different potentials often exhibit such capacitance. Usually, this capacitance effect is unplanned and unwanted because it appears as an extra capacitance element in a circuit or system. For this reason, it is usually referred to as *stray capacitance*. Sometimes stray capacitance effects are small and can be neglected; at other times the effects may be relatively large and can cause significant changes in a circuit's behavior. For example, at high frequencies the stray capacitance can shunt large amounts of signal energy which should actually be transferred to other points in the circuit. When stray capacitance is significant, it must either be reduced or its magnitude must be included into the analysis of the circuit or measurement-system design. Examples of situations that can give rise to stray capacitance are shown in Fig. 11-3.

Figure 11-3(a) shows two current-carrying wires that are also at different potentials. A capacitive effect is established by the difference of potential between them. Figure 11-3(b) shows how a capacitive effect occurs between the turns of a

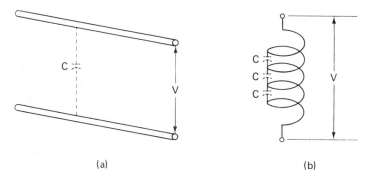

(a)                                    (b)

**Figure 11-3**  Stray capacitance effects: (a) stray capacitance between two current-carrying conductors at different potentials; (b) stray capacitance arising between the neighboring turns of a coil.

coil in an inductor. Because of the proximity and voltage drop between neighboring turns (a small but measurable drop), a stray capacitance results. A detailed discussion of how the stray capacitance effects can lead to capacitively coupled external noise interference in measurement systems is undertaken in Chapter 16.

## CAPACITOR CIRCUIT MODELS AND LOSSES

An ideal capacitor element stores but does not dissipate energy. It is a lossless element. However, a real capacitor always has some losses connected with its operation. Let us examine the structure of real capacitors to discover the sources of the major loss mechanisms.

If the dielectric separating the capacitor plates were a perfect insulator and if the leads and plates were made of perfectly conducting materials, there would be no energy dissipated by the capacitor during charging and discharging. However, since real dielectrics are not perfect insulators, they do cause some energy loss when a capacitor is operating in a circuit. This *dielectric loss* depends on how imperfect the dielectric is and on the frequency of the applied voltage.

In the special case of an applied dc voltage, a small current flows through the capacitor because of the few free charge carriers that exist in the dielectric. (The more the dielectric resembles a perfect insulator, the fewer are the free charge carriers.) Such a current is called *leakage current*. Polystyrene and mylar dielectrics possess the lowest leakage currents. Electrolytic capacitors have some of the highest leakage currents. Humidity and defects in the capacitor's encapsulation or packaging sometimes lead to additional leakage currents.

The other major losses of a capacitor involve so-called *resistance losses* or *plate losses*. These are due to the resistance of the material making up the plates and leads of the capacitor. At high frequencies the capacitor is partially charged and discharged at a high rate. Each time current flows into or out of a capacitor, it must flow through these conductors and lose some energy. In addition, at high frequencies the resistance of conductors can be much higher than their dc resistance value. Therefore, at such high frequencies the resistance loss effect can become quite significant.

These plate losses, together with the dielectric losses, show up as heat generated during a capacitor's operation. They must therefore be held to some reasonable level in order to avoid damaging the element by excessive heating.

The overall losses of an actual capacitor can be taken into account when creating an equivalent-circuit model of a capacitor for use in circuit analysis. One commonly used model is a resistor connected in parallel with an ideal capacitor. It is referred to as the *parallel model* [Fig. 11-4(a)]. The leakage current of a capacitor can be thought of as flowing through the resistor of this model. The lower the leakage current that exists for a given voltage, the larger is the *leakage resistance*, $R_p$. (A high leakage resistance is considered to be 100 M$\Omega$ or more. A low leakage resistance would be 1M$\Omega$ or less.)

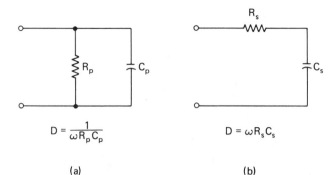

$$D = \frac{1}{\omega R_p C_p}$$

(a)

$$D = \omega R_s C_s$$

(b)

**Figure 11-4** Equivalent-circuit models of capacitors: (a) parallel; (b) series.

Another model that is also used to represent capacitors is the *series model*; it is shown in Fig. 11-4(b). It is not as easy to associate leakage current with this model, but the model is more useful than the parallel model for analyzing certain circuits. It is mentioned here because it is used in some of the circuits of the capacitance-measuring instruments which are discussed in a later section.

Both the parallel and the series equivalent-circuit models of a capacitor are frequency dependent. This means that the value of the elements used in them is liable to vary with the applied frequency. However, most capacitors have a range of frequencies over which $C$ and $R$ remain approximately constant (see Table 11-1). As long as the capacitor is used in this range, the simple models given in Fig. 11-4 can be used with confidence. (The values of $C$ and $R$ are determined by measurement.)

The dissipation or loss factor $D$ is also frequency dependent. This factor is defined as the ratio of the conductance to the capacitive reactance of a capacitor. For the parallel model, $D$ is found from

$$D = \frac{1}{\omega C_p R_p}\bigg|_\omega \tag{11-5}$$

Here $\omega$ is the frequency of the applied voltage, and $C_p$ and $R_p$ are the values of the elements of the parallel model measured at the applied frequency. For the series model, $D$ is found from

$$D = \omega C_s R_s\big|_\omega \tag{11-6}$$

where $C_s$ and $R_s$ are the values of the capacitance and resistance elements of this model measured at the applied frequency. Dissipation factor is also defined as the ratio of the amount of energy dissipated per half cycle to the average energy stored per half cycle. In a sense, this is a measure of the efficiency of a capacitor to store and then release energy. Typical values for commercial capacitors range from 0.001 to 0.0001. The lower the value, the better the capacitor. Some manufacturers use the term *quality factor*, Q, rather than *dissipation factor*, DF. These terms are simply reciprocals of each other.

In an ideal capacitor, $D$ would be zero, with $R_p$ being infinite or $R_s$ zero. In an actual capacitor, the larger the value of $D$, the larger the overall loss. Values of $D$ vary from about 0.1 in electrolytic capacitors to about $10^{-4}$ in polystyrene types.

**TABLE 11-1** VARIOUS CAPACITOR TYPES AND THEIR CHARACTERISTICS[a]

| Dielectric | Available capacitance values | Tolerances (%) | Leakages resistance (MΩ) | Maximum voltage ranges | Useful frequency ranges (Hz) |
|---|---|---|---|---|---|
| Mica (silvered) | 1 pF–0.1 μF | ±1 to ±20 | 1000 | 500–75 kV | $10^3$–$10^{10}$ |
| Ceramic (low-loss) | 1 pF–0.001 μF | ±5 to ±20 | 1000 | 6000 V | $10^3$–$10^{10}$ |
| Ceramic (high-$K$) | 100 pF–0.1 μF | +100 to −20 | 30–100 | 100 V or smaller | $10^3$–$10^8$ |
| Paper (oil-soaked) | 1000 pF–50 μF | ±10 to ±20 | 100 | 100 V to 100 kV | 100–$10^8$ |
| Polystyrene | 500 pF–10 μF | ±0.5 | 10,000 | 1000 V or smaller | 0–$10^{10}$ |
| Mylar | 5000 pF–10 μF | ±20 | 10,000 | 100 V to 600 V | 100–$10^8$ |
| Electrolytic | 0.47 μF–0.7 F | ±100 to −20 | 1 | 500 V or smaller | 10–$10^4$ |
| Air-variable | 10 pF (unmeshed) to 500 pF (meshed) | ±0.1 | | 500 V | |

[a] Adapted from B. D. Wedlock and J. K. Roberge, *Electronic Components and Measurements* (Englewood Cliffs, N.J.: Prentice Hall, 1969), p. 96.

In addition to the losses of a capacitor, there are other parameters that represent a departure in electrical behavior from that of an ideal capacitor structure. The most important of these is the *dielectric breakdown*. When the voltage across a dielectric exceeds a certain value, the bonds restraining the bound electrons of the material atoms are finally torn asunder. This results in a high current flowing through the capacitor, and it is called *dielectric breakdown*. The value of the electric field at which the breakdown takes place is called the *dielectric strength* of the material.

The maximum voltage a given capacitor can withstand is the product of the dielectric strength and the thickness of its dielectric layer. This is called the *breakdown voltage* of the capacitor. Operating a capacitor at rated voltage and core (internal) temperature establishes an *expected life* for the capacitor. While the expected life is often defined as the statistical time required to generate one failure in 25 units based on a 60 percent confidence level, this definition varies with different manufacturers. By operating the capacitor at a reduced voltage and core temperature, the life expectancy of the capacitor can easily be increased by three orders of magnitude. Core temperatures can be reduced by reducing ripple current and DF.

## CAPACITOR TYPES

Practical capacitors are built with various combinations of conductors and dielectrics. Families of capacitors are based on the type of dielectric used, such as mica, ceramic, paper, or oil.

Mica is a transparent, high-dielectric-strength mineral that is easily separated into uniform sheets as thin as 0.0001 inches. It has a high breakdown voltage and is almost totally chemically inert.

*Mica capacitors* are built in round, rectangular, or irregular shapes. They are constructed by sandwiching layers of metal foil and mica, as shown in Fig. 11-5(a). Sometimes silver is deposited on the mica in lieu of metal foil. The resulting stack of metal and mica sheets is firmly clamped and encapsulated in a plastic package.

Mica capacitors possess very small leakage current and dissipation factors. Available capacitance ranges are 1 pF to 0.1 μF, with tolerances of ±1 to ±20 percent. The capacitance is limited to this relatively small upper value because mica is not flexible enough to be rolled into tubes. As a result, the size of the mica capacitor structures cannot be markedly reduced.

There are two different types of *ceramic capacitors* being built: the low-loss, low-dielectric-constant type, and the high-dielectric-constant type (Fig. 11-6). The low-loss types have a very high leakage resistance (1000 MΩ) and can be used in high-frequency applications almost as well as mica capacitors.

The high-dielectric-constant types provide a large capacitance value in a small volume. However, their value of capacitance can change strongly with variations of temperature, dc voltage, and frequency. This is because the dielectric constants of high-$K$ capacitors are highly dependent on these variables. Thus, this type of capac-

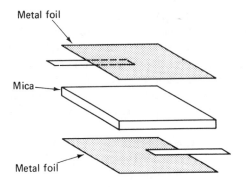

Metal foil

Mica

Metal foil

(a)

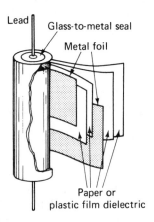

Lead

Glass-to-metal seal

Metal foil

Paper or
plastic film dielectric

(b)

**Figure 11-5**  Construction of mica,
paper, and plastic capacitors: (a) mica;
(b) paper and plastic capacitors.

itor is only suitable if an exact capacitance is not required (such as in circuit coupling or bypass applications). Capacitance values of the high-$K$ types range from 100 pF to 0.1μF. A typical tolerance range is +100 to −20 percent of its stated value.

The disk type of construction used to build ceramic capacitors is shown in Fig. 11-7. A ceramic disk or plate is coated with metal on both faces. Leads are attached to the metal and the resultant capacitor is packaged in a coating of plastic or ceramic to protect it from moisture and other environmental conditions. The capacitance value is printed directly on the body, or a color code is used. Ceramic capacitors possess no required voltage polarity.

*Paper capacitors* are the most widely used type of capacitors. Their popularity is due to their low cost and the fact that they can be built in a broad range of

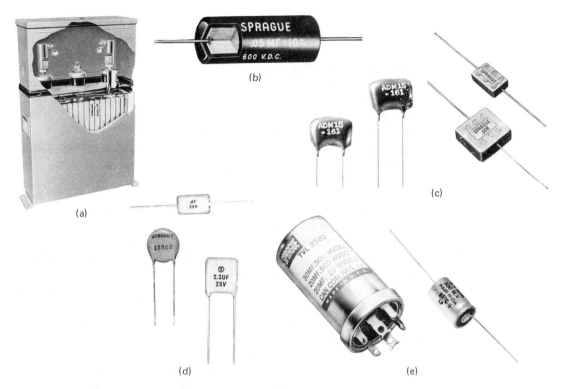

**Figure 11-6** Common types of capacitors: (a) oil-filled paper capacitor (Courtesy of Westinghouse Corp.); (b) cutaway view of plastic-film capacitor (Courtesy of Sprague Electric Co.); (c) mica units (molded and dipped types) (Courtesy of Sprague Electric Co. and Aerovox); (d) ceramic capacitors (Courtesy of Sprague Electric Co.); (e) electrolytic capacitors (Courtesy of Cornell-Dubiller Corp. and Sprague Electric Co.).

capacitance values (500 pF to 50 μF). Furthermore, they can be designed to withstand very high voltages. However, the leakage currents of paper capacitors are high, and their tolerances are relatively poor (±10 to 20 percent). These limitations restrict their use in some applications. If size permits, the capacitance value and voltage are usually printed on the capacitor body. For small units, a color code is used. When the color code is not used, a band (usually black) is often printed on the

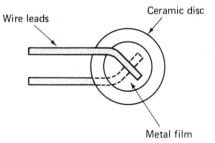

**Figure 11-7** Construction of ceramic disk capacitor.

tube nearest the lead that is connected to the outer metal sheet. This lead should always be connected to the circuit lead of lower potential.

Many paper capacitors are of a cylindrical shape because they are made by rolling a sandwich of metal and impregnated paper sheets into a tube. Axial leads are attached to each metal sheet, and the tube is encapsulated in waxed paper or plastic [Fig. 11-5(b)].

Various substances such as oil, wax, or plastic are used to soak the paper. If paper deposited with thin metal films is used rather than separate metal sheets, the volume per unit of capacitance can be reduced by 50 percent and the leakage current reduced by 90 percent. Unfortunately, this creates a resulting structure more prone to rupture by high-voltage transients.

*Plastic-film capacitors* are constructed in basically the same way as paper capacitors, except that a thin sheet of plastic (such as Mylar, Teflon, or polyethylene) is used as the dielectric. This dielectric improves the properties of the capacitor by minimizing leakage currents, even at temperatures of up to 150°–200°C. Their other characteristics are similar to those of paper units. However, the cost is higher for plastic units, so they are not usually used except when a paper capacitor cannot meet the design specifications. Commercial plastic-film capacitors are manufactured in ranges between 500 pF and 10 μF.

*Electrolytic capacitors* are usually made of aluminum or tantalum. The basic structure of the aluminum electrolytic capacitor consists of two aluminum foils, one of which is coated by an extremely thin oxide (Fig. 11-8). The oxide is grown on the metal by a process of applying a voltage to the capacitor; the process is called *forming*. The thickness of the oxide depends on the forming voltage. Between the foils is an electrolytic solution soaked into paper. This electrolyte is a conductor and serves as an extension of the nonoxidized metal foil. Since it is a fluid, the electrolyte can butt up directly against the oxide dielectric. The two oppositely charged plates are then effectively separated by only an extremely thin oxide film that possesses an extremely high dielectric constant.

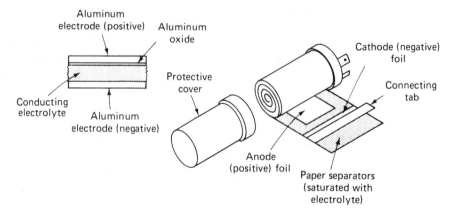

**Figure 11-8**   Construction of aluminum electrolytic capacitors.

Once the oxide is formed, the foils are rolled into a tube, and the piece of foil without the oxide is connected to the capacitor's exterior package. This lead serves as the negative connection to the capacitor. The other lead is marked by a plus on the capacitor body and *must* be connected to the positive terminal of the circuit in which it is used.

It needs to be strongly emphasized that the electrolytic capacitor should only be connected in a circuit with the proper polarities. If one connects the positive lead of the capacitor to the negative lead of a circuit, chemical action by the electrolyte will rupture the oxide dielectric and *destroy* the capacitor. (With reversed polarity, the oxide no longer acts like an insulator. As a result, a substantial leakage current can flow and disintegrate the oxide.[1]) In addition, as for other capacitors, the rated voltage must not be exceeded. For the largest capacitance values, the maximum voltage will be small because the oxide layer is so thin.

Electrolytic capacitors have the largest capacitance values per volume of element of any capacitor type. But they also possess large leakage current values. These properties limit their use to special applications. For example, in transistor circuits, large capacitances in a small volume are desirable, but leakage currents or exact capacitance values are not necessarily critical. Thus electrolytic capacitors are suitable for some of these circuits. Electrolytic capacitors are available in values that range from 1 to 500,000 $\mu$F. However, their corresponding leakage resistances are only about 1 M$\Omega$.

### Variable Capacitors

As with resistors, it is often necessary to be able to vary the value of a capacitor while it remains connected in a circuit. For example, it may be desired to tune the circuit of a radio receiver or an oscillator. *Variable capacitors* are available to fulfill such application requirements.

The *air-variable* capacitor is one such common variable type. It is constructed by mounting a set of metal plates (usually aluminum) on a shaft, and meshing them with a comparably shaped set of fixed metal plates (see Fig. 11-9). As the shaft is rotated, either more or less area (depending on the direction of rotation) between the adjacent and oppositely charged plates is created. This variation of the area changes the capacitance. (The larger the interleaved area, the greater the capacitance.) By designing the shapes of the plates appropriately, various capacitance versus shaft position curves can be achieved. For example, a linear or a square-root variation of capacitance can be obtained. Because the dielectric is air, the separation between the plates must be kept fairly large to ensure that they do not touch and discharge. (If dust or conducting debris falls in between capacitor plates, it can cause arcing and changes in capacitance values. Thus air-variable capacitors must

---

[1]If an oxide is grown on both metal foils of an electrolytic capacitor, the problem of proper polarity connections does not exist. However, the capacitance-to-volume ratio of the element is also reduced by half. These types of electrolytic capacitors are called *nonpolar electrolytic capacitors*, and they are not commonly used.

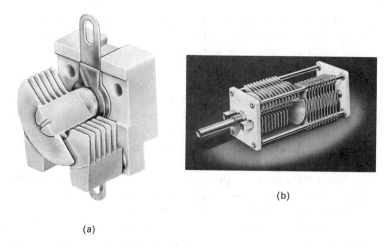

(b)

(a)

**Figure 11-9**   Air-variable capacitors. (Courtesy of E.F. Johnson Co. and James Millen Mfg. Co.)

be kept clean.) This limits capacitance values of air-variable capacitors to about 500 pF. Because air capacitors have such low leakage, they are used to build precision adjustable capacitors which act as standards for measuring small values of capacitance.

The trimmer capacitor is also a variable capacitor, but it is used primarily in circuits that need only one-time or infrequent tuning adjustments (as in the setting of the frequency range of a tuned amplifier). The trimmer capacitor is usually a mica capacitor that has a screw that clamps the metal-mica sheets. When the screw is tightened, the separation between the plates (and thus the capacitance) is adjusted. The overall range of trimmer capacitors is about 15 to 500 pF. Each individual unit has a small variable range (i.e., between 50 and 40 pF or 20 and 100 pF).

## COLOR CODING OF CAPACITORS

Capacitance values at one time were stamped directly on all the bodies of capacitors. However, the popularity of the color-coding scheme used on resistors led to the development of color-coding systems for capacitors. Today color coding is used on many capacitors that have small packages. It is seen most commonly on tubular paper, mica, and ceramic units. Figure 11-10 gives the color codes used for each of these various capacitor types.

## INDUCTORS AND INDUCTANCE

*Inductance* is that property of a device that reacts against a change in current through the device. *Inductors* are components designed for use in circuits to resist changes in current and thus serve important control functions.

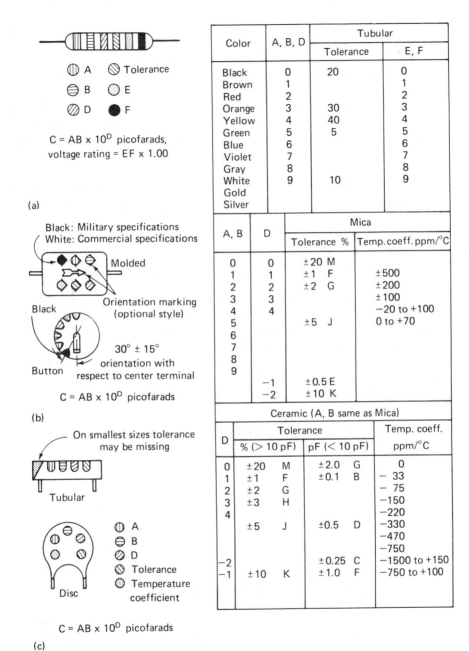

**Figure 11-10** Capacitor color codes: (a) tubular paper; (b) mica; (c) ceramic. [Adapted from B. D. Wedlock and J. K. Roberge, *Electronic Components and Measurements* (Englewood Cliffs, N.J.: Prentice Hall, 1969), p. 15.]

Inductor design is based on the principle that a varying magnetic field induces a voltage in any conductor in that field. Thus, a practical inductor may simply be a coil of wire as shown in Fig. 11-11(a). The current in each loop of the coil produces a magnetic field that passes through neighboring loops. If the current through the coil is constant, the magnetic field is constant and no action takes place. A change in the current, however, produces a change in the magnetic field. The energy absorbed or released from the changing magnetic field reacts against the change in current, and this is exhibited as an induced voltage (electromotive force, or emf), which is counter to the change in applied voltage. The inductor thus behaves as an impedance to ac current.

The counter emf is directly proportional to the rate of change of current through the coil ($V_L = L[di/dt]$). The proportionality constant is the inductance $L$, which has the unit of *henrys* (H).

In an ac circuit, as shown in Fig. 11-11(b), the inductor offers reactance to alternating current. The *inductive reactance* $X_L$ has the units of ohms and is given by

$$X_L = \omega L = 2\pi f L \qquad (11\text{-}7)$$

Note that inductive reactance, like capacitive reactance, is frequency dependent. For inductors, however, the reactance *increases* with increasing frequency. An inductor is said to be a short circuit to direct current since $X_L \to 0$ as $f \to 0$.

Actual inductors exhibit resistance as well as inductive reactance because of the resistivity possessed by the wires from which inductor coils are wound. The

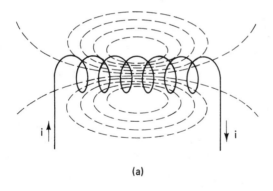

(a)

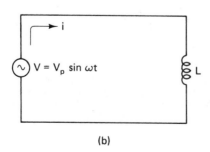

(b)

**Figure 11-11**   (a) Changing current in coil produces a changing magnetic field. The energy absorbed or released from the changing magnetic field reacts against the change in current. The coil thus exhibits *inductance* in ac circuits. (b) ac circuit in which an inductance is present.

resistance of an inductor, however, is rarely specified. Instead, a factor called the quality factor, $Q$, is used. It gives the ratio of the inductive reactance of the inductor to its resistance at a specific frequency, or

$$Q = \frac{\omega L(\omega)}{R} \qquad (11\text{-}8)$$

If $R$ were equal to zero, $Q$ would be infinite and such an inductor would exhibit ideal behavior. Thus the larger the value of $Q$, the more ideal is the inductor. (The best made inductors have values of $Q$ about 1000.) The $Q$ value of an inductor can be measured with inductance bridges or $Q$-meters.

## INDUCTOR STRUCTURES

Inductors are constructed by winding wire in various coil configurations. This restricts the magnetic field to the physical space around the inductor and creates the largest inductance effect per volume of element. (For the closely wound toroidal coil, the magnetic field is almost wholly confined to the space enclosed by the winding.)

The major factors that determine the magnitude of the inductance of a coil are

1. The number of coil turns
2. The type and shape of the core material
3. The diameter and spacing of the turns

The coils are usually wound around cores of ferromagnetic material because this makes magnetic flux density within the wound coil area vastly greater than if the core is air. The larger flux density allows an increase in the inductance of the structure. But this type of core also makes the inductor subject to eddy current and hysteresis losses.

For inductors shaped like those shown in Fig. 11-12, an approximate value of inductance can be calculated from Eq. (11-9) (as long as the current is not so large that the linear region of $B$ versus $H$ curve is exceeded).

$$L = \frac{\mu_r \mu_0 N^2 A}{l} \qquad (11\text{-}9)$$

In Eq. (11-9), $L$ is the inductance in henries, $\mu$ is the relative permeability of the core, $N$ is the number of turns,[2] $A$ is the area of one turn, and $l$ is the length of the coil.

---

[2]The term $N^2$ is used to calculate $L$ in most practical inductor structures because of the configuration of the coil winding. The coil is wound so that the magnetic flux of each loop is allowed to cut or intersect each of the other loops. In this manner, the changing current in each turn will cause an inductive effect in all the other turns.

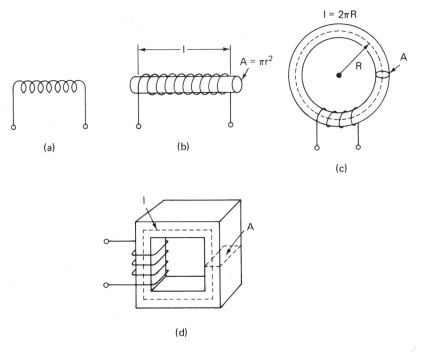

**Figure 11-12**   Various inductor configurations.

### Example 11-1

Given an inductor such as the one shown in Fig. 11-12(c), with $N = 100$ turns, $l = 6$ cm, and $r = 0.5$ cm. For cores of (a) air and (b) iron, find the inductance of each coil. Use $\mu_{r\,iron} = 1000$.)

**Solution.**
(a) Using $L = \mu_0 N^2 A /l$, where

$$l = 6 \text{ cm} = 0.06 \text{ m}$$

$$A = \pi r^2 = \pi(0.005 \text{ m})^2$$

$$= \pi(2.5 \times 10^{-5}) \approx 8 \times 10^{-5} \text{ m}^2$$

$$\mu_0 = 4\pi \times 10^{-7} = 12.6 \times 10^{-7} \text{ H/m}$$

then

$$L = \frac{(12.6 \times 10^{-7}) \times (10^4) \times (8 \times 10^{-5})}{6 \times 10^{-2}}$$

$$= 1.6 \times 10^{-5} \text{ H}$$

$$L_{\text{air}} = 16 \text{ μH}$$

(b) $L_{\text{iron}} = \mu_r L_{\text{air}} = 1000 \times 16 \text{ μH} = 16 \text{ mH}$

For *low-frequency* applications, large-valued inductors ($>5$ H) are commonly used. Laminated iron or silicon steel is used for the inductor core. For *high-*

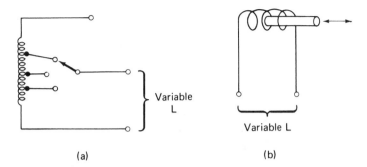

**Figure 11-13** Variable inductors: (a) tap switching; (b) moveable-core type.

*frequency* applications, much smaller inductors are encountered ($10^{-3}$ to $10^{-6}$ H) and powdered iron slugs and ferrites are selected as core materials.

Some applications call for variable rather than fixed inductors. Tuning circuits, phase shifting, and switching of bands in amplifiers sometimes require a variable inductance. Such inductors can be made in different ways. Figure 11-13 shows how inductance is varied in several commercial elements. The inductor shown in Fig. 11-13(a) can be varied by switching from one tap on the coil to another. In Fig. 11-13(b) a movable core is employed. As more of the core is inserted into the coil, the inductance increases. By appropriately varying the spacing of the coil windings, we can obtain a relatively linear variation of inductance with core insertion.

## TRANSFORMERS

Transformers are devices designed to transfer electric energy from one circuit to another. They achieve this transfer by utilizing a magnetic field that intersects both circuits. In addition to performing such energy transfers, transformers are also capable of delivering a different value of ac current or voltage at their output terminals than the value applied to their input terminals.

The transformer operates by using the electrical phenomenon of *mutual inductance*. Mutual inductance is the effect that occurs when the magnetic field of one element also influences other elements in its vicinity. The result of such magnetic coupling is that currents and voltages are induced in the nearby elements. Although mutual inductance may be an undesirable effect in some cases, the operation of a transformer depends on using this effect to its fullest extent.

The transformer consists of two coils (called the *primary* and the *secondary*) wound around a common core of magnetic material (Fig. 11-14). If a current flows in the primary winding, it sets up a magnetic field largely restricted to the magnetic core around which the primary is wound. If another winding (called the *secondary*) is also wound on the same core, the magnetic field will also link the secondary winding. If the current in the primary is steady (dc), it will not affect the secondary coil because the magnetic field will also be constant. In particular, no current will flow in the secondary coil.

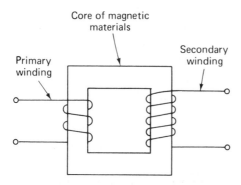

Core of magnetic
materials

Secondary
winding

Primary
winding

**Figure 11-14**   Diagram of a
transformer.

If the current in the primary is a changing (ac) rather than a steady current, the magnetic field in the core will also change. Since a changing magnetic field appears to a conductor as a moving magnetic field, the free charges in the conductor of the secondary coil experience a force. Since they are mobile, these free changes will move under the influence of the force, and a current will flow. In this manner, a changing current in the primary will cause a current to flow in the secondary of a transformer.

The current flow induced in the secondary also has a voltage associated with it. Faraday's law says that the magnitude of the voltage induced by the changing of magnetic flux in a coil of $N$ turns is given by

$$v = NK\frac{d\phi}{dt} = M\frac{di}{dt} \tag{11-10}$$

where $M$ is called the mutual inductance that exists between the coils.

In the ideal transformer, all the magnetic flux created by the primary coil also links the secondary. Then the voltage at the terminals of the secondary is dependent on the rate of change of current in the primary

$$V_2 = M\frac{di_1}{dt} \tag{11-11}$$

The ratio of the number of turns on the primary to the number on the secondary is an important quantity and is written as

$$\text{turns ratio} = \frac{N_p}{N_s} \tag{11-12}$$

In this equation, $N_p$ is the number of turns on the primary and $N_s$ is the number of turns on the secondary. It is the *turns ratio* that determines how much a transformer steps up or steps down a voltage.

In fact, the ratio of the voltage across the primary ($V_p$) to the voltage across the secondary ($V_s$) is equal to the turns ratio

$$\frac{V_p}{V_s} = \frac{N_p}{N_s} \tag{11-13}$$

Transformers can also connect two parts of a circuit without an *electrical* connection. This property is used in many instruments. Since only magnetic effects actually link the two parts, they remain *isolated electrically* from one another. This is a very useful property in such applications as restricting a high dc voltage level to one part of a system. When an isolation transformer is used, the ac component of a signal can be coupled between two parts of a circuit, while the dc level is kept from being transferred. (Remember that dc quantities are not passed by a transformer.)

## Types of Transformers

Transformers come in many shapes and sizes, depending on the specific application in which they are to be used. However, most transformers share the property of being wound on a ferromagnetic core and possessing one primary and one or more secondary windings. If there is a large transformer ratio, the high-current winding is usually wound of heavy-gauge wire to reduce resistance losses. The low-current winding is then wound of fine wire. For purposes of shielding, most transformers are enclosed in a suitable ferromagnetic or copper housing (depending on the frequency the transformer is designed to handle).

Two kinds of transformers are most commonly available as standard products: high-frequency pulse transformers and power transformers. High-frequency pulse transformers are for used coupling ac signals while isolating the dc levels of primary and secondary circuits.

Power transformers are used to step up and supply high voltages to the various parts of communication equipment or measuring instruments drawing 60-Hz power. Usually, a single power transformer can feed several different elements that require such voltages. Thus, power transformers can have one primary and several secondary windings.

Figure 11-15 shows a schematic of such a power transformer connected on its primary side to the 115-V, 60-Hz power line. There are several secondaries: two for 350 V, one for 10 V, and one for 60 V. The wire in these transformers is enamel- or plastic-insulated to prevent conduction or arcing between coil turns (particularly in the high-voltage secondary windings). Power transformers are also shielded to prevent the magnetic field that they generate from producing 60-Hz hum signals in nearby elements. Sometimes an electrostatic shield is also employed between the primary and secondary coils to prevent RF and noise voltages from entering through the power lines.

Autotransformers are an exception to the rule that no electrical connection exists between the primary and secondary windings of a transformer. In the autotransformer, the same coil serves as both the primary and secondary winding. The single winding has a *tap* that can be connected anywhere along the length of the winding. If the transformer is to be used to step down voltages, the entire length of the coil is used as the primary. The part between the tap and the bottom end acts as the secondary. If step-up action is required, the entire coil is used as the secondary. Special variable autotransformers known as *variacs* or *powerstats* are also available

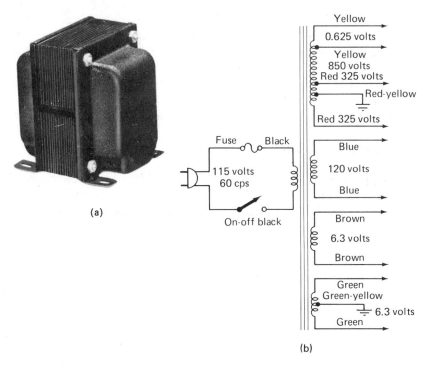

**Figure 11-15**   Power transformer used in an oscilloscope: (a) power transformer photograph; (b) schematic. (Courtesy of Micotran Co., Inc.)

for use where manual regulation may be required. Figure 11-16 is a cutaway photograph of a variac.

## IMPEDANCE

At dc, the resistance of a linear, two-terminal device is defined as the ratio of the voltage across it to the current through it, by Ohm's law ($R = V/I$). For sinusoidal ac, the ratio of the voltage to current is generally a complex number. The ac equivalent of Ohm's law in Cartesian form is

$$\frac{V}{I} = Z = R + jX \qquad (11\text{-}14)$$

where $Z$ is called the impedance of the device and $X$ is called the reactance. (We have already introduced $X_c$ and $X_L$, the capacitive reactance and inductive reactance, respectively.) The reciprocal of impedance is called the *admittance* of the device, and

$$Y = \frac{1}{Z} = G + jB \qquad (11\text{-}15)$$

**Figure 11-16** Cutaway of a Variac. (Courtesy of General Radio Corp.)

where $Y$ is the admittance, $G$ the conductance, and $B$ the susceptance of the device.

The impedance can be expressed in polar as well as Cartesian form and the relationships between them are

$$Z = R + jX = |Z|e^{j\theta} = |Z|(\cos \theta + j \sin \theta) \qquad (11\text{-}16)$$

where the impedance magnitude $|Z|$ is

$$|Z| = \sqrt{R^2 + X^2} \qquad (11\text{-}17)$$

and the impedance phase angle $\theta$ is

$$\theta = \tan^{-1}\frac{X}{R} \qquad (11\text{-}18)$$

Actual physical components are not ideal and actually possess all three impedance parameters (resistance, inductance, and capacitance). At any frequency an element possesses a complex impedance, which, however, may be simulated by two ideal circuit elements, an *equivalent* resistance and *equivalent* inductance or capacitance. If each element represents one term in the expression $R + jX$, they are assumed to be connected in series, since impedances of series elements are additive.

The impedance of an element or circuit may also be expressed as an admittance, where the two equivalent elements are $G$ and $B$, as in Eq. (11–15). In this case, the two elements are assumed to be connected in parallel.

Whether the two-equivalent-element impedance or admittance model is used to represent a device, the model with specific values for each element is accurate only at a single frequency.

## CAPACITANCE AND INDUCTANCE MEASUREMENTS

Although capacitance and inductance can be measured by indirect methods such as measuring $RC$ and $L/R$ time constants, these methods are generally hampered by a lack of accuracy. Consequently, most measurements of the capacitance and inductance of components (i.e., capacitors or inductors) are made by using bridge circuits, which can yield very accurate results.

The bridge circuit was introduced in Chapter 10 in the form of the Wheatstone bridge for measuring resistance. Recall that a condition of balance in the bridge circuit established a null reading in the bridge detector. At balance, the value of the unknown resistance could be computed from a knowledge of the other resistance values in the circuit.

The methods for measuring capacitance and inductance by the use of bridges are also based on the principle of establishing a null condition in a bridge circuit. The unknown value is calculated from the other elements of the circuit at balance.

### Bridge Circuits for Measuring Capacitance Values

The condition for balance of the Wheatstone bridge examined in Chapter 10 was found to be

$$R_x R_1 = R_2 R_3 \qquad (11\text{-}19)$$

where $R_x$ was the unknown value of resistance and $R_1$, $R_2$, and $R_3$ were the known values. If the resistances of the Wheatstone bridge are replaced by impedances of both a resistive and reactive nature (Fig. 11-17), and if an ac voltage is applied between points $A$ and $B$ of the circuit, the balance equation is, in general,

$$Z_x Z_1 = Z_2 Z_3 \qquad (11\text{-}20)$$

Since any impedance $Z$ can be expressed as a complex number

$$Z = R + jX \qquad (11\text{-}21)$$

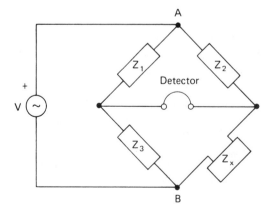

**Figure 11-17**  An ac impedance-bridge circuit.

then deriving an unknown $Z_x$ from the condition of balance is not quite as easy as finding $R_x$ in a purely resistive bridge. In fact, for a null condition to be achieved, it is necessary to specify two matching conditions: one for the resistive part of $Z_x$ and one for the reactive part

$$Z_x = R_x + jX_x = \frac{Z_2 Z_3}{Z_1} = \text{Re}\left(\frac{Z_2 Z_3}{Z_1}\right) + \text{Im}\left(\frac{Z_2 Z_3}{Z_1}\right) \qquad (11\text{-}22)$$

Two types of capacitance bridge circuits are commonly employed for measuring the capacitance values and the dissipation factors of a capacitor. If the dissipation factor of a capacitor is small $(0.001 < D < 0.1)$, the *series-capacitance comparison bridge* is used [Fig. 11-18(a)]. If $D$ is larger than this $(0.05 < D < 50)$, the *parallel-capacitance comparison bridge* is used [Fig. 11–18(b)].

For the *series-capacitance comparison bridge* the impedances of Eq. (11-22) are $Z_1 = R_1$, $Z_2 = R_2$, $Z_3 = R - j(1/\omega C_3)$ and $Z_s = R_s - j(1/\omega C_s)$. Substituting these impedances into Eq. (11-22) and separating the imaginary and real values, we find that when a null is established, $R_s$ and $C_s$ are calculated from the following relations:

$$R_s = \frac{R_2 R_3}{R_1} \qquad \text{and} \qquad C_s = C_3 \frac{R_1}{R_2} \qquad (11\text{-}23)$$

$C_s$ is the value of the capacitance we want to determine.

Note that if $R_2$ and $C_3$ are chosen as the fixed quantities, and $R_3$ and $R_1$ as the variable elements, we can achieve a null condition by varying $R_3$ and $R_1$ and then calculating $R_s$ and $C_s$. In this circuit the only variable elements are resistors, and no reactive elements have to be adjusted to achieve a condition of balance. This aids the user of the bridge in approaching the null value rapidly. In addition to $C_s$ the dissipation factor $(D)$ is also indicated by the bridge. Usually, a dial on the meter is calibrated so that the value of $D$ is automatically calculated and directly indicated.

The condition of balance of the parallel-capacitance comparison bridge is

$$R_p = \frac{R_1 R_3}{R_2} \qquad \text{and} \qquad C_p = \frac{C_1 R_2}{R_3} \qquad (11\text{-}24)$$

In this bridge, $R_2$ and $C_1$ are fixed and $R_3$ and $R_1$ are variable. Both $C_p$ and $D$ are read directly from the bridge settings at the condition of balance.

For measuring capacitors in circuits where the phase angle is very nearly 90°, the Schering bridge offers more accurate readings than either of the capacitance comparison circuits. This bridge is shown in Fig. 11-18(c).

It uses a parallel $RC$ network ($R_1$ and $C_1$) for $Z_1$, a resistance $R_2$ for $Z_2$, and a capacitance $C_3$ for $Z_3$. Therefore, for an unknown element in the place of $Z_4$, the conditions of balance for the Schering bridge are

$$R_x = \frac{R_2 C_1}{C_3} \qquad (11\text{-}25)$$

and

$$C_x = \frac{R_1 C_3}{R_2} \qquad (11\text{-}26)$$

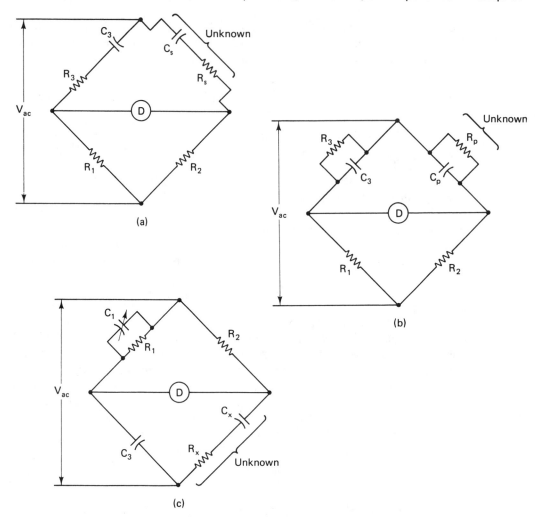

**Figure 11-18** Bridge circuits for measuring capacitance: (a) series comparison bridge; (b) parallel comparison bridge; (c) Schering bridge.

There are two types of bridge circuits most commonly used to determine inductance. The first, the *Maxwell bridge*, is best suited for measuring inductors that have a low $Q$ (i.e., $1 < Q < 10$). The second is the *Hay bridge*, and it determines $L$ most accurately when the $Q$ of an inductor is high (i.e., $10 < Q < 1000$).

The Maxwell bridge is shown in Fig. 11-19(a). It measures unknown inductance by comparison to a standard capacitance. The use of a capacitance as the standard element is advantageous because a capacitor is a compact element and is easy to shield.

A condition of balance in a Maxwell bridge exists when

$$L_x = R_1 R_3 C_2 \qquad (11\text{-}27)$$

$$R_x = \frac{R_1 R_3}{R_2} \qquad (11\text{-}28)$$

in these equations, $L_x$ is the value of the unknown inductance and $R_x$ is the corresponding resistance value of the element. We see from these equations that by choosing $C_2$ and $R_3$ as constants, we only have to vary $R_1$ and $R_2$ until a null is achieved. However, since $R_1$ appears in both Eq. (11-27) and Eq. (11-28), finding the two conditions of balance requires several adjustments. The common procedure for establishing a balance is to first determine $L_x$. Then the balance of $R_x$ is sought. In finding the condition of balance of $R_x$, we invariably disturb the condition of balance for $L_x$. We must then return to find $L_x$ again. After making several adjustments, we finally reach a balance for both conditions simultaneously.

The Hay bridge shown in Fig. 11-19(b) is best for measuring the inductance of high-$Q$ inductors. It also uses a standard capacitor as a comparison element for determining the unknown $L$. However, the capacitor is in a series with a resistor in one arm of the bridge, rather than in parallel. The equations of balance for the Hay bridge are

$$L_x = \frac{R_2 R_3 C_1}{1 + \omega^2 C_1 R_1^2} \qquad (11\text{-}29)$$

and

$$R_x = \frac{\omega^2 C_1^2 R_1 R_2 R_3}{1 + \omega^2 C_1^2 R_1^2} \qquad (11\text{-}30)$$

These equations appear more complex than the balance conditions for the other bridges considered. Also, the equations for $L_x$ and $R_x$ appear to be dependent

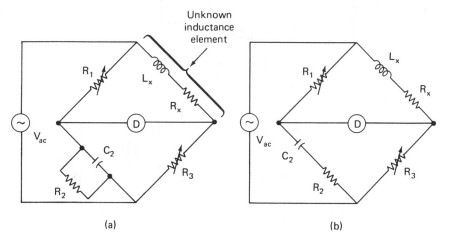

**Figure 11-19**  (a) Maxwell bridge; (b) Hay bridge.

on $\omega$. However, for the cases when $Q > 10$, the term with $\omega$ becomes less than $\frac{1}{100}$ and hence can be neglected. In such cases, the equation for $L_x$ becomes

$$L_x = R_2 R_3 C_1 \qquad (11\text{-}31)$$

For this reason, the Hay bridge is not as accurate if the value of $Q$ of the inductor being measured is less than 10.

### Commercial Capacitance and Inductance Bridges

Bridge-based instruments that measure only capacitance or inductance are commercially manufactured. Other instruments that can measure $R$, $L$, and $C$ are also available, and these are referred to as *universal bridges*.

The instruments that measure only capacitance usually contain two comparison bridge circuits: a series comparison type and a parallel capacitance type. Most such bridges contain an internal ac source at a single fixed frequency. A few others provide multiple internal frequencies. However, almost all contain provisions for connecting additional external ac sources so that the instrument can be used at other frequencies as well. Typical capacitance values that can be measured by using capacitance bridges range from 1 pF to 1000 $\mu$F to accuracies of 1 percent. Some extremely accurate bridges can measure capacitance values up to $\pm 0.1$ percent in accuracy. Examples of this type of bridge are shown in Fig. 11-20.

(a)

(b)

**Figure 11-20**   Commercial capacitance bridges: (a) GR-1680; (b) Boonton 75D. (Courtesy of General Radio Corp. and Boonton Electronics, Inc.)

Commercial inductance bridges are available that measure $L$ with values from nanohenries up to 1100 H. The basic accuracies of these instruments range from 0.1 to 1 percent. The Maxwell and Hay bridges are usually used as the bridge circuits that determine $L$.

The *universal impedance bridge* is an instrument designed to be able to measure $R$, $L$, and $C$ over a wide range of values. To be able to carry out these functions, the instrument has five or six built-in bridge circuits. These five bridge circuits are the Wheatstone bridge, the series and parallel capacitance bridges, the Maxwell bridge, and the Hay bridge. In addition to the internal, fixed-frequency excitation signal, they allow others, external ac sources to serve as excitation signals. Thus, operation at various frequencies can be accomplished. The specifications of three commercially available universal impedance bridges are given in Table 11-2. Figure 11-21 shows a photograph of two of the models listed in the table.

## Digital Capacitance Meters

Digital meters that determine capacitance by measuring the discharge time of capacitor through a known resistor are also commercially available. They can measure capacitance values from 1.0 pF to 2,000 µF at 0.1 percent accuracy.

In such instruments the unknown capacitor $C_x$ is connected to the meter and is charged to some known reference voltage. When this value is reached, the capacitor is allowed to discharge through a resistor of known value, $R_K$. The time required to discharge the capacitor to a second voltage value is measured by using a gate and counter circuit (similar to the one described in Chapters 5 and 8 in the sections that discussed digital voltmeters and frequency meters). The START and STOP signals that open and close the gate occur when the capacitor discharge begins (START) and when the second voltage reference value is reached (STOP). A clock of fixed frequency feeds pulses through the gate while it is open (in the interval between START and STOP signals), and the number of pulses is counted by a binary counter.

If the time constant of the $R_K C_x$ is large compared to the discharge time measured, the current through the known resistor during discharge, the discharge will remain sufficiently constant to allow the change in the charge on the capacitor to be accurately inferred. The capacitance according to Eq. (11-1) is found by dividing the change in the charge on the capacitor by the change in the voltage. Another technique used to measure capacitance is to apply a constant current to a capacitor for a specific time and then measure the voltage across the capacitor. The resultant voltage is inversely proportional to the value of the capacitor.

$$C = \frac{IT}{V} \quad \text{or} \quad V = \frac{IT}{C} \quad\quad (11\text{-}32)$$

This technique is used in many variable capacitance transducers that output a voltage that is proportional to capacitance. An example would be a liquid level transducer in which a change in the liquid level would result in a proportional change in output voltage.

**TABLE 11-2** SUMMARY OF THE SPECIFICATIONS OF THREE TYPICAL UNIVERSAL IMPEDANCE BRIDGES

| | R | C | L | Percent accuracy | D | Q | Internal source frequency |
|---|---|---|---|---|---|---|---|
| GR 1650-B | 1 mΩ–1.1 MΩ | 1 pF–1100 μF | 1 μH–1100 H | 1 | 0.001–50 | 0.02–1000 | dc, 1 kHz |
| John Fluke 710B | 10 mΩ–12 MΩ | 1 pF–1200 μF | 1 μH–1200 H | 1 | 0–1.05 | 0–1000 | dc, 1 kHz |
| Hewlett-Packard 4260-A | 10 mΩ–10 MΩ | 1 pF–1000 μF | 1 μH–1000 H | 1 | 0.001–50 | 0.002–1000 | dc, 1 kHz |

**Figure 11-21** Universal impedance bridge: HP 4274. (Courtesy of Hewlett-Packard Co.)

## Capacitance Measurement Using an ac Voltmeter

Although capacitance measurements made with a capacitance bridge are quite accurate, this instrument is not always available when an unknown capacitor needs to be measured. Therefore, a method that uses a high impedance ac voltmeter (e.g., with an input impedance of 10–11 MΩ) to determine the unknown capacitance is presented here. This type of measurement is limited to measuring capacitors with values of 0.001 μF or more, and it is good to only about 10 percent accuracy (because of uncertainties in the applied frequency and voltage and because of meter inaccuracies).

The unknown capacitor is connected in series with a resistor and the combination is put across an ac voltage source that is less than the rating of the capacitor (Fig. 11-22). Then the voltage across each element is measured separately. We first find $I$ (an rms value) from

$$I = \frac{V_R}{R} \tag{11-33}$$

where $R$ is the resistance of the resistor and $V_R$ is the rms voltage measured across the resistor. Then we find $C$ from

$$V_C = IX_C = \frac{I}{\omega C} = \frac{I}{2\pi f C} \tag{11-34}$$

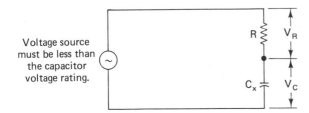

Voltage source must be less than the capacitor voltage rating.

CAUTION – This technique can only be used for nonpolarized capacitors.

**Figure 11-22** Circuit for measuring capacitance with an ac voltmeter.

or

$$C = \frac{I}{2\pi f V_C} \qquad (11\text{-}35)$$

where $V_C$ is the rms voltage measured across the capacitor.

### Measurement of Inductance with an ac Voltmeter

A quick measurement of inductance can be made by using an ac voltmeter and the 115-V ac voltage from the power line. The method is not nearly as accurate as measuring inductance with a bridge, but it does produce values that are adequate for many applications. Its chief advantage is that the measurement can be made with instruments that are commonly available in the electronics laboratory.

The method consists of connecting the unknown inductance in series with a variable resistor $R$ (as shown in Fig. 11-23). The total ac voltage is applied across the series connection, and the voltage across both elements is equalized. The equalization is carried out by first measuring the voltage across the inductor and then connecting the meter across $R_1$. The resistance is adjusted until the voltage across it is equal to the voltage measured across the inductor. When both voltages are equal, the impedances of the inductor and the resistor are also equal. Therefore, we can equate them and use the resulting relation to calculate the value of the inductance. The relation for finding $L$ is

$$L = \frac{\sqrt{R^2 - r^2}}{2\pi f} \qquad (11\text{-}36)$$

where $R$ is the value of the adjustable resistor at balance, $r$ is the measured value of the dc resistance of the coil, and $f$ is 60 Hz (or any other conveniently available frequency).

#### Example 11-2

An unknown inductor is connected in series with an adjustable resistor as shown in Fig. 11-23. The dc resistance of the inductor had previously been measured to be 100 Ω. When the series connection is hooked up across the 60-Hz power line, a condition of

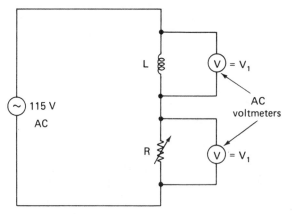

**Figure 11-23**   Measurement of an unknown inductance with ac voltmeters.

equal voltage across both elements exists when $R$ is set to 3200 $\Omega$. Find the value of the inductance.

**Solution.**   Using Eq. (11-35), we find that

$$L = \frac{\sqrt{R^2 - r^2}}{2\pi f} = \frac{(3200^2 - 100^2)^{1/2}}{6.28 \times 60} = \frac{3150}{377} = 8.4 \text{ H}$$

## COMPLEX IMPEDANCE MEASUREMENTS

When it is desired to measure the complex impedance of a component or circuit, a determination of the magnitude of the impedance $|Z|$ and the phase angle $\theta$ will allow the impedance to be expressed in polar form [see Eq. (11-16)]. Below about 100 MHz, it is usually sufficient to measure the voltage across and current through a device to determine $|Z|$. The phase difference can be obtained in a number of ways, including the use of an oscilloscope to display a Lissajous pattern. On the other hand, special instruments, called vector impedance meters, are especially designed to perform such measurements of $|Z|$ and $\theta$ over a wide range of frequencies quite easily.

### Vector Impedance Meters

The impedance of even simple electrical components becomes almost impossible to calculate theoretically at high RF frequencies (i.e., 10 MHz or more). Rather than attempting to make such calculations, a measurement of the quantity is usually made. The vector impedance meter is an instrument designed to measure the magnitude and phase of the impedance at the frequency of interest. The element or circuit of interest is merely connected to the inputs of the meter, the desired frequency is dialed in, and the magnitude and phase angle of the impedance are displayed by the two front-panel meters. Vector impedance meters are made to cover either AF or RF frequencies. AF meters have ranges of 5 Hz to 500 kHz and 1 $\Omega$ to 10 M$\Omega$. The ranges of RF vector impedance meters are 5 to 100 MHz and 1 $\Omega$ to 100 k$\Omega$.

Vector impedance meters perform their measurement functions by passing a known current of constant amplitude through a complex impedance. They then measure the magnitude of the voltage across the impedance to get $|Z|$. A phase meter internal to the instrument determines the phase angle between the voltage and current. Because of shunting impedance, however, it is difficult to drive a constant current through a high impedance. Therefore, on the high-impedance ranges of vector impedance meters, a constant voltage is applied, and the resultant current is proportional to the admittance magnitude.

### Q Measurements

The quality factor $Q$ of a coil was defined as the ratio of the reactance of the coil to its resistance or

$$Q = \frac{\omega L}{R} \tag{11-37}$$

where $\omega$ is the test frequency ($\omega = 2\pi f$) and $R$ is the effective series resistance of the coil. The effective resistance often differs markedly from the dc resistance because of such ac-associated resistance effects as eddy currents and the skin effect. Therefore, $R$ varies in a highly complex manner with frequency. For this reason $Q$ is rarely calculated by determining $R$ and $L$. Instead, $R$ is determined indirectly from a measurement of $Q$.

One way in which $Q$ can be measured is by using the commercial inductance bridge. But since the circuits of inductance bridges are rarely capable of producing an accurate measurement when $Q$ is high, special meters designed to yield accurate values are built for measuring $Q$.

The circuit of the common $Q$-meter is shown in Fig. 11-24. A variable-frequency voltage source (with a very low impedance) is used to drive a series connection of a capacitor and the inductor under test. Since the resistance of the inductor is in series with its inductance, this connection makes a series $RLC$ network. By varying the frequency of the voltage source, we can find the resonant frequency $\omega_0$ of the circuit. At resonance the capacitive reactance and the inductive reactance of the circuit cancel, leaving only the effective resistance $R_E$ of the inductor to limit the current. Thus, at resonance the current in the entire circuit (rms value) will be found from

$$|I| = \frac{|V_1|}{R_E} \tag{11-38}$$

Since the current must have the same magnitude in the entire circuit, the current in the capacitor, $I_C$, will also be equal to

$$|I| = |I_C| = \omega_0 C |V_2| \tag{11-39}$$

Therefore, in connection with the rapid increase of $I$ at resonance, a high-impedance voltmeter placed across the capacitor will detect a sharp increase in $|V_2|$ at resonance. We also know that at the resonant frequency $\omega_0$,

$$\omega_0^2 = \frac{1}{LC} \tag{11-40}$$

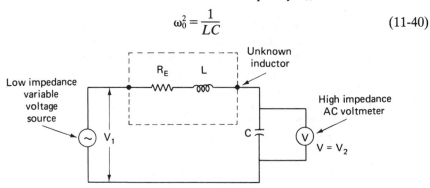

**Figure 11-24**    The circuit of a $Q$-meter.

Therefore, by combining Eqs. (11-38), (11-39), and (11-40), we see that

$$\left|\frac{V_2}{V_1}\right| = \frac{\omega_0 L}{R_E} = Q \tag{11-41}$$

In commercially available $Q$-meters, a continuously variable oscillator with a very low output impedance acts as the voltage source. The meter also contains a high-impedance VTVM and a variable capacitor. The scale of the VTVM is cali-brated to indicate $Q$ directly. A separate scale is often provided for low $Q$ (i.e., between 0 and 10). The frequency range of the oscillators is typically 50 kHz to 50 MHz, with special oscillators available for lower and higher frequencies. The accu-racy of such $Q$-meters varies from about 1 to 5 percent, depending on the magni-tude of $Q$ and the resonant frequency.

## PROBLEMS

1. Find the capacitance of a capacitor structure if a charge of 100 μC is deposited on the plates of the structure when 15 V are applied across it.

2. How much charge is deposited on the plates of a 300-pF capacitor if 250 V are applied across it?

3. A parallel-plate capacitor has a capacitance of 10 μF. What will be the effect on the capacitance of this component if:
   (a) The area of the plates is tripled?
   (b) The spacing between the plates is increased from 1 to 5 mm?
   (c) If the oil-soaked paper dielectric is replaced by a mica dielectric?

4. If the capacitance of an air capacitor is 600 pF and an unknown dielectric is placed between the plates, its capacitance increases to 3240 pF. Identify the unknown material used as the dielectric.

5. If capacitors are connected together in parallel with one another, the capacitance of the entire connection is given by the expression
$$C_t = C_1 + C_2 + C_3 + \cdots + C_n$$
If the capacitors are connected in series, the capacitance is given by the expression
$$\frac{1}{C_t} = \frac{1}{C_1} + \frac{1}{C_2} + \cdots + \frac{1}{C_n}$$
For capacitors whose individual values are 16 μF, 32μF, and 40μF, find the capacitance value of the three capacitors connected
   (a) In parallel
   (b) In series

6. Find the total capacitance value of the connection shown in Fig. P11-1.

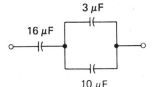

16 μF   3 μF   10 μF

**Figure P11-1**

7. Explain why the current flowing in an inductor does not rise immediately to its maximum value when a voltage is applied across the inductor.

8. **(a)** An induced voltage of 3 V is developed in a coil when the current in it is changing at a rate of 0.7 A/s. What is the inductance of the coil?
   **(b)** What would be the inductance of the above coil if the induced voltage were
      (1) 8?
      (2) 0.06 V?
      (3) 0.0001 V?

9. If a coil with 75 turns has an inductance of 250 μH, how many turns of the coil must be removed for it to have an inductance of 200 μH?

10. For the electromagnet shown in Fig. P11-2, find the flux density in the core.

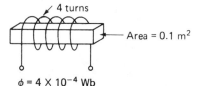

4 turns

Area = 0.1 m²

$\phi = 4 \times 10^{-4}$ Wb                    **Figure P11-2**

11. Find the inductance of the inductor shown in Fig. P11-3.

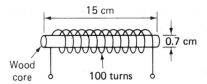

15 cm

0.7 cm

Wood core       100 turns               **Figure P11-3**

12. Find the inductance of the inductor shown in Fig. P11-4.

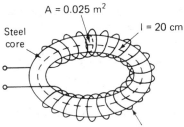

A = 0.025 m²

l = 20 cm

Steel core

500 turns        **Figure P11-4**

13. What is the effect of the following actions on the inductance of a coil?
    **(a)** Reducing the number of turns by one-third
    **(b)** Using an iron core rather than an air core

14. What is the reactance of a 50-mH coil at
    **(a)** 3000 Hz?
    **(b)** 3000 kHz?

15. At what frequency will a 325 μH coil have a reactance of 1500 Ω?

16. What happens to the value of inductive reactance if
    (a) The inductance is doubled?
    (b) The frequency is tripled?
    (c) The frequency is reduced to one-half?
17. Why are inductors and transformers often enclosed in metal containers?
18. What types of metals are usually used for such containers?
19. What is the significance of the $Q$ of an inductor?
20. If a 350-$\mu$H inductor has a resistance of 80 $\Omega$ at 4 MHz, what is the $Q$ of the coil?
21. For the transformer shown in Fig. P11-5, find the magnitude $V_2$ of the induced voltage.

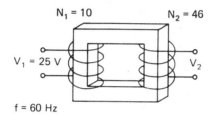

$N_1 = 10$        $N_2 = 46$

$V_1 = 25$ V        $V_2$

$f = 60$ Hz        **Figure P11-5**

22. What is the value of $L_1$ in Fig. P11-6 if $E_0 = O$?
23. What is the value of $R_4$ in Fig. P11-6?
24. Find the $Q$ of the $R_4/L_1$ network in Fig. P11-6.

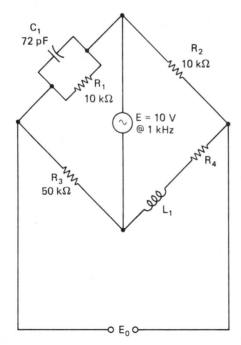

$C_1$
72 pF

$R_2$
10 k$\Omega$

$R_1$
10 k$\Omega$

$E = 10$ V
@ 1 kHz

$R_4$

$R_3$
50 k$\Omega$

$L_1$

$E_0$        **Figure P11-6**

# REFERENCES

1. Mullin, W.F., *ABC's of Capacitors*. Indianapolis, Ind.: Howard W. Sams, 1967.
2. Bukstein, E., *ABC's of Coils and Transformers*. Indianapolis, Ind.: Howard W. Sams, 1968.
3. Stout, M. B., *Basic Electrical Measurements*, 2nd ed. Chap. 9–10, 12–14. Englewood Cliffs, N.J.: Prentice Hall, 1960.
4. Oliver, B. M., and Cage, J. M., ed., *Electronic Measurements and Instrumentation*. Chap. 9. New York: McGraw-Hill, 1971.
5. Cooper, W. D., *Electronic Instruments and Measurement Techniques*. Chaps. 8 and 10. Englewood Cliffs, N.J.: Prentice Hall, 1978.
6. Kaufman, M. and Seidman, A. H., eds., *Handbook for Electronic Engineering Technicians*. Chap. 2–5. New York: McGraw-Hill, 1976.

# DC Signal Sources

All circuits and electrical equipment require sources of electric power. Some require ac power; others require dc power. Electronic amplifiers that use FETs, transistors, and operational amplifiers all require *dc power* to allow them to amplify electric signals (whether these signals are in dc or ac form). Since amplifiers are utilized in oscilloscopes, electronic voltmeters, radio transmitters and receivers, hearing aids, and a host of other electronic equipment, dc power sources are required for operating such instruments. However, the electric power commonly delivered from generating stations to the home and laboratory is 60-Hz ac power (see Chapter 9 for more details on how electric power is distributed). In order to get dc power for the electronic devices described, another source must be used or the available ac power must be converted to dc form. The *battery* is the most common alternative source of dc power. It utilizes energy from electrochemical reactions to supply this power. On the other hand, the device used to convert the 60-Hz line power to dc power at the desired current and voltage levels is called a *dc power supply*. The power supply can be constructed as a separate piece of laboratory equipment. It may also be designed to be an integral component of equipment that requires dc power but is operated from the ac power line.

## BATTERIES

Batteries were the first available sources of dc power and were therefore used almost exclusively to power early electronic circuits (such as amplifiers or radios). Alessandro Volta of Italy was honored by having the units of electric potential named after him because of his pioneering work that led to the development of the battery in the late 1700s. His first battery in 1800 consisted of alternate zinc and silver discs separated by moist cardboard and was known as the "Voltaic pile." It was not until 1859 that the modern automotive battery was developed by Gaston Planté.

Even though batteries are relatively simple devices, their chemical reactions are very complex. Scientists today still debate over the chemical reaction that occurs in the common zinc-carbon battery used in many flashlights. Since batteries must be periodically replaced or recharged, most electronic equipment that is not portable uses the dc power supply to generate any needed dc voltages. Advances in electronic components that require only pA to operate, however, for example, have allowed more equipment to be powered by batteries. Transmitters that send signals from transducers to signal processing equipment over twisted pairs of wires are being replaced by battery-operated VLSI chips that function as both transmitters and signal conditioners. The conditioned signal is transmitted over an optical fiber (as described in Chapter 17). Such electronic systems can be powered by advanced new batteries such as a lithium-sulfur dioxide battery. This type of battery will last 5 years before requiring servicing and can have a shelf life of 20 years.

Many batteries function because of the difference in the work function of the metals used as the battery electrodes, even though the energy is stored in the electrolyte. For example, the potential difference (difference in work function) of the zinc-carbon cell is about 1.5 volts and that of the lead storage battery is 2.2 volts. This difference in potential is called the cell voltage. A nickel-cadmium cell consists of a negative electrode of cadmium, a positive electrode of nickel hydroxide, and an aqueous solution of potassium hydroxide as the electrolyte. There are three classes of lithium cells based on their electrochemical system: (1) the solid cathode and a liquid electrolyte, (2) the solid cathode and a solid electrolyte, and (3) the soluble cathode and liquid electrolyte. Some of the properties of commercially available batteries are summarized in Table 12-1.

The reactions taking place at an electrode sometimes yield a gas as their product. If this gas forms bubbles that cling to the electrode surface, they can have a deleterious effect on the battery operation. Gas is a poor conductor, and if it surrounds the electrode, it will prevent further ions from wandering near to it. The chemical reactions and hence the current flow in the battery will slow down. Such gas formation is called *polarization* and is counteracted by substances called *depolarizers*. Depolarizers react with the gas to form a liquid, keeping the electrolyte in its well-conducting state.

Each configuration of two electrodes and electrolyte is called a *cell*. Batteries are made of one or more cells. (The term battery is derived from "battery of cells.") The circuit symbol for a single-cell battery is ⊣⊢ . Batteries consisting of two or more cells in series are shown as ⊣⏐⏐⊢. When two or more cells are connected in series, the voltage across the entire connection is equal to the sum of the voltages of each of the individual cells. When cells of equal voltage are connected in parallel, the voltage output of the assembly will still be the same as the voltage of one cell.

**TABLE 12-1** CHARACTERISTICS OF COMMERCIAL BATTERIES

| Principle chemical systems | General characteristics | Nominal cell voltage (v) | Practical energy density (WH/kg) |
|---|---|---|---|
| Zinc-air | Flat discharge characteristic | 1.4 | 310 |
| Lithium-sulfur dioxide | Flat discharge characteristic | 3.0 | 275 |
| Lithium-manganese dioxide | Flat discharge characteristic | 3.0 | 175 |
| Alkaline-manganese dioxide | Sloping discharge characteristic | 1.5 | 130 |
| Zinc-carbon ($Zn/MnO_2$) | Sloping discharge | 1.5 | 75 |
| Mercuric-oxide ($Zn/HgO$) | Flat discharge characteristic | 1.2 | 110 |
| Silver oxide ($Zn/Ag_2O$) | Flat discharge characteristic | 1.5 | 130 |
| Nickel-cadmium ($Cd/Ni(OH_2)$) | Good high rate | 1.2 | 35 |
| Lead-acid ($Pb/PbO_2$) | Low cost | 2.0 | 35 |

However, the current output of the assembly will be equal to the sum of the individual current outputs.

Cells are categorized in different groupings. One distinction is made between *wet* cells and *dry* cells. The difference between the two types is that the electrolyte in wet cells is completely liquid, whereas the electrolyte in dry cells is mixed with other substances to form a semiliquid or moist paste. The common zinc-carbon flashlight battery is a dry cell that has an electrolyte of ammonium chloride and zinc chloride. The lead storage (automobile) battery is a wet-cell battery that uses sulfuric acid ($H_2SO_4$) diluted in water as its electrolyte.

A further distinction is made between *primary* and *secondary* cells. In a *primary cell*, a decomposed electrode cannot be restored to use again by charging. The chemical reactions that led to its decomposition are irreversible. A *secondary cell* possesses the property that once the battery is discharged and its electrodes are partially decomposed into other compounds, it can be restored to its initial chemical state by means of another external energy source (i.e., a power supply battery charger). The external source recharges the battery by passing an electric current between the terminals of the battery in a direction opposite to the battery's normal current-flow direction.

The size of the battery determines the total amount of energy it can deliver (the more electrolyte and electrode material, the longer the energy-generating reactions can be sustained). The quantity termed the *capacity* of a battery indicates the number of *ampere-hours* (A-h) a battery can deliver before its terminal voltage drops below some designated level. A 1.2 A-h battery, however, will not deliver 1.2 amperes for 1 hour. Instead, a battery will deliver current for varying times depending on the discharge rate. Most manufacturers specify a discharge time that is based on a discharging rate of $C/20$ (where $C$ is a number equal to the A-h rating). Always ask the manufacturer how the A-h rating was determined. For example, in a sealed lead-acid battery rated 1.2 A-h, a discharge rate of $C/20$ (1.2 A/ 20 = 60 mA) will allow the battery to function for 20 hours [Fig 12-1(a)] if the temperature is a constant 20°C. A decrease in temperature will reduce markedly the A-h capacity of a battery as shown in Fig. 12-1(b).

Another way to measure a battery's capacity is by specifying the total amount of energy (in joules) it can provide before its voltage drops below a specified value. Using this measure, a D-type (flashlight) zinc-carbon cell can deliver $2 \times 10^4$ J under conditions of moderate discharge (i.e., no more than about 100 mA is drawn from the battery at any time). An automobile battery can provide about 100 times as much energy as a zinc–carbon flashlight battery.

Most batteries have a limited *shelf life*. This quantity is usually defined as the period of time required to reduce battery voltage to a specified percentage (usually 90 percent) of its original voltage value if the battery is not used.

A graph showing the discharge characteristics of various batteries is given in Fig. 12-2(a). The shelf life of a battery is very temperature dependent [Fig. 12-2(b)] as is the energy density per unit volume [Fig. 12-2(c)]. Battery selection must be based upon many factors in addition to the A-h rating. Some battery features that must be considered are physical size, charge/discharge characteristics, life cycle cost, duty cycle, shelf life, failure mode, custom design costs, environmental hazards, cost per watt-hour of service time, and maintenance costs. The decision on the type of battery to use should not be a last minute decision in a product design. Instead, it must be carefully considered at the beginning of the design task. A pacemaker is a prime example of an application in which all aspects of a battery have to be carefully considered (because of the critical nature of a battery failure).

### Battery Internal Resistance

The battery is used as a voltage source and is treated as such in analyzing a circuit. However, the electrical characteristics of actual batteries do not exactly equal those of an ideal voltage source. The main difference between them is that an ideal voltage source is defined as having zero resistance, whereas an actual battery always exhibits some resistance. This resistance exists because the electric current generated in a battery must first flow in the electrolyte and electrodes before reaching the external circuit. Since these component parts of a battery are not perfect conductors, the current is presented with an impedance before it gets to the external circuit. This impedance is called the *internal resistance* of the battery. In the circuit model used to represent actual batteries, the internal resistance is shown as a resistor in series with the ideal voltage source [Fig. 12-3(a)].

The value of the internal resistance varies with the condition and age of the battery. For example, the internal resistance of a freshly prepared zinc–carbon dry

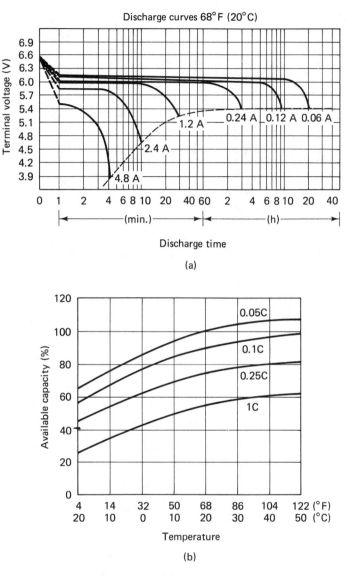

**Figure 12-1** Characteristics of a 6 V, 1.2 A-h sealed lead-acid battery, which is commonly called a gel cell: (a) discharge characteristics; (b) effects of temperature and discharge rate on available capacity.

cell is about 0.05 Ω, while the internal resistance of the same cell that has aged a year without use may be 100 Ω or more. When the internal resistance of a battery becomes too large, the voltage drop across this resistance becomes so great as to render the battery useless as a voltage source.

To measure the internal resistance of a battery experimentally, we can use a circuit such as the one shown in Fig. 12-3(b). The resistor $R$ should have a resistance

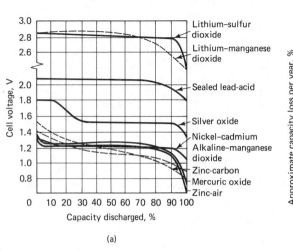

(a)

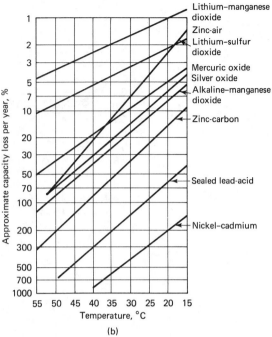

(b)

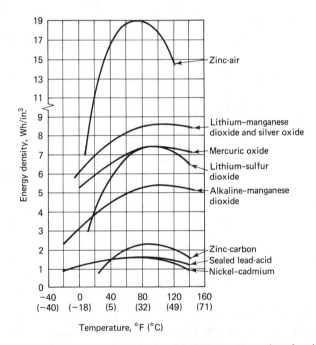

Figure 12-2  Characteristics of major batteries: (a) discharge properties of major batteries; (b) capacity loss as a function of temperature; (c) energy density as a function of temperature. [Sources for the graphs are from *Handbook of Batteries and Fuel Cells*, edited by David Linden (New York: McGraw-Hill, 1984) and Duracell, Inc.]

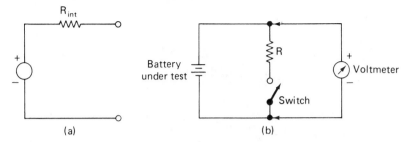

**Figure 12-3** (a) Equivalent-circuit model of a battery; (b) circuit used for measuring the internal resistance of a battery.

of about 10 Ω and a power rating of $\frac{1}{2}$ W (or more). For accurate results, a digital voltmeter should be used to perform the measurement (although any other volt-meter can also be used).

The voltage across the battery terminals with the switch open is measured first. This voltage is called the *open-circuit voltage* of the battery ($V_{oc}$). Then the switch is momentarily closed, and the voltage is measured again. The second volt-age value is referred to as the *loaded voltage*, $V_L$ (or terminal voltage). The internal resistance of the battery $R_{int}$ is then found from the expression

$$R_{int} = R\left(\frac{V_{oc}}{V_L} - 1\right) \tag{12-1}$$

### Example 12-1

If a battery connected as shown in Fig. 12-3(b) is found to have a $V_{oc}$ of 1.50 V and a $V_L$ of 1.41 V, what is the internal resistance of the battery (assuming that $R = 10$ Ω)?

**Solution.**  Using Eq. (12-1), we obtain

$$R_{int} = R\left(\frac{V_{oc}}{V_L} - 1\right) = 10\left(\frac{1.50}{1.41} - 1\right)$$

$$= 10(1.06 - 1)$$

$$= 0.6\ \Omega$$

### Example 12-2

If the same battery has aged and now has an internal resistance of 100 Ω, what will be the value of $V_L$ if we use the measuring circuit of Fig. 12-4 (with $R = 10$ Ω)?

**Solution.**  The open-circuit voltage of the battery $V_{oc}$ remains at 1.50 V. Then using Eq. (12-1) and solving for $V_L$, we obtain

$$R_{ins} = R\left(\frac{V_{oc}}{V_L} - 1\right)$$

$$V_L = \frac{V_{oc}}{1 + R_{ins}/R}$$

$$= \frac{1.5}{1 + 100/10} = \frac{1.5}{11}$$

$$= 0.14\ V$$

Positive terminal
binding post

Cover
plastic coated
insulation board

Negative terminal
binding post

Inner seal asphalt

Expansion chambers

Seal support washer

Carbon
electrode

Paste coated
pulpboard separator

Depolarizing
mix

Zinc can
outside surface
asphalt coated

Chipboard
jacket

(a)

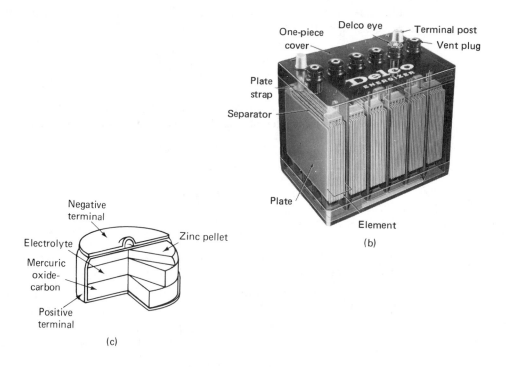

One-piece
cover

Delco eye

Terminal post

Vent plug

Plate
strap

Separator

Plate

Element

(b)

Negative
terminal

Zinc pellet

Electrolyte

Mercuric
oxide-
carbon

Positive
terminal

(c)

**Figure 12-4** Construction of various battery types: (a) zinc-carbon dry cell (Courtesy of Union Carbide Corp.); (b) lead storage automobile battery (Courtesy of Delco-Remy, Division of General Motors); (c) mercury battery.

### Common Battery Types

The following batteries are the types of batteries that are most likely to be encountered in measurement applications: (1) zinc–carbon, (2) zinc–air, (3) lithium, (4) silver–oxide, (5) mercury, (6) nickel–cadmium, (7) manganese–alkaline, and (8) lead–acid. Let us briefly introduce their most important characteristics and indicate where they are most likely to be used.

The *zinc–carbon battery* is a primary dry cell developed by Georges Le Clanche in 1868. When freshly prepared, it has a terminal voltage of about 1.55 V. Dry-cell batteries of the zinc-carbon type are manufactured with voltage values of 1.5, 3, 6, 7.5, 22.5, 25, 67, and 90 V. Such batteries are very commonly used because of their low cost and general applicability. However, they do suffer from the following disadvantages. First, their capacity is very dependent on the rate of discharge. Second, they are hampered by temperature limitations (at temperatures below freezing they are quite ineffective). Third, since the zinc electrode forms part of the outer wall of the cell, its dissolution weakens the cell structure. Sometimes this will cause the cell to rupture and spill its corrosive contents. Therefore, equipment should never be stored with such dry cells installed. Figure 12-4(a) shows a zinc-carbon cell.

A *zinc–air battery* is unlike any other battery in that oxygen from the air is the active cathode material. This primary battery has the highest energy density of commercial batteries and approximately 10 times that of a lead–acid battery. In addition, it can be stored for up to 10 years before being activated. Because of its small size for a given capacity it is becoming widely used in portable medical equipment such as hearing aides, heart monitors, and pagers. Once the battery is activated it must be used within one year in order to obtain its rated A-h capacity. Their current output can be as high as 7 A, but additional output can be obtained by connecting such batteries in series, parallel, or series-parallel.

The term *lithium battery* applies to many electrochemical systems based upon a lithium anode. The positive cathode can be manufactured from a dozen or so different materials. Most lithium batteries will operate from $-40°$ to $71°C$. This battery can be either a primary or secondary type. The primary types can have a shelf life as long as 20 years. The lithium/sulfur dioxide is the preferred battery for many military applications. The energy density of this battery is approximately 4.7 times that of a lead–acid battery. It is replacing most of the exotic military cells such as the zinc–silver oxide and lithium/thionyl chloride cells. The acceptance of this battery type for many consumer products, such as cameras, hearing aides, and industrial process control, has increased its production and lowered its cost. As a secondary battery, the lithium-molybdenum disulfide system has an energy density of 100 watt-hours/kilogram (WH/kg), which is four times that of the nickel–cadmium battery. Because of its high reliability and long shelf life, the lithium–iodide battery is used to power most implanted cardiac pacemakers. It has an expected life of 10 years when discharged at 7 μA.

The *silver–oxide battery* has an extremely flat discharge voltage characteristic, long shelf life, and wide temperature range. It has a self-discharge rate of 6 percent per year compared to 1 percent for lithium batteries. An advantage of this battery over the mercury battery is its higher cell voltage of 1.5 volts. Higher material costs, however, can offset the cell voltage advantage of this primary battery.

The *mercury battery* shown in Fig. 12-4(c) is a primary cell that has the advantage of keeping a relatively constant terminal voltage (1.35 to 1.4 V) throughout its useful life. This voltage stability makes the mercury battery especially attractive for use in those devices which require a specific voltage to operate properly. Occasionally, mercury batteries are also used as voltage references in measurement circuits. When a mercury battery is exhausted, its internal resistance rises drastically (within 2 to 3 h of the end of its 150-h life). The capacity in ampere-hours exceeds that of a comparably sized zinc–carbon cell by about a factor of 3. The chief disadvantages of mercury batteries are their greatly decreased capacity at temperatures below 40°F. These batteries are based on systems that have either cadmium or zinc for an anode. The Cd battery (also called mercad) has a cell voltage of only 0.85, and zinc is 1.2. Both types are primary and have good capacity per unit volume. These batteries are used in cameras, instruments, and watches.

*Nickel–cadmium batteries* are secondary cells that are easily recharged with a simple battery charger. They are very well suited for use as power sources in battery-operated appliances. Under normal use, these batteries can be recharged several hundred times, and they are also unharmed by large loads, overcharging, being left discharged, and low temperatures. Their terminal voltage is about 1.25 V, and they can be made in many different sizes and capacities. Although the initial cost of the nickel–cadmium batteries is high, their cost per use drops to a very low sum as they are recharged many times. In addition to the nickel-cadmium battery there is a nickle-hydride battery which has approximately 175 percent more capacity than a nickel-cadmium battery for a given size. For example, the AA size nickel-hydride battery has a capacity of 1070 mA hr which amounts to an energy density 175 Whr/L (54 Whr/kg). A main advantage of the nickel-hydride battery is that it does not contain any toxic materials or have the charge/discharge memory that is associated with the nickel-cadmium battery. Both batteries are usually charged using a constant current source that is rated C/10 (C is the Ahr capacity of the battery). Faster charges can be made up to C, but a temperature sensor (or thermal cutout) must be used to stop the charging before the battery temperature reaches 45°C. A constant current trickle charge is usually done at C/20, and a float charge is usually done at C/30. Both types of batteries can be charged and discharged from 400 to 500 cycles.

The *manganese–alkaline battery* can be made as either a primary or secondary cell, but the primary type is more common. This battery is very similar in construction and terminal voltage (1.5 V) to the carbon-zinc cell, but has a lower internal resistance, longer shelf life, and larger energy capacity than that of the carbon-zinc cell. In addition, alkaline cells can operate at temperatures as low as −40°F. Although their initial cost is greater, the larger capacities of alkaline cells make their cost per ampere-hour lower than that of carbon–zinc cells.

The *lead storage battery*, which is commonly used in automobiles is made of secondary wet cells. An example of such a battery is shown in Fig. 12-4(b). Each lead cell has a terminal voltage of about 2.2 V when fully charged. The 6- and 12-V requirements of automobile electrical systems are met by placing three or six cells in series. Because the lead battery is made of secondary cells, it can be recharged to its initial state hundreds of times after partial discharge. However, lead batteries are quite bulky and require some maintenance in order to give proper service.

There is also the sealed lead–acid battery which is widely used as a secondary battery. There are two general categories, the gel cell (in which the electrolyte is a polymeric gel) and the starved electrolyte (which has little or no free fluid electrolyte). Both types have a wider operating temperature range than the NiCd battery. The sealed lead–acid battery is a very cost effective alternative when weight and size are not major factors in the selection of a battery type.

A battery charger is a device capable of supplying direct current to the battery for the purpose of returning the battery to the chemical state of high energy. There are three basic techniques commonly used by such devices to charge batteries: constant-voltage, constant current, and tapered current. Combinations of these techniques are also used and are known as two-step systems. Characteristic curves for these systems are shown in Fig. 12-5.

Most batteries are damaged if repeatedly overcharged and also have a short-

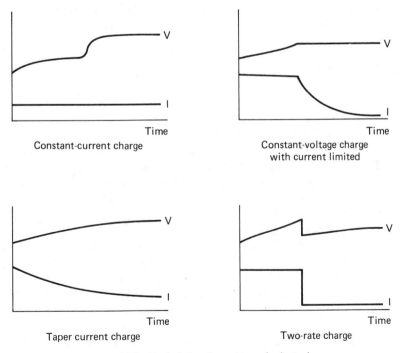

**Figure 12-5**   Typical charging patterns for batteries.

ened life if they are repeatedly deeply discharged (i.e., discharged until virtually no energy remains stored). Secondary batteries (batteries in which the electrochemical process is reversible) require a charger that is tailored to the battery type and to the end use. A *float charger* is used, for example, for instrumentation, emergency backup batteries in hospitals, emergency lighting systems in buildings, telephone systems, and electrical substations. A float charger would not be appropriate for a portable data logger because it would take too long to recharge the batteries.

Batteries are temperature sensitive, and, therefore, a constant voltage charger must compensate for changes in terminal voltage as the battery temperature changes. That means that the constant voltage charger output must also be altered as the temperature is changed. A battery can remain connected indefinitely to such a properly compensated voltage charger without being damaged. On the other hand, the output of a constant current charger is not altered as the temperature changes. Therefore, if a battery is connected too long to a constant current charger, it can be damaged by overcharging. As a result, the charging time when a constant-current charger is used is dependent on the discharge state of the battery.

Standby or backup battery charging is usually performed by either trickle charging or float charging. Batteries connected to trickle chargers are not connected to a load until the primary source fails, whereas batteries being float charged are also connected to the load. Trickle chargers supply just enough current to the battery to overcome internal battery losses. *Float charge systems* provide sufficient current to a battery so that the battery will not discharge more than 10 percent (except when the source that drives the charger itself fails).

## DC POWER SUPPLIES

To convert the readily available 115-V, 60-Hz electric power to the dc form required for the operation of electronic devices, a dc *power supply* is used. The name may be misleading because the power supply does not actually generate power; it only converts ac power to an approximate dc voltage of current. Power supplies can be designed to provide either *constant voltage* (CV) or *constant current* (CC). Some general-purpose laboratory units, however, can be operated in either the constant-voltage or constant-current mode (but only one mode at a time, not both). Such supplies are referred to as constant-voltage/constant-current (CV/CC) models. When power supplies are operated to deliver a constant voltage (which is by far the most commonly used mode), their function is to exhibit the characteristics of a *constant-voltage source*. That is, once the voltage setting of the CV supply is chosen, the selected output voltage should remain constant with time and constant in response to variations in the demands for output current. Figure 12-6(a) shows the current versus voltage characteristic which an ideal CV power supply should exhibit. We see that the output voltage $V_{out}$ remains constant for all the values of output current that must be provided by the supply ($I_{out}$). Since Figure 12-6(a) indicates that the ratio $\Delta V_{out}/\Delta I_{out}$ is zero, an ideal CV power supply would have an

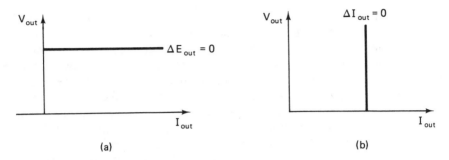

**Figure 12-6** (a) Ideal constant-voltage (CV) power supply output characteristic;
(b) ideal constant-current (CC) power supply output characteristic.

output resistance of zero ohms. Typical CV power supplies are commercially available with output resistances of less than $0.01\ \Omega$ at 60 Hz.

For the occasional application that requires a source of constant current
(rather than a constant voltage), a CC power supply must be used. Just as a CV
supply alters its output current in response to load resistance changes, the CC
supply changes its output voltage whenever the load resistance changes. The $I$–$V$
characteristic exhibited by an ideal CC power supply is given in Figure 12-6(b). It
shows that a constant current, $I_{out}$, is supplied by the CC source regardless of the
voltage output required of it. This means that the ratio $\Delta V_{out}/\Delta I_{out}$ is infinitely large
and that the output resistance of a CC supply is infinite. Practical CC supplies have
output resistances greater than 1 M$\Omega$. To understand how both CV and CC supplies
are capable of converting ac power to dc quantities, let us examine Fig. 12-7.

From this figure we see that the incoming 115-V ac voltage is first stepped up
or down by a transformer. The initial voltage change enables the power supply to

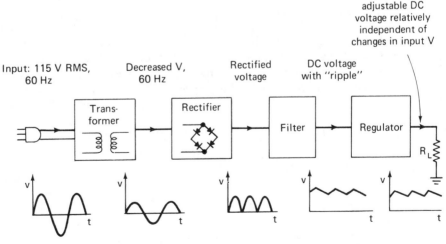

**Figure 12-7** Block diagram of dc-voltage power supply operation.

deliver voltages which may be greatly different from the 115-V power-line voltage. Next, the transformed voltage (still at 60 Hz) is fed to a rectifier (discussed in Chapter 4). The rectifier converts the ac voltage to a rectified dc voltage. (The output of the rectifier shown in Fig. 12-7 is a full-rectified waveform, although half-wave and peak rectifiers are also available.) The rectified waveform is then fed a filter that smooths out the variation or *ripple* in the rectified waveform. Filters can eliminate most, but not all, of the ripple in a signal. Therefore, the output of a filter is a dc quantity with a small residual ac component superimposed on it (Fig. 12-8). However, if sufficient filtering has been used, the magnitude of the remaining ac component is small enough that it will not impair the operation of the device that is drawing power from the supply. The magnitude of the remaining ripple voltage in the output of a power supply is expressed in terms of its rms value. Typical commercial supplies have ripple voltages in their outputs which range from less than 0.1 mV to about 10 mV. Power supply ripple can be displayed and measured using an oscilloscope. As discussed in Chapter 6, with the Input Coupling selector of the oscilloscope set to its ac position, the relatively small amplitude of the ripple signal is easily observed even though the ripple is actually superimposed on a much larger dc level.

The filtered waveform could now be employed as the dc output of the power supply, but it becomes more useful with the addition of one more modification. As is, the value of the voltage output is fixed by the input voltage level and the transformer construction. There is no way to adjust the magnitude of the output level. Furthermore, any variations in the power-line voltage or load current could vary the dc output level. For instruments that depend on a constant voltage level, this variation (possibly greater than the variation owing to the remaining ripple) might be too large to allow proper functioning. For such reasons, a regulation device must be installed to allow adjustments of the output level and help keep the output value constant once the level is chosen. With the help of the *regulator*, the supply output is a dc voltage that is adjustable over the designed range.

The ability of a power supply to keep a voltage constant once an output level is chosen is referred to as its *regulation* ability. This quantity is expressed by the percentage of change the output will undergo for each 1 percent of change of the input voltage. Another way of expressing regulation is to specify the maximum percentage change in output voltage for a particular change in input voltage (usually from

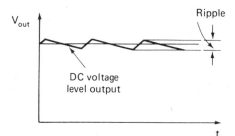

Figure 12-8   A dc waveform with ripple.

105 V to 125 V). Typical values of regulation ability in various power supplies allow about 0.05 percent change in the output for a variation of 20 V (105 V to 125 V) at the input.

Because instrumentation circuits generally require very small currents, power supplies can be placed inside an instrument. Devices that have made this economical are integrated circuit (IC) voltage regulators such as the LM317. There are over a hundred types of IC regulators from which to choose. Most of them are very simple to use and well documented by the manufacturers. A typical circuit for the LM317 is shown in Fig. 12-9. Note that the input voltage can be as high as 37 volts. Because of the losses in the regulator, the input voltage ($V_{in}$) must be at least 2 volts higher than the output voltage ($V_{out}$). The voltage between $V_{out}$ and $V_{in}$ is a precision 1.25 V that requires at least 5 mA to operate properly. This current is set by $R1$. Since the ADJ terminal sources very little current (100μA), all of the current passing through $R1$ must also pass through $R2$. By Kirchhoff's voltage law, the output voltage is

$$V_{out} = iR1 + iR2 \qquad V_{out} - V_{ADJ} = 1.25 \text{ volts}$$

$$V_{out} = \frac{V_{out} - V_{ADJ}}{R1}(R1) + \frac{V_{out} - V_{ADJ}}{R1}(R2) = 1.25\left(1 + \frac{R2}{R1}\right) \qquad (12\text{-}2)$$

By replacing $R2$ with a variable resistor, the output voltage can be made adjustable.

It is very important that the power loss in the regulator be maintained below the manufacturer's recommendation. The power lost in the regulator is

$$P_{lost} = (V_{in} - V_{out})I_{load} \leq \text{manufacturer's rating} \qquad (12\text{-}3)$$

Capacitor $C2$ can range from 1 to 1000 μf, depending on the load requirements. Capacitor $C1$ is sized to limit the input ripple to the regulator. Remember that the input cannot be less than 3 volts above output voltage. $C1$ can be very small (0.1μf) if the input has very little ripple. This would be the case if $V_{out}$ is 5 volts and $V_{in}$ is from a 12 V dc battery. The ripple rejection ratio for a typical three-terminal regu-

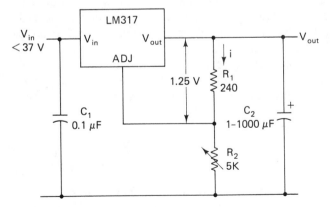

**Figure 12-9** Typical three-terminal regulator circuit.

lator is 65 dB. That is, the output ripple would be 1778 times smaller than the input ripple as defined by

$$\text{output ripple} = \text{input ripple}/\text{antilog}[0.05(\text{ripple rejection ratio}_{dB})] \quad (12\text{-}4)$$

### Safe Operation and Current Limiting of Power Supplies

When using a constant-voltage power supply, the current requirement of the load can vary. Under the most extreme circumstances of a short circuit in the load, a high current delivered from the power supply could severely damage either the load circuits or the power supply itself. To guard against such an occurrence, *electronic current limiters* are included in the regulator circuits. They limit the maximum current flowing in the output, regardless of the output voltage required of the supply. By doing this, they ensure that the power supply is being operated safely (in the sense that it is kept from burning out). Some power supply models feature a variable-current-limiting control that allows the operator to set the limiting current to any value within range.

### Additional Control Features of Power Supplies

There are some applications that require a power supply to provide power to a point that is at a fairly distant location from the supply itself. Thus wires must carry the power from the supply to that point. When high currents are required by such a distant load, the voltage at the load may differ from the voltage at the output terminals of the power supply. This change occurs because the wires connecting the supply output to the load may drop a measureable percentage of the supply voltage before it gets to the load. Therefore, the load voltage will be dependent on the connections to the load as well as the output of the supply. The regulation at the load may not be as good as at the power supply output. To overcome this problem, a voltage reading can be taken at the load and this value can be fed back to the supply by an additional pair of wires usually called "remote sense." Similar to the way that the voltage output of the supply is regulated internally, this voltage monitor at the load can control the output of the supply. It will maintain the desired voltage at the load regardless of the drop that occurs in the connections. This feature is available in some supplies and is called *remote sensing*.

Another feature, *remote programing*, provides the capability of controlling the supply output by means of a remotely varied voltage or resistance. This feature is also used to allow rapid switching between preset output values.

### dc Power Supply Specifications

**1.** *dc output.* Describes the range of dc voltages or currents available from a particular power supply.

**2.** *ac input.* Describes the characteristics of the ac voltage required to drive the power supply. Usually, the required ac input is 115 V ac ±10 percent, 50–63 Hz. However, some supplies are built to also operate over other voltages and frequencies.

**3.** *Load regulation.* The change in the dc output voltage resulting from a change in the load resistance from zero (short circuit) to the value that results in the supply's maximum rated output voltage. Typical values are 0.001 to 1 percent.

**4.** *Line regulation.* The change in the dc output voltage of the supply resulting from a change in the input line voltage from its lowest to its highest values (usually the maximum change allowed is from 105 V to 125 V). Typically, 0.001 to 1 percent.

**5.** *Ripple and noise.* Describes the rms value of the ac component that remains unfiltered and superimposed on the dc output. Typically, 50 μV to 1 mV.

**6.** *Output impedance.* For a supply designed as a constant-voltage source the output impedance should be very small ($\rightarrow 0$). For a constant-current supply, the output impedance should be very large ($\rightarrow \infty$).

**7.** *Temperature rating.* The temperature range over the supply can operate and remain within its capabilities.

### How to Use a Power Supply

Figure 12-10 shows a typical instrumentation power supply that is capable of supplying four isolated outputs. Each of these outputs can be controlled from either the front panel or by a signal from an IEEE-488 bus. The dc power supply operates from a three-terminal 120 vac outlet. The *LINE* switch turns ON power, permitting programing of the supply. The supply can be used as either a constant voltage supply (with overcurrent protection) or as a constant current supply (with overvoltage protection). Remote sensing is also available. In remote sensing, the power

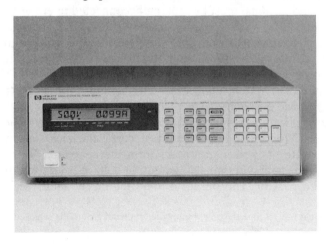

**Figure 12-10** Typical commercial power supply with IEEE-488 bus. (Courtesy of Hewlett-Packard Co.)

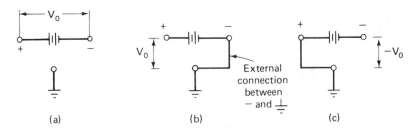

**Figure 12-11**  How to make connections to power supply output: (a) ungrounded $V_o$ between + and − terminals; (b) with extra ground wire connected as shown, a positive voltage $V_o$ relative to ground exists between (+) and (−) terminals; (c) with extra ground wire connected as shown, a negative voltage $V_o$ relative to ground exists between (+) and (−) terminals.

supply senses the voltage at the load and adjusts the output of the supply to compensate for any voltage drop on the wires supplying the load. To configure the power supply as a constant voltage source, select and output (1–4) by pressing *OUTPUT SELECT* and then entering a digit (1–4). When selecting the constant voltage mode, VSET is pressed, and then the desired voltage is entered by using the *ENTRY* keys. The maximum allowable current for output 4 is set by pressing OCP (overcurrent protection) and then entering the maximum current by using the *ENTRY* keys (Fig. 12-10). For example, the display shows 50.0V on output 4 and a maximum current of 0.099 A (Fig. 12-10). The equivalent circuit of the supply as seen from the plus and minus terminals is shown in Fig. 12-11 (a).

The equivalent circuit shows that the ground terminal is connected to the third wire of the power-line cord through the chassis, but it is not connected internally to either the plus or minus terminal. Thus the voltage between the plus and minus terminals is floating.[1] If a load is connected between the plus and ⏚ terminals alone, the output voltage of this connection will be zero. To supply a voltage output relative to ground, the ground terminal must be separately connected to either the plus or minus terminal, depending on the polarity desired [see Fig. 12-11(b) and (c)]. On the other hand, if it is necessary to float the power supply voltage off ground, only connections to the plus and minus terminals need be made [Fig. 12-11(a)]. Floating the power supply output is necessary for example, when the supply is used in a measurement system that employs single-point grounding to reduce ground loop interference problems. (See Chapter 16, especially Fig. 16-9, for an example of such an application.) Most regulated low-voltage power supplies will operate up to at least 300 V off ground. One limiting factor is the mica washers, which, on most units, separate the power transistors from the heat sink. With a proper connection made to the supply terminals, the output voltage is adjusted by means of the *Voltage Adjust* control. This output is indicated on the meter scale.

---

[1]Note that this ability to float (i.e., isolate the power supply output from ground) is achieved by the use of an isolation transformer internal to the power supply circuitry.

The range of the meter scale is controlled by the *Meter Range* switch. The *Short Circuit Current* control indicates the maximum output current of the supply.

The following guidelines should also be followed to ensure proper operation of the power supply:

**1.** Make sure that connections to the supply terminals are made with connectors (i.e., spade lugs) that can be slipped onto the power supply binding-post nut and then tightly clamped down. Casual, clip-lead connections will lead to supply performance degradation because the contact resistance of such poor connections can easily be 100 times the output resistance of the power supply. Thus the supply will appear to have an output resistance far greater than its actual value.

**2.** When more than one load is to be connected in parallel to a power supply, Fig. 12-12(a) shows the correct method for making such a connection [and also a frequently used, but incorrect method-Fig. 12-12(b)]. By correctly connecting each load with a separate set of leads to the supply output terminals [Fig. 12-12(a)], variations in the current drawn by a single load will not be mutually conductively coupled to the other parallel connected loads. When incorrectly connected [Fig. 12-12(b)], these current variations can cause large current spikes to appear across the other loads.

If the other parallel loads happen to be digital circuits, problems such as false triggering of these circuits may occur. Digital circuits are often connected as shown in Fig. 12-12(b). To prevent the voltage spikes that occur on the power bus from affecting individual chips (and thus causing false triggering), capacitors should be connected between plus and minus as closely as possible to every chip. Even with the connection shown in Fig. 12-12(a), capacitors should be positioned as closely as possible to each load to eliminate voltage transients that occur when the loads switch on or off.

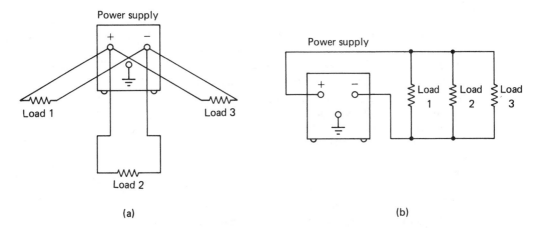

(a)                                                      (b)

**Figure 12-12** (a) Correct method of connecting several loads in parallel to a single power supply; (b) incorrect method of connecting several loads in parallel to a power supply.

**3.** Always turn off the power supply before connecting or disconnecting a load. In addition, always use an ohmmeter to check a potential load for a shorted condition before connecting it to a power supply. Although this may seem to be time consuming, some power supplies are not protected for short circuits, and damage to the supply can result if a shorted load is connected to the power supply.

**4.** Ground loops, as discussed in Chapter 16, can easily be established when power supplies are connected to test setups. Sizable currents may flow in these loops and will lead to degradation of the power supply ripple and noise performance. As will be explained in Chapter 16, precautions should be taken to prevent ground loops from being created (i.e., by using single-point grounding, differential amplifiers, and guarding techniques).

## PROBLEMS

1. Define the following terms.
   (a) Ion
   (b) Electrolyte
   (c) Electrode
2. What is the difference between
   (a) A primary and a secondary cell?
   (b) A wet and a dry cell?
   (c) A cell and a battery?
3. What is the ampere-hour rating of a battery that can provide 500 mA for 50 h?
4. What current will a battery with an ampere-hour rating of 100 theoretically provide for 15 h?
5. What is meant by polarization of a cell? How is polarization reduced in the zinc-carbon cell?
6. Three cells whose open-circuit voltages are each 2.50 V are connected in series. What is the open-circuit voltage of the series connection? If the same three cells were connected in parallel, what would be the open-circuit voltage of the parallel connection?
7. What are two functions of the zinc can in the zinc–carbon dry cell shown in Fig. 12-4(a)? What is the disadvantage of this type of construction?
8. What type of cells would be best suited for the following applications?
   (a) Constant voltage, light loads
   (b) Low temperatures (below 0°F)
   (c) Portable appliances
   (d) Long life, moderate initial cost
   (e) High current values
9. A cell has an open-circuit voltage of 1.50 V and a terminal voltage of 1.42 V at a load of 200 mA. What is the internal resistance of the cell?
10. An automobile battery has an open-circuit voltage of 12.63 V and a total internal resistance of 0.012 $\Omega$. What is the terminal voltage of the cell when it is delivering 175 A to the starter motor?

11. What is the purpose of a power supply, and why must ac power be rectified for use in electronic circuits?

12. What is the *ripple factor*? How is it determined?

13. What is meant by the regulation of a power supply?

14. If a power supply has a no-load voltage output of 100 V and a 96 V output under full load, what is the percentage regulation of the supply?

15. List the major precautions which must be observed when using a standard cell.

16. A standard cell is found to have an open-circuit voltage of 1.01892 V. When a 1 M$\Omega$ resistor is connected across the cell terminals, the terminal voltage of the cell drops to 1.01874 V. Find the internal resistance of the cell.

17. How much does the open-circuit voltage of the normal standard cell change from its value at 20°C if it is operated at 0°C?

## REFERENCES

1. Coombs, C., ed., *Basic Electronic Instrument Handbook*. Chap 39. New York: McGraw-Hill, 1972.

2. Malmstadt, H. V., Enke, C. G., and Toren, E. C., *Electronics for Scientists*. Chap. 2. New York: W.A. Benjamin, 1962.

3. *Power Supply Handbook*. Berkeley Heights, N.J.: Hewlett-Packard, 1970.

4. Stout, M. B., *Basic Electrical Measurements*, 2nd ed. Chap. 6. Englewood Cliffs, N.J.: Prentice Hall, 1960.

5. Kaufman, M. and Seidman, A.H., eds., *Handbook for Electronic Engineering Technicians*. Chaps 16 and 17. New York: McGraw-Hill, 1976.

6. David Linden, ed., *Handbook of Batteries and Fuel Cells*. New York: McGraw-Hill, 1984.

# 13

# *AC Signal Sources*

For the process of providing power for electronic instruments and measurements, both dc and ac power sources are needed. In Chapter 12 we discussed some devices that are used as sources of dc power. This chapter examines the operation and use of instruments that produce time-varying signals. Although the term *ac signal source* can be used to describe the entire family of such instruments, more specific classifications are used for identifying particular subcategories. The classifications are based on the type of output signals that can be generated by the particular signal source.

The subcategories of ac signal sources that we examine in this chapter follow:

1. Oscillators
2. Sweep-frequency generators
3. Pulse generators
4. Function generators

With the advent of low-cost, easy-to-use microprocessors, as well as digital-to-analog and analog-to-digital convertors, many instruments are being designed with multi-function instead of single-function capabilities. Such broad-based instruments are commonly called *function synthesizers* (Fig. 13-1). They are capable of out-putting any of the major waveforms including sine waves, square waves, pulses, triangular waves, and ramps. Multi-function sweep generators can provide either a

**Figure 13-1** Synthesizer/function generator: HP model 3325B. (Courtesy of Hewlett-Packard Co.)

linear or logarithmic frequency sweep. Some of the additional features they offer include variation in phase, dc offset, and amplitude. Control of the instruments can be either manual or by a signal through the IEEE-488 bus connection located on the rear of the instrument.

## OSCILLATORS

Oscillators are instruments that generate sinusoidal output signals. Although there are also other ac signal sources capable of producing sinusoidal outputs, we usually reserve the term *oscillator* for those instruments which are designed to produce sine wave signals exclusively. Most oscillators are capable of generating sinusoidal signals whose frequencies and amplitudes are adjustable over specific ranges. However, there are also a few fixed-frequency oscillators available.

The basic principles involved in the operation of oscillators will be introduced with the help of the analogy of the child's swing. Although not all oscillators operate on a principle exactly analogous to the child's swing, the analogy remains quite instructive.

If a child's swing is initially at rest, a substantial amount of energy must be expended to start it oscillating. However, once the desired amplitude of oscillation is reached, a much smaller amount of energy must be added per cycle to sustain a constant amplitude of oscillation. This energy must be supplied to the swing at appropriate times during the oscillation cycle in order for it to be effective in sustaining the oscillation. If we stop adding energy altogether, the oscillatory motion of the swing will continue for some time. But the frictional and air-resistance losses during each cycle will gradually diminish the amplitude of the oscillations until the swing finally comes to rest. Therefore, to keep the swing oscillating, energy must be continually added to overcome the losses of the system.

An oscillator also produces an output voltage whose magnitude oscillates back and forth between positive and negative values, much like the positions of the moving swing. In those oscillators whose operation resembles that of a swing, a natural sinusoidal voltage variation takes place within a portion of the oscillator circuitry (when that portion is initially excited). The frequency of the oscillation is determined by the values of the electrical components comprising the oscillatory portion of the circuit. The natural voltage oscillation is then amplified by an electronic amplifier, and the resulting amplified signal is used as the output of the oscillator. However, a small portion of the amplified output is diverted and returned to the oscillatory part of the circuit. (This procedure of returning a portion of the signal to its source is called *feedback*.) By feeding back a fraction of the output signal so that it is in phase with the natural voltage oscillations, the losses that would lead to a diminishing voltage amplitude can be counteracted. If the energy contained in the feedback signal is just equal to the energy lost during each oscillatory cycle, the amplitude of the voltage oscillations in the oscillatory circuit is kept constant. The frequency of the output of the oscillator can be varied by changing the values of the electrical components that make up the oscillatory portion of the circuit.

From this discussion it can be inferred that an oscillator must contain the following three elements (Fig. 13-2):

1. Oscillatory element or circuit
2. Amplifier
3. Feedback path to provide the energy to *regenerate* the oscillating signal within the oscillatory element

We should not think, however, that the oscillator is a perpetual motion machine; neither does it violate the law of conservation of energy. The amplifier portion of the oscillator is actually a device that converts dc power into the ac power contained in its output signals. This dc power must be supplied by some external energy source such as a battery or power supply.

### Oscillator Types

The frequency spectrum over which oscillators are used to produce sine wave signals is extremely wide (from less than 1 Hz to many hundreds of gigahertz). However, no

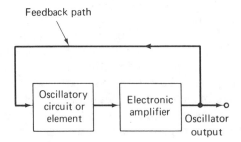

**Figure 13-2**  Oscillator block diagram.

single oscillator design is practical for producing signals over this entire range. Instead, a variety of designs are used, each of which generates sine wave outputs most advantageously over various portions of the frequency spectrum. In this section we discuss some of these oscillator types.

The details of their electronic circuitry will not be covered. Instead, the emphasis of the discussion will be to identify the most commonly used designs and provide a description of their chief characteristics. By being exposed to such a catalog of the major types of oscillators, the reader should be able to recognize and select appropriate instruments for various measurement applications. On the other hand, for those readers interested in the electronic circuit aspects of specific oscillators, a list of references containing this information is provided at the end of the chapter.

Oscillators that use *inductance–capacitance* (LC) circuits as their oscillatory elements resemble the child's swing in their operation rather closely. There is a particular frequency at which such circuits are resonant (that is produce a natural sinusoidal variation). Such LC oscillators are very popular for producing high-frequency (RF) outputs (e.g., 10 kHz to 100 MHz). The most widely used LC oscillator designs are the *Hartley* and the *Colpitts* oscillators. Although they are slightly different from one another in their electronic circuitry, these types of oscillators have virtually identical frequency ranges and frequency-stability characteristics. However, LC oscillators as a whole are not well suited for producing low-frequency sine wave outputs. This is due to the fact that the components necessary for constructing low-frequency LC resonant circuits are too bulky and heavy. Therefore, *resistor–capacitor* (RC) oscillators are usually used for generating low-frequency sine waves (from 1 Hz to about 1 MHz).

The two most common RC oscillators are the *Wien-bridge* and *phase-shift* types. (The basis of their operation is somewhat different than that described by the child's swing analogy.) The Wien bridge design is used in almost all oscillators that produce signals in the audio-frequency range (20–20,000 Hz). This type of oscillator is simple in design, compact in size, and remarkably stable in its frequency output. Furthermore, its output is relatively free from distortion. However, the maximum frequency output of typical Wien bridge oscillators is only about 1 MHz.

*Phase-shift oscillators* also employ a simple circuit and produce sine wave outputs that are quite distortion free. Their principle advantage over Wein bridge oscillators is that they have a wider frequency range (up to several thousand kHz). But they also have the disadvantage of not being as frequency stable as Wein bridge oscillators.

Other less frequently used oscillators include the crystal oscillator and the negative-resistance oscillator. *Crystal oscillators* use a prizoelectric crystal to generate a sinusoidal signal of constant frequency. The output frequency is extremely stable with time but cannot be tuned (adjusted). Therefore, crystal oscillators are used only in applications requiring a fixed frequency of high stability. *Negative-resistance oscillators* are used primarily to produce very high frequency signals. Table 13-1 summarizes the frequency ranges of the most common oscillator types.

**TABLE 13-1** COMMON OSCILLATOR TYPES

| Type | Approximate frequency ranges |
|------|------------------------------|
| Wien bridge (RC) | 1 Hz to 1 MHz |
| Phase-shift (RC) | 1 Hz to 10 MHz |
| Hartley (LC) | 10 kHz to 100 MHz |
| Colpitts (LC) | 10 kHz to 100 MHz |
| Negative-resistance | >100 MHz |
| Crystal | Fixed frequency |

## Output Impedance of Oscillators

When we described the operation of the battery in Chapter 12 we noted that it did not possess the same electrical characteristics as the ideal dc voltage source. The major difference between the two devices was that the battery contained some internal resistance, while the resistance of the ideal voltage source was zero. The internal resistance of the battery was represented in its circuit model by a resistor in series with an ideal dc voltage source.

Similarly, the circuit model of the oscillator can be represented by an ideal ac voltage source in series with a resistor, $R_G$ [Fig. 13-3(a)]. The origin of this resistance value arises in connection with the electronic circuitry of the instrument; in the absence of a reactive component, it is the *output impedance* of the oscillator. This impedance is designed to remain constant as the frequency of the output is varied. In AF oscillators, $R_G$ usually equals 600 $\Omega$, while in RF oscillators (and in signal generators and pulse generators) $R_G$ usually equals 50 $\Omega$. The 600 $\Omega$ value is used in AF oscillators because the characteristic impedance of audio-frequency communications systems (e.g., telephone circuits) have been standardized at 600 $\Omega$.

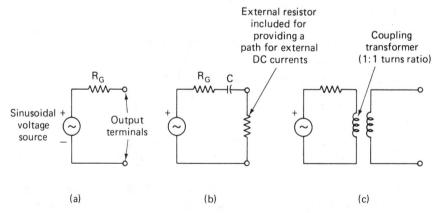

**Figure 13-3** Oscillator outputs: (a) oscillator equivalent circuit (output dc-coupled); (b) ac-coupled output; (c) transformer-coupled output.

Similarly, RF oscillators and pulse generators have 50 Ω output impedances because RF signals and pulses are transmitted along coaxial cables (when not being propagated through space). Such cables typically have a characteristic impedance of 50 Ω.

Why is it so important that the output impedance of an oscillator be equal to the characteristic impedance of the system to which it is connected? The answer lies in the concept of impedance matching, which was discussed in Chapter 3. To transfer the maximum power from the generator to the load, the impedances of the generator and the load must be equal (*matched*). The rated power output of an oscillator is therefore conditional on its being connected to its rated load impedance. This factor is an important consideration in many measurement applications where the full rated power of the oscillator is required by the test circuit.

If the oscillator is connected to a load whose value does not equal (match) its output impedance, the value of the maximum output voltage (as well as the maximum power output) will change.

To see the effect of the load impedance on the voltage, we use the circuit model of the oscillator shown in Fig. 13-3(a). From this model we see that, if a load whose impedance value is equal to $R_x$ is connected to the oscillator, the voltage, $V_o$, across $R_x$ can be found from

$$V_o = \frac{R_x}{R_G + R_x} V \qquad (13\text{-}1)$$

We note from Eq. (13-1) that as $R_x$ approaches infinity (open circuit), $V_o$ approaches $V$ in magnitude. On the other hand, if $R_x$ becomes very small, $V_o$ approaches zero. Finally, if $R_x = R_G$ (matched load), $V_o$ is one-half the maximum voltage output.

When a load is connected to the oscillator, current flows through the load. As defined in Chapter 3, the drawing of current from a circuit is called *loading*. Since a progressively smaller impedance connected across an oscillator will cause more and more current to be drawn from it, we say that this action causes the oscillator to become more *loaded down*. In some oscillators, excessive loading introduces severe distortion of the output signal. In others, even a maximum loading (i.e., a short-circuiting of the output leads) should not cause serious distortion. However, if the oscillator is to be used to drive a circuit whose input impedance is considerably smaller than $R_G$, and if there is any doubt about the ability of the oscillator to operate properly under severe loading, the output signal should be monitored with an oscilloscope for distortion.

## Selection of an Oscillator

When an oscillator is being chosen for a particular application, the requirements of the task at hand should be compared with the performance capabilities of the oscillator. Examples of some of the most common requirements an oscillator is asked to satisfy include the following.

**1.** *Frequency range.* The oscillator should be able to supply an output signal whose upper- and lower-frequency limits exceed those required by the measurement.

**2.** *Power and/or voltage requirements.* The measurement may have a specific voltage or a specific power requirement. The oscillator should be able to produce the pertinent quantity with a magnitude large enough to fulfill the requirement.

**3.** *Accuracy and dial resolution.* The accuracy of an oscillator specifies how closely the output frequency corresponds to the frequency indicated on the instrument dial. Dial resolution indicates to what percentage of the output frequency value the dial setting can be read.

**4.** *Amplitude and frequency stability.* The *amplitude stability* is a measure of an oscillator's ability to maintain a constant voltage amplitude with changes in the frequency of the output signal. *Frequency stability* determines how closely the oscillator maintains a constant frequency over a given time period. Sometimes the frequency stability is included in the accuracy specifications of the oscillator.

**5.** *Waveform distortion.* This quantity is the measure of how closely the output waveform of the oscillator resembles a pure sine wave signal. Sometimes the oscillator is used as a source in a test, which measures the tendency of a circuit to distort a sine wave signal. In such tests, the distortion produced by the oscillator should be much less than the anticipated distortion due to the circuit under test.

**6.** *Output impedance.* The aspects of the output impedance of oscillators were discussed in the preceding section.

Integrated circuits (IC) are available for incorporating signal generators directly on printed circuit boards. They are capable of being either modulated or operated as sweep generators. These ICs are used in instrumentation transmitters. The frequency can be dependent on the magnitude of a dc voltage signal from a transducer, in which case the signal generator is called a voltage controlled oscillator (VCO). Outputs can be sine wave, square wave, triangular wave, or multiple wave as shown in Fig. 13-4. The sine wave generator has a distortion as low as 1 percent, and the triangular wave generator has at least 0.1 percent linearity.

## SWEEP-FREQUENCY GENERATORS

*Sweep-frequency generators* are instruments that produce a sine wave output whose frequency is automatically varied (swept) between two chosen frequencies. One complete cycle of the frequency variation is called a *sweep*. The rate at which the frequency is varied can be either linear or logarithmic, depending on the particular instrument design. However, the amplitude of the signal output is designed to stay constant over the entire frequency range of the sweep. Figure 13-5 shows a block diagram of the major components of a sweep-frequency generator.

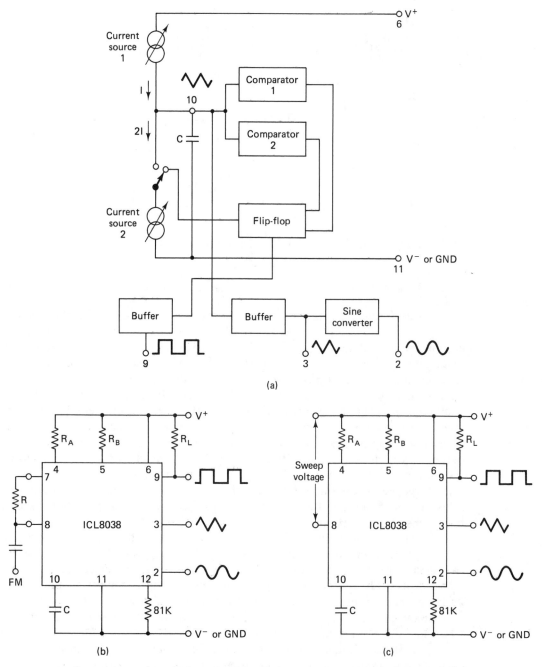

**Figure 13-4** Integrated circuit waveform generator/voltage–controlled oscillator (VCO), Intersil type ICL8038: (a) block diagram; (b) frequency-modulated generator; (c) VCO or sweep generator.

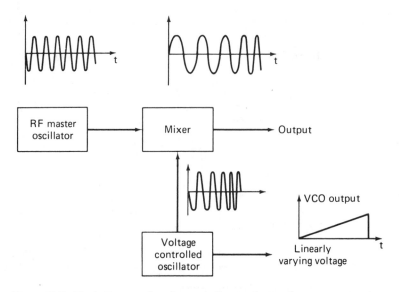

**Figure 13-5**  Block diagram of an electronically tuned sweep-frequency generator.

Sweep-frequency generators are used primarily to measure the responses of amplifiers, filters, and electrical components over various frequency bands. For example, when amplifier circuits or filters are built, they must be tested to ensure that their responses over a specific band of frequencies (bandwidth) meet the design requirements. Performing the measurement of the bandwidth over a wide frequency range with a manually tuned oscillator would be a time-consuming task. In addition, since every change is design requires that the measurement be repeated, many hours could be spent performing this measurement by the end of the project. By using a sweep-frequency generator, a sine wave signal that is automatically swept between two selected frequencies can be applied to the circuit under test, and its response versus frequency can be displayed on an oscilloscope or X-Y recorder. Therefore, the measurement time and effort is drastically reduced. Sweep generators can also be used in the same capacity to check and repair amplifiers used in televisions and radar receivers.

The major component of the sweep generator is a master oscillator (usually an RF type) with several operating ranges selected by a range switch. The frequency of the signal generator's output signal can be varied by either a mechanical or electronic process. In the mechanically varied models, the frequency of the master oscillator output signal is changed (tuned) by a motor-driven capacitor.

In the electronically tuned types, the frequency of the master oscillator is kept fixed, and a second oscillator is used to provide a varying frequency signal. The second oscillator contains an element whose capacitance value depends on the voltage applied across it. This element is used to vary the frequency of the sine wave output of the second oscillator. As a result, this second oscillator is called a *voltage-controlled oscillator*. The output of the VCO is then combined with the output of the master oscillator in a special electronic device called a *mixer*. The output of the mixer is a sine wave whose frequency depends on the difference between the frequencies of the two applied signals. Thus if the master oscillator frequency is fixed at 10.00 MHz and the variable frequency is varied from 10.01 MHz to 42 MHz, the output of the mixer will be a sine wave whose frequency is swept from 10 kHz to 32 MHz. The sweep rates of sweep generators can be adjusted to vary from 1000 to 0.01s per sweep. A voltage that varies linearly (or logarithmically) according to the sweep rate can be used to synchronously drive the $X$ axis of an oscilloscope or $X$-$Y$ recorder. (In the electronically controlled sweep generators, the same voltage that drives the VCO serves as this voltage.)

The frequency of various points along the frequency-response curve can be interpolated from the values of the end frequencies if we know how the frequency is varied (i.e., linear or logarithmic). For more accuracy, markers can be used. (Markers are pulses that appear along the frequency-swept output at accurately known frequencies.)

As an example of a typical sweep-frequency generator, let us examine the sweep-frequency specifications of the HP3325 shown in Fig. 13-1. This instrument can be swept from 1 μHz to 20.999 999 999 MHz. The sweep can be either linear or logarithmic, and the Start, Stop, and Mkr (marker) frequencies are entered in the Sweep Linear/Log section of the front control panel. Sweep times from 0.01 to 1000 s are available and may be either continuous or discrete with a resolution of 0.01 s. For log sweeps the minimum frequency is 1 Hz. Harmonic distortion of an output sine wave varies from −25 to −65 dB. In addition to frequency variation, the output wave can be varied in amplitude, phase, and dc offset.

## PULSE GENERATORS

Pulse generators are instruments that are designed to produce a periodic train of equal-amplitude pulses [Fig. 13-6(a)]. In pulse generators, the duration of the *on* time of a pulse may be independent of the time between pulses. However, if the pulse train has the property of being *on* 50 percent of the time and *off* 50 percent of

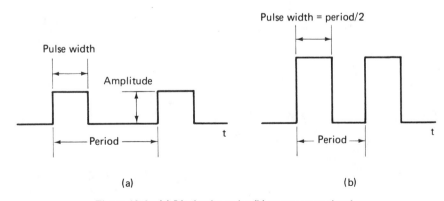

(a)   (b)

**Figure 13-6**   (a) Ideal pulse train; (b) square-wave signal.

the time, the waveform is called a *square wave* [Fig. 13-6(b)]. Square wave generators can be considered to be a special class of pulse generators.

To describe the output of pulse generators and the applications in which they are used, the terminology associated with pulses must be introduced. The first group of such terms denotes the characteristics of ideal rectangular pulses. The second group of terms provides measures of the deviation from the shape and periodicity of ideal pulses.

The terms that characterize a train of ideal periodic pulses include the following ones:

1. *Period.* The time (in seconds) between the start of one pulse and the start of the next. The frequency (or *pulse repetition frequency*) is inversely related to the period.
2. *Amplitude.* The peak voltage value and polarity of the pulse.
3. *Pulse width.* The time duration of the pulse (in seconds).
4. *Duty cycle.* The ratio of the pulse width to the period (expressed in percent of the period). Square waves have a duty cycle of 50 percent.

However, real pulses and pulse trains only approximate the characteristics of their ideal counterparts. As a consequence of this fact, some additional terms are used to describe the nonideal aspects of real pulses. These terms and their definitions are listed below and are illustrated in Fig. 13-7.

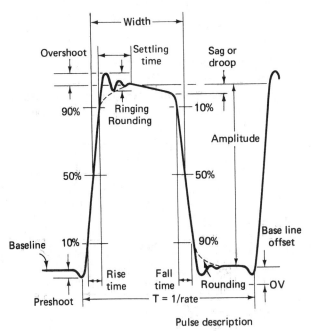

Pulse description

**Figure 13-7**   Actual pulse. (Courtesy of Hewlett-Packard Co.)

1. *Rise time.* ($t_r$) the time (in seconds) it takes for the pulse to increase from 10 to 90 percent of its amplitude.

2. *Fall time.* ($t_f$) the time (in seconds) it takes for the pulse to decrease from 90 to 10 percent of its amplitude.

3. *Overshoot.* The extent (in percent of amplitude) to which the pulse surpasses its correct value during the initial rise.

4. *Ringing.* Oscillation occurring (in percent of pulse amplitude) as a result of overshoot.

5. *Sag* or *droop.* Any decrease (in percent of pulse amplitude) in the pulse amplitude taking place during the pulse width.

6. *Jitter.* Specifies the maximum variation in period from one cycle to the next (in terms of the percentage of the period).

7. *Baseline.* The reference dc level at which the pulse starts.

8. *Settling time.* Time required for the overshoot to be within a specified percent of pulse amplitude.

Pulse generators are designed to produce pulses that approximate ideal pulses as closely as possible. High-quality pulses ensure that any distortion in the output pulse from a test circuit is due to the test circuit alone. The amplitude, pulse width, and period of the generated pulses are usually adjustable over various ranges. The duty cycle is also made adjustable; but if the power contained by each pulse is large, the maximum duty cycle must be kept small. As the maximum duty cycle of a pulse generator is reached, the pulse waveform becomes irregular or the width of the pulse no longer increases.

The source or generator impedance of pulse generators is 50 $\Omega$. This value is chosen so that pulse generators will be matched to the cables which transmit the output pulses of the generator (the cables are coaxial cables which have a 50 $\Omega$ characteristic impedance). Such matching is necessary because loads connected to pulse generators (at the other end of the coaxial cables) are not always matched to the source impedance or cable impedance. Any such *mismatching* causes a part of the pulse to be reflected back to the pulse generator along the coaxial cable. Since the cable and the generator are matched, the reflected signal is completely absorbed upon returning to the generator. If this total absorption did not take place, a portion of the pulse would be re-reflected, and spurious pulses would appear to be generated from the pulse generator (Fig. 13-8). A typical pulse/function generator is shown in Fig. 13-9. The operator can control the period, amplitude, symmetry, burst length, duty cycle, baseline offset, and start phase of the output pulse. In addition, the pulse polarity can be selected and the generator can be either manually or externally triggered.

Some pulse generators allow *manual triggering*. That is, single pulses are produced only upon the manual operation of a front-control-panel pushbutton switch. In the *paired pulse* (or double pulse) mode, two successive pulses are produced in each period. The first pulse is the same as the undelayed pulse, while

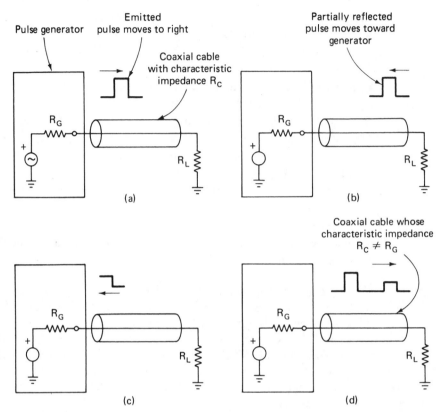

**Figure 13-8** The importance of matching the impedance of a pulse generator and the pulse-transmission cable. (a) Pulse emitted by pulse generator is transmitted to the load along a coaxial cable. Coaxial cable has a characteristic impedance $R_C = R_G$. Load has impedance $R_L \neq R_G$. (b) Pulse is partially reflected by the load $R_L$ because $R_L \neq R_C$. Reflected pulse travels toward the left on coaxial cable. (c) Since $R_C = R_G$, the reflected pulse is absorbed when it returns to the generator. (d) If $R_G \neq R_C$, the reflected pulse would be re-reflected and would propagate to the right again along the cable. The pulse would appear at the load in addition to the desired pulses.

the second is the delayed pulse. In the burst mode, the generator is turned *on* and pulses (0–1999 for the model shown in Fig. 13-9) are produced by the generator only during a brief period (which is controlled by an internally generated gating waveform, which in turn is triggered by an external signal).

Besides the features included in the ordinary pulse generators described above, other options are available on special instruments. One such feature is found in so called *digital word generators*. These instruments allow a pattern of several pulses to be produced and then a length of time before this same pattern is re-

**Figure 13-9** Pulse/function generator: HP model 8116A. (Courtesy of Hewlett-Packard Co.)

peated. As the price of such instruments is increased, a greater flexibility in adjusting the pulse amplitudes and widths within a given word is made available.

When selecting a pulse generator for use in a specific application, it is wise to consult the specifications of the generator to determine if it will be adequate for making the desired measurements. Listed below are the important characteristics of general-purpose pulse generators, together with ranges and specifications available in commercially available instruments. Note that no single instrument will be able to offer operation over all ranges and within the limits described.

**1.** *Pulse period.* Can be as long as 10 s and as short as 1 to 2 ns, with accuracies of 3 percent and jitter better than 0.05 percent.

**2.** *Pulse delay.* Can be as long as 1 s and as short as 1 to 2 ns, with accuracies of 3 percent and jitter better than 0.05 percent.

**3.** *Pulse duration.* Same as delay.

**4.** *Double pulse generation.* Minimum separation between pulses, 1 to 2 ns.

**5.** *Rise times and fall times.* 0.5 ns to seconds, with accuracies of 3 percent and linearity within 1 percent of ideal ramp. Often, rise times and fall times can be separately controlled.

**6.** *Pulse amplitude.* Up to 100 V with accuracies of 3 percent and pulse abberations (including preshoot, overshoot, ringing, and flatness) of no more than 3 percent.

**7.** *Gating and triggering.* Compatible with semiconductor-logic levels.

**8.** *Output impedance.* Typically 50 Ω.

**9.** *Baseline offset.* Up to ±10 V with accuracies of 0.1 percent.

**10.** *Duty cycle.* 10–90 percent.

**11.** *Start phase.* ±90°.

## FUNCTION GENERATORS

A function generator, (Fig. 13-10), is a signal source that has the capability of producing several different types of waveforms as its output signal. Most function generators can generate sine waves, square waves, and triangular waves over a wide range of frequencies. Other models are capable of generating pulse, ramp, trapezoid, or sawtooth waveforms as well as the three common types of waveforms mentioned. The frequency range of a function generator is generally 0.001 Hz to 20 MHz. Some units are also able to interact with automatic test equipment via microprocessor control and the IEEE-488 bus. Finally, function generator manufacturers plan to continue increasing their instruments' versatility by including features such as sweep-frequency generation, complex waveform synthesis, and high-resolution synthesis as available features in their models. Because function generators can produce a wide variety of waveforms and frequencies, they are becoming the "bread and butter" signal generators of the electronic laboratory. In fact, each of the wave shapes they produce is particularly suited for a different group of applications. The uses of the *sine wave output* were described in the earlier section on oscillators. The *square wave* signal can be employed for testing electronic amplifiers and the transient responses of other circuits. Since low-frequency square waves are composed of a wide frequency range of sinusoidal components, they provide a unique measurement capability for the testing of amplifier circuits.[1] In other words, the response of an amplifier circuit to a square wave input signal yields the same type of data about its electrical characteristics as if it were tested sequentially with sine wave inputs of many different frequencies. To use the square

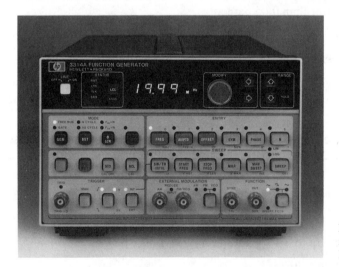

**Figure 13-10**   Function generator: HP model 3314A. (Courtesy of Hewlett-Packard Co.)

[1]For a more elaborate discussion of this point, see any mathematics or circuits text which covers Fourier analysis. The section in Chapter 8 entitled "Harmonic Analysis and Spectrum Analyzers" also has a brief discussion on some of the introductory aspects of Fourier analysis.

wave for this type of measurement, it must be applied to the input of the circuit under test. The output of the circuit is displayed on an oscilloscope.

The triangular wave and sawtooth wave outputs of function generators are commonly used for those applications which require a signal that increases (or decreases) at a specific linear rate. They are also useful for driving sweep oscillators in oscilloscopes and the $X$ axis of $X$-$Y$ recorders. Many function generators are also able to generate two different waveforms simultaneously (from different output terminals, of course). This can be a useful feature when two generated signals are required by a particular application. For example, a triangular wave and a sine wave of equal frequencies can be produced at the same time. If the zero crossings of both

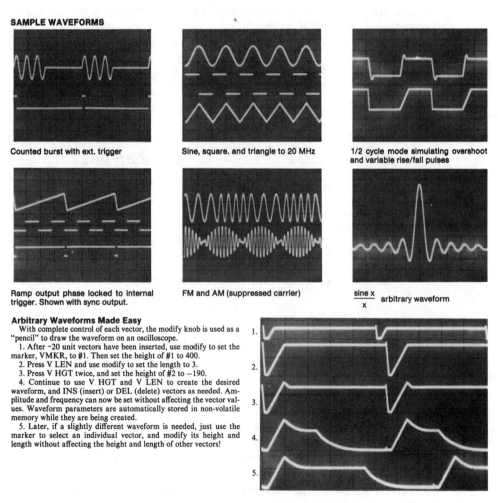

**SAMPLE WAVEFORMS**

Counted burst with ext. trigger

Sine, square, and triangle to 20 MHz

1/2 cycle mode simulating overshoot and variable rise/fall pulses

Ramp output phase locked to internal trigger. Shown with sync output.

FM and AM (suppressed carrier)

$\dfrac{\sin x}{x}$ arbitrary waveform

**Arbitrary Waveforms Made Easy**

With complete control of each vector, the modify knob is used as a "pencil" to draw the waveform on an oscilloscope.

1. After ~20 unit vectors have been inserted, use modify to set the marker, VMKR, to #1. Then set the height of #1 to 400.

2. Press V LEN and use modify to set the length to 3.

3. Press V HGT twice, and set the height of #2 to −190.

4. Continue to use V HGT and V LEN to create the desired waveform, and INS (insert) or DEL (delete) vectors as needed. Amplitude and frequency can now be set without affecting the vector values. Waveform parameters are automatically stored in non-volatile memory while they are being created.

5. Later, if a slightly different waveform is needed, just use the marker to select an individual vector, and modify its height and length without affecting the height and length of other vectors!

1.

2.

3.

4.

5.

**Figure 13-11** Typical waveforms output by HP 3314A function generator. (Courtesy of Hewlett-Packard Co.)

waves are made to occur at the same time, a linearly varying waveform is available which can be started at the point of zero phase of a sine wave.

Another important feature of some function generators is their ability to *phase-lock* to an external source. This means that when the *phase-lock* features are employed, each cycle of the waveform from a function generator will bear a fixed phase relationship to an applied external signal. (This phase relationship is also adjustable.) One example of how this feature is used is when a sine wave output of a function generator is phase-locked to another sine wave from a second function generator. If this other sine wave is the second harmonic of the first, the summation of the two sine waves at different amplitudes and phase shifts can produce a wide variety of unusual waveforms. In addition, if the function generator is connected to an accurate frequency standard (and the phase-lock feature is being employed), all the output waveforms from the function generator will have the same frequency, stability, and accuracy as the standard. Figure 13-10 shows a photograph of a typical function generator. Arbitrary waveforms and standard waveforms that can be generated by the HP 3314A function generator are shown in Figure 13-11.

A typical set of the most important specifications of a function generator is given in Table 13-2.

TABLE 13-2   TYPICAL FUNCTION-GENERATOR SPECIFICATIONS

| | |
|---|---|
| Frequency range | 0.001 Hz to 20 MHz |
| Frequency stability | 0.05% |
| Sine wave distortion | −55 dB below 50KHz |
| | −40 dB above 50KHz |
| Square wave transient response | Rise and fall time < 9 ns |
| Output amplitude (open circuit) | 10 V p-p |
| Output impedance | 50 Ω |
| Output waveforms | Sine, square, triangle, ramp, pulse, arbitrary, AM and FM modulated, counted burst |

# PROBLEMS

1. Briefly describe the various types of output signals that are generated by each of the following ac signal sources.
   (a) Oscillators
   (b) Sweep-frequency generators
   (c) Function generators
2. Qualitatively describe the operation of the oscillator. From where does the oscillator draw its power to overcome the losses present in its oscillating circuitry?
3. Explain why AF and RF oscillators usually possess output impedances of 600 Ω and 50 Ω, respectively.

**4.** Describe what effect the value of the load impedance connected to the output of an oscillator will have on the maximum voltage that can be put out by the oscillator.

**5.** Define the following terms used to specify the performance of an oscillator.
(a) Dial resolution
(b) Frequency stability
(c) Range
(d) Amplitude stability

**6.** Refer to other texts and then describe the processes of amplitude modulation and frequency modulation in more detail.

**7.** If a pulse has the following period and width, calculate its duty cycle.
(a) Pulse width = 4 μs, period = 7μs
(b) Pulse width = 12 ms, period = 2 s

**8.** Explain why the output impedance of a pulse generator and the impedance of the cable along which the output pulses of the generator are transmitted should have the same values.

**9.** Describe what is meant by the capability of a function generator to phase lock to an external source.

**10.** Define the following terms in reference to the description of a pulse.
(a) Rise time
(b) Fall time
(c) Overshoot
(d) Ringing
(e) Jitter

**11.** Using a block diagram, design a test setup that utilizes an oscilloscope to calibrate the frequency of an oscillator.

## REFERENCES

1. Hewlett-Packard, *Instrumentation Catalog*. Palo Alto, Calif.: Hewlett-Packard, 1982.
2. Coombs, C., *Handbook of Basic Electronic Instrumentation*. Chap 28–30. New York: McGraw-Hill, 1972.
3. Cooper, W. D., and Helfrick, A. D., *Electronic Instrumentation and Measurement Techniques*, 3rd ed. Englewood Cliffs, N.J.: Prentice Hall, 1985.
4. Millman, J., and Taub, H., *Pulse, Digital, and Switching Waveforms*. New York: McGraw-Hill, 1965.

# 14

# *Electrical Transducers*

*Transducers* are broadly defined as devices that convert energy or information from one form to another. They are widely used in measurement work because not all quantities that need to be measured can be displayed as easily as others. A better measurement of a quantity can usually be made if it can be converted to another form which is more easily or accurately displayed. For example, the common mercury thermometer converts changes in temperature to changes in the length of a column of mercury. Since the change in the length of the mercury column is rather simple to measure, the mercury thermometer becomes a convenient device for sensing changes in temperature. On the other hand, the actual temperature change is not as easy to display directly.

The purpose of this chapter is to introduce some of the transducers that convert physical quantities into electrical signals (as well as a few transducers that convert them to nonelectrical forms). Electrical transducers (i.e., microphones, loudspeakers, etc.) make up the vast majority of transducers in use today. The reason for their popularity is that electrical signals possess many desirable measurement traits. The first of these is that electrical instrumentation is so highly developed that there are usually several different methods for converting most physical quantities into electrical signals. Next, if weak signals are converted to an electrical form, they can be faithfully amplified until their amplitudes become large enough to be easily displayed. Finally, there are display and recording devices (including digital computers) that can follow very-high-frequency variations in electrical signals. Thus a nonelectrical quantity which also has a high-frequency variation (e.g., the vibrations of a solid) can be converted to an electrical form and accurately monitored.

When we discuss the various types of transducers, a means for classifying them must be selected. For the most part, our method will be to classify transducers according to the type of quantity to which they are designed to respond.[1] The two exceptions to this procedure will involve the discussion of the *strain gauge* and the *linear variable differential transformer*. These two transducers are used for such a wide variety of measurements that we will treat them as multipurpose devices (rather than as single-purpose transducers which are capable of converting only one or two physical quantities to an electrical signal). However, the discussion of this chapter is not meant to be an examination of *all* possible electrical transducers. Instead, its aim is to introduce several of the most commonly used varieties and thereby alert the reader to some of the ways that measurement possibilities can be expanded through the use of transducers.

## Role of Transducers in Measurement Systems

There are, in general, three major elements which are common to most measuring systems (Fig. 14-1). The first of these is the *detecting element* (or *sensor*). The purpose of the detecting element is to respond to the magnitude (or changes in the magnitude) of the quantity being measured. The response of the sensor takes the form of an output signal whose magnitude is proportional to the magnitude of the quantity being measured.

The second element is the *signal modifier*. This element receives the output signal of the detecting element and modifies it by amplification or by suitable shaping of its waveform. When the signal emerges from the signal modifier, it should be in a form that is appropriate for display or recording. The third element of measurement systems is the *display* or *recording device*. In electrical systems, display or recording devices include such instruments as meters, cathode-ray tubes, chart recorders, tape recorders, *X-Y* recorders, and digital computers.

If the measurement system is one in which a nonelectrical quantity is to be measured by converting it into an electrical form, an electrical transducer is used as the detecting element.[2] If the electrical transducer produces a signal without re-

---

[1]Another method groups transducers according to the type of electrical parameter that undergoes changes in the process of producing the transducer output signal.

[2]Sometimes the initial transducing element of an electrical measurement system converts the measured quantity into a nonelectrical form. In such cases, the nonelectrical output signal of the first transducer can be fed to a second transducer. This second transducer is an electrical transducer which converts the signal to an electrical form. One example of such a two-transducer arrangement is the Bourdon tube and linear variable differential transformer (LVDT), which is often used to measure fluid pressure. The Bourdon tube converts fluid pressure to a mechanical displacement. An LVDT converts mechanical displacements to an ac electrical signal. Thus if the LVDT is connected to the Bourdon tube, pressure can be converted to an ac electrical signal.

A fine reference which presents a more complete exposition of the subject than is possible in the present chapter is a text by H. N. Norton, *The Handbook of Transducers for Electronic Measuring Systems* (Englewood Cliffs, N.J.: Prentice Hall, 1969).

**Figure 14-1**   Elements of a general measurement system.

quiring an electrical excitation, it is called an *active* transducer. If the transducer is capable of producing an output signal only when it is used in connection with an excitation source, the transducer is called a *passive* transducer. A complete transducer system includes the transducer and the voltage excitation source if one is required.

### Guidelines for Selecting and Using Transducers

When a measurement of a nonelectrical quantity is to be undertaken by converting the quantity to an electrical form, an appropriate transducer (or combination of transducers) for carrying out this conversion must be selected. The first step in the selection procedure is to clearly define the nature of the quantity that is to be measured. This awareness should include a knowledge of the range of magnitudes and frequencies that the quantity may be expected to exhibit. When the problem has thus been stated, the available transducer principles for measuring the desired quantity must be examined. If one or more transducer principles are capable of producing a satisfactory signal, we must decide whether to use a commercially available transducer or undertake to build the transducer. If commercially manufactured transducers are available at a suitable price, the choice will probably be to purchase one of them. On the other hand, if no transducers are made which can perform the desired measurement, one may have to design, build, and calibrate his own device.

When the specifications of a particular transducer are examined, the following points should be considered in determining its suitability for a particular measurement:

**1.** *Range.* The range of the transducer should be great enough to encompass all the expected magnitudes of the quantity to be measured.

**2.** *Sensitivity.* To yield meaningful data, the transducer should produce a sufficient output signal per unit of measured input.

**3.** *Loading effects.* Since transducers will always consume some energy from the physical effect under test, it must be determined either that this absorption is negligible, or that correction factors can be applied to compensate the readings for the loss.

**4.** *Frequency response.* The transducer must be able to respond to the maximum rate of change in the effect under observation.

**5.** *Electrical output format.* The electrical form of the transducer output signal must be compatible with the rest of the measurement system. For example, a dc output voltage would not be compatible with an amplifier that can respond only to ac signals.

**6.** *Output impedance.* The transducer output impedance must have a value that makes it compatible with the next electrical stages of the system. If an impedance incompatibility exists, additional signal modifying devices may have to be added to the system to overcome this problem.

**7.** *Power requirements. Passive* transducers require external excitation. Thus, if passive transducers are to be used, it is necessary to ensure that appropriate electrical power sources for operating them are available.

**8.** *Physical environment.* The transducer selected should be able to withstand the environmental conditions to which it may be subjected while making the test. Such parameters as temperature, moisture, and corrosive chemicals might damage some transducers but not others.

**9.** *Errors.* The errors inherent in the operation of the transducer itself, or those errors caused by environmental conditions of the measurement, should be small enough or controllable enough that they allow meaningful data to be taken.

Once the transducer has been selected and incorporated into the measurement system design, the following guidelines should be observed to increase the accuracy of the measurements:

**1.** *Transducer calibration.* The transducer output should be calibrated against some known standards while it is being used under actual test conditions. This calibration should be performed regularly as the measurement proceeds.

**2.** Changes in the environmental conditions of the transducer should be monitored continuously. If this procedure is followed, the measured data can later be corrected to account for any changes in environmental conditions.

**3.** By artificially controlling the measurement environment, we can reduce possible transducer errors. Examples of artificial environmental control include the enclosing of the transducer in a temperature-controlled housing and isolating the device from external shocks and vibrations.

## STRAIN GAUGES

On March 22, 1939, Professor Arthur C. Ruge of MIT received a letter from the university patent committee that stated " ... the committee does not feel that the commercial use is likely to be of major importance. . . . " The best minds in engineering felt that the strain gauge was not useful at the time it was conceived. Today the *strain gauge* is one of the most commonly used electrical transducers. Its popularity stems from the fact that it can detect and convert *force* or *small mechanical displacements* into electrical signals. Since many other quantities such as torque pressure, weight, and tension also involve force or displacement effects, they can also be measured by strain gauges. Furthermore, if the mechanical displacements to be measured have a time-varying form (such as vibrational motion), signals with frequencies of up to 100 kHz can be detected.

Strain gauges are so named because when they are strained they change in resistance. Strain is defined as a change in length of a material owing to an externally applied force or stress:

$$\epsilon = \Delta L / L \tag{14-1}$$

where   $\Delta L$ = change in length owing to applied force
$L$ = original length

To maintain linear operation of a transducer, the applied force must not strain any of the material in a strain gauge–based transducer beyond 50 percent of its elastic limit (the elastic limit is the maximum strain that a material can undergo and still return to its original length once the stress is removed). When strain is applied to a strain gauge, its resistance will change ($\Delta R$) in accordance with

$$\Delta R = \rho \frac{\Delta L}{A} \tag{14-2}$$

where   $\rho$ = resistivity of material ($\Omega$-m)
$\Delta L$ = change in the effective length of conductor
$A$ = area of conductor

Most strain gauges are manufactured such that the resistance varies linearly with changes in the length. Changes in either resistivity (as a function of stress) or area are usually minimal and thus ignored. The unstrained resistance is usually 120 or 350 ohm. These values were originally chosen because they were the critical dampening resistance of the galvanometers used with strain gauges. Since very few instruments today use galvanometers for measuring the output of bridge circuits, the resistance of strain gauges must no longer be kept at these values. As a result, it is being increased to match the requirements of integrated circuit instrument amplifiers, to reduce power requirements, and to increase the signal-to-noise ratio. As a result the unstrained resistance is being increased to values as high as can be manufactured. These values in the state-of-the-art platinum (92 percent) tungsten (8 percent) strain gauges are about 1 k ohm, and such gauges have a gauge factor ($K$) of 4.5. where $K$ is defined by

$$K = \frac{\dfrac{\Delta R}{R_g}}{\dfrac{\Delta L}{L}} \tag{14-3}$$

where   $K$ = gauge factor (usually $\approx 2$ but as high as several hundred)
$\Delta R$ = change in gauge resistance
$R_g$ = gauge resistance
$\Delta L$ = change in gauge length
$L$ = gauge length

**Example 14-1**

Assume a 120 ohm strain gauge is attached to the cylindrical surface of a 12 oz can designed to hold a carbonated beverage. When the can is filled and pressurized, it causes the walls of the can to expand, resulting in a strain of the metal can material and, therefore, the strain gauge. If the gauge factor $K$ is 2.00 and the resultant strain after the filling and pressurization process strain is 243$\mu$E, calculate the change in resistance of the strain gauge. Solution

$$K = \frac{\frac{\Delta R}{R_g}}{\frac{\Delta L}{L}} = 2.00 = \frac{\frac{\Delta R}{120}}{243 \times 10^{-6}}$$

$$\Delta R = 0.0583 \text{ ohms}$$

A basic strain gauge pattern is shown in Fig. 14-2. The large end loop areas (small resistance) reduce the effects of the transverse strain (strain perpendicular to the gauge length). Gauge length can vary from a few millimeters to several inches depending on the length of area over which the strain will be averaged. Concrete and other nonhomogeneous materials require very long gauges, whereas roots of gears (Fig. 14-3) require very small gauges. A gauge should be bonded to a surface that is subjected to uniform strain since the strain is averaged over the length of the gauge. The use of plastic models and polarized light permit strain patterns to be observed. The patterns can also be determined by using a computer and finite element analysis software. An example of a strain pattern and placement of strain gauges on a transducer are shown in Fig. 14-4. The transducer is designed for a full bridge (defined in a later section) in which two gauges are subjected to tension and two are subjected to compression.

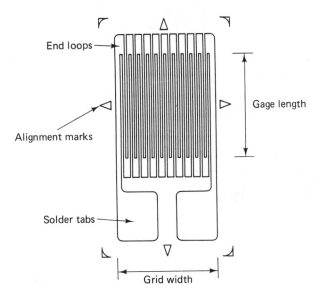

End loops

Alignment marks

Gage length

Solder tabs

Grid width

**Figure 14-2** Typical strain gauge pattern. (Courtesy of Micro Engineering II.)

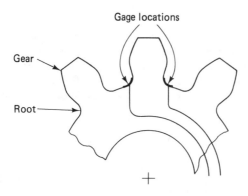

**Figure 14-3** Strain gauge placement for determining the stress at the root of a gear tooth.

Strain gauges are constructed by bonding a very thin film of metal to a backing such as polyimide or glass fiber–reinforced phenolic resin. The gauge pattern is then etched by using processes similar to those used in the etching of printed circuit board wiring patterns. The metal films usually used to form the gauge are constantan, karma, manganin, or platinum-tungsten. Each gauge is trimmed to the final resistance. A finished gauge is attached to the area that will be strained by special adhesives such as nitrocellulose, cyanoacrylate, phenolic, or epoxy. The temperature coefficients of expansion of the material to be analyzed and the strain gauge should be matched as closely as possible to minimize temperature effects. Some strain gauges must remain attached for years (such as those attached to airplane surfaces for FAA tests), while others are attached for very short one-time impact tests (which may destroy the specimen under test).

Because the *resistance change* of a strain gauge is generally very small, Wheatstone bridges are usually used to measure the change in resistance. In addition, since the *resistance of the gauge itself* is small, special bridge connections are required to eliminate the resistive effects of the lead wires (Fig. 14-5). As long as the length and temperature of both lead wires remain the same, the error signal $e_0$ (in volts) will not be affected by any lead wire resistance changes, because the resistance is equally divided between the bridge legs. In a balanced bridge

$$\frac{R_1}{R_2} = \frac{R_{L1} + R_3}{R_{L2} + R_4} \tag{14-4}$$

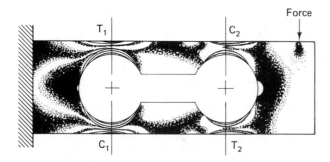

**Figure 14-4** Constant stress contour line and strain gauge location on a transducer. Strain lines are created by using a plastic model and polarized light.

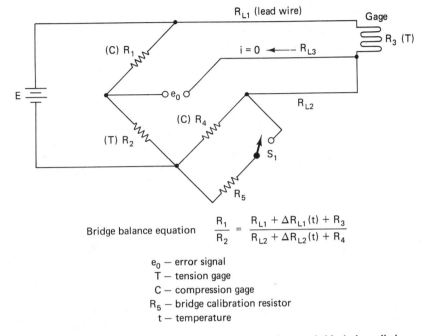

$$\text{Bridge balance equation} \quad \frac{R_1}{R_2} = \frac{R_{L1} + \Delta R_{L1}(t) + R_3}{R_{L2} + \Delta R_{L2}(t) + R_4}$$

$e_0$ — error signal
T — tension gage
C — compression gage
$R_5$ — bridge calibration resistor
t — temperature

**Figure 14-5** Three-wire circuit for single active gauge (quarter bridge) also called a lead-wire-compensated connection.

and, in the unbalanced bridge, the error signal $e_0$ is found from

$$e_0 = i \frac{\Delta R}{4 + \dfrac{\Delta R}{R}} \qquad (14\text{-}5)$$

where $\Delta R =$ is the change in resistance of the strain gauge due to a stress (Chapter 10 describes the circuit equations of both the balanced and the unbalanced Wheatstone bridge in more detail).

The lead length of the third conductor, $R_{L3}$, does not affect the error signal $e_0$ because no current flows in this lead, and, hence, no voltage drop is produced across $R_{L3}$. Resistor $R_4$ can be an unstrained gauge that is placed adjacent to $R_3$ so that temperature effects will be minimized. It can also be a gauge that is strained opposite (tension versus compression) to gauge $R_3$ but at the same temperature. A resistance can be placed in shunt with any of the bridge arms for calibration purposes. For example, when $S_1$ in Fig. 14-5 is closed, the parallel combination of $R_4$ and $R_5$ simulates a compressive load, since the parallel resistance is smaller than $R_4$ alone. The simulated strain is found from

$$\text{simulated strain } (\epsilon) = \frac{R_4}{K(R_4 + R_5)} \qquad (14\text{-}6)$$

where $K =$ gauge factor

**Example 14-2**

If the gauge factor of a 120 ohm strain gauge is 2.00 and the calibrating resistor is 59,880 ohms, find the simulated strain.

**Solution.**

$$\text{simulated strain } (\epsilon) = \frac{120}{2.00(120 + 59880)} = 1000 \ \mu\epsilon$$

The three most common configurations of the bridge circuits used with strain gauges are the quarter, half, and full bridges (Fig. 14-6). In full bridges there are four gauges, and these alternate in tension and compression, resulting in four times the output signal of a quarter bridge. (Note that the $R_4$ should not be mounted at a

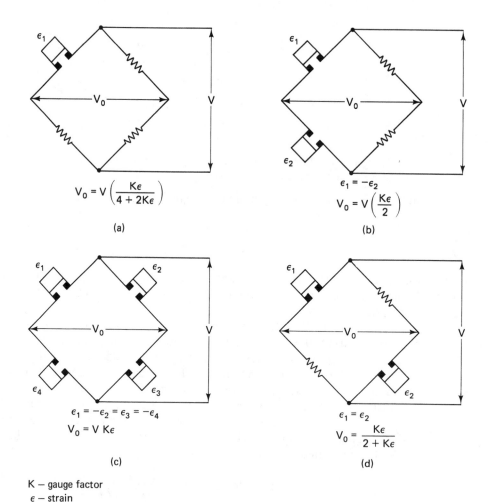

$$V_0 = V\left(\frac{K\epsilon}{4 + 2K\epsilon}\right)$$

(a)

$$\epsilon_1 = -\epsilon_2$$
$$V_0 = V\left(\frac{K\epsilon}{2}\right)$$

(b)

$$\epsilon_1 = -\epsilon_2 = \epsilon_3 = -\epsilon_4$$
$$V_0 = V\,K\epsilon$$

(c)

$$\epsilon_1 = \epsilon_2$$
$$V_0 = \frac{K\epsilon}{2 + K\epsilon}$$

(d)

K — gauge factor
$\epsilon$ — strain

**Figure 14-6** Common bridge configurations for strain gauges: (a) quarter bridge; (b) half bridge; (c) full bridge; (d) half bridge.

right angle to $R_3$ when $R_4$ is used to provide temperature compensation). This is because $R_4$ is subjected to reduced stress based upon Poisson's ratio (which is the ratio of lateral strain to axial strain). For steel this strain can be as high as 30 percent of the strain in $R_3$.

Strain gauges can also be mounted to appropriate mechanical structures to create a variety of transducer devices that can measure such physical variables such as tension, compression, pressure, acceleration, and differential pressure (velocity) (Fig. 14-7).

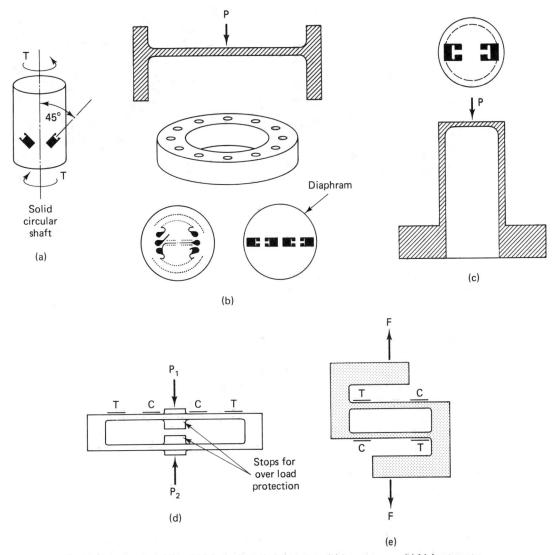

**Figure 14-7**  Typical strain gauge transducers: (a) torsion; (b) low pressure; (c) high pressure; (d) differential pressure; (e) tension or compression (scales).

Detail A

**Figure 14-8**   Integrated circuit strain gauge mass flow sensor. (Courtesy of Micro Switch, a Honeywell Division.)

Semiconductor strain gauges are considerably more sensitive than metal film strain gauges, and their gauge factors are usually greater than 200. Because of their increased sensitivity, such semiconductor strain gauges can measure much smaller strains. Unfortunately, the gauges are much more sensitive to temperature fluctuations than are metal film gauges. (Nevertheless, temperature-compensated circuits should be used with both metal film and semiconductor gauges). Semiconductor strain gauges are incorporated into three-dimensional integrated circuits for measuring both pressure and mass flow. Flows as low as one CC/min can be measured, and pressures as low as 2.5 kPa can be measured. A temperature-compensated pressure and mass flow transducer is shown in Fig. 14-8. In addition to strain gauges, pressure transducers also use ion-implanted piezoresistive elements in a bridge configuration (Fig. 14-9).

## LINEAR VARIABLE DIFFERENTIAL TRANSFORMERS

The second general-purpose transducer is the linear variable differential transformer (abbreviated LVDT). Like the strain gauge, it produces an electrical signal

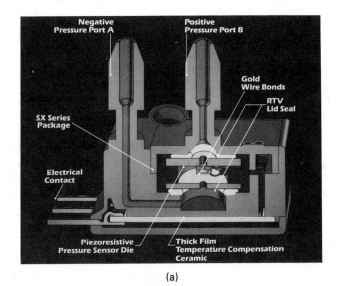

(a)

(b)

**Figure 14-9** Integrated circuit piezoresistive pressure sensor: (a) schematic; (b) typical assembly. (Courtesy of SenSym, Inc.)

that is linearly proportional to mechanical displacements.[3] The displacements detectable by LVDTs are relatively large compared to those detectable by strain gauges. Thus, LVDTs are suitable for use in applications where the displacements are too large for strain gauges to handle. (Conversely, strain gauges are generally more suitable for the smaller displacements.) For example, LVDTs can detect displacements that range from microinches to inches. Since the LVDT can also be connected to other transducers whose outputs are a mechanical displacement, they are often used together with other transducers (as well as alone).

The LVDT senses displacements by the motion of the ferromagnetic core within a special transformer [see Fig. 14-10(a)]. The transformer has one primary winding and two secondary windings. All three of these windings are wound on the same hollow insulating tube. The primary winding is wound at the center of the tube, and the two secondary windings (which have an equal number of turns) are connected in series-opposition. This means that if the mutual coupling between each secondary winding and the primary winding is equal, the voltage, $V_o$, across the secondary winding wires will be zero, even when the primary is excited by an ac signal.

If the ferromagnetic core is centered (with respect to the length of the transformer), the mutual coupling between each secondary winding and the primary will be equal. As long as this condition exists, $V_o = 0$. However, if the core is moved from this centered position, the mutual coupling between each secondary winding and the primary will no longer be equal. For example, if the core in Fig. 14-10(b) is moved to the right, the mutual coupling between secondary winding 2 and the primary will increase, while the mutual coupling between secondary winding 1 and the primary will decrease. A shift of the core position to the left will have the opposite effect. As a result of the changes in the mutual coupling, the voltage, $V_o$, across the output wires connected to the secondary windings will no longer be zero. Instead (for small displacements of the core), this output voltage will be linearly proportional to the magnitude of the displacement.

The *sensitivity* of an LVDT is rated in mV/V/0.001 in. Its actual output voltage is thereby found by multiplying the sensitivity by the input voltage and displacement. Since the output is directly proportional to the applied primary voltage, very good power supply regulation is required. Any variation in either frequency or voltage will appear in the output signal. A typical input signal ranges from 1 to 10 V ac and from 1 kHz to 100 kHz. An advantage of the LVDT is its ability to change the output signal level without modifying the single-conditioner circuitry. Instead, either the frequency or input voltage level of the primary input signal is changed.

Linearity of an LVDT is defined as the maximum deviation of the output curve from a best-fit straight line passing through the origin expressed as a percentage of the nominal output. For example, if the output of an LVDT is 5.00 volts at a

---

[3]Angular displacements can be measured by a similar device called a rotary variable differential transformer (RVDT).

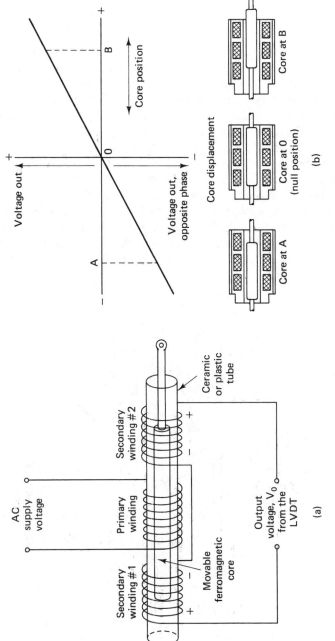

**Figure 14-10** Linear variable differential transformer (LVDT): (a) schematic; (b) output voltage as a linear function of position of core in an LVDT. (Courtesy of Schaevetz Engineering.)

displacement of 0.250 inch, and the maximum deviation of the output curve from the straight line through the origin is 0.006 volt, then the linearity is

$$\text{linearity} = \frac{\text{deviation}}{\text{output}} = \frac{0.006}{5.00} = 0.12\%$$

As the primary input voltage is increased, the harmonic distortion also increases, necessitating care in designing the primary voltage oscillator. For maximum power to be delivered by the LVDT, the input impedance of the signal conditioning circuit must match the output impedance of the LVDT. Signal conditioning of both the primary and secondary voltage is available in VLSI form. One example is the NE5520 by Signetics, which includes a sinewave oscillator (4 percent distortion), a synchronous demodulator with feedback, and an amplifier.

The length to range ratio can be very large. For example, a ±0.050 inch range LVDT is 1.12 inches long, and a ±10 inch range LVDT is 30 inches long. This large ratio is impractical when measuring the large distances encountered in numerical control machinery, such as in lathes and robotic equipment. For these systems, digital positioners that employ either optical or magnetic encoders are used. In an optical system, light is passed through an encoded linear optical scale to a photo-sensitive receiver, (Fig. 14-11). Encoder lengths of 10 feet with a resolution of 0.1

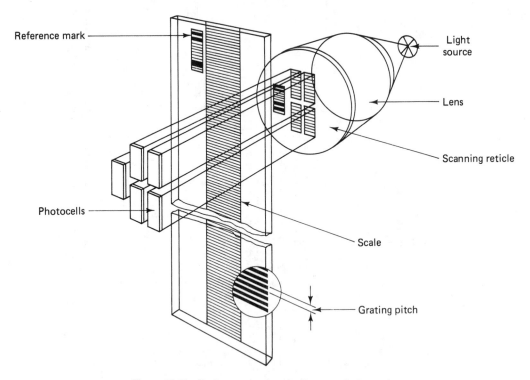

**Figure 14-11**   Basic components of a linear optical encoder.

μm inches are possible. In a magnetic system a rod or tape as long as 100 feet is magnetically encoded in a manner similar to that in a dual-track computer disk. Because of the long lengths, error compensation is incorporated about every 200 mm. To increase resolution, an interpolation process is performed on two quadrature signals. The code is read in a manner similar to the method used to read computer disks. The automotive and robotics industries are continuting to incorporate more and more linear sensors in their control systems.

### LVDT Accelerometer

As an example of the wide variety of uses in which the LVDT is utilized, we will examine the LVDT accelerometer. Such accelerometers are used to detect earthquakes and to measure missile accelerations. (Other accelerometers which measure shock and vibration primarily use piezoelectric crystal transducers. Piezoelectric crystals are discussed in a later section.)

A form of the LVDT accelerometer is shown in Fig. 14-12. We see that the magnetic core of the LVDT is connected by two cantilever springs to a larger external piece of equipment. If this piece of equipment is accelerated in the direction shown, the core undergoes a force proportional to the acceleration and therefore bends the cantilever springs. The change in the position of the core is thereby proportional to the acceleration, and this shift yields a voltage signal which is also proportional to the acceleration.

A typical missile accelerometer has a rated output voltage of 2 $V_{rms}/g$ with a 115-V, 400-Hz input voltage.

### Other Position and Velocity Transducers

Although the strain gauge and the LVDT produce accurate indications of position and velocity, there are other devices that also find use as transducers of these quantities. The two most common ones are the linear-motion potentiometer and the linear-motion variable inductor. These two transducers are simpler, cheaper, and easier to use than the strain gauge and the LVDT. However, their accuracies and sensitivities are not as high, and this limits their use in many cases.

The linear-motion potentiometer is shown in Fig. 14-13. We see that it is a variable-resistance device whose resistance value is changed by the motion of the slider along the resistance element. The slider is connected to an arm that couples the motion being measured to the transducer. If the centered position of the slider corresponds to the zero-value position, a resistance change will accompany any change of position in either the positive or negative direction. A Wheatstone bridge can be used to measure these resistance changes.

The linear-motion inductor works on a similar principle as the linear-motion potentiometer. However, the moving element in this device is a magnetic core which is placed inside the coil of an inductor. As the core is moved in relation to the coil, the inductance value of the inductor changes. The change in inductance can be monitored by an impedance bridge to indicate changes in position.

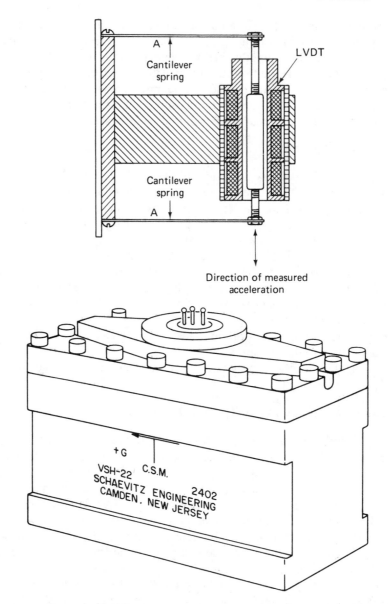

**Figure 14-12**   LVDT accelerometer. (Courtesy of Schaevetz Engineering.)

## FLUID-PROPERTY TRANSDUCERS (PRESSURE AND FLOW RATE)

Since they both act like fluids in many respects, the most common properties of liquids and gases can usually be monitored by using the same type of transducers. This is especially true of the properties of pressure and flow rate.

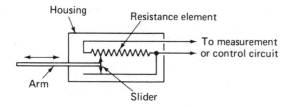

**Figure 14-13**   Linear-motion potentiometer.

## Fluid Pressure

The quantity, *pressure*, can be described in many ways, If the value of pressure is being described relative to vacuum, this type of pressure is called *absolute pressure*. When the value of pressure is compared to the absolute pressure of air at sea level, the type of pressure is referred to as *relative pressure*. If the pressure of interest is the difference of the pressure of two fluids (or the difference of the pressures of the same fluid in different parts of a system), what is being described is the *differential pressure*.

Pressure can be electrically measured if it directly changes an electrical parameter (such as capacitance). It may also be measured if it produces a mechanical displacement. The mechanical displacement can then be made to activate a linear-displacement transducer, thereby causing an electrical signal.

A variable-capacitor transducer is shown in Fig. 14-14(a). It is very similar in operation to the capacitor microphone (to be described later). The reference pressure of the transducer of this example can be atmospheric pressure (for relative pressure measurement), a vacuum (for absolute measurements), or a fluid from the second pressure of interest (for differential pressure measurements).

A metal diaphragm within the capacitor transducer moves closer or farther away from a rigid plate and thereby causes a change in the capacitance of the structure. If the capacitance value is made part of an oscillator circuit, the frequency

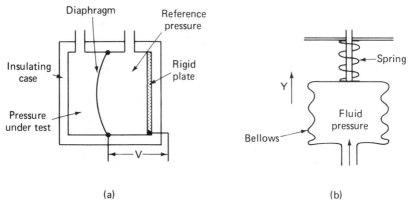

(a)                                                 (b)

**Figure 14-14**  Fluid-pressure transducers: (a) capacitor pressure transducer; (b) bellows pressure transducer.

of the oscillator will change as the capacitance value changes. The frequency changes can be monitored to indicate the pressure change.

The capacitor pressure transducer is one of the most rugged and accurate transducers available for measuring pressure. It can be built to respond to a wide range of pressure values as well as to high-frequency pressure changes.

The devices used to convert pressure into a mechanical displacement are made in many ways. We will mention only a few of the most common ones. The first is a flexible bellows like the one shown in Fig. 14-14(b). The fluid is allowed to enter the bellows, and its pressure extends them in the $Y$-direction. In low-pressure bellows, the external spring shown in Fig. 14-14(b) is not present. Then the springiness of the bellows alone is used to resist the pressure. For higher pressures, an external spring is used to add its restraining force to the force against the pressure. The extension of the bellows due to the force moves a rod which is connected to a position transducer. The position transducer converts the displacement to an electrical signal. Depending on the design of the bellows and springs, relative and absolute pressures can be measured with this device.

Another common pressure-to-displacement transducer is the Bourdon tube (shown in some of its various forms in Fig. 14-15). The Bourdon tube is a flat, hollow tube that is curled into a spiral or helical shape. When a fluid under pressure is introduced into it, the tube tries to straighten out. The extent of the straightening is proportional to the pressure. For low pressures, a simple shape like that shown in Fig. 14-15(a) is used. For higher pressures, the tube is wound into a helical shape [Fig. 14-15(b)]. Since the output of the Bourdon tube is a mechanical displacement, the end of the tube must be connected to an additional electrical transducer that will convert the displacement to an electrical signal.

The *pressure cell* and *pressure transducer* shown in Fig. 14-16 use strain gauges to measure the effects of pressure on a sealed metal tube. In the pressure cell, the fluid causes a sealed tube to expand. A strain gauge mounted on the surface of the tube senses the extent of the expansion by changing its resistance value. The *pres-*

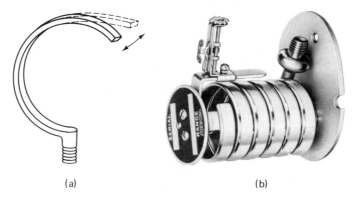

|          (a)          |          (b)          |

**Figure 14-15**   Bourdon tubes: (a) simple Bourdon tube; (b) helical- spiral Bourdon tube. (Courtesy of Foxboro Co.)

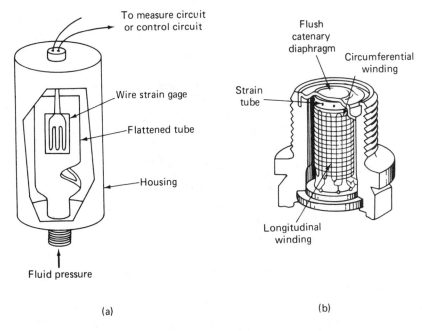

To measure circuit
or control circuit

Wire strain gage

Flattened tube

Housing

Fluid pressure

(a)

Flush
catenary
diaphragm

Circumferential
winding

Strain
tube

Longitudinal
winding

(b)

**Figure 14-16**  (a) Pressure cell; (b) pressure transducer.

*sure transducer* is a cylindrical tube with strain gauges attached to its circumference. The pressure is applied to a diaphragm at one end of the cylinder. This pressure causes the tube to contract lengthwise while increasing in diameter. The increased diameter causes the circumferentially attached gauges to change their resistances. This type of transducer is used to measure the compression capability of the cylinders in automobile and other internal-combustion engines.

## Fluid Flow Transducers

There are many types of instruments used to measure the rate of flow of fluids. The three we discuss in this section are the turbine flowmeter, the magnetic flowmeter, and the hot-wire anemometer.

    *Turbine flowmeters* are probably the most commonly used type of flowmeter (Fig. 14-17). They provide a direct method for measuring both liquid and gas flow rates. They are also particularly useful for remote monitoring and aircraft applications. The turbine flowmeter consists of a rotor mounted in a pipe through which the fluid flows. The flowing liquid causes the rotor to turn. The greater the rate of flow, the faster is the speed of rotation. The vanes of the rotor are metal, and a *magnetic pickup* mounted on the wall of the pipe senses the passing of each vane as an electrical pulse. The frequency of the pulses is proportional to the rate at which the rotor is spinning and hence to the rate of flow of the liquid. Turbine flowmeters

**Figure 14-17**   Turbine flowmeter.
(Courtesy of Cox Instruments Div.)

are available for liquid flow rates from less than 0.01 gallon per minute (gpm) to over 35,000 gpm.

The *magnetic flowmeter* is a transducer used to measure the flow of electrically conductive fluids. It has the advantage of not presenting any sort of obstruction to the fluid flow during the measurement. This type of flowmeter operates on the principle that a voltage is induced in a conductor when the conductor moves in a magnetic field. Since the voltage depends on the rate at which the conductor moves through the magnetic field, the strength of the voltage can be used as an indication of the rate of flow of the liquid.

The *hot-wire anemometer* is a fine resistance wire heated by a current passing through it. If a cooler fluid flows past the wire, heat is removed from it by the fluid. The rate of heat transfer varies with the type of fluid, but it also tends to vary as the square root of the velocity at which the fluid flows past the wire. If the current in the wire is kept constant, the change in resistance due to the cooling will yield a voltage signal. Because the diameter of the wire element can be made very small, the device can be made very sensitive and responsive to high-frequency changes in the flow rate. One of its chief uses is in aerodynamic research.

## TEMPERATURE TRANSDUCERS

A wide variety of transducers are used to measure temperature. Some of them convert temperature directly to an electrical signal, and others must be used in combination with an electrical transducer to convert the temperature indication into an electrical form. The most common temperature transducers include

1. Bimetallic strips
2. Thermocouples

**3.** Resistance-temperature detectors (RTDs)

**4.** Thermistors

**5.** Semiconductor temperature sensors

**6.** Radiation pyrometers

Each is best suited for a particular application or range of temperatures.

## Bimetallic Strip

The *bimetallic strip* is made up of two strips of different metals welded together. Because of the difference in the coefficients of thermal expansion of the two metals, a heating of the entire strip will cause one of the metals to expand in length more than the other. Since the strips are welded together along one entire edge, the complete strip will bend in the direction of the metal that expands the least. The extent of the bending is directly proportional to the degree of temperature change. If one end of the strip is firmly clamped while the other end remains free, the extent of the bending can be used to indicate temperature change. This is done by attaching a position transducer (such as an LVDT) to the free end of the strip and calibrating its displacement due to the temperature change.

Bimetallic strips are actually used more frequently as controlling devices than as temperature-indicating devices. In this role they are used most commonly as the thermostats which control the on-off switches of heating furnaces, and automotive automatic chokes. As pointed out in Chapter 3, bimetallic strips are also used in some forms of circuit breakers. Overload currents cause the strips to bend and break the circuit connections (see Fig. 3-9).

## Thermocouples

Thermocouple operation is based on the physical principle that if two dissimilar metal wires are joined together and the point of joining is heated (or cooled), a voltage difference appears across the two unheated ends. This principle (called the *Seebeck Effect*) was discovered in 1821 by T. J. Seebeck. The magnitude of the resultant voltage difference due to the Seebeck Effect is quite small (on the order of millivolts). For example, a type K thermocouple develops approximately 0.04 mV/°C. Nevertheless, the voltage difference is directly proportional to the temperature difference that exists between the heated junction and the cooler ends. If a sufficiently sensitive detector is used, temperature difference can be measured with a thermocouple. Because such small voltages are produced, the electronic signal conditioning circuitry used with thermocouple must eliminate both common mode signals and noise caused by electric and magnetic fields. Design of circuitry with this capability can be a real challenge for an engineer in an industrial environment.

The combinations of metals most commonly used for constructing thermocouples are as follows: iron and constantan, Chromel (alloy of nickel and chromium) and Alumel (alloy of aluminum and nickel), and platinum and rhodium–

**TABLE 14-1     SOME COMMON THERMOCOUPLES**

| Junction materials | Typical useful temperature range (°C) | Voltage swing over range (mV) | ANSI designation |
|---|---|---|---|
| Platinum–6% rhodium/platinum–30% rhodium | 38 to 1800 | 13.6 | B |
| Tungsten–5% rhenium/tungsten–26% rhenium | 0 to 2300 | 37.0 | (C) |
| Chromel/constantan | 0 to 982 | 75.0 | E |
| Iron/constantan | −184 to 760 | 50.0 | J |
| Chromel/Alumel | −184 to 1260 | 56.0 | K |
| Platinum/platinum–13% rhodium | 0 to 1593 | 18.7 | R |
| Platinum/platinum–10% rhodium | 0 to 1538 | 16.0 | S |
| Copper/constantan | −184 to 400 | 26.0 | T |

platinum. Table 14-1 lists a number of standard thermocouples, their useful temperature range, the voltage swing over that range, and their ANSI (American National Standards Institute) designations.

Table 14-2 is a list of the standard colors used for thermocouple insulations. The negative wire is either solid red or red with a trace of the positive wire color. Extension wires are manufactured specifically for each type of thermocouple. These are used to connect a thermocouple to a remote instrument. Their part number usually has an X suffix; for example, EPX is the positive extension wire for a type E thermocouple.

Because of their small size, thermocouples are fast devices and thus suit applications that emphasize speed of response. In addition, they function from cryogenic to well above jet-engine exhaust temperatures, are economical and rugged, and have long-term stability. Thermocouples are available in four basic types of junctions: bare wire, bare wire beaded, insulated junction, and grounded junction, (Fig. 14-18). The type of junction selected depends upon the required response time, the temperature, and the unique stresses associated with the environments in which the thermocouple is to be used (such as corrosive atmospheres, mechanical abrasion, and moisture). Most thermocouples are also placed inside a thermowell for protection from the environment.

**TABLE 14-2     THERMOCOUPLE WIRE COLOR CODING**

| Type | Positive wire color | Negative wire color | Overall insulation jacket |
|---|---|---|---|
| E | Purple | Red-purple trace | Purple |
| J | White | Red-white trace | Black |
| K | Yellow | Red-yellow trace | Yellow |
| K | Green | Red-green trace | White |
| R or S | Black | Red-black trace | Green |
| T | Blue | Red-blue trace | Blue |

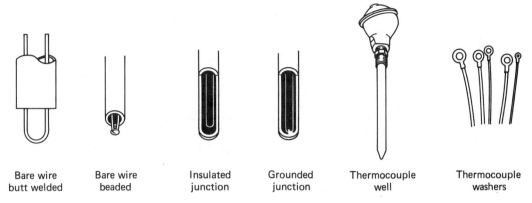

Bare wire butt welded   Bare wire beaded   Insulated junction   Grounded junction   Thermocouple well   Thermocouple washers

**Figure 14-18**  Thermocouple types and hardware.

Figure 14-19 is a comparative plot of thermocouple output as a function of temperature, referred to a 0°C fixed-temperature reference junction. Figure 14-20 shows a circuit for making a measurement with a thermocouple, using an ice bath to maintain the reference junction at 0°C.

Because thermocouples generate output voltages that are so small (only tens

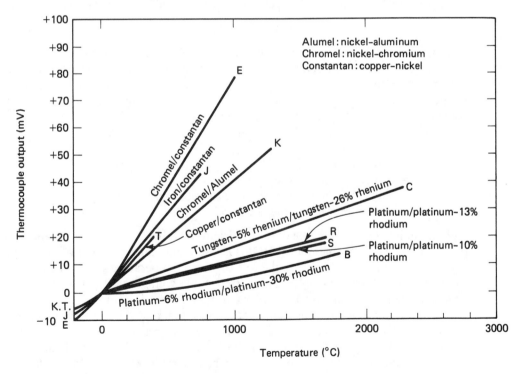

**Figure 14-19**  Output characteristics of thermocouples.

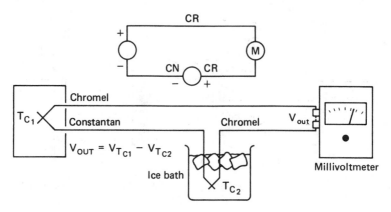

**Figure 14-20** Simple temperature measuring circuit using an ice bath at the reference junction. Thermocouple measurements are inherently differential.

of microvolts per degree), they provide a challenge to the signal conditioning electronics circuits to allow accurate resolutions of less than 1°C. In addition, the linearity of the relationship between the temperature and output voltage in many thermocouple types is poor, albeit predictable and repeatable. The temperature–voltage relationship can usually be closely approximated by an equation of the form

$$V_o = AT + BT^2 \tag{14-6}$$

when the reference junction temperature is 0°C. If $A$ and $B$ are known, linearization techniques that allow corrections to the output voltage to be made result in output voltage curves that are linear to within 1 percent. Linearization techniques applied to thermocouples are discussed in references 4, 5, and 9 of Chapter 17. Better conversion accuracy can be obtained by using the power series polynomial

$$T = a_0 + a_1 x + a_2 x^2 + a_3 x^3 + a_4 x^4 + \cdots + a_n x^n \tag{14-7}$$

where   $T$ = temperature in °C
   $x$ = thermocouple voltage
   $a$ = polynomial coefficients unique to each thermocouple
   $n$ = maximum order of the polynomial

and the National Bureau of Standard coefficients in Table 14-3. A look-up table in a programable read-only memory (PROM) is also a popular method of converting a digitized signal from a thermocouple to a digitized temperature scale. The analog-to-digital converters that convert the thermocouple voltages to digital signals must have a resolution of 1 microvolt in order to obtain a sensitivity of 0.1°C.

Another serious handicap of thermcouples is that they require a reference to a known temperature for use in absolute temperature measurements; that is, the circuitry must compare the output of the "signal" thermocouple with that of a similar "reference" thermocouple. Of course, the latter *must* be held at a known temperature.

**TABLE 14-3** NATIONAL BUREAU OF STANDARDS THERMOCOUPLE COEFFICIENTS

| | TYPE E | TYPE J | TYPE K | TYPE R | TYPE S | TYPE T |
|---|---|---|---|---|---|---|
| | Nickle-10% Chromium (+) Versus Constantan (−) | Iron (+) Versus Constantan (−) | Nickel-10% Chromium (+) Versus Nickel-5% (−) (Aluminum Silicon) | Platinum-13% Rhodium (+) Versus Platinum (−) | Platinum-10% Rhodium (+) Versus Platinum (−) | Copper (+) Versus Constantan (−) |
| | −100°C to 1000°C* ±0.5°C 9th order | 0°C to 760°C* ±0.1°C 5th order | 0°C to 1370°C* ±0.7°C 8th order | 0°C to 1000°C* ±0.5°C 8th order | 0°C to 1750°C* ±1°C 9th order | 160°C to 400°C* ±0.5°C 7th order |
| $a_0$ | 0.104967248 | −0.048868252 | 0.226584602 | 0.26362917 | 0.927763167 | 0.100860910 |
| $a_1$ | 17189.45282 | 19873.14503 | 24152.10900 | 179075.491 | 169526.5150 | 25727.94369 |
| $a_2$ | −282639.0850 | −218614.5353 | 6723.4248 | −48840341.37 | −31568363.94 | −767345.8295 |
| $a_3$ | 12695339.5 | 11569199.78 | 2210340.682 | 190002E+10 | 8990730663 | 78025595.81 |
| $a_4$ | −448703084.6 | −264917531.4 | −860963914.9 | −4.82704E+12 | −1.63565E+12 | −9247486589 |
| $a_5$ | 1.10866E+10 | 2018441314 | 4.83506E+10 | 7.62091E+14 | 1.88027E+14 | 6.97688E+11 |
| $a_6$ | −1.76807E+11 | | −1.18452E+12 | −7.20026E+16 | −1.37241E+16 | −2.66192E+13 |
| $a_7$ | 1.71842E+12 | | 1.38690E+13 | 3.71496E+18 | 6.17501E+17 | 3.94078E+14 |
| $a_8$ | −9.19278E+12 | | −6.33708E+13 | −8.03104E+19 | −1.56105E+19 | |
| $a_9$ | 2.06132E+13 | | | | 1.69535E+20 | |

Temperature Conversion Equation: $T = a_0 + a_1x + a_2x^2 + \cdots + a_nx^0$

Nested Polynomial Form: $T = a_0 + x(a_1 + x(a_2 + x(a_3 + x(a_4 + a_5x))))$ (5th order)

Source: Practical Temperature Measurements, Application Note 290, August 1980. (Courtesy of Hewlett-Packard.)

Providing such a suitable temperature reference and minimizing the effects of unwanted thermocouples can prove challenging. Techniques include physical references (ice-point cells at 0.1°C, which are accurate and easy to build but cumbersome to maintain); ambient-temperature reference junctions (acceptable as long as the temperature variation in the vicinity of the reference junction is smaller than the desired resolution of the temperature being measured); and electronic cold-junction compensators, which provide an artificial reference level and compensate for ambient-temperature variations in the vicinity of the reference junction. That is, an offsetting circuit measures the ambient temperature at the reference junction and adds a voltage to the thermocouple output equal to the voltage expected to be developed by the reference, but of opposite polarity. This is easily done by using a temperature-compensated low-voltage reference diode such as the LM113 (which has highly predictable properties) or by using a monolithic temperature sensor such as the LT1025 manufactured by Linear Technology (Figure 14-21), which is specifically designed for cold junction compensation. The net output of the thermocouple circuit when used with either of these devices is a voltage whose value is equivalent to the voltage that would be produced if the reference junction were at 0°C.

There are two empirical laws of thermocouples that permit the analysis of most practical thermocouple circuits. These can be stated as follows:

**1.** The output voltage from a thermocouple is unaffected by the temperature of the wire between the thermocouple junction and the reference junction. This principle permits the wires between the thermocouple junction and the reference

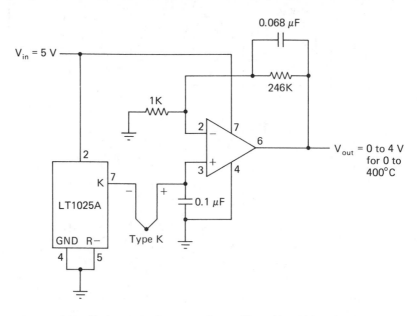

**Figure 14-21** Single supply thermocouple amplifier with cold junction compensation.

junction to be routed throughout various temperature regions of a plant without affecting the output voltage.

**2.** A third metal may be connected to either side of a thermocouple without affecting the output voltage so long as both junctions of the third metal are at the same temperature. This principle allows a measuring device such as a voltmeter with copper terminals and wire to be inserted into an iron-constantan thermocouple circuit for measuring the voltage without affecting the voltage. This also says that the thermocouple junction itself can be either soldered with a third metal or welded and the output voltage will not be affected.

### Resistance-Temperature Detectors (RTDs)

In 1821, Sir Humphrey Davy discovered that metals change their resistance value with temperature. Nearly 50 years later, Sir William Siemens suggested the use of platinum be used as the element for a resistance thermometer. Platinum is still used today for high-accuracy resistance thermometers. The change in resistance of a metal makes it possible to measure temperature by measuring the change in resistance of a current-carrying device. The *resistance-temperature detector* (RTD) is a device whose operation is based on this effect. The platinum resistance temperature detector (PRTD) is used today as an interpolation standard from the boiling point of oxygen ($-182.962°C$) to the boiling point of antimony ($630.74°C$). The classical construction of an RTD consists of a coil of fine copper, nickel, or platinum wire attached to a supporting framework. This device configuration was proposed by C. H. Meyers in 1932.

For the industrial environment, the measuring element consists of a bifilar winding around a ceramic bobbin that is encased in a coating of molten glass. The coefficient of expansion of the bobbin and the measuring element must be precisely matched to prevent strain-induced changes in resistance. A commercially available RTD is shown in Fig. 14-22. The tube that surrounds the measuring element is made from a high-thermal–conductivity material to allow fast response to temperature changes. Metal film RTDs are also used in the industrial environment. They are constructed by depositing a thin film coating on a ceramic substrate. A pattern in

**Figure 14-22** Resistance temperature detector. (Courtesy of Thomas A. Edison Instrument Division, McGraw-Edison Co.)

this film is then etched and laser trimmed to a precise resistance value. For very low temperature work (i.e., less than 50°K), carbon resistors are used. Copper and nickel are used when economics require reduced cost, but at a penalty of reduced linearity. The resistance value of RTDs can range from 10 ohms for bird-cage models to several thousand ohms for metal film devices. The most popular value, however, is 100 ohms at 25°C.

Wheatstone bridges are used to sense the resistance changes arising in resistance thermometers. The bridges are usually calibrated to indicate the temperature that caused the resistance change rather than the resistance change itself. The three-wire bridge, [Fig. 14-23(a)] must be used to reduce the effects of lead resistance. The third wire carries no current, and therefore its resistance does not affect the output voltage. An improved technique that is less susceptible to noise uses a four-wire system consisting of a constant current source and a digital voltmeter [Fig. 14-23(b)]. Since the digital voltmeter draws very little current, lead resistance does not have to be considered when a four-wire RTD temperature measuring system is being designed. Noise is a reduced factor because of the constant current source.

Several practical precautions must be observed when installing an RTD. The size of the current source must produce the proper output voltage from the RTD, yet it must not cause self-heating in the RTD. A 100-ohm platinum film RTD can rise 1°C in still air with a current source of only 5 mA. As a result, current sources of

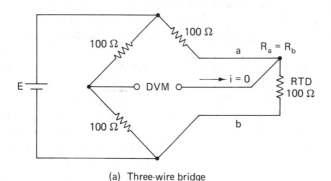

(a) Three-wire bridge

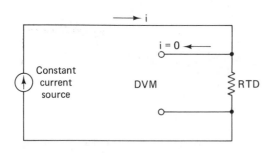

(b) Four-wire measurement

**Figure 14-23**  RTD measuring circuit schematics: (a) three-wire bridge; (b) four-wire measurement.

**TABLE 14-4**  TYPICAL RTD PROPERTIES

| Material | Temperature range (°C) | TC (%/°C)@25°C |
|----------|------------------------|----------------|
| Platinum | −200 to +850 | 0.39 |
| Nickel | −80 to +320 | 0.67 |
| Copper | −200 to +260 | 0.38 |
| Nickel–iron | −200 to +260 | 0.46 |

less than 5 mA are typically used. Large RTDs can also cause thermal shunting, thereby causing the actual measured temperature to change. Finally, any changes in metal (e.g., platinum to copper) can result in thermal emf offsets, which require compensation.

Even though the RTD is more linear than the thermocouple, it still requires curve fitting. The coefficients for a 20th-order polynomial are available from most manufacturers of RTDs.

Table 14-4 compares the temperature ranges and temperature coefficients of resistivity (TCs) of commonly used RTD materials.

## Thermistors

Thermistors are devices that also measure temperature through a changing resistance effect. However, the resistance of materials from which thermistors are made[4] decreases with increasing temperature from about −100°C to +300°C. In some thermistors, the decrease in resistance is as great as 6 percent for each 1°C of temperature change (although one percent changes are more typical).

The decrease in resistance that takes place in thermistors involves the chemical bonding properties of electrons in semiconductor materials. In these materials the valence electrons are locked in covalent bonds with their neighbors. As the temperature of the thermistor is increased, the thermal vibrations of its atoms break up some of these bonds and release electrons. Since the electrons are no longer bound to the specific atoms in the lattice, they are able to respond to applied electric fields by moving through the material. These moving electrons add to the current in the semiconductor, and the material appears to have a smaller resistance.

Because the change of resistance per degree of temperature change in thermistors is so large, they can provide good accuracy and resolution when used to measure temperatures between −100°C and +300°C. If an ammeter is utilized to monitor the current through a thermistor, temperature changes as small as ±0.1°C can be detected. If the thermistor is instead put into a Wheatstone bridge, the measuring system can detect temperature changes as small as ±0.005°C.

Thermistors are most commonly made in the form of very small beads. This shape and others are shown in Fig. 14-24. Because of their small size, they can be inserted into regions where other larger temperature-sensing devices might not fit.

[4]Semiconductor ceramics consisting of a mixture of metallic oxides (such as manganese, nickel, cobalt, copper, and iron) are used as the materials from which thermistors are made.

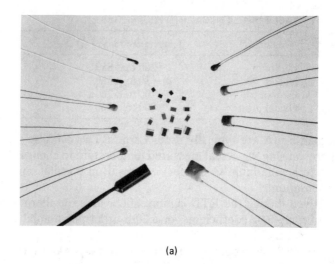

(a)

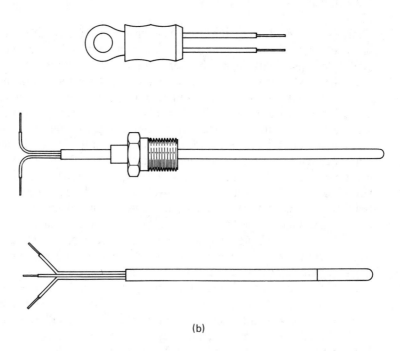

(b)

**Figure 14-24** Thermistors: (a) surface mount and glass-beaded thermistors; (b) commercial thermistor assemblies. (Courtesy of Alpha Thermistor, Inc., San Diego, Calif.)

    The change in resistance of thermistors in response to changes in temperature is inherently nonlinear, as shown in Fig. 14-25. An individual thermistor curve, however, can be very closely approximated by using the Steinhart-Hart equation (Eq. 14-8) and data either provided by the manufacture or obtained by direct measurement.

$$\frac{1}{T} = A + B \ln R + C (\ln R)^3 \tag{14-8}$$

where    $T$ = degrees Kelvin
           $R$ = resistance of thermistor
   $A, B, C$ = curve-fitting constants

The constants $A$, $B$, and $C$ are found by selecting three data points from the manufacturer's data and solving the three simultaneous equations that result when the three data points are substituted into Eq. (14-8). The result is an equation that approaches a $\pm0.02°C$ curve fit. A less accurate approximation of the resistance can be obtained by using the expression

$$R = R_0 e^{\beta\left(\frac{1}{T} - \frac{1}{T_0}\right)} \tag{14-9}$$

where    $R$ = resistance at $T(°K)$
        $R_0$ = resistance at $T_0(°K)$
        $\beta$ = curve-fitting constant (2000–4000)

    Thermistors are used as the sensor for low temperature controllers, linearizers for resistance circuits, moisture detectors, and air flow detectors. The resistance of a thermistor can vary from 50 $\Omega$ to 2 M$\Omega$ ohms at 25°C. Thermistors are available with negative temperature coefficients (NTC) and positive temperature coefficients (PTC), with those possessing an NTC being much more prevelant in the commercial market.

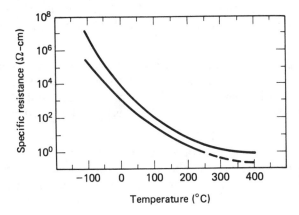

**Figure 14-25** Resistance versus temperature characteristic of two typical thermistor materials.

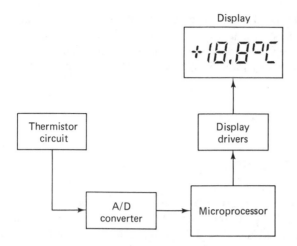

**Figure 14-26** Block diagram of microprocessor (or EPROM) based thermistor to display circuit.

Figure 14-26 is a block diagram of the method used to convert a thermistor output to a digital display of temperature. The microprocessor can be replaced by an EPROM. In this case the output from the A/D converter is the address of the code in the EPROM that corresponds to a particular temperature. This code is directly output to the display driver, which is simply a buffer or inverter, depending upon whether the display is common cathode, common anode, or LCD matrix.

Special linearized thermistor networks can be built and are available commercially in single-package form (e.g., from Yellow Springs Instrument Co.). Such products consist of two or more thermistors and fixed resistors and provide a nearly linear voltage-versus-temperature output over moderate (−30 to +100°C) temperature ranges.

### Semiconductor Temperature Transducers

There are many types of temperature-sensing transducers that take advantage of the physical properties of semiconducting materials. We discuss three of the most popular types: bulk semiconductor resistors, semiconductor diodes, and integrated circuits (i.e., the AD590).

**1.** The simplest semiconductor temperature transducers are merely pieces of *bulk silicon*. They are inexpensive, reasonably linear from −65 to 200°C (±0.5 percent), and have a positive temperature coefficient of 0.7%/°C. Physically they look like $\frac{1}{4}$-W resistors and their nominal resistance ranges from 10 Ω to 10 kΩ. Like RTDs, silicon resistors may be used in bridge circuits.

**2.** *Semiconductor diodes* are used to measure temperature because their junction potential is proportional to the diode temperature. In silicon diodes a change in junction potential of about 2.2 mV/°C is manifested. Diode temperature sensors are inexpensive, provide fast response, and are useful over temperature ranges from about −40 to −150°C.

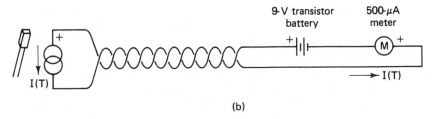

(a)

(b)

**Figure 14-27** (a) AD590 as a two-terminal device; (b) simple implementation of the AD590.

**3.** Integrated circuits that provide a measurement of temperature are also available commercially. A very popular example of such a device is Analog Devices' AD590. This sensor is a two-terminal device [Fig. 14-27(a)] in which the current passing through the device is numerically equal to the absolute temperature (over the temperature range −125 to −200°C. That is, if the temperature of the device is −55°C (218.2°K), a current of 218.2 μA flows through the AD590. If the device is at 0°C (273°K), 273 μA of current flows through the device, and so on. The excitation source required by the AD590 is just a voltage supply ranging from +4 to +30 V. Figure 14-27(b) shows how the AD590 can be simply used in a remote measurement application.

The AD590 has found wide application because it possesses the advantages of ease of use, high linearity (i.e., better than ±1 percent), and low-voltage excitation. Furthermore, it does not require bridge circuitry. Finally, since the output is in current form, long leads may be used without introducing errors owing to voltage drops or induced noise voltage.

## Radiation Pyrometers

*Radiation pyrometers* are devices that sense temperature by measuring the optical radiation emitted by hot bodies. The higher the temperature to which a body is heated, the higher is the dominant frequency of radiation it emits. This means that as the temperature of a body increases to a point where it begins to emit visible light, the heated surface will first have a dull red color. As the body gets hotter and more incandescent, its surface becomes progressively less red and more white.

To detect the emitted radiation, the radiation pyrometer does not have to be placed in the furnace or region it is measuring. It is merely necessary to point it at the heated surface of interest to make the measurement.

The *disappearing-filament* type of pyrometer uses a heated-wire filament to provide a radiant temperature standard. An electric current passing through the filament provides an accurate heating method. When the filament is heated to the same temperature that exists at the surface being examined, the filament image is no longer visible because it has the same color as the surface (Fig. 14-28). Since the current through the filament is a known quantity, the pyrometer can be calibrated to yield the temperature of the surface from this current value. Because a body begins to emit visible light when heated to about 775°C, a disappearing-filament pyrometer can measure temperature from this point to about 4200°C.

In the *brightness* type of pyrometer, the radiation from the heated surface being examined is collected by a lens and focused onto a thermistor or thermocouple. The net radiation from a heated ideal blackbody follows the Stefan-Boltzmann Law, which states that the total radiation from a black body is given by

$$W_{\text{total}} = \sigma T^4 \tag{14-10}$$

where    $W$ = total radiation expressed in watts/cm$^2$
  $T$ = absolute temperature of the hot body, °K
  $\sigma$ = Stefan-Boltzmann constant, $5.672 \times 10^{-12}$ W cm$^{-2}$ deg$^{-4}$

The problem with applying this equation is that it is difficult to find an ideal blackbody in an industrial environment (e.g., red objects and blue objects at the

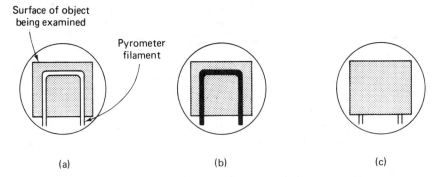

**Figure 14-28**  Principle of disappearing-filament optical pyrometer: (a) Too much current through filament raises it to a higher temperature than the surface. (b) Filament of wire too cold. Not enough current is being passed through it. (c) Filament "disappears" when it reaches the same temperature as the surface being examined.

same temperature do not emit the same amount of radiation). Emissivity is defined as the radiation efficiency of an actual surface in comparison with that of an ideal blackbody radiator. Consequently, to take such emissivity differences into consideration, Plank's equation for graybodies is used.

$$W = \frac{\epsilon C_1}{\lambda^5 (e^{C_2/\lambda T} - 1)} \tag{14-11}$$

where  $W$ = radiation power in watts per unit area

 $\lambda$ = radiation wavelength in microns

 $T$ = target temperature in degrees Kelvin

 $C_1 = 3.7405 \times 10^4$ when area is in square centimeters and wavelength in microns.

 $C_2 = 1.43879 \times 10^4$ when square centimeters and microns are used.

 $\epsilon$ = emissivity or emission efficiency for the graybody surface.

By using two detectors that are filtered to accept radiation in two independent adjacent windows of the spectrum, $\lambda 1$ and $\lambda 2$, an output voltage ratio can be obtained that is independent of the material emissivity. This is called *two-color (ratio) infrared thermometry*. The output voltage of the ratio is defined as follows:

$$R = \frac{E_1}{E_2} = \frac{\lambda_2^5}{\lambda_1^5} \times \frac{(e^{C_2/\lambda_2 T} - 1)}{(e^{C_2/\lambda_1 T} - 1)} \tag{14-12}$$

This equation is usually solved by a microcomputer in the instrument that processes the incoming infrared radiation. The optical system that is associated with an instrument must be able to accurately define the area being observed. Because the response time of the sensors is considerably slower than a microprocessor, several transducers can be multiplexed by one controller (Fig. 14-29).

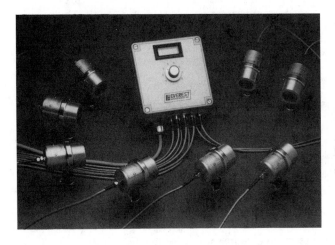

**Figure 14-29** Eight infrared transducers that have internal signal conditioning and are connected to a multiplexer. (Courtesy of Everest Interscience Inc.)

## LIGHT AND RADIATION TRANSDUCERS

The spectrum of electromagnetic radiation extends from radio waves (less than 10 Hz) to gamma rays ($10^{20}$ Hz or higher). The very low frequency radio waves have the longest wavelengths; the gamma rays the shortest. In between these extremes are all the other categories of electromagnetic radiation, including light (see Fig. 14-30). For our purposes of classification, light will be defined to include the radiations belonging to the infrared, visible light, and ultraviolet portions of the electro-magnetic spectrum.

In this section we are interested primarily in those transducers that can sense light radiation and convert it into an electrical form. The general class of light-radiation transducers, also known as *phototransducers*, is used to detect the presence and intensity of light under various circumstances. In fact, many photo-transducers can be made much more sensitive to light radiation than the human eye is. The three primary types of light-to-electrical energy transducers are

1. Photoemissive devices
2. Photoconductive devices
3. Photovoltaic devices

Each of these types possesses special advantages over the others.

### Photoemissive Light Sensors

Photoemissive light sensors are so named because they contain materials whose surfaces emit electrons when struck by light radiation. The electrons are emitted

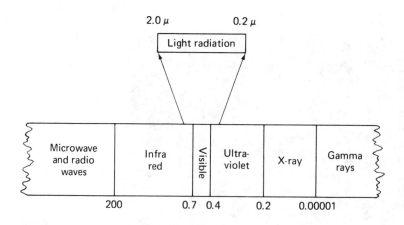

**Figure 14-30**  The electromagnetic spectrum.

when the photons of the incident light are able to transfer enough energy to the electrons to break them free from both their atomic bonds and the forces of the entire material lattice. Materials in which this phenomenon takes place easily enough to produce many electrons when struck by visible light are called *photoemissive* materials. Because the photoemissive material is usually housed in a glass tube, photoemissive devices are also often known as *phototubes*. Figure 14-31 shows the basic principle behind the operation of phototubes.

The surface of a specially shaped cathode is coated with a photoemissive material (such as cesium-antimony). The cathode (now called a *photocathode*) is housed in a sealed glass tube along with another electrode called the anode. A voltage is created between the photocathode and the anode (with the anode having the positive voltage level). When light strikes the photocathode, the electrons emitted from the surface are attracted and collected by the positive anode. The stronger the intensity of the light incident on the photocathode, the more electrons it emits. Therefore, the magnitude of the current flowing in the circuitry connected to the electrodes of the tube is directly proportional to the intensity of the light incident on the photocathode.

Before going on with the description of the various types of photoemissive tubes, one precautionary note on their use should be sounded here. Because phototubes suffer from a condition known as *phototube fatigue*, the data obtained with the use of phototubes are often misinterpreted. Phototube fatigue is the loss of sensitivity of a photoemissive surface when it is subject to constant illumination by an intense source of light. The time constant of this effect varies from tube to tube, but it is on the order of $\frac{1}{2}$ h. The mistakes are made in interpreting the cause of the decreased intensity of the output signal of the phototube. Instead of attributing the

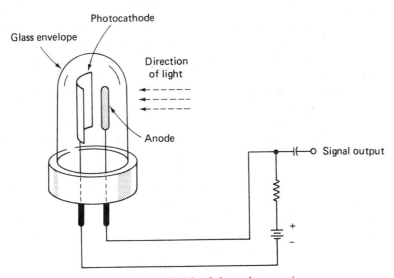

**Figure 14-31**   Principle of phototube operation.

resultant decrease to the characteristics of the tube, the cause for the decrease is sought as a change in the magnitude of the quantity being measured. One way to avoid having this effect occur is to use a mechanical chopper ahead of the tube. By chopping the light beam, the chopper converts the constant light intensity to a pulsating form. This greatly reduces the effect of phototube fatigue.

There are three commonly used types of phototubes. The first is the *vacuum phototube* shown in Fig. 14-31. In this type a vacuum exists within the glass tube. When light strikes the photocathode of the tube, electrons are emitted from its surface. If a sufficient voltage exists between the anode and the photocathode, the resultant current is almost linearly dependent on the intensity of the light. In fact, the response of vacuum phototubes is linear over such a wide range of light levels that they are used as standards in light-comparsion measurements. In addition, the response time of phototubes to the incident light is so rapid that they are suitable for applications where very short duration light pulses must be observed.

The second type of phototube is the *gas-filled phototube*. In this device, the tube that houses the photocathode and anode is filled with an inert gas (such as argon) at a very low pressure. As electrons are emitted from the photocathode, they are also accelerated toward the anode by a voltage difference. While in transit between the electrodes, these electrons collide with the argon gas atoms. If the energy of an electron is high enough, the collision ionizes (i.e., tears away electrons) the argon atoms and therefore creates a positive ion and additional free electrons. The electrons are attracted by the anode and the positive ions by the cathode. As a result, a greater current appears to flow between the anode and photocathode. Because of the multiplication effect of charge carriers arising from the collisions, the resulting current collected by the electrodes is often large enough that it does not require amplification. This makes gas phototubes very simple and inexpensive devices.

The relatively slow motion of the positive ions toward the cathode makes the response of gas phototubes to variations in the intensity of the incident light relatively slow. Therefore, gas phototubes are only suitable for applications where such a slow response time is not a hindrance. They are chiefly used to reproduce the sound tracks for movie films because their response-times are sufficiently rapid for this task.

The last type of phototube is the *photomultiplier tube* (PM) tube. These devices are probably the most widely used types of light detectors. Their outstanding characteristic is that they can detect very low level light intensities.

The ability to detect very small intensities of light is because photomultiplier tubes are actually amplifying devices. In Fig. 14-32 we see that the incident light beam is first made to strike a photoemissive surface in the same manner as in the ordinary vacuum phototube. However, the emitted electrons are not drawn immediately to an anode. Instead, they are attracted (by a voltage difference) to another electrode called a *dynode*. The dynode emits *secondary electrons* when struck by an electron beam. Thus each original photoelectron is accelerated by an electric field and knocks several (i.e., three to six) secondary electrons out of the dynode. There

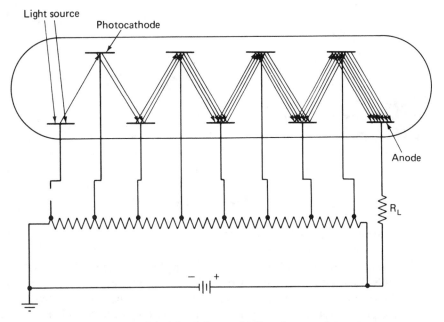

**Figure 14-32**   Principle of photomultiplier-tube operation.

are usually ten dynodes in a PM tube, and each one is designed to form electric field lines which guide the secondary electrons emitted by the previous dynode to itself. [Each dynode is at a higher potential ($\approx 100$ V) than the preceding one.] Thus the secondary electrons are *multiplied* in number at each dynode, and the final burst is collected by the anode. In this way, a multiplication factor of over $10^6$ is achieved in commercially available tubes. Such amplifications allow PM tubes to detect the event of even a single electron being emitted from the photocathode. The response time of PM tubes is also very rapid, and frequencies of up to hundreds of megacycles can be followed by them. However PM tubes are generally not suitable for detecting infrared radiation because materials are not photoemissive in response to infrared radiation.

Night vision imaging systems (Fig. 14-33) consist of a semiconductor photocathode made from gallium arsenide (GaAs) that is bonded to a glass faceplate. The GaAs photocathode has a sensitivity of 1000 $\mu$A/lumen. Electrons are emitted from the photocathode in proportion to the intensity of the incident light. The electrons are then accelerated by an electric field to a microchannel plate, which consists of millions of fiberoptic channels. The plate (which is approximately 1 mm thick) emits secondary electrons due to an emissive material on the fibers. Following the plate is a phosphor screen that emits light that is proportional to the number and velocity of the impinging electrons. This type of night imaging system can intensify both visible and near-infrared light.

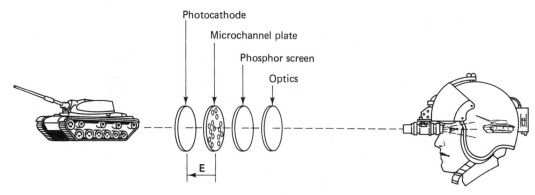

Photocathode

Microchannel plate

Phosphor screen

Optics

E

**Figure 14-33**    Principle of night vision.

## Photoconductive Light Detectors

Photoconductive light detectors are basically light-sensitive resistors. They are also called light dependent resistors (LDR), photoresistors, and junctionless detectors. The major difference between the photoresistor and the photomultiplier tube is that in a photoresistor an electron is excited into the conduction band rather than being removed from the material. Typical curves are shown in Fig. 14-34. Visible light detectors are generally made from the semiconductor materials cadmium sulfide (CdS) or cadmium selenide (CdSe), both of which have a bandgap energy, $E_g$, of approximately 2.42 eV. There are many infrared detector materials. Some of the major materials are lead sulfide (PbS), lead selenide (PbSe), indium antimony (InSb), and mercury- or copper-doped germanium (Ge). In all cases, the thermally induced current must be much smaller than the optically induced current. To achieve this, some detectors must be cooled with liquid nitrogen.

The resistance of CdS and CdSe devices decreases as a result of the creation of electron-hole pairs when the energy of incident photons is greater than the bandgap energy, $E_g$. The freed electrons are available as charge carriers in the conduction band. Some of the infrared detectors such as the mercury and copper compound devices operate by incident photons ionizing an acceptor state. The ionization energy required for copper germanium Ge devices is only 0.04 eV.

Photoconductors are constructed by applying a thin layer of the semiconductor material on a substrate of ceramic or silicon. The dark resistance varies from 10 K$\Omega$ to 200 M$\Omega$, depending on the device. The ratio of the dark-to-light resistance can be as high as 10,000. Each cell is sensitive to different wavelengths. Their peak spectral response varies from 0.5 to 2.2 $\mu$m, with CdS near 0.6 $\mu$m and CdSe near 0.75 $\mu$m. The switching times of these devices are relatively slow. They range from 1 to 100 ms.

A medical application of a photoconductor is the plethysmograph (or pulse detector) used by athletes (Fig. 14-35). The amount of light from the LED that is reflected to the photoconductor is inversely proportional to the volume of material

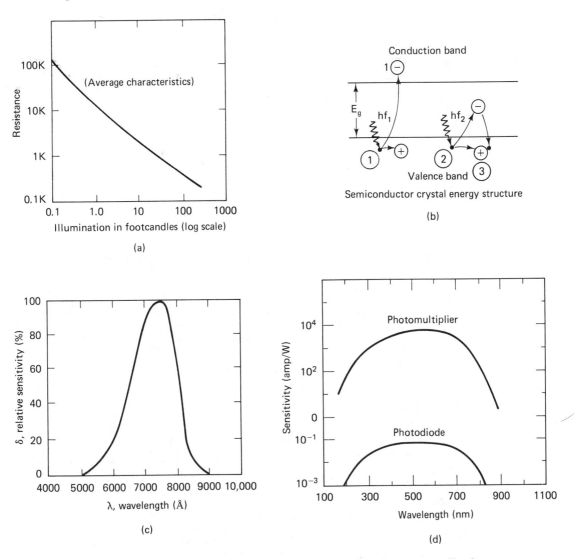

**Figure 14-34** Characteristics of photoconductive light detector: (a) resistance versus illumination; (b) photoeffect; (c) spectral response of CdSe; (d) comparison of photomultiplier and photodiode sensitivity.

the light passes through. An arterial pulse will increase the artery volume, thus increasing the photoconductor's resistance. A constant current source through the photoconductor results in a voltage pulse applied to a rate counter and display. It is important that the spectral responses of the LED and photoconductor be matched for maximum transfer efficiency.

Photoconductive cells are often used as the photosensitive elements in photo-

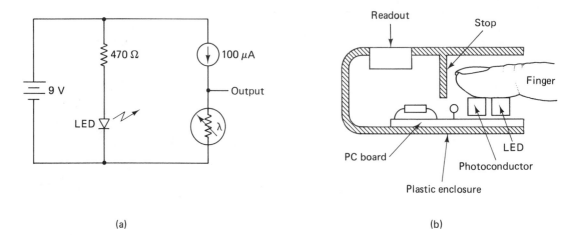

(a)                                                                                (b)

**Figure 14-35**   Pulse detector: (a) schematic; (b) mechanical arrangement.

electric relays and proximity switches, as devices that measure light intensity, as automatic iris control in cameras, and as part of the automatic control that activates home security and street lights at dusk. They are also used as the sensor in part counters and intrusion alarms. Typical curves for devices sensitive in the infrared region are shown in Fig. 14-36.

Figure 14-37 shows a cutaway view of a photoconductive device.

### Photodiodes

Photodiodes are devices that have the same electrical characteristics as conventional PN-junction diodes when they are not subjected to illumination. That is, their forward biased resistance is low and their reverse-biased resistance is high. When diodes are illuminated by light of the proper frequency, however, the reverse-bias current of the diode increases with the illumination intensity. Their characteristic I-V curves are similar in shape to those of a transistor, [Fig. 14-38(a)]. Their structures can be those of a PN junction, a PIN (a PN structure with an intrinsic layer between the P and N layers), an avalanche photodiode (APD), or a Schottky barrier diode. The subtle differences in the characteristics of the photodiodes offered by different manufacturers should be examined as part of the selection process. Some of these characteristics include the size, noise, temperature limits, light-level limits, and speed. PIN photodiodes are designed for infrared radiation detection in high frequency fiberoptical systems. Their rise time is typically 1.5 ns, and they exhibit their best linearity when operated with a current amplifier as shown in Fig. 14-38(c). Logarithmic operation can be obtained when the photodiode is operated at zero bias with a very high impedance amplifier.

When a photodiode is not biased with a voltage, but is nevertheless illuminated, it acts as a voltage source (with polarity of the p-layer being positive with

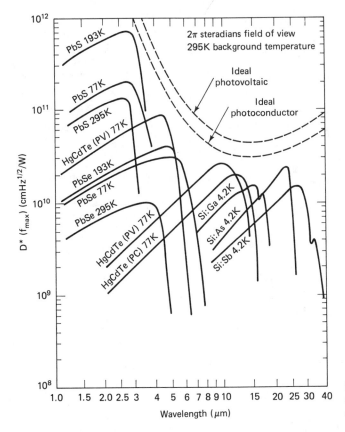

Figure 14-36 Spectral response of commercially available detectors. (Courtesy of EG&G Judson.)

respect to that of the n-layer). In this mode of operation, the photodiode is usually called a *photovoltaic diode* or *solar cell*. For silicon cells the typical short circuit current is 80 mA/cm$^2$ and the open circuit voltage is 0.6 volts. Their efficiency ranges from 5 to 31 percent in a light intensity of 35–50 W/cm$^2$. The cost of such devices is almost directly proportional to their efficiency. The materials used to construct solar cells are usually silicon or selenium. Other materials such as gallium

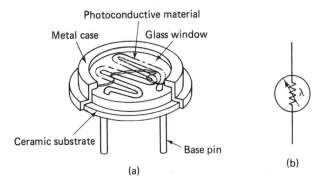

Figure 14-37 Photoconductive cell: (a) cutaway view; (b) symbol.

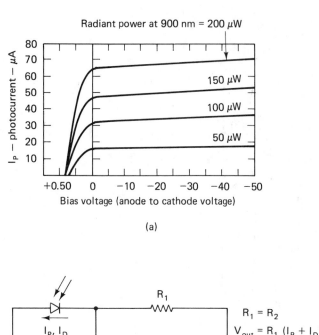

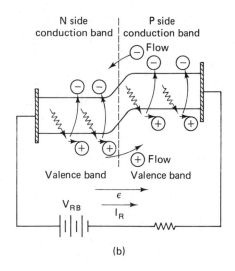

(a)

(b)

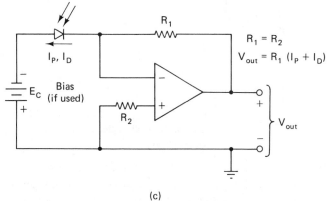

$R_1 = R_2$

$V_{out} = R_1 (I_P + I_D)$

(c)

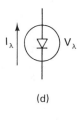

(d)

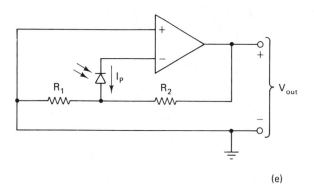

$$V_{out} = \left(1 + \frac{R_2}{R_1}\right) \cdot \frac{kT}{q} \cdot \ln\left(1 + \frac{I_P}{I_S}\right)$$

where $I_S = I_F \left(e^{\frac{qV}{kT}} - 1\right)^{-1}$ at $0 < I_F < 0.1$ mA

(e)

**Figure 14-38** Photo diode properties: (a) characteristic curves; (b) photo effect; (c) schematic for linear operation; (d) symbol; (e) schematic for logarithmic operation.

arsenide, cadmium sulfide, and indium arsenide are less commonly used. The most efficient cells use a multijunction structure that absorbs a broader part of the incident light spectrum than a single cell. For example, the upper cell is sensitive to blue light and has an efficiency of 27 percent, while the lower cell is sensitive to red light and has an efficiency of 4 percent. A typical characteristic curve for a solar cell is shown in Fig. 14-39. The shaded area is the maximum power available from the cell. Because sunlight is not monochromatic, the energy gap of the semiconductor should be as small as possible to allow the largest percentage of sunlight photons to be absorbed.

### Phototransistors

Phototransistors have a photodiode as the reverse biased-collector base junction. The base terminal can be left floating, or it can be biased. The diode current is the reverse-bias leakage current, which increases as a result of photo-stimulation. Owing to transistor action, the increase in base current will be amplified from 100 to 1000 times. To increase sensitivity, the base area is enlarged. Since the diode dark current is also amplified, the signal-to-noise ratio is not increased. The sensitivity can be increased another order of magnitude by using a Darlington pair instead of a single transistor. The response of the phototransistor is a function of both the light intensity and the wavelength. The same sensitivity can be obtained by using a photo-sensitive FET that uses the gate-to-channel junction as the photodiode. Also available are photothyristors, which are called light-activated SCRs or LASCRs. Phototransistors are available in a variety of packages. The TO-18 is shown in Fig. 14-40. These devices are designed for industrial applications such as light modulators, security systems, shaft position encoders, end of tape detectors, and disk hole detectors.

By combining an infrared emitting diode and a phototransistor, an optocoupler/isolator is formed. The emitting diode and the receiving photo transistor must be matched for wavelength response to maximize the *transfer ratio* (defined as the ratio of $I_c$ to $I_{diode}$). Most manufacturers express this ratio as a percentage (such as 150 percent). For this example transfer ratio, the collector current is only 1.5 times the diode current. This percentage can be as high as 500, but it is usually less than 100. External stages of amplification are usually necessary when designing devices with such photoelectric switches and solid state relays. A typical characteristic curve and application schematics are shown in Fig. 14-40. The optoisolator is usually packaged in a six-pin DIP package.

### X-Ray and Nuclear Radiation Transducers

X-ray and nuclear radiation sensors utilize some of the same principles of operation as light-radiation sensors. For example, one type of X-ray sensor is like a photo-conductive device, while another is made of a material that emits visible light when struck by X rays. The intensity of the emitted visible light in this latter type is

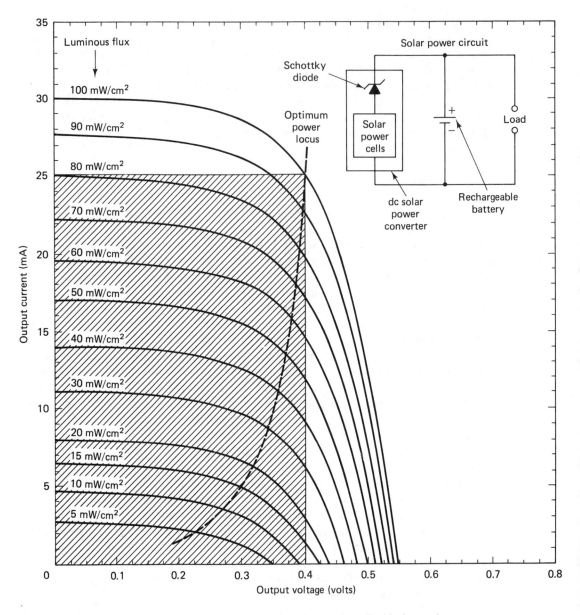

**Figure 14-39** Typical output characteristic of a silicon solar cell with the maximum power rectangle shaded in.

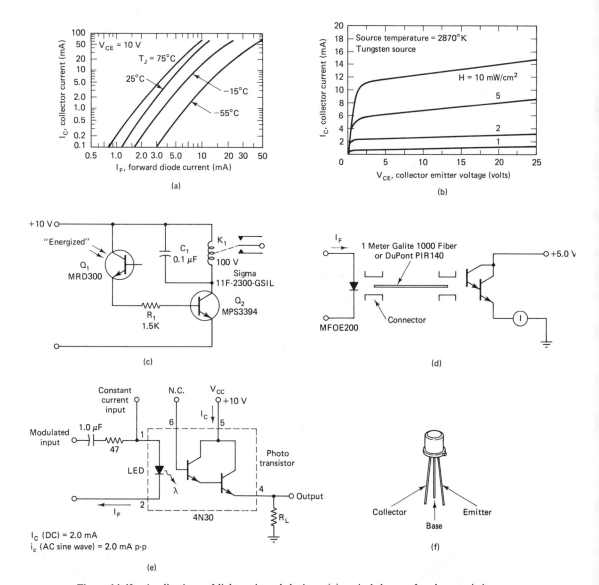

**Figure 14-40** Applications of light-activated devices: (a) typical dc transfer characteristics for a 4N29 and 4N30 optu-isolator; (b) typical characteristics for a phototransistor; (c) light-activated relay; (d) fiber-optic connection; (e) modulated output; (f) phototransistor in a TO-18 package.

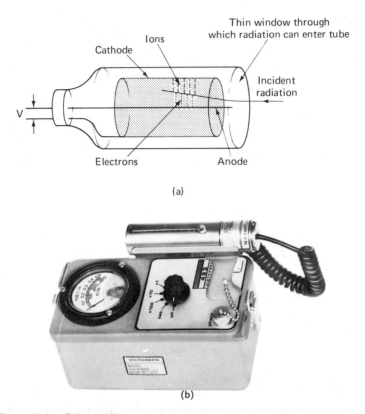

(a)

(b)

**Figure 14-41** Geiger-Müller radiation detector: (a) schematic; (b) Victoreen model 493. (Courtesy of Victoreen Instrument Co.)

proportional to the intensity of the incident X rays. The strength of the glow is then measured by a phototube. X-ray sensors are used to monitor the thickness of materials which are manufactured in sheet form. They are also used for locating flaws in metal structures and other construction materials, and in determining liquid levels in sealed tanks.

The most common nuclear radiation sensors are the *Geiger–Müller* tube and the *scintillation counter*.[5] The Geiger–Müller tube is a sealed tube filled with an inert gas (such as argon). Within the tube is a cathode in the shape of a long cylinder, as well as a long wire anode placed along the axis of this cylinder (Fig. 14-41). At one end of the glass tube is an extremely thin window through which

---

[5]There are many other types of nuclear radiation detectors that find application in special radiation-measurement systems. They include the ionization chamber (similar to the Geiger–Müller tube), the cloud chamber, the Cerenkov counter, neutron detectors, and radiation dosimeters. Rather than discuss them further in this book, the reader is asked to refer to other texts (listed as references at the end of the chapter) that have been written specifically on radiation detection. These books cover the operation and uses of the devices in extensive detail.

radiation can enter the tube. If a high voltage ($\approx$900 V) exists between the anode and the cathode, each burst of nuclear radiation (in the form of beta particles or gamma rays) that enters the tube will ionize some of the argon atoms. The ionized particles will rush toward the electrodes of the tube. During their voyage to the electrodes, the electrons and ions are accelerated and collide with other argon atoms, which are in turn ionized. When the burst of ions is collected by the electrodes, a pulse of current is caused in the circuit to which the tube is connected.

As the pulse passes through the connecting circuitry, the voltage between the anode and cathode drops below the value which causes further ionization by collision. The gas in the tube returns to its deionized state, and current in the connecting circuit ceases. Then the voltage between the cathode and anode rises again, and the tube is ready to sense the next radiation burst. The "dead time" between bursts is about 100 to 200 $\mu$s.

The number of pulses occurring in a given time is a measure of the strength of the radiation in the region near the tube. The pulses can be used to cause audible clicks from a loudspeaker or they can be counted by an electronic counter to yield a cumulative value over a specific time. Typical GM meters can detect up to 15,000 events (or counts) per minute.

The *scintillation counter* is a device that uses a photomultiplier tube to count light flashes that are produced by certain crystals when struck by nuclear radiation. These crystals (such as zinc sulfide or soluim iodide) produce a brief *scintillation* of light each time such an event occurs. These emitted flashes are reflected by mirrors and are carried to a photomultiplier tube through fiberoptic *light pipes*. The photomultiplier tubes convert and amplify the feeble light flashes to electrical pulses which have sufficient magnitude to be measured. The number of pulses the photomultiplier tube puts out is a measure of the intensity of the nuclear radiation.

Because of the amplifying ability of the photomultiplier tube, *scintillation counters* are much more sensitive than Geiger–Müller tubes in measuring nuclear radiation.

## PROBLEMS

1. List five devices that act as nonelectrical transducers [i.e., the mercury thermometer (temperature to dimensional length transducer)].

2. List the series of steps that you would take if you were required to select a transducer for a particular measurement application.

3. Are the following transducers active or passive transducers? Why?
   - **(a)** Strain gauge
   - **(b)** LVDT
   - **(c)** Thermocouple
   - **(d)** Phototube
   - **(e)** Solar cell
   - **(f)** Carbon microphone
   - **(g)** Piezoelectric crystal

4. Comparing the three types of strain gauges (bonded metal, unbonded metal, and semiconductor), list the advantages and disadvantages of each type.

5. A semiconductor strain gauge has a length of 3 cm and an initial resistance of 2.3kΩ. An applied force causes a 0.8 change in its length and a change in resistance of 530Ω. What is the GF of the strain gauge?

6. How does temperature affect the operating characteristics of strain gauges?

7. Describe the operation of the linear variable differential transformer (LVDT).

8. An LVDT has an output of 5 V when the displacement is 0.036 in. Determine the sensitivity of the device in mV/0.001 in.

9. Choose the most suitable temperature transducer for measuring the temperature in each of the following applications.
   (a) Rapidly changing temperatures
   (b) Very small temperature variations about 40°C
   (c) Very high temperatures (>1500°C)
   (d) Highly accurate temperature measurements
   (e) Wide temperature variation
   (f) Application calling for a rugged, accurate, temperature-sensing device

10. Describe the differences in the principles of operation of photoemissive, photo-conductive, and photovoltaic transducers.

11. What is phototube fatigue, and how can it be avoided?

12. A capacitive-type pressure transducer has two metal plates that are 4 cm in diameter and 0.5 cm apart. A 200-psi pressure on one plate will cause the separation between the plates to decrease by 0.04 cm. If no pressure is applied to the capacitor, it has a capacitance of 400 pF. Determine the capacitance value if 200 psi is applied to the transducer.

## REFERENCES

1. Norton, H. M., *Handbook of Transducers for Electronic Measurement Systems*. Englewood Cliffs, N.J.: Prentice Hall, 1969.

2. Tektronix, *Use of Transducers*. Beaverton, Ore.: Tektronix, 1971.

3. Cobbald, R.S.C., *Transducers for Biomedical Measurements: Principles and Applications*. New York: John Wiley, 1974.

4. *Introduction to Transducers for Instrumentation*. Los Angeles, Calif.: Statham Instruments, n.d.

5. Coombs, C., ed., *Basic Electronic Instrument Handbook*. Chap. 6. New York: McGraw-Hill, 1972.

6. Sheingold, D.H., ed., *Transducer Interfacing Handbook*. Norwood, Mass.: Analog Devices, 1981.

7. Jones, K.A. *Introduction to Optical Electronics*. New York: Harper & Row, 1987.

8. *Practical Temperature Measurements, Application Note 290*. Palo Alto. Calif.: Hewlett-Packard, *Corp.*, *Palo Alto, Ca.*, 1980.

# 15

# *Electronic Amplifiers*

One of the primary reasons why electrical methods are so widely used in scientific instrumentation is that very weak electrical signals can be *amplified* to the point where they can directly activate indicating or recording devices and be thereby measured. The *electronic amplifier* is the component of electrical systems that provides the necessary power to the signals to make such measurements feasible. The purpose of this chapter is to describe the general characteristics and terminology associated with electronic amplifiers, especially those amplifiers that are most commonly used in electrical measuring instruments.

The level of the discussion is such as to introduce the qualitative aspects of the most common amplifier types and list some of their important applications. The quantitative electronic design principles of amplifier circuits are not stressed. Even a basic coverage of such circuit design aspects would require far more space than this text can devote to the subject.[1] As a result we treat the amplifier as a device that somehow amplifies the signals that are fed into it.

One important amplifier type that is covered more fully (but still in a very introductory fashion) is the *operational amplifier*. These amplifiers are extremely versatile and are finding use as key building blocks in a wide variety of modern measurement instrumentation systems.

[1]Readers desiring more information on the electronic design aspects of amplifiers should consult any of the standard texts on electronics.

## GENERAL PROPERTIES OF AMPLIFIERS

The primary function of an amplifier is to increase, or amplify, the signals that are fed to its inputs. Although amplifiers are also used in some cases to isolate one part of an electrical system from another, our overall convern will be with the amplifying properties. Thus we will view amplifiers as instruments that are designed to receive a signal ($a_{in}$) and produce a replica of this signal with an amplitude that has been multiplied by some factor, $K$. The value of the output signal of the amplifier ($a_{out}$) may therefore be expressed as

$$a_{out} = Ka_{in} \qquad\qquad (15\text{-}1)$$

From this relation we see that $K$ (which is known as the *gain* of the amplifier) is equal to the ratio of $a_{out}$ to $a_{in}$. $K$ is a function of the amplifying device, the bias level, and the circuit configuration. If $K > 1$, it is evident that the amplifier increases the value of signals fed to its inputs. The circuit symbol for an amplifier is $-\!\boxed{K}\!\!>$ .

If $a_{in}$ and $a_{out}$ are expressed in terms of the voltage value of the input and output signals, respectively, the ratio of their magnitudes yields the *voltage gain* of the amplifier. On the other hand, if $a_{in}$ and $a_{out}$ are expressed in terms of the power contained in the input and output signals, their ratio yields the *power gain* of the amplifier. These two gain values are often quite different from one another.

The frequency range over which an amplifier is designed to amplify input signals with a constant gain is known as the *bandwidth* of the amplifier. This quantity is formally defined as the interval between those frequencies where the *power gain* of the amplifier has dropped to one-half of its midfrequency value (Fig. 15-1).

In many cases, the voltage gain of an amplifier is specified rather than the

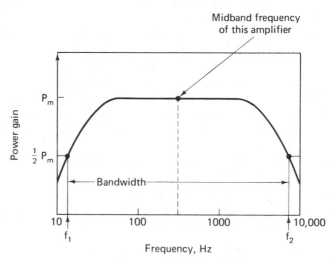

**Figure 15-1**   Amplifier bandwidth.

power gain. However, the bandwidth is not the frequency interval between the two frequencies at which the *voltage* gain falls off to one-half its midfrequency value. Instead, since the voltage value of the half-power points is 0.707 $V_m$ (where $V_m$ is voltage value at midfrequency), the bandwidth is defined as the frequency interval between the points where the voltage gain drops to 0.707 of its midband value.

Normally, the bandwidth specification adequately describes the range of frequencies over which an amplifier can amplify signals with suitable fidelity. That is, it seems reasonable to assume that if the frequency components of a signal fall within the bandwidth of an amplifier, the signal will be amplified without distortion. However, it was shown (and an example was given in Chapter 8) that nonsinusoidal periodic waveforms are composed of a series of harmonics that extend to infinite frequency. Thus an attempt to display nonsinusiodal waveforms with an amplifier of finite bandwidth will cause some distortion in the amplified signal. Thus as an aid in determining how much distortion an amplifier will cause when it amplifies an arbitrary signal, it would also be useful to have a measure of the ability of the amplifier to respond to instantaneous signal changes. Such a measure is given by the *rise time*, $\tau_r$, of the amplifier.

The *rise time* concept, as applied to pulses, was introduced in Chapter 13. In an analogous fashion, we can define the rise time, $\tau_r$ of an amplifier. That is, $\tau_r$ is the time required for the output voltage of an amplifier to move between the 10 and 90 percent points of its final value when a step function is applied to its input (see Fig. 15-2).

Although perfect (ideal) step-function generators do not exist, there are some electronic instruments that can produce output waveforms that have rise times which are much smaller than the rise times of the amplifiers being tested (e.g., some pulse generators are capable of producing signals with rise times shorter than $10^{-10}$ s). When such devices are used to test an amplifier, the rise time of the amplifier is defined as the amplifier response to that applied step function. Rise-time tests also serve to demonstrate an important point about amplifiers: *The output signal of an*

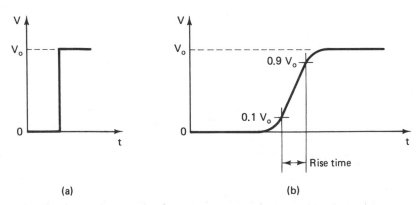

(a)  (b)

**Figure 15-2** (a) Ideal step function applied to amplifier; (b) response of amplifier output to ideal step functions of part (a).

*amplifier will not change voltage levels any faster than the amplifier rise time, no matter how short the rise time of the input signal.*

Rise times also add together as the square root of the sum of the square of the individual rise time involved:

$$\tau_{r_{TOT}} = \sqrt{\tau_{r_1}^2 + \tau_{r_2}^2} \tag{15-2}$$

For example, if an applied test signal has a rise time $\tau_{rs}$, which equals the rise time $\tau_{ra}$ of the amplifier to which it is applied, the observed output signal will have a rise time $\tau_{r_{TOT}}$ which is given by

$$\tau_{r_{TOT}} = \sqrt{\tau_{r_S}^2 + \tau_{r_A}^2} = \sqrt{2}\,\tau_{r_S} \qquad (\text{since } \tau_{r_S} = \tau_{r_A})$$

or

$$\tau_{r_{TOT}} = 1.4\tau_{r_S}$$

Note that in this case the observed rise time of the input signal (as expressed by $\tau_{r_{TOT}}$) would have been in error by 40 percent.

Figure 15-3 is a chart that shows the percentage error present in the rise time of an observed waveform $\tau_{r_{TOT}}$ for various ratios of $\tau_{r_S}/\tau_{r_A}$. For example, the figure indicates that if the rise time of the input signal is seven times as long as the amplifier rise time $(\tau_{r_S}/\tau_{r_A} = 7)$ the error of the displayed rise time $\tau_{r_{TOT}}$ is only 1 percent.

Although rise time is a more general specification of amplifier frequency response than bandwidth, the two quantities are nevertheless closely related. In general, the product of the $\tau_{r_A}$ (in nanoseconds) and bandwidth (in megahertz) produces a number whose value lies between 0.33 and 0.35

$$\tau_{r_A} \times \text{bandwidth} = 0.35 \tag{15-3}$$

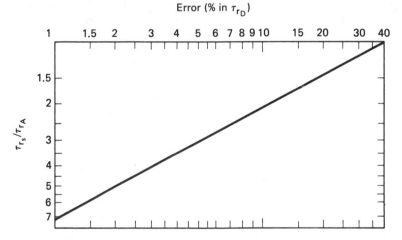

**Figure 15-3** Percent by which the output or displayed time rise $\tau_{rD}$) exceeds the input signal rise time depends on the ratio of the amplifier or scope rise time to the input signal rise time.

For example, oscilloscopes with bandwidths of 15 MHz will have rise times of approximately

$$\tau_{r_A} = \frac{0.35}{15 \text{ MHz}} \approx 23 \text{ ns}$$

According to Fig. 15-3, such oscilloscopes are capable of displaying waveforms with rise times of 69 ns ($\tau_{r_s}/\tau_{r_A} = 3$) with an accuracy of 5 percent.

One broad distinction between types of amplifiers is based on their low frequency response capabilities. If an amplifier is able to produce a constant dc output signal in response to a constant dc input, it is classified as a *dc amplifier*. The bandwidth of dc amplifiers extends from dc to the frequency at which the voltage gain of the amplifier drops to 0.707 of the voltage gain at dc. All other types of amplifiers are known as *ac amplifiers* [Fig. 15-4(a)].

Ac amplifiers are sometimes further divided into categories according to the band of frequencies over which they can amplify signals. If the amplifier is designed to be able to amplify a wide band of frequencies, it is called an *untuned amplifier* [Fig. 15-4(b)]. On the other hand, if an amplifier is designed to be able to amplify frequencies within a very narrow band, the amplifier is known as a *tuned amplifier* [Fig. 15-4(c)].

Untuned amplifiers are subdivided even further into the following classes: *audio-frequency* (AF) *amplifiers*, designed for amplifying signals [30 Hz to 15,000

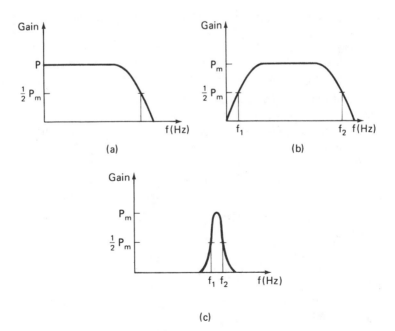

**Figure 15-4** Frequency characteristics of various amplifiers: (a) dc amplifier; (b) untuned ac amplifier; (c) tuned ac amplifier.

Hz]; *radio-frequency* (RF) *amplifiers* (capable of amplifying high-frequency waves, typically 500 kHz or higher. Such signals are found in radios, radar, and other applications); and *video amplifiers*, designed for amplifying the wide band of signals used in television receivers, typically from 30 Hz to 4 MHz.

Dc and ac amplifiers both have their advantages. Ac amplifiers tend to be less expensive than dc amplifiers, particularly if high gain is required. Thus, if dc amplification is not needed, ac amplifiers are usually a more sensible choice.

Dc amplifiers are required when dc and very low frequency signals must be amplified. Weak dc voltage signals are produced by such sources as thermocouples, photovoltaic devices, and strain gauges. Many bioelectrical phenomena also generate weak dc and very low frequency signals. A dc amplifier must be used to amplify such signals. In addition, the oscilloscope becomes a much more versatile instrument when equipped with dc amplifiers.

Dc amplifiers have a tendency, however, to suffer from the problems of *offset voltages* and *drifts*. The *offset voltage* is a small dc voltage that appears at the output of the amplifier, even when no voltage is applied to its inputs. To correct this voltage error (which would be zero in an ideal dc amplifier), a small compensating voltage must be applied to the input of the amplifier. *Drift* is the rate at which the output voltage of an amplifier changes due to temperature variations or aging of the amplifier components. Because of drift, the zero level of dc amplifiers must be periodically checked and adjusted in order to get accurate readings.

### Miscellaneous Specifications of Amplifier Operation

In addition to the characteristics of amplifiers described in the preceding section, there are other specifications which also provide a measure of how well amplifiers perform certain functions. Some of the more common of these are listed below.

**1.** *Linearity.* The output of an *ideal* amplifier is directly proportional to its input; that is, a plot of the output versus input is a straight line. (This means that in an ideal linear amplifier the output signal is an amplified replica of the input signal, except for phase.) The extent to which this ideal is approached in an actual amplifier is specified by the linearity of the amplifier. Linearity is usually described either in terms of a percentage of the output or a percentage of the full-scale value. Nonlinearities in an amplifier lead to *distortions* of the output signal.

**2.** *Amplifier output.* Describes the maximum voltage, current, or power output of an amplifier.

**3.** *Input/output impedance.* An amplifier should have a high input impedance to avoid *loading* the signal source (or, if applicable, the previous amplifier stage). When the amplifier output is to be connected to another amplifier, the output impedance of the first one should be low (for *efficient* power transfer). If the output is to be connected to a recording or indicating device, the output impedance should match the input impedance of the recording device (for *maximum* power transfer).

**4.** *Noise.* One source of distortion in amplifiers, besides nonlinearity, is *noise*. This quantity can be divided into two classes: (1) unwanted signals that are generated externally and enter the amplifier from the outside and (2) unwanted signals that are generated by the components of the amplifier itself. The greater the magnitude of the noise, the more the desired signal will be distorted and obscured. The noise signal created by the amplifier itself is usually listed in the specifications of the amplifier and is expressed in terms of an rms voltage value.

### Measuring Amplifier Gain and Bandwidth

The gain and bandwidth of an amplifier can be measured using standard laboratory equipment. To begin with, a suitable load resistance $R_L$ should be connected across the amplifier output terminals. An oscillator provides the input signal to the amplifier. An oscilloscope is used to observe both the input and output signals of the amplifier.

To measure the amplifier *voltage gain* $K_v$, a sinusoidal signal from the oscillator at the frequency of interest (often near the midband frequency) is fed to the amplifier. The output signal is monitored by the oscilloscope to ensure that the level of the input signal is small enough so that output clipping does not occur. The input and output signal voltages are measured and $K_v$ is calculated by using Eq. (15-1).

To measure the *bandwidth* of the amplifier, the input signal to the amplifier from the oscillator is kept at a constant amplitude but varied in frequency by using a sweep frequency generator (Chapter 13). The output voltage of the amplifier is measured at various frequencies. The gain of the amplifier at these points is usually expressed in decibel from, and the 0-dB reference level is specified as the output voltage at a given midband reference frequency (usually 400 or 1000 Hz). The bandwidth of the amplifier will be the range between those two frequencies at which the voltage gain of the amplifier has dropped to 0.707 ($-3$ dB) of its midband reference value. The $-3$ dB frequency is also defined as the half-power frequency from Chapter 2, Eq. 2-3

$$dB = 10 \log \frac{P_{out}}{P_{in}} = 20 \log \frac{V_{out}}{V_{in}}$$

$$dB = 10 \log \frac{1}{2} = 20 \log \frac{1}{\sqrt{2}} = -3$$

Note that the input signal amplitude should be measured at each frequency of interest and adjusted if necessary to ensure that it remains constant. A sweep frequency generator will automatically maintain a constant amplitude input signal as it sweeps the frequency.

## DIFFERENTIAL AMPLIFIERS

There is a special type of amplifier, called the *differential amplifier*, which is designed to amplify the difference between the voltage values of two input signals ($V_1$ and $V_2$). In such amplifiers, each of the two signals is applied to one of the input

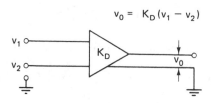

$$v_0 = K_D(v_1 - v_2)$$

**Figure 15-5**  Ideal differential amplifier.

terminals of the amplifier (as shown in Fig. 15-3). This means that the term $a_{in}$ of Eq. (15-4) becomes

$$a_{in} = V_1 - V_2 \tag{15-4}$$

From this relation, it follows that for a differential amplifier, Eq. (15-4) written as

$$a_{out} = V_{out} = K_D(V_1 - V_2) \tag{15-5}$$

In this equation, $K_D$ is called the *differential gain* of the amplifier. (The reason for using this nomenclature will become clearer as we discuss the operation of differential amplifiers.)

We also note from Fig. 15-5 that in a differential amplifier neither of the input terminals is grounded. As a result, the differential amplifier can be used to amplify the difference between the voltages of two ungrounded points in a circuit. This makes differential amplifiers suitable for amplifying the signal output of such devices as the strain gauge bridge shown in Fig. 15-6. In Chapter 6 we also saw how oscilloscopes equipped with differential amplifiers are able to make use of this feature to measure nongrounded voltages. However, the fact that differential amplifiers can amplify nongrounded voltages is not the only characteristic that makes them so useful for certain applications. Some of these other unique characteristics will be examined as we continue with the discussion.

Since an ideal differential amplifier is designed to amplify only the difference between the voltage values applied to its inputs, two input signals of equal magnitudes should yield an output voltage of zero. As a result, the component of a signal which is *common* to both signals applied to the input of a differential

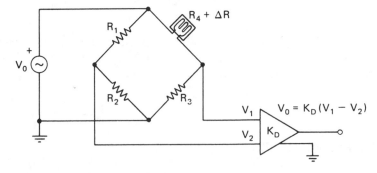

**Figure 15-6**  Use of a differential amplifier to measure the two ungrounded voltages of a Wheatstone bridge.

amplifier should not be amplified. This component is called the *common-mode component* ($V_C$) and is defined as being equal to

$$V_C = \tfrac{1}{2}(V_1 + V_2) \tag{15-6}$$

The *difference component*, $\Delta V$, which is the component we desire to be amplified, is defined as

$$\Delta V = V_1 - V_2 \tag{15-7}$$

**Example 15-1**

If (a) $V_1 = 5$ and $V_2 = 3$, and if (b) $V_1 = 4 + 3 \sin \omega t$ and $V_2 = -5 + \sin \omega t$, find $V_C$ and $\Delta V$ for each case.

**Solution.**

(a) If $V_1 = 5$ and $V_2 = 3$, then

$$V_C = \tfrac{1}{2}(V_1 + V_2) = 4$$

and

$$\Delta V = V_1 - V_2 = 2$$

(b) If $V_1 = 4 + 3 \sin \omega t$ and $V_2 = -5 + 5 \sin \omega t$, then

$$V_C = -\tfrac{1}{2} + 4 \sin \omega t$$

$$\Delta V = 9 - 2 \sin \omega t$$

Actual differential amplifiers only approach the ideal characteristic of solely amplifying the $\Delta V$ component. In reality, instead of producing an output voltage according to Eq. (15-7), they also produce an output signal component that is proportional to $V_C$. That is, the the output signal of an actual differential amplifier is given by the equation

$$V_{\text{out}} = K_D(V_1 - V_2) + K_C \frac{V_1 + V_2}{2} = K_D \, \Delta V + K_C V_C \tag{15-8}$$

where $K_C$ is called the *common-mode gain*. This common-mode gain is unwanted, and amplifier designers seek to minimize its value. It is this component that causes the offset seen on an oscilloscope when the (A-B) mode is used and no input signal is present. As a method of specifying how closely an actual differential amplifier approaches the characteristics of its ideal counterpart, a quantity called the *common-mode rejection ratio* is used. The common-mode rejection ratio (CMRR) is defined to be the ratio of the differential gain ($K_D$) to the common-mode gain ($K_C$) of the amplifier:

$$\text{CMRR} = \frac{K_D}{K_C} \tag{15-9}$$

An ideal differential amplifier would have an infinite CMRR. In practice, a well-designed commercially available differential amplifier is one with a CMRR of

1000 or more. CMRR is often expressed in terms of voltage decibels (dB) as well. In that case, the relation

$$CMR(dB) = 20 \log_{10} CMRR \qquad (15\text{-}10)$$

is used to convert the ratio (CMRR) to a voltage decibel value (CMR). Thus a differential amplifier which is specified as having a CMR of 80 dB has a CMRR of 10,000.

**Example 15-2**

A differential amplifier receives two signals, $V_1$ and $V_2$, both of whose magnitudes are 10 V. The CMR of the amplifier is 90 dB and its differential gain is $K_D = 100$. Find the output voltage of the amplifier.

**Solution.**   Using Eq. (15-9), we determine the CMRR and $K_C$:

$$CMR = 20 \log_{10} CMRR$$

$$90 = 20 \log_{10} CMRR$$

$$4.5 = \log_{10} CMRR$$

$$CMRR = 31,000$$

Then

$$CMRR = \frac{K_D}{K_C}$$

or

$$K_C = \frac{10^2}{3.1 \times 10^4} = 0.0032$$

Then using Eq. (15-8), we can find $V_O$:

$$V_O = K_D(V_1 - V_2) + K_C \frac{V_1 + V_2}{2} = K_D(0) + 0.0032(10)$$

$$= 0.032 \text{ V} = 32 \text{ mV}$$

If the CMRR of a differential amplifier is high, the amplifier is able to reject hum (and other common-mode noise or interference) that appears simultaneously and in phase at the amplifier inputs. For example, if

$$V_1 = V_{signal} + V_{noise} = V_S + V_{noise}$$

and

$$V_2 = V_{noise}$$

then from Eq. (15-8) we can write

$$V_{out} = K_D\left(\Delta V + \frac{K_C}{K_D}V_C\right) = K_D V_S + \frac{K_D V_{noise}}{CMRR} + \frac{K_D V_S}{2CMRR} \qquad (15\text{-}11)$$

Equation (15-11) shows that if the CMRR of a differential amplifier is large, the component of voltage due to noise is largely rejected. This feature of differential

**TABLE 15-1** DIFFERENTIAL AMPLIFIER CIRCUIT-RESISTANCE TOLERANCE VERSUS COMMON-MODE GAIN OF DIFFERENTIAL AMPLIFIERS

| Resistor tolerance (%) | 5 | 2 | 1 | 0.5 | 0.1 |
|---|---|---|---|---|---|
| Average common-mode gain. $K_C$ | 0.1 | 0.04 | 0.02 | 0.01 | 0.002 |

amplifiers makes them extremely valuable for observing low-level signals in the presence of common-mode noise.

The CMRR of a differential-amplifier circuit is, in practice, determined by how well various resistors required in the amplifier circuit can be matched in value. Specifically, the magnitude of the unwanted common-mode gain, $K_C$, will increase in proportion to the difference in value between such critical circuit resistors. For example, Table 15-1 shows the average realizable values of $K_C$ for various resistance tolerances. We see that if resistors with 5 percent tolerance values are used in a differential-amplifier circuit, $K_C$ will be equal to 0.1. On the other hand, if the tighter tolerance of 0.1 percent is imposed on the differential-amplifier circuit resistors, $K_C$ is reduced to $K_C = 0.002$. Since the differential gain, $K_D$, is essentially not effected by such resistor value mismatching, we can see from Eq. (15-9) that the CMRR will be reduced as circuit resistor values become less perfectly matched.

The CMRR capability of a differential amplifier is also reduced if the signals to its two inputs arrive from the signal source through paths that do not have the same impedance value (Fig. 15-7). Such CMRR degradation is said to be due to *impedance unbalance*, and when an amplifier is configured in such a system, this lower exhibited CMRR is known as the *apparent CMRR* of the amplifier. The effect of impedance unbalance is to allow a small proportion of the common-mode signal $V_{cm}$ to appear as a difference component $\Delta V$ to the amplifier (which then applies to this unwanted noise component the full differential gain).

A common application in which impedance unbalance due to unequal electrode impedances is encountered is in the measurement of temperature by use of thermocouples. As discussed in Chapter 14, thermocouples employ two wires of dissimilar metal (i.e., one wire might be iron, the other constantan) whose junctions are subjected to two different temperatures. A voltage signal proportional to the difference of the temperatures is produced. Since the thermocouple output signals

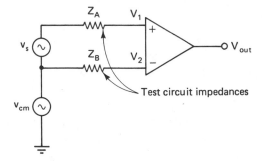

**Figure 15-7** Differential amplifier receiving a signal at $(+)$ and $(-)$ terminals along paths containing impedances $Z_A$ and $Z_B$. If $Z_A \neq Z_B$, the amplifier CMRR is degraded.

are in the millivolt range, they must be amplified prior to display or other signal processing. Since the wires that make up the thermocouple are of dissimilar metals, their resistivities will also be different. Thus the equivalent impedances of each of the thermocouple wires (i.e., $Z_A$ and $Z_B$ in Fig. 15-7) may not be equal. The amplifier will perceive such differing impedances as an impedance unbalance being presented to its inputs.

Impedance unbalance can also result from unequal values of capacitance to ground in the input cables. Such unequal capacitances will lead to unequal impedances that tend to become even more unequal as higher-frequency signals are measured.

### Example 15-3

A differential amplifier with a CMRR of 100,000: 1 has test circuit impedances $Z_A = 150 \ \Omega$ and $Z_B = 0$ connected to its inputs, together with a voltage source $v_s$ as shown in Fig. 15-7. If the input impedance of the amplifier is $Z_{in} = 10^7 \ \Omega$ and the differential gain is $K_D = 2 \times 10^4$, and $v_{cm} = 1.0$ V, find the apparent CMRR of the amplifier.

**Solution.**   Since $Z_B = 0$, $v_2$ appearing at the $(-)$ terminal of the amplifier is 1.0 V. Since $Z_A = 150 \ \Omega$, $v_1$ at the $(+)$ terminal is found from

$$v_1 = \frac{v_{cm} Z_{in}}{Z_{in} + Z_A} = \frac{(1.0)10^7}{10^7 + 150} = 0.999985 \text{ V} = 999{,}985 \ \mu\text{V}$$

since 15 $\mu$V of $v_{cm}$ appears across the 150-$\Omega$ resistor. Now this results in a difference component $\Delta v = 15 \ \mu$V appearing across the amplifier inputs. With $K_D = 2 \times 10^4$, this 15-$\mu$V signal will appear as 300 mV at the output, while an additional 200 mV will appear at the output due to the CMRR of the amplifier. Thus a maximum of 500 mV of common-mode signal $v_o$ may appear at the output even when $\Delta V = 0$. An equivalent differential signal that would produce this same output would be

$$\Delta V = \frac{v_o}{K_D} = \frac{500 \text{ mV}}{2 \times 10^4} = 25 \ \mu\text{V}$$

But since the amplifier will still produce $v_o = 500$ mV when $V_{cm} = 1.0$ V and $\Delta v = 0$, we can find $K_C$ (apparent) from

$$K_C \text{ (apparent)} = \frac{V_o}{V_{cm}} = \frac{0.5 \text{ V}}{1.0 \text{ V}} = 0.5$$

Thus the apparent CMRR is

$$\text{CMRR (apparent)} = \frac{K_D}{K_C \text{ (apparent)}} = \frac{2 \times 10^4}{0.5} = 4 \times 10^4$$

and the CMRR is seen to be reduced from 100,000:1 to 40,000:1.

From Example 15-3 it is clear that the output of a differential amplifier is degraded from that of an ideal differential amplifier (which would exhibit complete rejection of common-mode interference signals) by impedance unbalance as well as by its finite CMRR (determined largely by imperfect matching of circuit resistance values). It is thus desirable either to attempt to minimize the impedance unbalance, or to reduce the magnitude of the common-mode voltage $v_C$ at the signal source ground through *input guarding*. The technique of common-mode voltage reduction at the signal source ground point by input guarding is discussed in detail in Chapter 16. As a general rule, when selecting an operational amplifier for improved performance, do not select an amplifier with a higher CMRR but rather select an amplifier with a higher gain or increase the feedback signal. The reason that this is the preferred approach is that the CMRR error term is usually several orders of magnitude smaller than the dc gain error term.

### Measuring CMRR

To measure the common-mode rejection ratio (CMRR) of a differential amplifier, a convenient differential voltage level is selected. This voltage, $\Delta V$, is fed to the amplifier and also carefully measured. The output voltage $V_o$ for this $\Delta V$ is also measured. Next, the two inputs of the amplifier are tied together and used as a single, common-mode input. An input signal voltage $V_C$ is applied to this common-mode input and is increased in amplitude until the same output voltage $V_o$ is obtained as with the measured $\Delta V$. From Eq. (15-9) we can show that

$$\text{CMRR} = \frac{V_C}{\Delta V}\bigg|_{V_o = \text{constant}}$$

and our measured values of $V_C$ and $\Delta V$ will allow us to calculate CMRR from this equation.

## OPERATIONAL AMPLIFIERS

Operational amplifiers (op-amps) are basically high-gain dc amplifiers with differential inputs and a single-ended output. The ideal op-amp responds only to the difference voltage between the two input terminals. The general symbol for op-amps is shown in Fig. 15-8. The single-ended input types (i.e., one input terminal grounded) can be treated as a special case where the $(-)$ input terminal of the general type is grounded [Fig. 15-8(b)]. The major difference between op-amps and

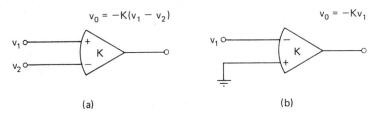

**Figure 15-8**  (a) General symbol for an operational amplifier; (b) operational amplifier with one input grounded (i.e., single-ended input).

ordinary dc differential amplifiers is that operational amplifiers are designed to be used with external feedback networks. That is, a portion of the output signal of the operational amplifier is fed back to its input through various feedback paths, depending on the specific function of the op-amp. It is because of this feature that the op-amp can be used to perform such an extraordinary number of different functions.

The term *operational amplifier* originated because high-gain, dc feedback amplifiers were first developed to perform the mathematical *operations* of addition, subtraction, and integration in analog computers. This name has clung to them even though their use has spread to a vast number of measurement and control applications outside of analog computing.

The op-amp is being manufactured in complete units from integrated circuits as well as from discrete components. In either case the designer of instrument systems can incorporate complete op-amps into his system. Because of the ease of incorporating such complete building blocks into instrumentation systems, the design and construction of new measurement systems has become a much simpler task.

## Characteristics of Operational Amplifiers

The op-amp, without any feedback paths connected to it is described as being operated in an *open-loop* mode (i.e., the feedback loop is not closed). The ideal characteristics of the op-amp in this open-loop mode are

1. Gain $= \infty$
2. Bandwidth $= \infty$
3. Input impedance $= \infty$
4. Output impedance $= 0$
5. Output signal, $V_o = 0$ when $V_2 = V_1$

Actual op-amps, of course, cannot meet these ideal open-loop specifications. Instead, they are designed to approximate these requirements as closely as possible. Actual op-amps have open-loop gains ranging from $10^3$ to $10^9$ (typically $10^5$) and a flat response (i.e., constant gain) from zero frequency out to several kilohertz. The input impedance is $10^5$ to $10^{12}$ $\Omega$, while the output impedance is about 25 to 50 $\Omega$. Since actual op-amps do not have an infinite input impedance, their input current is not zero. This small current causes an offset voltage on the output that can be partially compensated for by inserting a resistor $R_C$ as shown in Fig. 15-9. The value of the compensation resistor $R_C$ for the amplifiers shown in Fig. 15-9 is

$$R_c = \frac{R_{in} R_f}{R_{in} + R_f} \qquad (15\text{-}12)$$

Resistor $R_C$ can be omitted whenever the offset does not adversely affect the ouput voltage. Offset, however, is very important and cannot be ignored when op-amps are used in the signal conditioning circuit of small signal such as those from thermo-couples, strain gauges, and bridge circuit outputs near null. A simple test that can determine if an op-amp is operating in its linear region is to measure the voltage difference between the noninverting (+) and inverting (−) terminals. The voltage difference should be zero. This implies that the two terminals are always at the same voltage for linear operation. By using this fact and the fact that the input current is zero, most op-amp circuits can be analyzed by using node equations. For example, the voltage gain of the circuit in Fig. 15-9 (a) can be determined as explained in the following paragraph.

Since the plus terminal of the operational amplifier is connected to ground, its voltage must be zero with respect to ground. This forces the minus terminal to ground, commonly called a virtual ground. The input current can then be calculated by using Ohm's law.

$$i_{in} = \frac{V_{in}}{R_{in}}$$

The input current can only flow through the feedback resistor $R_f$; therefore,

$$V_{out} = -i_{in}(R_f) = -\left(\frac{V_{in}}{R_{in}}\right)R_f \qquad (15\text{-}13)$$

The op-amp is invaluable as a general-purpose amplifier primarily because of its high gain. This characteristic makes it possible to utilize external feedback connections with it in such a way that the *overall gain or the characteristics of the output signal from the amplifier depend primarily on the values of the elements in the feedback path*. The passive electrical components that are usually used to make up

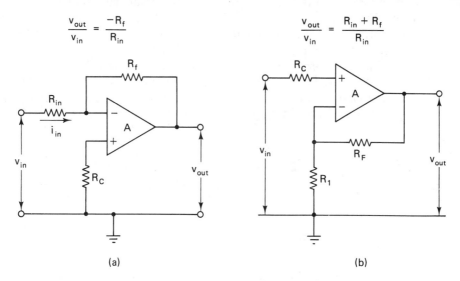

(a)

(b)

**Figure 15-9** (a) Inverting amplifier configuration; (b) noninverting amplifier configuration.

the elements of the feedback path have accurately known and stable values. Therefore, the overall output characteristics of the op-amp can be accurately controlled, *independently* of the properties of the amplifying element itself.

Once a feedback path is connected to the basic op-amp shown in Fig. 15-8, the amplifier is said to be operating in a *closed-loop mode* (see Fig. 15-9). The two basic closed-loop amplifying circuits in which op-amps can be connected are (1) the *inverting amplifier* configuration and (2) the *noninverting amplifier* configuration. Another very important circuit, which is actually a combination of these two basic circuits, is the *differential-amplifier* configuration.

The *inverting-amplifier* configuration is shown in Fig. 15-9(a). In this circuit, the (+) input is connected to ground, the signal is applied to the (−) input through $R_{in}$, and feedback is returned from the output through $R_f$. Gain of an ideal op-amp connected in the inverting configuration is

$$\frac{V_{out}}{V_{in}} = \frac{-R_f}{R_{in}} \tag{15-14}$$

From Eq. (15-14) we see that the output voltage $V_{out}$ is negative (as long as $V_{in} > 0$, of course) and that the gain can be varied by adjusting $R_{in}$ or $R_f$.

The *noninverting-amplifier* configuration is shown in Fig. 15-9(b). In this circuit the output voltage is seen to have the same polarity as the input voltage, the input signal is applied to the (+) terminal, and the gain can be shown to be

$$\frac{V_{out}}{V_{in}} = \frac{R_{in} + R_f}{R_{in}} \tag{15-15}$$

From Eq. (15-13) we see that the lower limit of the gain of the noninverting configuration occurs when $R_f = 0$, and this minimum gain is thus *unity*. In the ideal case, the input impedance of this configuration is infinite. The minimum gain characteristic and the high-input impedance of the noninverting configuration are used in the voltage-follower circuit, Fig. 15-10.

The *voltage-follower* circuit has this name because the output $V_o$ is equal in sign and magnitude to the input voltage. The function of this circuit is to isolate the output from the input of the device. Consequently, the input impedance can be very high (up to $10^{14}$ Ω) and the output impedance very low (less than 1 Ω). With this

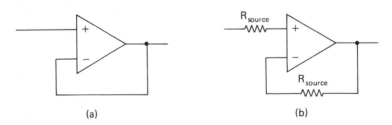

(a)     (b)

**Figure 15-10** Voltage follower: (a) common circuit; (b) with current compensation.

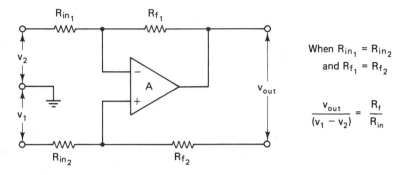

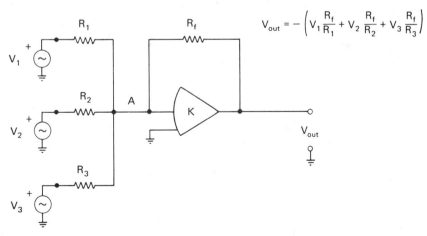

When $R_{in_1} = R_{in_2}$
and $R_{f_1} = R_{f_2}$

$$\frac{V_{out}}{(v_1 - v_2)} = \frac{R_f}{R_{in}}$$

**Figure 15-11**   Differential-amplifier configuration.

capability, less than 1 pA can be drawn from a circuit whose voltage is being measured by an instrument with such a voltage follower as its input element.

The *differential-amplifier* configuration, shown in Fig. 15-11, is a combination of the inverting and noninverting configurations. It has signals applied to both inputs and uses the natural differential-amplifier characteristics of the op-amp. The gain for the differential component of the input signal, $\Delta v = v_1 - v_2$, is

$$K_D = \frac{v_{out}}{\Delta v} = \frac{R_f}{R_{in}} \tag{15-16}$$

If the ratio $R_{f_1}/R_{in_1}$ is equal to the ratio $R_{f_2}/R_{in_2}$, the gain for the common-mode component of the input signal, $v_{cm} = (v_1 + v_2)/2$ will be zero (since by definition, the amplifier has no gain for equal signals applied to both inputs). Thus an ideal op-amp and perfectly matched resistors would have an infinite CMRR.

If additional input branches containing resistances are connected to point $A$ of Fig. 15-12, the output signal can be shown to be equal to

$$V_{out} = -\left( V_1 \frac{R_f}{R_1} + V_2 \frac{R_f}{R_2} + V_3 \frac{R_f}{R_3} \right)$$

**Figure 15-12**   Summing network.

$$V_{\text{out}} = -\left(V_1 \frac{R_f}{R_1} + V_2 \frac{R_1}{R_2} + V_3 \frac{R_f}{R_3}\right) \tag{15-17}$$

If $R_1 = R_2 = R_3 = R_f$ in this expression, then

$$V_{\text{out}} = -(V_1 + V_2 + V_3) \tag{15-17a}$$

Thus the negative of the sum of the signals applied to the inputs of this circuit is provided at the output. We note that to get a positive output value, a unity-gain inverting amplifier would have to be added in series to the summing amplifier.

If the resistor $R_f$ in the feedback loop of Fig. 15-9 is replaced by a capacitor $C_f$, the op-amp becomes an *integrator*, Fig. 15-13. From Eq. (15-14) we find that the voltage output of an operational amplifier with such a feedback path is

$$v_{\text{out}} = -\frac{1}{RC}\int v_{\text{in}}\, dt \tag{15-18}$$

The integrator is a low frequency circuit in which the frequency should not exceed

$$f < \frac{1}{2\pi R_s C} \tag{15-19}$$

There are a large number of other circuits that can be built by using operational amplifiers, including precision voltage sources, logarithmic amplifiers, voltage-to-current sources, dc power supply regulators, charge amplifiers, and voltage limiters. Readers interested in using op-amps as building blocks to create such circuits should consult the texts devoted to op-amp applications. One of the best is *The IC Op-Amp Cookbook*, by Jung (Reference 3).

### Real Operational Amplifiers and Their Limitations

As we noted in the introductory section on op-amp characteristics, real op-amps can only approximate the characteristics of the ideal devices. In this section we study these deviations from the ideal in more detail. We will see that in many applications,

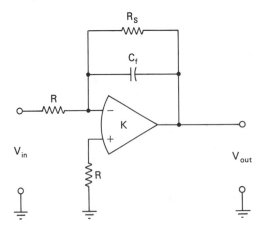

**Figure 15-13** Integrator.

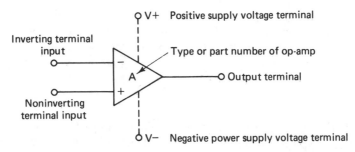

**Figure 15-14** Standard operational amplifier symbol.

the differences between ideal and real behavior have a negligible effect on the results. In other cases, sufficient error from the ideal result will be caused by behavior of real op-amps.

First let us specify the standard op-amp schematic signal (Fig. 15-14). The basic symbol is the triangle, which in electronic usage is generally understood to imply amplification. The inputs are at the base of the triangle and the output at the apex of the triangle. The noninverting (+) input is usually drawn as the lower of the two inputs. The type or part number of the op-amp is listed in the center of the triangle, but if the circuit is a general one (merely indicating an op-amp), the symbol used is A. Power supply leads are drawn extending above and below the triangle as shown in Fig. 15-14. These leads, however, may not always be drawn (in the interests of simplicity), but since an amplifier always requires dc power for operation, their presence is always implied.

The important parameters of an ideal versus a typical real op-amp are shown in Table 15-2. Let us briefly discuss each one and point out where the deviations from the ideal device become significant.

• *Open-loop gain.* A principal source of error in op-amp applications may arise from the fact that the amplifier open-loop gain is not infinite. In reality, the open-loop gain of actual op-amps differs from the ideal in two respects: (1) the dc gain is

**TABLE 15-2** TYPICAL PARAMETERS OF IDEAL VERSUS AN ACTUAL OPERATIONAL AMPLIFIER

| | Parameter | Ideal | Actual |
|---|---|---|---|
| Open-loop gain | $A_{VOL}$ | $\infty$ | $10^5$ |
| Open-loop 3-dB bandwidth | $BW_{OL}$ | $\infty$ | 10 Hz |
| Input impedance | $Z_{in}$ | $\infty$ | $10^5$–$10^{12}$ $\Omega$ |
| Output impedance | $Z_{out}$ | 0 | 50 $\Omega$ |
| Input offset voltage | $V_{OS}$ | 0 | 2 mV |
| Input bias current | $I_{OS}$ | 0 | 50–300 nA |
| Common-mode rejection ratio | CMRR | $\infty$ | $10^4$–$10^5$ |
| Slew rate | SR | $\infty$ V/$\mu$s | 1 V/$\mu$s |
| Rated output current | $I_{out(max)}$ | $\infty$ | 100 mA |
| Rated output voltage | $V_{out(max)}$ | $\infty$ | $\pm 10$ V |
| Maximum common-mode voltage | $V_{cm(max)}$ | $\infty$ | $\pm 10$ V |

not infinite, and (2) the bandwidth is not infinite. The effect of the noninfinite dc gain on the actual closed-loop gain of the inverting amplifier configuration [Fig. 15-9(a)] is to reduce the actual gain from its ideal value. To keep the error due to this effect below 0.1 percent, the following relationship must be obeyed:

$$\frac{A_{V_{OL}} R_{in}}{R_f} \geq 10^3 \qquad (15\text{-}19)$$

where $A_{V_{OL}}$ is the open-loop gain of the op-amp. This limitation is not severe since $A_{V_{OL}}$ is typically well in excess of $10^4$. Thus, if Eq. (15-19) can be obeyed, the effect of finite $A_{V_{OL}}$ does not result in significant difference between ideal and real behavior.

• *Open-loop 3-dB bandwidth*. As was mentioned, the open-loop gain drops off from the full dc value at some low frequency (most often near 10 Hz in general-purpose [compensated] op-amps). This roll-off continues at 20 dB per decade of frequency (see Fig. 15-15) until the unity-gain frequency $f_t$ is reached (typically at 1 MHz). To allow a larger closed-loop bandwidth, the closed-loop gain must be reduced. This fact of life is expressed by the relationship that the product of the closed-loop gain $A_v$ and the bandwidth BW yields a gain-bandwidth product (GBP), which is constant in an amplifier:

$$GBP = A_v \cdot BW = \text{constant} \qquad (15\text{-}20)$$

For example, if an op-amp has an $A_{VOL} = 10^5$ and $BW_{OL} = 10$ Hz, then GBP $= 10^6$. If it is desired that this amplifier have a closed-loop bandwidth of 10 kHz, its maxi-

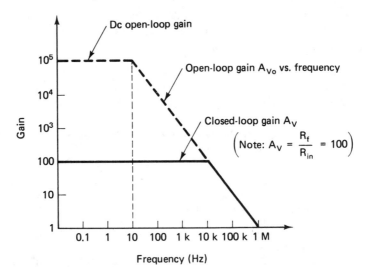

**Figure 15-15** Typical plot for a *compensated* operational amplifier connected in the inverting-amplifier configuration showing open-loop voltage gain ($A_{v_o}$) versus frequency, and closed-loop gain ($A_v$) for the amplifier versus frequency (for the case when $A_v = 100$).

mum closed-loop gain will be limited to 100 (see Fig. 15-15). This also implies that if the GBP of an op-amp can be increased, greater bandwidths for an identical closed-loop gain will be possible. Techniques used for properly extending the GBP of op-amps (to allow operation at higher frequencies) must be sought in other references since they are beyond the scope of discussion of this book. However, for many applications, there is no need to extend the frequency response of the op-amp beyond that which is nominally available. Thus for such "ordinary" applications it is best to select op-amps that are of the *compensated* type and do not need additional components (i.e., the μ741 is a widely used compensated op-amp).

• *Input impedance.* The effect of a finite (rather than an infinite) input impedance is to reduce the loop gain of the amplifier. However, if the amplifier input resistance $r_{in}$ is greater than 10 times the output impedance of the feedback network (i.e., $r_{in} > 10 \times (R_{in} \| R_f)$), the error due to input resistance loading will be negligible. Note that the input resistance of op-amps ranges from about $10^6$ Ω for general-purpose models to $10^{13}$ Ω in FET-input devices.

• *Output impedance.* Although a typical value of output impedance is 50 Ω, in the closed-loop operation mode this value drops to less than 0.1 Ω, certainly small enough to ignore in most applications.

• *Output offset voltage.* When an op-amp is used as a dc amplifier, $v_{out}$ should be zero when the input voltage $v_{in}$ is zero. In actuality, a dc voltage still appears at the output of real op-amps even when no input voltage signal is present. This *output offset voltage* can seriously limit the accuracy of the op-amp. There are two principal components which make up the output offset voltage, each caused by a different effect: (1) input offset voltage and (2) input bias current.

*Input offset voltage* $v_{io}$ is a small, relatively constant, but temperature-dependent voltage that exists between the inputs terminals of an op-amp even when no input signal is present. It is caused by imperfect matching of component characteristics within the input stage of the op-amp. It can be treated as a small voltage in series with one of the inputs. Therefore, unless it is balanced out, $v_{io}$ will be amplified by the same factor as the input signal, and $v_{out}$ will not be zero even when $v_{in} = 0$. In most amplifiers, provision is made for allowing adjustment of $v_{io}$ to zero at some constant temperature (i.e., 25°C) with an external potentiometer. However, since temperature changes will alter the state of balance in the input stage, some input voltage $\Delta v_{io}$ may reappear due to such temperature drift (i.e., $\Delta v_{io}$ can be from 0.1 to 75 μV/°C depending on the specific op-amp). If minimum total drift (after nulling of the input offset voltage) is required, an amplifier with a low offset drift specification should be selected for the application.

*Input bias current* $i_b$ is current that flows into the inputs of a nonideal (i.e., $Z_{in} \neq \infty$) amplifier due to leakage currents, gate currents, and so on, of the amplifier components. It is present even when $v_{in} = 0$ and may range from a high of 1.0 μA to a low of 1.0 pA (in very high input impedance FET op-amps). The voltage drop generated by $i_b$ across the input resistances $R_{in}$ and $R_f$ causes a voltage $v_b$ to appear at the amplifier inputs according to

$$v_b = i_b \frac{R_{in}R_f}{R_{in} + R_f} \tag{15-21}$$

As with $v_{io}$, $v_b$ is amplified by the same factor as the input signal. To minimize $v_b$ (and thus reduce any error in $v_{out}$ caused by $i_b$), (1) the parallel resistance of $R_f$ and $R_{in}$ must be small, or (2) an op-amp with low $i_b$ must be used, or (3) an equivalent resistance, $R_{ef} = (R_f$ in parallel with $R_{in})$ must be connected to the $(+)$ input configuration shown in Fig. 15-9(a). The latter technique causes a voltage $v_b(-)$ to appear at the $(+)$ input, and it serves to cancel $v_b$ due to $i_b$ appearing at the $(-)$ terminal. Thus if $i_b(+) = i_b(-)$, the offset voltage due to $i_b$ can be made reasonably small. (Note that $i_b$ is also quite sensitive to temperature changes, and the op-amp spec sheet for $\Delta i_b / \Delta T$ should be consulted if this problem is suspected.)

• *Slew rate*. Another important potential source of error due to the characteristics of real op-amps is inability of the amplifier to change the value of its output voltage instantaneously. That is, the *slew rate* (or maximum rate of change of the amplifier output voltage) is finite. For example, if a square-wave or sine-wave input is applied to an amplifier, distortion in the output will occur under conditions of slew-rate limiting (Fig. 15-16). As we see in Fig. 15-16, after the application of a full-scale square-wave input, the amplifier output slews, or changes at its maximum rate, until it reaches the required output value. Under slew-rate-limited response, the amplifier output voltage will change at a maximum slew rate (SR) of

$$\mathrm{SR} = \frac{\Delta V}{\Delta t} \tag{15-22}$$

The slew rate is usually specified in terms of volts per microsecond (V/μs), and can vary from 1 V/μs to 1000 V/μs. The relation between slew rate maximum

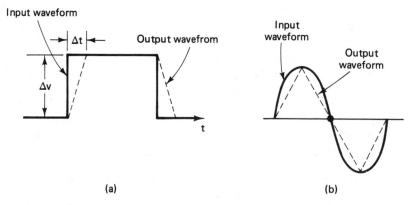

(a)   (b)

**Figure 15-16** Examples of distortions in amplifier output waveforms due to conditions of slew-rate limiting: (a) square-wave input; (b) sine-wave input.

amplitude of a sinusoidal output $(V_p)$ and maximum undistorted frequency $(f_m)$ is given by

$$SR = 2\pi f_m V_p \qquad (15\text{-}23)$$

Thus if an op-amp output is to have an amplitude of 2 V, and its slew rate is 5 V/$\mu$s, the maximum input sinusoidal frequency that will produce no distortion in the output signal is 796 kHz.

• *Common-mode rejection ratio.* The CMRR of op-amps is typically $10^5$ (CMR = 100 dB) at +5 V, and $10^4$ (CMR = 80 dB) at +10 V.

• *Rated current and voltage outputs and maximum common-mode voltages.* The rated output voltage of op-amps is somewhat less than that of the power supply voltages (which are usually from about $\pm 12$ V to $\pm 18$ V). The rated current output varies considerably with the specific op-amp, and may range from 5 to 70 mA. The maximum common-mode voltage rating is defined as the maximum peak common-mode voltage that will produce less than a 1 percent error at the output (typically $\pm 10$ V).

## INSTRUMENTATION AMPLIFIERS

Many industrial and laboratory applications call for the measurement of low-level analog signals originating at remote sources (i.e., thermocouples, strain gauges, current shunts, and biological probes). These signals must be amplified by devices possessing adequate gain, input impedance, CMRR, and stability. Typically, *instrumentation amplifiers* are selected for such applications, since they are amplifiers especially designed to fulfill these requirements. An introduction from an article by J. Ruskin of Analog Devices defines instrumentation amplifiers and their general characteristics quite elegantly:[2]

> An instrumentation amplifier is a precision differential voltage gain device that is optimized for operation in an environment hostile to precision measurement. The real world is characterized by deviations from the ideal; temperature fluctuates, electrical noise exists, and voltage drops caused by current through the resistance of leads from remote locations are dictated by the laws of physics. Furthermore, real transducers rarely exhibit zero output impedance and nice neat zero-to-ten-volt ranges. Induced, leaked or coupled electrical interference (noise) is always present to some extent. In brief, even the best "cookbook" must be taken with a grain of salt.
>
> Instrumentation amplifiers are intended to be used whenever acquisition of a useful signal is difficult. IA's must have extremely high input impedances because source impedances may be high and/or unbalanced. Bias and offset currents are low and relatively stable so that the source impedance need not be constant. Balanced

[2]Riskin, J., *A User's Guide to IC Instrumentation Amplifiers* (Norwood, Mass.: Analog Devices, 1978).

differential inputs are provided so that the signal source may be referenced to any reasonable level independent of the IA output load reference. Common mode rejection, a measure of input balance, is very high so that noise pickup and ground drops, characteristic of remote sensor applications, are minimized.

Also included in this category of amplifiers is the important subclass of *isolation amplifiers* (IAs).

As preamplifiers, IAs are capable of extracting small differential signals from large common-mode voltages. In analog-to-digital systems, their output signal amplitudes are adapted to satisfy the necessary input signal requirements of the receiving A/D converters (usually 5 or 10 V, full scale). They are also employed in special current measurement applications, that is, by amplifying the small voltages that appear across low-resistance shunts inserted in high-voltage lines.

Let us examine three different IA types:

1. The single op-amp connected in the differential-amplifier configuration
2. The three-op-amp IA type
3. Commercially built IAs

We shall see that the three types vary both in complexity and in their effectiveness in performing the required amplification tasks.

Single op-amps can be used as IAs if they are connected in the differential-amplifier configuration shown in Fig. 15-8. Op amps used in such fashion provide the simplest type of IA, but their CMRRs are limited to a maximum value equal to the intrinsic CMRR of the op-amp. (From Table 15-2 we note that the intrinsic CMRR of single op-amps ranges from about $10^4$ to $10^5$.) However, if the gain-setting external, resistors of the amplifier circuit ($R_{in_1}$ and $R_{in_2}$ and $R_{f_1}$ and $R_{f_2}$ of Fig. 15-8) are not perfectly matched, the actual CMRR (or apparent CMRR) of the single op-amp IA will be degraded (as indicated by the information given in Table 15-1). In addition, this IA type is very vulnerable to even further CMRR degradation from source impedance unbalance due to the relatively low input impedance of suitable op-amps (about $10^6\ \Omega$). Nevertheless, for the many applications for which the apparent CMRR of single op-amp IAs is adequate, this simple circuit can satisfactorily be used.

The second type of IA is built from three op-amps and appropriate resistors (Fig. 15-17). Such an IA is more complex, but provides improved performance (because of its higher input impedance and CMRR) over the single-op-amp type.

Typically, the three-op-amp type will exhibit a CMRR about 10 times larger than a single amplifier type (assuming both are designed with identical component tolerances and op-amp model types). In addition, the higher input impedance of the three-op-amp type makes it less susceptible to impedance unbalance CMRR degradation. For a great many laboratory measurement applications, the three-op-amp IA provides excellent performance. Let us discuss its behavior in a bit more detail.

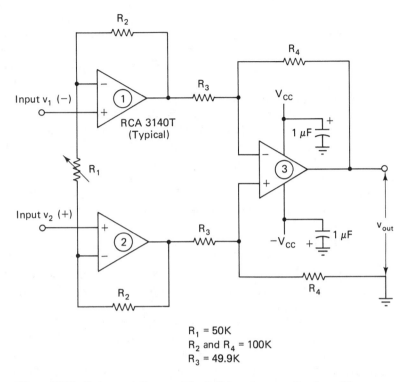

R₁ = 50K
R₂ and R₄ = 100K
R₃ = 49.9K

**Figure 15-17**   Instrumentation amplifier built by using operational amplifiers.

Operational amplifiers ① and ② of the three-op-amp type (Fig. 15-17) are connected in the voltage-follower configuration introduced in Fig. 15-10(b). These two voltage-followers serve to establish the very high input impedance that this type of IA presents to input signals. The two voltage followers then drive operational amplifier ③, which is connected in the differential-amplifier configuration to provide gain and common-mode voltage rejection. The output voltage of such an instrumentation amplifier is

$$v_o = \left(\frac{R_4}{R_3}\right)\left(1 + \frac{2R_2}{R_1}\right)(V_2 - V_1) \qquad (15\text{-}24)$$

When the values of $R_2$, $R_3$ and $R_4$ are kept equal to better than 0.1 percent, it is possible to obtain CMRRs of about 100 dB. Impedance unbalances of up to 1 kΩ can also be tolerated without serious degradation of CMRR. With the typical ±15-V power supplies, the amplitude limit for common-mode voltages for such three-op-amp IAs is approximately ±10 V. Note that the three-op-amp IA circuit shown in Fig. 15-17 can provide gains as high as 1000. If it is desired to be able to vary the gain with a single adjustment, some additional circuit complexity is required. See Reference 3 for additional details as to how this can be achieved.

The third type of IA that we shall discuss is sold commercially as a compact, single-packaged part, ready for use. Such IAs generally contain modified circuits of the three-op-amp IA just discussed. The highest performance commercially built IAs can provide another tenfold improvement in both the CMRR and input impedance over those of the three-op-amp types. In addition, the gain of these IAs can be adjusted by using a single external gain-setting resistor.

One example of a commercially available IA is Analog Devices' AD521J. It is a low-cost commercially built IA with an input impedance of $3 \times 10^9$ $\Omega$ and the minimum CMRR of $3 \times 10^4$ (at a gain of 100 and 1 k$\Omega$ source impedance unbalance). Gain can be preset from 0.1 to 1000 by using two external resistors. Somewhat more costly commercially available IAs can provide in-circuit CMRR values up to $10^6$ (CMR = 120 dB) and input impedances of up to $10^{13}$ $\Omega$. Table 15-3 provides a summary comparison of three different IA types.

### isolation Amplifiers

Isolation amplifiers are a special subclass of instrumentation amplifiers. They are actually dc differential amplifiers equipped with input circuit-guard shields and therefore their input circuits are ohmically separated (electrically isolated) from both the output circuit and the power supply of the amplifier. As a result, leakage and ground loop currents flowing through circuits containing isolation amplifiers are restricted to very small values (i.e., less than 10 $\mu$A). In addition, isolation amplifiers offer the capability of protecting measurement system components from very large voltages (up to 5000 V) encountered in some industrial environments. Thus they find application in the following measurement tasks for which ordinary IAs are inadequate:

1. In medical electronic equipment applications, where for safety reasons it is mandatory that leakage current levels passing through patients connected to medical electronic equipment be restricted to extremely small values (i.e., <10 $\mu$A).
2. When the common-mode voltages in the circuits under test are in the range

**TABLE 15-3**  INSTRUMENTATION AMPLIFIER COMPARISON

|  | Maximum CMRR | $Z_{in}$ ($\Omega$) | Advantages |
|---|---|---|---|
| Single-op-amp type | $10^4$ | $10^6$ | Simplicity |
| Three-op-amp type | $10^5$ | $10^9$ | Moderate performance, adequate for many applications, easy to design and build |
| High-performance commercial IA | $10^6$ | $10^{13}$ | High performance, adjustable gain, ready to use |

500 to 5000 V (i.e., in electric power plants and other high-voltage industrial process control systems).

3. In applications where a large impedance unbalance exists, but when a large CMRR is still required.

The amplifier input circuit of the isolation amplifier is isolated by enclosing it in a floating guard shield. (See Chapter 16 for additional information on input guarding.) The amplified input signal is coupled to the amplifier output circuit by an isolation element (either an isolation transformer or an optoelectronic coupler). The operating power required by the amplifier is also coupled into the shielded input circuits through an isolation transformer.

The *optoelectronic coupler* as an isolating element is shown in Fig. 15-18(a). A light-emitting diode and phototransistor are mounted very close to one another in a single package. Light from the diode, caused by current $i_1$, falls on the phototransistor, giving rise to a current $i_2$. The information contained by current $i_1$ is therefore transferred to the photodetecting element (the phototransistor) without any electrical coupling path. Light is the coupling link. Typically, the coupling capacitance that exists between the LED and phototransistor of optoelectronic couplers is $\approx 1$ pF. Optoelectronic couplers have the advantage over transformer isolation couplers of being able to operate from dc up to about 10 kHz. However, they do not provide as much electrical isolation or linearity.

Transformer isolation coupling devices, as shown in Fig. 15-18(b), use magnetic flux coupled through shielded transformers to pass signal information from the input to the output of the amplifier. Although transformer coupling devices can be built to exhibit improved linearity and a very much smaller coupling capacitance $C_c$ between input and output than optoelectronic couplers (i.e., $C_c = 0.1$ pF versus 1.0 pF), they do not function at dc or very low frequencies, and have an upper frequency limit of about 1 kHz.

Low-cost isolation amplifiers offer low capacitance between input and output

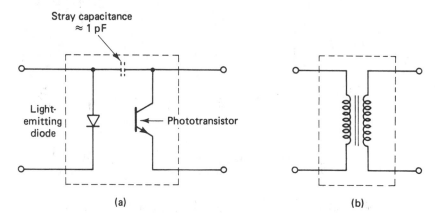

(a)                                                                (b)

**Figure 15-18**    (a) Optoelectronic isolator; (b) transformer isolation coupler.

circuits (<10 pF), high CMR (115 dB at 60 Hz), and high maximum common-mode voltage ratings (to 5 kV). Analog Devices' AD284J is an example of a commercially available isolation amplifier. It offers a minimum CMR of 115 dB with 5 kΩ of impedance imbalance between input and ground, an adjustable gain of 1 V/V to 10 V/V, a maximum of 2.0 μA rms leakage current at 115 V ac, and the ability to withstand 2.5-kV (continuous) and 5-kV (pulse) common-mode voltages. It also exhibits 78 dB of CMR between the low side of the signal input and the guard. (This

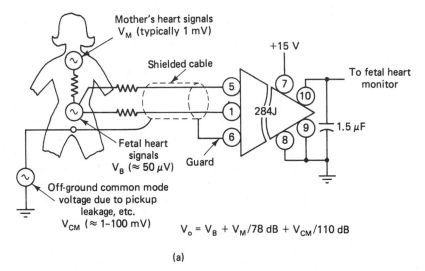

$$V_o = V_B + V_M/78 \text{ dB} + V_{CM}/110 \text{ dB}$$

(a)

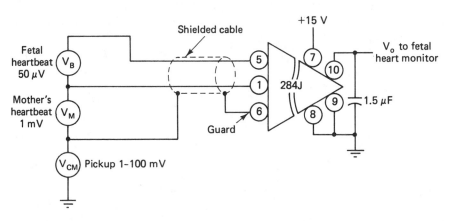

Amplifier's 78 dB input-to-shield CMR separates fetal heartbeat
from mother's, while 110-dB input-to-ground CMR attenuates 60-Hz pickup

(b)

**Figure 15-19**   Fetal heart monitoring with isolation amplifier.

specification needs to be considered in applications where the guard cannot be connected directly to the low side of the signal source.) Figure 15-19(a) illustrates how an AD284J is used to monitor the heartbeat of an unborn child. Figure 15-19(b) shows an equivalent-circuit model of this measurement application and indicates how the 78-dB CMR between the input electrodes and the guard shield allows the fetal heartbeat signal (50 μV) to be separated from the mother's heartbeat signal (1 mV). The 110 dB of CMR from input to ground screens out common-mode signals due to 60-Hz pickup and other interference.

## PROBLEMS

1. Define the following terms.
   (a) Voltage gain $A_v$
   (b) Power gain $A_p$
2. Explain the difference between dc and ac amplifiers.
3. Demonstrate that the two definitions of bandwidth of an amplifier (i.e., as defined in terms of power gain and in terms of voltage gain) are equivalent.
4. If the magnitudes of the input and output voltages of an amplifier are given as listed below, calculate the voltage gain of the amplifier.
   (a) $v_i = 2$ V, $v_o = 37$ V       (b) $v_i = 75$ mV, $v_o = 13$ V
   (c) $v_i = 50$ μV, $v_o = 42$ mV       (d) $v_i = 25$ mV, $v_o = 21$ mV
5. If a 50-mV signal appears at the output of a dc amplifier when no signal is applied to its input, is this an unexpected event? Why or why not? How can the magnitude of this output signal be reduced?
6. A differential amplifier has a gain of $A_v = 120$. When the common-mode gain $A_c$ is being measured, we get $v_i = 3$ V and $v_o = 15$ mV. Calculate the CMRR and the CMR of this amplifier.
7. Determine the output voltages for the op-amp circuits shown in Fig. P15-1.

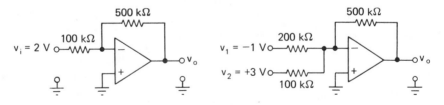

**Figure P15-1**

8. Since an amplifier is basically designed to multiply the magnitude of an input signal by some constant amount, what is the advantage of using an op-amp with a feedback path to accomplish essentially the same purpose?
9. Explain why the property of having a high input impedance is often a desirable characteristic in amplifiers.

**10.** List some of the applications that make the op-amp such a versatile building block in the design of measurement instrumentation.

## REFERENCES

1. Millman, J., and Halkias, C.C., *Integrated Electronics*. New York: McGraw-Hill, 1972.
2. Irvine, R. G., *Operational Amplifier Characteristics and Applications, 2nd ed.*, Englewood Cliffs, N.J.: Prentice Hall, 1987
3. Jung, W., *IC Op-Amp Cookbook*. Indianapolis, Ind.: Howard W. Sams, 1976.
4. Diefenderfer, A.J., *Principles of Electronic Instrumentation*. Chap. 9. Philadelphia, Pa.: W. B. Saunders, 1979.
5. Strong, P., *Biophysical Measurements*. Chap. 19. Beaverton, Ore.: Tektronix, 1970.
6. Garret, P.H., *Analog Systems for Microprocessors and Minicomputers*. Chap. 2. Reston, Va.: Reston Publishing, 1978.
7. *Isolation and Instrumentation Amplifier Design Guide*. Norwood, Mass.: Analog Devices, n.d.
8. Berlin, H.M, and Getz, Frank C., Jr., *Principles of Electronic Instrumentation and Measurement*. Columbus, Ohio.: Charles E. Merrill, 1988.

# 16

# *Interference Signals and Their Elimination or Reduction*

Many electrical measurements involve the detection and measurement of low-level signals (defined as signals with amplitudes of less than 100 mV). Usually, these low-level signals are amplified by a part of the measuring system so that they can be more easily displayed or otherwise utilized.

Unfortunately, the electrical laboratory environment contains many sources of electric and magnetic energy that can induce extra, unwanted (interference) signals in the wires that carry low-level signals. These unwanted signals are then amplified together with the signal being measured. Under some circumstances, the magnitudes of such induced signals can become so large that they distort or obscure the signal of interest and lead to inaccurate or meaningless measurement results. Therefore, as a part of making accurate measurements, the sources of interfering signals need to be identified and measures taken to eliminate or minimize the interference.

Unwanted noise signals can be generated external to the measurement system. These signals may then be coupled into the system in various ways and thereby become interference signals. Noise signals may also arise from the inherent operation of the devices and components that make up the system, and such signals are considered to be internally generated. However, there is general agreement among experimenters that the externally generated interference signals are more serious in limiting the maximum sensitivity of most measurements. Therefore, we direct the emphasis of our discussion to these more serious external interference signals and to the methods developed for their suppression. However, a brief introduction to internally generated interference signals is also be undertaken at the end of the chapter.

*External interference signals* can be classified according to the physical phenomena that are responsible for their generation or transmission. The five major types of external interference signals are therefore known as

1. Capacitive (or electrically coupled) interference
2. Inductive (or magnetically coupled) interference
3. Electromagnetic interference
4. Conductively coupled interference
5. Ground-loop (or common-mode) interference

When a conductor possesses a net electric charge, an electric field exists in the space around the conductor. If the conductor carries a current, the region surrounding the conductor contains a magnetic field. If the charge or current varies with time, the associated fields at any point in the surrounding space will also vary; that is, they will shrink and grow in magnitude in proportion to the strength of their source. In the regions nearby the conductor, these fields will pulsate in synchronization with the time variation of the current or charge. Such synchronous fields are known as the *near* (or *induction*) fields. Capacitive and inductive interference signals are caused by such *near fields*.

At points far from the conductor, the near electric and magnetic fields become negligible and plane-wave propagation of electromagnetic energy takes over. This region is called the *far field*. Radiated electromagnetic waves can cause ac signals of the same frequency as their originating source in any suitable conductors they encounter. Such induced signals are thus called *electromagnetic* interference.

Interference signals in ground loops can arise from near-field effects and far-field effects, as well as other electrical phenomena.

## CAPACITIVE (ELECTRICALLY COUPLED) INTERFERENCE

Although unconnected physically, nearby conductors are coupled electrically by the capacitance between them (in the same way that the parallel plates of a capacitor are coupled). Thus a voltage change occurring in one conductor is coupled to other conductors nearby. The amount of capacitance (see Chapter 11) between the conductors determines the degree of coupling between them.

In measurement systems, the *objects* of capacitive interference are the low-level signal transducers and low-level signal-carrying conductors. The *sources* of capacitive interference are conductors that have large varying voltages, typically with little or no current flow. In the usual measurement environment, such sources include fluorescent light bulb fixtures, and unconnected wall or ceiling power outlets. These sources exhibit voltages which vary from $+163$ to $-163$ V (peak value) at

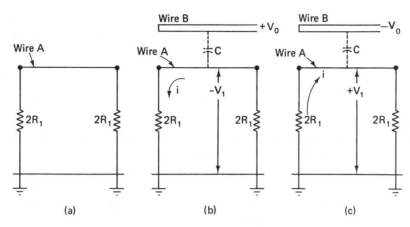

**Figure 16-1** Capacitive interference.

60 Hz. As long as the outlets remain unconnected, virtually no current flows in the wires, but the voltage continues to vary from +163 to −163 V. Since such sources are present in virtually all laboratories, capacitive interference signals at 60 Hz can be appreciable (in varying degrees of severity) in nearly every measurement situation.

In fact, 60-Hz noise is so pervasive that it is even referred to by its own name, *hum*.[1] Other typical sources of higher-frequency capacitive interference include electric arcs and oscillators and signal generators being used in the proximity of the test setup.

As an example of how capacitive effects induce interference signals in a low-level-signal wire, consider wire $A$ shown in Fig. 16-1(a). Let us assume that it is a low-level signal wire connected to ground by resistor of value $2R$, at each of its ends. As long as there is no current flowing in the wire, it remains at zero potential (the same potential as the ground point to which it is connected). Let us then place another wire [wire $B$ of Fig. 16-1(b)] near wire $A$. The voltage of wire $B$ varies from $+V_0$ to $-V_0$ volts at some specific frequency. When the voltage of wire $B$ equals $+V_0$, electrons are drawn up to wire $A$ from the ground plane due to the capacitive effect between wires $A$ and $B$. The charges must move through the resistors to get to wire $A$, and moving charges constitute a current flow. This current, $i$, flowing through the resistors will cause the wire to acquire some nonzero voltage according to

$$-v = -iR_1$$

Similarly, when wire $B$ is at $-V_0$ volts, electrons are repelled from wire $A$ and will move away from it. A current thereby flows through the resistors again, this

[1]All interference signals caused by the voltage or current variations in the 60-Hz power-line wires are called *hum*. They get this name from the fact that if the interference is amplified and fed to a loudspeaker, the resulting sound is a low hum. The sound is that of a 60-Hz audio tone. The 120-Hz ripple associated with typical full-wave rectifiers is also recognized as "hum."

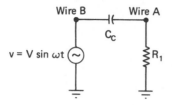

**Figure 16-2**   Model for calculating the magnitude of capacitive interference between two wires.

time in the opposite direction. The potential of wire $A$ also reverses and becomes equal to $-V_1$ when wire $B$ is at voltage $-V_0$.

Thus, as the voltage of wire $B$ varies from $+V_0$ to $-V_0$, the voltage induced in wire $A$ will vary from $-V_1$ to $+V_1$. The magnitude of $V_1$ will depend on the capacitance that exists between wires $A$ and $B$, the value of the resistors connecting wire $A$ to the ground, the value of $V_0$, and the frequency of the variation of the voltage $V_0$. This information can be utilized to develop a circuit model of the events occurring in the system (Fig. 16-2). In the model, wire $B$ is connected to a voltage source that varies at a frequency of $\omega$ radians per second. A capacitive effect exists between wires $A$ and $B$, and this is represented by a coupling capacitor of value $C_c$.

If we know the frequency of $V_0$ and its magnitude, the value of $R_1$, and the magnitude of the interference signal induced in wire $A$, we can find $C_c$.

### Example 16-1

A test lead is connected to an instrument that has a 2-M$\Omega$ input impedance and to a circuit that has a 2-M$\Omega$ output impedance (Fig. 16-3). If a 1-mV rms signal is induced in the lead from nearby 60-Hz power wires, what is the capacitance that must exist between the lead and the power wires?

**Solution.**   Using the model shown in Fig. 16-3, we see that the rms voltage of point $A$ will be 1 mV. $R_1$ of the model is 1 M$\Omega$ for this case. Then the current flowing in the circuit can be found from

$$I_{\text{rms}} = \frac{V_{A\,\text{rms}}}{R_1} = \frac{1\ \text{mV}}{1\ \text{M}\Omega} = \frac{0.001\ \text{V}}{10^6\ \Omega} = 10^{-9}\ \text{A}$$

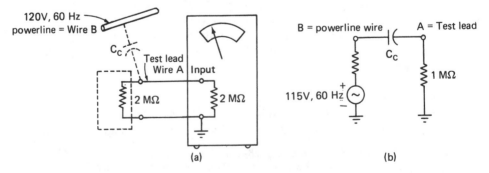

**Figure 16-3**   (a) Instrument with 2 M$\Omega$ input impedance connected to test circuit having an output impedance of 2 M$\Omega$; (b) equivalent-circuit model of (a).

The capacitance between conductors $A$ and $B$ is found from

$$V_{B\,\text{rms}} = I_{\text{rms}}Z = I_{\text{rms}}\left(R^2 + \frac{1}{\omega^2 C_c^2}\right)^{1/2} \tag{16-1}$$

Since in this case $1/\omega C_c \gg R$, (actually, in this step we are assuming that the value of $C$ is so small that $\omega C_c = 377C$ is such a tiny number that $1/\omega C_c \gg 10^6$; the final answer bears out the validity of this assumption) we can neglect $R$ in Eq. (3-7) and write

$$V_{B\,\text{rms}} \approx \frac{I_{\text{rms}}}{\omega C_c} \tag{16-2}$$

or

$$C = \frac{I_{\text{rms}}}{\omega V_{B\,\text{rms}}} = \frac{10^{-9}}{2\pi(60)\times(120)} = 2\times 10^{-14}\,\text{F} = 0.02\,\text{pF}$$

Example 16-1 illustrates the fact that even very small capacitive effects can potentially induce potentially significant interference signals in high-impedance circuits. In actual practice the magnitude of the source-coupling capacitance varies widely, but 0.2 pF is probably within an order of magnitude of its typical value. The example is also relevant to instrument practice because the input impedance of oscilloscopes, electronic voltmeters, and most amplifiers is 1 MΩ or more. If such instruments are used to measure low-level signals in high-impedance circuits, the test leads will behave very much like the test lead (wire $A$) examined in Example 16-1. We can see that such leads will be very prone to picking up capacitive interference from any nearby conductors possessing a time-varying potential.[2] In addition to the 60-Hz signal, Eq. (16-2) shows that as the frequency of the source of the interference increases, the induced current due to this capacitive pickup also becomes larger. Therefore, conductors carrying high-frequency signals are also liable to be sources of capacitive interference.

In summary, we see that unwanted capacitive pickup is increased as

1. The input impedance of the instrument increases.
2. The capacitance between the interfering source and test lead increases.
3. The voltage between the interfering source and test lead increases.
4. The frequency of the voltage in the interfering source increases.

To prevent low-level signal sources and the wires that carry low-level voltage signals, from being influenced by capacitive effects, several courses of action are possible. The most common countermeasure is to surround such wires with an electrostatic shield. The shield consists of a metal enclosure surrounding the low-level signal source and a braided metal sleeve which surrounds the two test leads

---

[2] In Chapter 6, in the section "Oscilloscope Errors," there is further discussion of such 60-Hz pickup by an oscilloscope. There is also a description of an easy method for demonstrating how this interference is coupled to the scope and displayed on the scope screen.

[Fig. 16-4(a)]. This type of shielding is effective because external static electric fields cannot penetrate an enclosure surrounded by an electrical conductor. However, when the signal source is grounded at one end, the metal shield surrounding the test signal cable should also be connected to ground at the signal source ground point through a low-impedance path. This will ensure that the shield will remain at a potential very close to zero [Fig. 16-4(b)]. Thus, although there is a relatively large coupling capacitance between the low-level signal wire and the surrounding shield, the small potential difference existing between them will keep any noise picked by the shield from causing a significant signal on the signal lead.

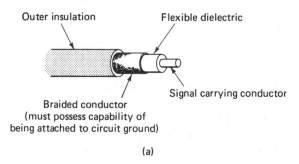

(a)

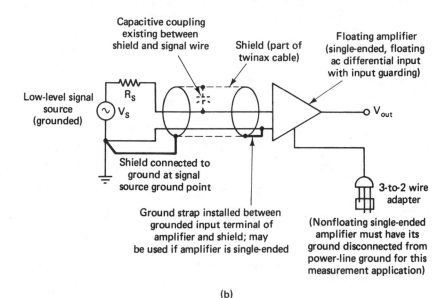

(b)

**Figure 16-4** (a) Sketch of Twinax (two signal leads) shielded (enclosed) in an axial cable; (b) shielding low-level signal wires against external capacitive interference through the use of shielding and proper grounding.

As we emphasize again in the following sections on inductive interference and ground-loop interference, the shield lead and the signal source should be connected to earth ground at the same single point, and the amplifier must be disconnected from ground (i.e., floated) to prevent establishment of a ground loop [Fig. 16-4(b)]. If we were to create a ground loop by connecting the signal ground lead and the amplifier ground lead to the ground plane at separate points, ground-loop interference signals (i.e., common-mode voltages and inductively coupled ground currents) would degrade the measured signal.

## INDUCTIVE INTERFERENCE AND SHIELDING

Capacitive interference involves electrostatic fields that exist between conductors at different potentials. Inductive interference stems from magnetic fields that are associated with current-carrying conductors. The current creates a magnetic field in the space surrounding the conductor. If the current changes with time, the magnetic field will also change. If there is a closed conducting path (a loop) nearby, the magnetic field that intercepts the loop will cause a current to be induced and flow around the loop. If this loop is part of the measurement system, the induced current will represent an inductively caused interference signal. The magnitude of the induced current will depend on the strength of the magnetic field, the frequency of its variation, and the area of the loop. (The loop area is a factor because a large loop will encompass more magnetic flux than will a smaller one.) In addition, the current will depend on the resistance of the loop. If the resistance of the current path forming the loop is high, the magnetic fields usually encountered in the laboratory environment will induce only insignificantly small currents. However, inductively coupled interference signals can still cause problems in high-impedance measurement setups. If a low-impedance loop is formed by a shield conductor and a portion of the ground plane (i.e., if the shield conductor is connected to ground at two different points), large currents can be induced in this path. The capacitance that exists between the shield and the low-level signal wires (typically 80 pF/m between the outer shield and inner conductor of a coaxial cable) will still cause the interference signal to be coupled to the measured signal.

There are also instruments adversely affected by external magnetic fields in other ways. These instruments include electromechanical meter movements (especially the iron vane and electrodynamometer movements) and cathode-ray tubes (CRTs). The meter movements detect current with the help of rather weak magnetic field effects, and these magnetic effects can be disturbed by additional magnetic fields in their vicinity. Unwanted changes in the magnetic fields of the movements can cause inaccurate meter readings. (The basic meters discussed in Chapter 4 are sensitive to such magnetic interference.) The electron beam in the CRT can be deflected by magnetic fields, and such deflections can lead to misleading shifts in the position of the beam.

The most common sources of inductive interference in the laboratory follow:

(1) the current flowing in power cord wires in the instruments themselves and in the surrounding benches, walls, and ceilings; (2) inductors and power transformers; (3) large currents flowing in ground loops; (4) current surges caused by power being delivered to large motors (of nearby elevators, air-conditioning units, or subway trains, for example); and (5) abruptly changing currents that occur in the operation of, for instance, motors, relays, automobile ignitions, and lighting dimmers.

Inductive interference is the most difficult to shield against. Therefore, it is often easier to attempt control of the interference at the source. There are several techniques that can be used to limit such interference at the source. First, since the magnetic flux produced in any closed path is proportional to the sum of the current enclosed by the path, by running the signal wires, as well as the power and return leads of an instrument, as a twisted pair, the external magnetic field surrounding the power cord is reduced [Fig. 16-5(a)]. Twisting the wires reduces the area between the conductors, and, as the area is reduced, the induced current is minimized.

Second, since a magnetic field will concentrate itself in a ferromagnetic material, by enclosing transformers and other magnetic-field sources in ferromagnetic

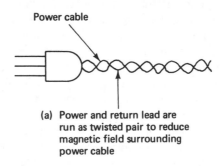

Power cable

(a) Power and return lead are run as twisted pair to reduce magnetic field surrounding power cable

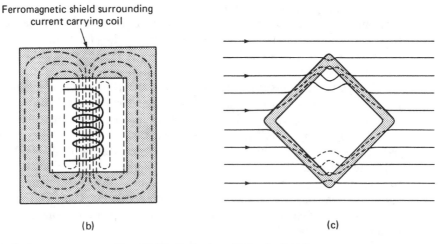

Ferromagnetic shield surrounding current carrying coil

(b)                                              (c)

**Figure 16-5**   Reduction of inductive interference.

enclosures, the magnetic field outside the enclosure is weakened [Fig. 16-5(b)]. Third, ground loops carrying large circulating currents must be prevented from being established. Finally, the effect of the magnetic field can be reduced by increasing the distance between the source and object of the magnetic interference and by reducing the area of the loop intersecting the interfering magnetic field. Thus, interfering sources such as relays and motors should be kept far from the measurement setup. In addition, it is good practice to keep the low-level signal leads as far apart as possible from the (twisted) power cable or digital signal lines. Where this is not totally possible, the wires should be oriented so that signal wires and digital lines or power cables are allowed to cross only when they are at right angles to one another.

Once the magnetic fields of the sources are reduced by the methods described above, magnetic shielding for further reduction becomes practical. The procedure again is to use a metal with a high magnetic permeability (to attenuate the magnetic field, which would otherwise intercept the sensitive device or wire). Thus, magnetic fields that cannot be reduced at their sources (such as the magnetic field of the earth) are made to bypass the elements enclosed within the shielded volumes, and their effects are minimized [Fig. 16-5(c)].

## ELECTROMAGNETIC INTERFERENCE AND SHIELDING

At high frequencies, a part of the energy associated with the fluctuating current or charge in a conductor is radiated away from it in the form of electromagnetic (EM) radiation. This phenomenon is specifically used to generate radio waves for communication and radar applications. However, it has become common parlance to refer to any EM waves that have frequencies comparable to radio or radar waves as *radio-frequency* (RF) waves, whether they are actually signals from radio or radar transmitters or not. Besides radio waves, there are many other sources (both manufactured and natural) that produce RF signals. However, in measurement systems, *all* types of RF signals are considered to be sources of unwanted EM interference. As a result, sensitive circuits must be protected from all RF signals no matter what their source might be.

Human-made sources of RF signals include gas discharges in fluorescent lights and X ray tubes; arching in electric motors, generators, switches, and relays; and high-frequency oscillations in pulse circuits, discharge circuits, and oscillators. (Of course, radio and TV transmitters can also be sources of RF interference.) Natural sources of RF radiation include lightning and other electrical atmospheric phenomena, and cosmic radiation (including solar disturbances).

Since EM waves have both electric and magnetic field components, interference effects comparable to those generated by capacitive and inductive pickup can be caused by EM radiation. If the circuits of a measuring system are susceptible to one of these kinds of pickup more than the other, the EM waves will predominantly produce that type of interference in the circuit. Furthermore, both inductive and capacitive interference effects increase as the frequency of the disturbing effect increases. Since EM waves are usually high-frequency signals, this makes them even more likely to induce interference. For these reasons, a nearby source of EM radiation is likely to cause interference in the widest variety of circuits. Fortunately, it is also possible to effectively shield against EM radiation.

EM shields make use of the fact that EM radiation must simultaneously contain electric and magnetic fields in order to propagate independently through space. If we eliminate either the electric or the magnetic component of an EM wave, the other component is also halted. Therefore, since a shield designed to eliminate electrostatic fields can be made quite easily, this type of shield is also used to eliminate the electric-field component of an Em wave. Without the electric component, the magnetic field cannot continue to propagate and is thereby extinguished, too. Thus, an enclosure consisting of a good electric conductor connected to the ground through a low-impedance path will provide shielding against EM interference. (Note that special rooms that are completely surrounded on all sides by a continuous conductor are often used to house test setups that are highly sensitive to external interference.) When constructing such enclosures, care must be taken to prevent any discontinuity or holes that are of a size that is as large as a significant part of the wavelength of the signal to be rejected. Thus metallic wire mesh is often used as a shield material. If holes in the shield are required for ventilation of an instrument, it is good practice to keep all uncovered holes less than about 3 mm in diameter. Larger holes should be covered with a fine wire mesh. Where there are joints in the shield or across doors, continuous electrical and mechanical bonds must be used. Use of RF gaskets at the joints and doors also helps to assure low-impedance contact along the joints.

If a known source of EM radiation is in the vicinity of the measurement setup, it is a wise policy to turn off the source while making measurements or to enclose it in a housing made of a good conductor which is tied to ground. Such a grounded conductor will stop the EM radiation from propagating away from the source and causing interference in circuits that are close enough to be affected by the emitted radiation.

## CONDUCTIVELY COUPLED INTERFERENCE

Interference can also be caused by electrical fluctuations or signals that originate in other electrical devices connected in the same circuit as the measuring instrument. Since such interference signals are coupled to the measurement circuits directly through electrical conductors (i.e., the wires or cables of the circuit), such interference is known as *conductively (or resistively) coupled interference*.

Three of the most common causes of conductively coupled interference in measurement systems follow: (1) the presence of a *common-impedance ground path* in the measurement system, (2) conductively coupled interference introduced into the system through the *power transformers of the measurement instruments*, and (3) *power supplies that are incorrectly connected to parallel loads.* Let us examine each of these situations in more detail.

**1.** A *common-impedance ground path* in a measurement circuit is present when a number of different circuit elements or instruments are connected to ground through the same conductor. This will cause conductively coupled interference in the system components connected to the common-ground path. The problem can become significant if the path possesses a substantial impedance to earth ground. As an example, consider the case where an analog-signal ground and a digital-signal ground are both connected to the same earth-grounded conductor. This conductor then acts as an impedance that is common to both the analog and the digital ground paths. If the impedance of the earth-grounded conductor is substantial, it is highly likely that switching transients from the digital logic circuits will be coupled to the wires carrying the analog signals.

To reduce interference arising from common-impedance paths, the following procedures should be followed. First, all low-level analog signal grounds should be kept separate (i.e., disconnected from) power grounds and digital signal grounds until the analog, digital, and power grounds are finally all tied to the single designated point on the systems earth grounded conductor (see, for example, Fig. 16-9). Second, if a common-impedance ground path must be utilized by more than one component of a measurement system, the impedance of the common path should be kept as small as possible. Third, the connection of all the ground leads of a measurement system should, whenever possible, be made at a designated single point where the common impedance of the system ground path to earth ground has the smallest value. This last statement implies that a better system ground path than that afforded by the power-line ground (i.e., the third wire) should be utilized in systems where sensitive measurements are being made. The third wire of the power-line is often inadequate as a system ground for two reasons. The power-line ground wire is usually of small diameter, and it is run all over a building before it is finally connected to a water pipe or other earth-grounding point. Thus the third-wire becomes an example of the previously described (and unwanted) common-impedance ground path (of relatively high impedance) for the components connected to it. Therefore, when designing systems that measure low-level signals, it is good practice not to rely on the grounding system of a building, but to establish a separate, low-impedance path to ground which is independent of the power-system ground. If this new ground path is used, however, care must be taken to ensure that instruments are not inadvertently reconnected to the power-line ground through the third prong of the power plug. Use of 3-to-2 wire plug adapters and isolation transformers allows disconnection of the third prong of the instrument power cords from the power system ground. Failure to observe this precaution will lead to the establishment of ground loops that will lead to a myriad of additional measurement problems for the experimenter.

**2.** The circuit consisting of the ac power line, together with the power transformer of a measurement system instrument, provides another path through which conductively coupled interference can be introduced into the measurement system. The sources of this interference include current spikes and fluctuations in the power-line voltage which occur when motors, temperature-controlled furnaces, and so on, are operated or turned on and off. Such interference is transmitted by the power lines throughout the common power distribution circuit. If a measurement setup is connected to this same power distribution circuit, these noise spikes will thereby be conductively coupled to the ac power inputs of the measuring instruments operating from the power lines. A large capacitance (typically 1000 pF) may exist between the primary and secondary windings of the power transformers of electronic instruments (i.e., oscilloscopes, amplifiers, etc.), and thus the current spikes can be strongly coupled to the internal power conductors of the instrument. Further coupling of this interference to the wires carrying the measured (input) signal can occur through capacitive or inductive effects, or through the degradation of the power supply output also caused by this interference. One technique used to suppress this type of interference is to build the instrument power transformers with capacitive interference shielding (called Faraday shielding) between their primary and secondary windings. This shielding consists of conducting (metal) foil and permits magnetic energy to pass, but attenuates capacitive interference. The conducting foil is connected to the power-line ground. Thus the electrostatically induced currents are carried to ground through the foil rather than through the circuitry of the transformer secondary winding capacitance of the transformer (i.e., from 1000 pF to less than 1.0 pF). The problem of the conductively coupled noise spikes is thereby effectively reduced. A second (and quite obvious) technique for reducing power transformer conductively coupled interference is to remove known sources of such interference from the local ac power distribution network.

**3.** Power supplies, when connected incorrectly to measurement setups, are the third cause of conductively coupled interference. As shown in Fig. 13-10(b), when two or more loads are incorrectly connected in parallel to the same power supply, fluctuations in the current drawn by one load will be conductively coupled to the other loads. This problem is overcome by connecting the parallel load correctly to the supply [Fig. 13-10(a)].

**4.** If the range of frequencies of the measured signal are known, unwanted conductively coupled signals of other frequencies can be minimized by proper use of electronic filters in the measurement system. Since the subject of filtering, however, is beyond the scope of this text, further information on filter design for noise reduction will have to be found by consulting other sources. Several useful references that discuss the matter are cited at the end of the chapter.

## GROUND-LOOP (COMMON-MODE) INTERFERENCE

Ground loops and ground-loop interference are a frequent and serious source of problems in many electronic measurement systems. In the earlier discussions on various types of external interference, we noted that ground loops often contribute to the strengthening of noise signals that arise from other forms of interference. It was also emphasized that the effects of ground loops must be considered when methods to suppress capacitively, inductively, and conductively coupled interference are employed. Ground loops can also cause interference difficulties on their own. Since they are often subtle and little-understood entities, ground loops are also frequently blamed for unexplained interference signals. Thus, a thorough discussion that explores the origins and effects of ground loop currents, as well as methods used to eliminate them, will be very useful.

*Ground loops* are closed electrical paths in which the sections of the path consist of the ground wires of a system and the ground plane. Ground loops are created whenever the ground conductor of an electrical system is connected to the ground plane at different points (Fig. 16-6). Since the ground wires of most systems and the ground plane are usually low-impedance conducting paths, ground loops as a whole are conducting paths of low impedance. Thus if even small voltage differences exist between any points along the loop, large currents will flow in them.

The two principal causes of current flow in ground loops are:

1. Differences in potential between the points of the ground plane to which the ground terminals are connected
2. Inductive pickup, owing to stray magnetic and RF fields

The electric power distribution systems that supply our electric power use earth ground as their reference. As a result, currents flow in the earth from these systems. Because power systems are used so extensively throughout the world, it can be assumed that these earth currents (at 60 Hz) flow almost everywhere. Such currents cause a potential drop (voltage) to exist between separate points on the earth (typically 1–10 V rms). If a measurement system is *grounded to earth at two separate points of the earth's surface*, it must be assumed that a potential drop will appear between these two points. The two ground connections will have caused a ground loop to be established, and large amounts of unwanted 60-Hz current will probably flow in this loop. If the shield of a measuring system constitutes part of such a loop, the shield itself becomes a source of interference (rather than the barrier to noise that it is designed to be).

*Inductive pickup* by a ground loop occurs the same way as inductive pickup in any closed conducting loop. Since a ground loop has a relatively large area and very low impedance, inductive pickup from stray magnetic or electromagnetic fields in

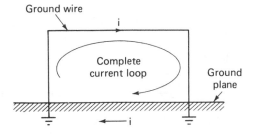

**Figure 16-6**   Ground loop.

the measurement environment can occur quite easily. Orthodox shielding methods are not as effective against inductive pickup in ground loops because the shielding conductors and the ground plane themselves form a part of the loop.

### Ground Loops Established by Capacitive Coupling

In our definition of the ground loop we said that a ground loop is created whenever a *closed electrical path*, consisting of ground wires (or shields) of a system and the ground plane, exist. In Fig. 16-5(a) we indicated that such a path can consist of an ohmic conducting path and the ground plane. However, owing to capacitive coupling, a *closed electrical path* can also exist even if there is no apparently complete ohmic conducting path (i.e., even though the measurement system is ohmically connected to ground only at a single point). For example, let us examine an amplifier, as shown in Fig. 16-7, with an input signal ground connected to earth ground at point ① and a shield surrounding the amplifier that is not connected to earth ground. Note that there is still a complete current path (loop) from the ground lead within amplifier ① to a second ground point. ④, through the mutual capacitances connecting the ground lead and shield ($C_{12}$) and the shield and the ground plane ($C_{23}$). Thus a ground loop through path ①–②–③–④–① does indeed exist. If an ac potential difference exist between the ground points ① and ④, a current will flow in this ground loop.

### Common-Mode Noise Voltage

In many measurement systems the source of the signal $v_s$ being measured is grounded at one point [point ① in Fig. 16-8(a)], and the amplifier to which it is connected is grounded at a different point [point ③ in Fig. 16-8(a)]. The shield of the measurement system is also usually grounded but may not be connected to the

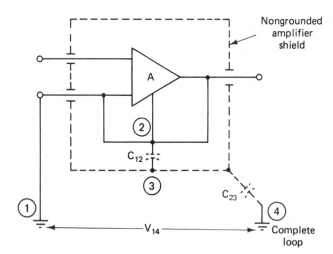

**Figure 16-7**  Ground-loop path around ①–②–③–④–①, which includes the capacitances $C_{12}$ and $C_{23}$ and the ground plane and shield. Note that there is no ohmic conductive path between points ② and ④.

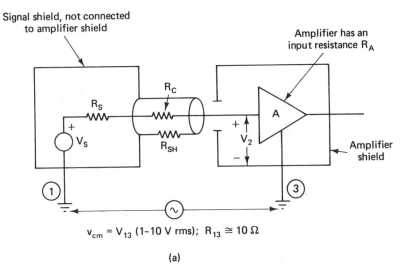

$$v_{cm} = V_{13} \text{ (1--10 V rms); } R_{13} \cong 10 \, \Omega$$

(a)

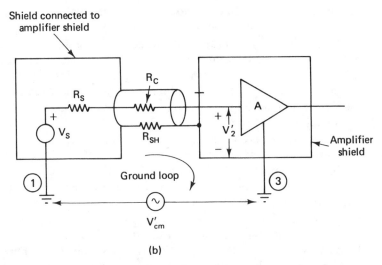

(b)

**Figure 16-8** (a) Common-mode voltage, $V_{cm} = V_{13}$, in this example, causes the input voltage of the amplifier to be $V_2 = V_s + V_{cm}$. If $V_{cm}$ is 1 to 10 V (rms) and $V_s$ is small, the desired signal will be obscured by the larger common-mode voltage. (b) Attempting to reduce the value of $V_{cm}$ appearing at the amplifier inputs by connecting the cable shield to the amplifier shield merely results in another common-mode voltage interference signal $V'_{cm}$ due to the ground loop established as shown.

amplifier shield. Thus the source, the signal lead, and the amplifier are all fully shielded from external capacitively coupled and EM interference. However, the voltage appearing across the amplifier input $v_2$ will be the sum of the source voltage $v_s$ and the potential difference between the two ground points ① and ③, $v_{13} = v_{cm}$

or $v_2 = v_s + v_{cm}$. If $v_s$ has a small amplitude (such as the 10- to 50-mV output signals typical of many transducers, $v_{cm}$ (the common-mode voltage) can completely obliterate $v_s$. For example, if the ground resistance between ground points ① and ③ of Fig. 16-8(a) is $R_{13} = 10$ MΩ, and if $R_s = 1000$ V, $R_c = 10$ Ω, and the amplifier input impedance $R_A$ is 10 MΩ, $v_2$ can be calculated to be approximately

$$v_2 = \frac{R_A}{R_A + R_s + R_{13} + R_c}(v_{cm} + v_s) = \frac{10^7(v_{13} + v_s)}{10^7 + 1000 + 10 + 10} \approx \frac{10^7}{10^7}(v_{cm} + v_s)$$

or

$$v_2 \approx v_{cm} + v_s$$

If an attempt is made to reduce $v_{cm}$ by connecting the source shield to the amplifier shield, a ground loop is produced [Fig. 16-8(b)]. Large ground currents, due to a combination of inductive pickup and the ever-present $v_{cm}$, will flow in the loop. Thus the new interference common-mode voltage $v'_{cm}$ may have a larger amplitude than the original common-mode voltage $v_{cm}$.

### Reduction of Ground-Loop-Induced Interference Signals

There are several techniques that can be used to eliminate or reduce the interference signals caused by ground loops and common-mode voltages. They include the following ones: (1) single-point grounding, (2) use of differential-input amplifiers, (3) input guarding, and (4) use of battery-powered instruments. We shall see that the use of differential-input amplifiers equipped with input guard shields (i.e., a combination of methods (2) and (3) provides the most complete solution for reduction of interference signals due to ground loops and common-mode voltages.

Prior to the widespread availability of differential amplifiers in the 1950s, single-point grounding implemented with heavy conductors was the most effective technique used for combating ground-loop interference problems. The *single-point grounding* method for eliminating ground-loop interference relies on the fact that current cannot flow in any path unless a complete loop exists. Therefore, to reduce ground-loop interference, we must avoid establishing any complete ohmic ground-loop paths and break up any such loops that already exist. The best way to ensure that no effective ohmic ground-loop paths exist is to design the measuring system so that only one point of the system is ever ohmically connected to ground. In that way, although other closed loops through the ground plane may still exist (due to capacitive coupling), their circulating currents will be much smaller than those in a low-impedance complete ohmic loop. In systems for which single-point grounding provides an effective reduction in ground-loop interference signals, single-ended rather than differential-input amplifiers can be safely used. Single-ended amplifiers are less complex and costly to employ than differential-input amplifiers.

Figure 16-9 shows an example of a measurement system that uses the single-point grounding method. All the ground leads are connected to the system's earth-ground connection at the same point. The best practice is to locate this single

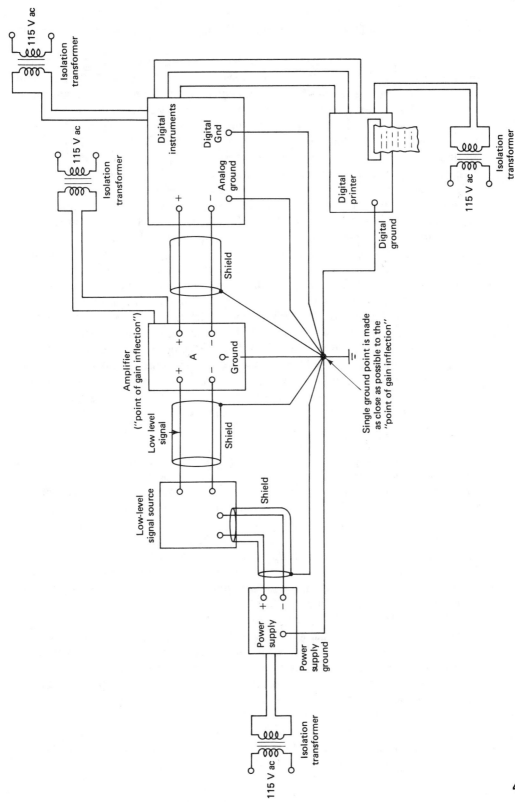

**Figure 16-9** Example of a measurement system using single-point grounding. Note that there are no ohmically complete ground loops and that no common-impedance ground paths exist.

469

ground point closest to the instrument in the system that is most susceptible to the interference (i.e., the point of gain inflection). For example, in Fig. 16-9 this would be the point where the amplifier is connected to the system's earth-grounded conductor. By using single-point grounding, conductively coupled interference due to a common-impedance path is also minimized. Note in Fig. 16-9 that the power supply, amplifier, and digital instruments have their power-line inputs isolated from ground by isolation transformers. These system components are therefore each connected to ground only at the single grounded point of the system.

In many systems, however (especially those in which the signal source and amplifier are separated by some distance), single-point grounding techniques may be either impractical or impossible. In addition, we have already discussed how systems utilizing the single-point grounding techniques still contain one or more ground loops that arise as a result of capacitive coupling. An example is shown in Fig. 16-10. The ground loop is the path made up of the power-line ground, power transformer primary-to-secondary winding capacitance, $C_{12}$, and the signal ground (the path connecting points ①–②–③–① of the simple, single-point grounded system shown in Fig. 16-10). This ground-loop path can still introduce an unwanted common-mode voltage in series with the measured signal $v_s$. For example, in Fig. 16-10 (with $v_{13} = 10$ V, $R_1 = 2\Omega$, and $C_{12} = 3000$ pF) at 60 Hz a current of 5 $\mu$A (and thus a common-mode voltage of 10 $\mu$V) will still be present as a noise signal $v_{cm}$ in series with $r_s$ at the input of the amplifier. For some low-level signal measurements, this noise voltage may still be excessively large. In such systems, use of an amplifier with a *differential input* together with the use of *input guarding* will probably be the answer to this ground-loop problem [Note that Fig. 16-4(b) shows a variation of the simple measurement circuit of Fig. 16-10. It also uses the single-point grounding method, but with the ground connections made in a modified way. If a single-ended amplifier is used in the circuit of Fig. 16-4(b), a ground-loop path including chassis-to-ground capacitive coupling (typically 3000 pF) will be created. This will cause a small common-mode voltage signal to appear at the amplifier inputs, comparable to one caused by the ground loop of Fig. 16-10.]

The use of amplifiers equipped with differential inputs is an alternative technique employed in efforts to eliminate or reduce many common-mode interference problems. The detailed characteristics of such differential amplifiers are explored in more detail in Chapter 15. Instruments designed to measure low-level signals (i.e., sensitive oscilloscopes and meters) are therefore generally equipped with differential amplifiers. Common-mode interference is minimized in such instruments because the differential amplifier is designed to amplify only the difference between the voltages applied to its two inputs. As shown in Fig. 16-11, the voltage at the (+) input of the ideal differential amplifier input is $v_s - v_{cm}$, the voltage at the (−) input is $v_{cm}$, and the output voltage is $K_D[(v_s + v_{cm}) - (v_{cm})] = K_D v_s$, where $K_D$ is the differential gain of these amplifiers.

The capability of a differential amplifier to reject common-mode voltages is specified by its common-mode rejection ratio (CMRR). As defined further in Chapter 15, CMRR is the ratio of the amplitude of the common-mode signal to the

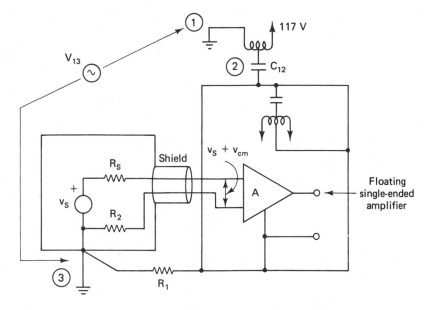

**Figure 16-10** Example of the limitations of the single-point grounding technique. Ground-loop path through ①—②—③—① causes a small common-mode voltage still to be applied to the inputs of the single-ended amplifier of this simple, single-point grounded system. If the measured signals are in the microvolt range, this residual common-mode voltage may still produce an error large enough to render the single-point grounding technique ineffective.

amplitude of the equivalent differential signal that would produce the same output from the amplifier. For example, if a 1.0 V signal were to be simultaneously fed to the two inputs of a differential amplifier, and the output voltage from these common-mode voltages was equivalent to that produced by a 10-$\mu$V differential signal, the CMRR of the amplifier would be 1.0 V/10 $\mu$V = 100,000:1. The larger the value of CMRR, the better will the amplifier be able to reject common-mode voltages.

As is discussed in Chapter 15, the differential amplifier remedy to common-mode problems works best when the test circuit impedances $Z_A$ and $Z_B$ (as shown Fig. 15-5) are equal (balanced) and are much less than the differential amplifier

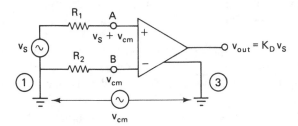

**Figure 16-11** Differential amplifier. In an ideal differential amplifier, only $V_s$ is amplified while $V_{cm}$ is rejected or $V_{out} = K_D V_s$.

input impedance. If a large difference (i.e., an impedance unbalance) exists between the values of $Z_A$ and $Z_B$ of Fig. 15-5, or if the signal source impedance is greater than a few hundred ohms, the differential amplifier CMRR will be reduced. (See Example 15-3 for a detailed examination of how source-impedance unbalance leads to degradation of the CMRR in differential amplifiers.) In addition, the common-mode voltages encountered may sometimes exceed the maximum permissible voltage values that can be applied to the specific amplifier inputs. If any of the conditions noted above are encountered, the use of a differential amplifier alone will not necessarily produce a satisfactory resolution of the common-mode ground-loop problem. The technique of *input guarding* may also have to be employed. Proper use of a differential amplifier with a guarded input provides the most effective solution for eliminating common-mode problems. In fact, many measurement applications that are considered almost routine today are possible only because of the application of this technique. Before it was available, single-point grounding had to be relied upon, and we have already noted that it is not effective enough for reducing common-mode interference problems in many sensitive measurement systems.

## INPUT GUARDING TO REDUCE GROUND-LOOP INTERFERENCE

*Input guarding* is a technique in which the entire measuring or input-circuit section of a differential amplifier is enclosed within a metallic guard. The guard shield is also insulated from the rest of the system (and ground), as shown in Fig. 16-12. An isolation device (such as an isolation transformer or optoelectronic isolator) is used to couple the signal from the input circuit of the amplifier to its output circuit. The entire "amplifier-guard shield-isolation device" assembly is referred to as an *isolation amplifier* (see Chapter 15).

The amplifier chassis, amplifier output shield, equipment enclosure, and low side of the amplifier output are grounded normally (i.e., they are all connected to the system power-line ground). This step serves to ensure personnel safety and stabilize the recording or display system (to which the amplifier provides its output) with respect to system ground. The isolated amplifier input guard shield is then connected to the signal cable shield, a step that extends the amplifier guard shield out along the entire length of the signal cable. (Note once again that the amplifier guard shield is isolated from the amplifier chassis and the measurement instrument enclosure.) Finally, the cable shield is connected to the signal ground point. Care must be taken that this is the only point at which the cable shield and the signal ground wires make contact with ground. It was pointed out earlier that the common-mode voltage appearing at the amplifier inputs has essentially the same value as at the signal ground point. Thus, by connecting the guard and cable shields to the signal ground point, the singal–cable, amplifier–guard shield circuit becomes driven to that common-mode voltage as well. Since the guard shield circuit is thereby kept at essentially the same potential as the low-level signal wires, capacitive coupling of any ground currents flowing in the guard shield to the signal leads will be kept to a minimum.

Let us describe in detail how input guarding will reduce common-mode interference in an otherwise difficult measurement situation. If a signal source is grounded, as shown in Fig. 16-12, and the source ground-leg resistance $R_2$ is small ($\sim 2\ \Omega$), a small common-mode voltage ($\sim 10\ \mu V$) will still occur at the amplifier input leads even if a floating amplifier is used. In some measurements, this small common-mode voltage value may be tolerable, and thus the use of a floating, single-ended amplifier, and single-point grounding could be effectively used.

Frequently, however, signal sources that produce small output voltage signals (10 to 30 mV) and which possess a larger source resistance value (typically $\approx 1\ k\Omega$) are encountered. Strain gauges, unbalanced Wheatstone bridge signals (as shown in Fig. 10-12), and themocouples bonded directly to grounded equipment are examples of such sources. In attempting to measure the output signals of these sources with floating, single-ended amplifiers, common-mode voltages in excess of 5 mV would probably be encountered (leading to inaccuracies in the measurement data). Use of an input-guarded differential amplifier would be a better choice. Most of the common-mode currents that would ordinarily flow in the ground-loop path would be diverted into the alternate ground-loop path of the cable- shield circuit (i.e., from the path of the low-level signal wires and $R_2$, ①–②–③–④–⑤–⑥–①, to the cable-guard shield path, ①–⑦–④–⑤–⑥–①). Such shunting would occur

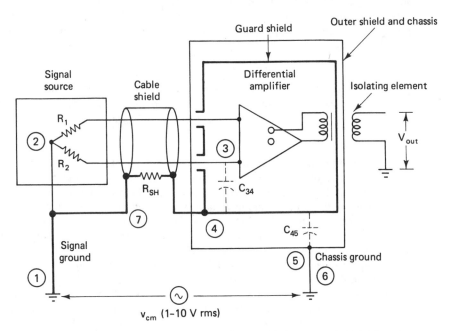

**Figure 16-12** Differential amplifier equipped with input guard shield (isolation amplifier).

because $R_{SH}$ (10 Ω) is much smaller than $R_2$ (1 kΩ) and $C_{34} = C_{45}$. As a result, the $v_{cm}$ appearing at the amplifier inputs is reduced to ~100 μV, a negligible small common-mode signal for the CMRR capability of the differential amplifier.

Thus we see that even with difficult common-mode measurement problems, the guarded differential-amplifier shunts ground-loop currents away from the measurement circuit, but in the process does not introduce interference into the low-level signal leads. Note, however, that if the signal source is itself floating, the guard circuit should be connected to ground as close to the transducer as possible, and only at one point.

## General Rules for Using Input-Guarded Amplifiers

**1.** Connect the amplifier guard shield to the signal cable shield. (Ensure that the signal cable shield is insulated from the chassis ground and the amplifier ground.) This step extends the internal amplifier guard shield out along the entire length of the signal cable.

**2.** Connect the signal cable shield to the low or shielded side of the signal source. This ensures that the amplifier guard and the signal cable shield are kept at the same potential as the low side of the signal source.

**3.** Connect the signal cable shield to the same ground point as the signal source is grounded. Do not allow the signal cable shield or the signal cables themselves to contact "ground" at any other point. If the signal is *not grounded* (i.e., floating), ground the signal cable shield as near as possible to the transducer and at only one point.

**4.** Ground the amplifier chassis, the low side of the amplifier output, and the shield of the output cable to the power-line ground of the amplifier. This step serves to keep the measurement system ground and the recording or display system ground at the same potential, and also ensures personnel safety.

### Example 16-2

A measurement system with a grounded thermocouple and a digital voltmeter (DVM) whose measuring circuit is nonguarded and floating but whose chassis is grounded is shown in Fig. 16-13(a). A circuit model for the system is shown in Fig. 16-13(b), including the common-mode voltage between the signal and instrument grounds and the coupling capacitances between the thermocouple leads and the instrument ground. In the model, $R_1$ and $R_2$ are the thermocouple lead resistances, and $R_{in}$ is the input resistance of the DVM. For the values of the components given in the model [Fig. 16-13(b)], calculate the common-mode voltage $V_m$ that appears across DVM inputs. (Note that the stray capacitance between the high terminal of the amplifier and the chassis, $C_{s1}$, is much smaller than the capacitance between the low terminal of the amplifier and chassis, $C_{s2}$. The reason for this is that the high side is physically a wire or narrow conductor, while the low side is a plane, or large metal area, in proximity to the power ground of the measuring system.

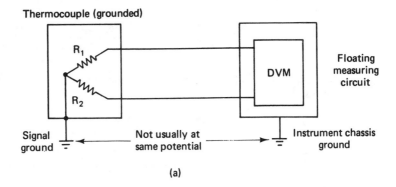

(a)

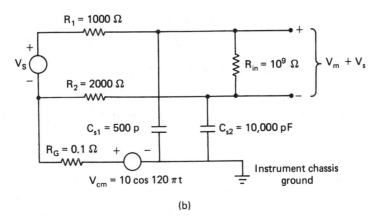

(b)

**Figure 16-13** Measurement system with a nonguarded floating differential-amplifier-driven meter: (a) measurement system; (b) circuit model of system.

**Solution.** The common-mode voltage $V^+_{cm}$ cm at the (+) terminal of the DVM in Fig. 16-13(b) with respect to instrument chassis ground is

$$V^+_{cm} = \frac{X_{cs1} \| (R_{in} + X_{cs2})}{R_1 + (X_{cs1} \| (R_{in} + X_{cs2}))} \times V_{cm} \simeq \frac{(X_{cs1} \| R_{in}) V_{cm}}{R_1 + (X_{cs1} \| R_{in})}$$

where $X_{cs1}$ is the capacitive reactance of $C_{s1}$ at $\omega = 2\pi(60)$, and since $R_{in} + 1/\omega C_{s2} \simeq R_{in}$ [when $\omega = 2\pi(60)$, $R_{in} = 10^9\ \Omega$, and $C_{s2} = 10,000$ pF]. Now since

$$X_{cs1} = \frac{1}{\omega C_{s1}} \simeq \frac{1}{2\pi(60)5 \times 10^{-10}} = 5 \times 10^6\ \Omega$$

which is much less than $R_{in}$, $(X_{cs1} \| R_{in}) \simeq X_{cs1}$. Thus,

$$V^+_{cm} \simeq \frac{X_{cs1}}{R_1 + X_{cs1}} \times V_{cm} = \frac{(1/\omega C_{s1})(V_{cm})}{R_1 + 1/\omega C_{s1}} = \frac{V_{cm}}{\omega R_1 C_{s1} + 1}$$

Then

$$V_{cm}^- = \frac{X_{c_s2}}{R_1 + X_{c_s2}} \times V_{cm} = \frac{(1/\omega C_{s2})(V_{cm})}{R_2 + 1/\omega C_{s2}} = \frac{V_{cm}}{\omega R_2 C_{s2} + 1}$$

Therefore, the voltage due to the common-mode interference, $V_m$, which appears across $R_{in}$ is

$$V_m = V_{cm}^+ - V_{cm}^- = V_{cm}\left(\frac{1}{\omega R_1 C_{s1} + 1} - \frac{1}{\omega R_2 C_{s2} + 1}\right)$$

$$= V_{cm} \times \frac{\omega R_2 C_{s2} - \omega R_1 C_{s1}}{1 + \omega R_1 C_{s1} + R_2 C_{s2} + (\omega^2 R_1 R_2 C_{s1} C_{s2})}$$

Now $\omega_1 R_1 C_{s1}$, $\omega_2 R_2 C_{s2}$, and $\omega^2 R_1 R_2 C_{s1} C_{s2}$ are all very much less than 1, so they can be neglected in the denominator of this expression. Thus $V_m$ is approximately equal to

$$V_m = V_{cm}(\omega R_2 C_{s2} - \omega R_1 C_{s1})$$

$$= 1200\pi (R_2 C_{s2} - R_1 C_{s1}) \cos 120\pi t$$

For the component values of this specific circuit (Fig. 16-13) we find that

$$V_m = (73.5 \text{ mV}) \sin 120\pi t$$

For a typical thermocouple or strain gauge signal that has an amplitude of 10 to 30 mV, the common-mode noise signal present in the system will obscure the desired signal.

**Example 16-3**

If the system of Example 16-2 is equipped with an input guard, as shown in the circuit model of Fig. 16-14, calculate the common-mode signal $V'_m$ which now appears across the DVM inputs.

**Solution.** With the input guard installed, the voltage at point ⓐ of Fig. 16-14 with respect to ground is

$$V_a = V_{cm} \frac{X_{C_{SH}}}{R_G + X_{C_{SH}}} = \frac{V_{cm}}{\omega R_G C_{SH} + 1}$$

Now the common-mode voltages that appear at the inputs of the DVM with respect to point ⓐ are approximately

$$V_{cm}^{+'} \simeq \frac{V_{cm} - V_a}{\omega R_1 C_{s1} + 1}$$

and

$$V_{cm}^{-'} \simeq \frac{V_{cm} - V_a}{\omega R_2 C_{s2} + 1}$$

or the common-mode voltage that now appears across the inputs of the DVM is $V'_m = (V_{cm}^{+'} - V_{cm}^{-'})$ or

$$V'_m \simeq (V_{cm} - V_a)(R_2 C_{s2} - R_1 C_{s1}) \sin 120 \pi t$$

Note that since $V_a$ for the circuit of Fig. 16-14 is $V_a \simeq 0.99999 V_{cm}$, the common-mode

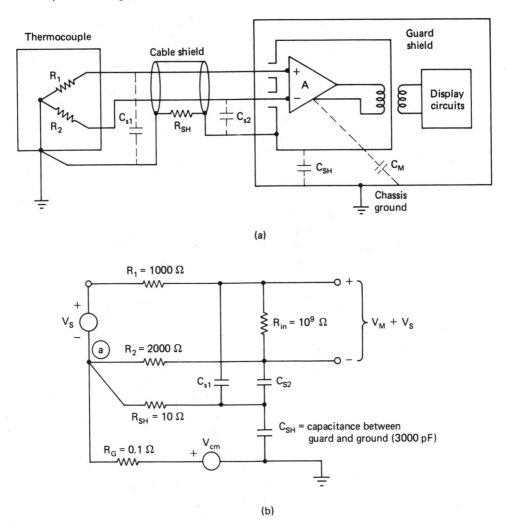

**Figure 16-14** Measurement system identical to that shown in Fig. 15-14, except that the amplifier is now surrounded by an input amplifier guard shield: (a) measurement system; (b) equivalent-circuit model of system (ignoring the effect of $C_m$).

voltage at 60 Hz appearing across the DVM inputs is reduced to 0.001 percent of the common-mode voltage without a guard:

$$V'_m \approx 0.7 \ \mu V$$

In practice, imperfections in the guard shield will result in some capacitance $C_M$ directly between the measurement circuit and the instrument chassis ground [Fig. 16-14(a)]. Thus the common-mode noise signal at the instrument inputs will actually be reduced from $V_m = 73.5$ mV to about $V'_m = 10 \ \mu V$.

## INTERNAL NOISE

As was discussed in the introduction to this chapter, externally generated interference is the major source of noise in most measurement systems. If, however, external noise is sufficiently suppressed, the noise generated internally in all electronic equipment and components will predominate. Such internal noise therefore limits the ultimate measurement sensitivity that can be achieved. Most measurement systems are designed to maintain an internally generated noise level below the sensitivity at which the system's instruments are intended to operate. Therefore, the internal noise is not noticeable at the output. If operation is extended (by amplification, for example) to lower signal levels, that is, down to the limit where the signal approaches the internally generated noise level, further sensitivity cannot be achieved without reducing the internal noise level.

Noise is random, but it has stationary properties, that is, it has consistently measurable amplitude, distribution, and frequency spectra. This means that noise can be characterized by a root-mean-square (rms) value.

$$\text{total noise} = \sqrt{(V_{\text{source1}_{\text{rms}}})^2 + (V_{\text{source2}_{\text{rms}}})^2 + \cdots}$$

For example, find the total noise, if the noise from several sources is as follows:

$$V_{\text{rms}} = 0.12 \ \mu V \qquad 10 \text{ Hz to } 100 \text{ Hz}$$

$$V_{\text{rms}} = 0.25 \ \mu V \qquad 100 \text{ Hz to } 1 \text{ kHz}$$

$$V_{\text{rms}} = 0.32 \ \mu V \qquad 50 \text{ Hz to } 2 \text{ kHz}$$

$$\text{total noise } (V_{\text{rms}}) = \sqrt{(.12 \times 10^{-6})^2 + (.25 \times 10^{-6})^2 + (.32 \times 10^{-6})^2} = 0.423 \ \mu V_{\text{rms}}$$

The total noise in an operational amplifier is provided in the form of a curve by the manufacturer. This curve is usually labeled *Broadband Noise for Various Bandwidths*. There is also a curve for flicker or $1/f$ noise, and it is usually labeled *Input Noise Voltage as a Function of Frequency*. White noise, that is noise which is independent of frequency, is the noise level that does not change with frequency. The major sources of internal noise in measurement systems are the following: (1) Johnson or thermal noise, (2) shot noise, (3) flicker, or $1/f$ noise, and (4) popcorn noise.

*Thermal (Johnson) noise* arises from the random motion of electrons in resistive materials due to thermal agitation. Because it is produced by thermal motion, the magnitude of the noise it generates increases with temperature. Since thermal noise has a white power density spectrum (i.e., the noise power density is independent of frequency), its magnitude is also proportional to the bandwidth of the measurement system (and to the resistance of the material in which it originates). Thermal noise is, therefore, minimized by (1) reducing the system bandwidth through filtering, (2) selecting of components and preamplifiers with low-noise characteristics, and (3) cooling those parts of the system's circuitry (typically the input stages) that are the largest contributors of noise in the system. Thermal noise is defined by

$$E_n = \sqrt{4kTR(f_H - f_L)}$$

where  $k$ = Boltzmann constant = $1.38 \times 10^{-23}$ J/K
$T$ = Temperature in degress Kelvin
$R$ = Resistance value in ohms
$f_H$ = Highest operating frequency
$f_L$ = Lower operating frequency

### Example 16-4
Assume that a resistor is operating over a bandwidth of 100 kHz ($f_H - f_L$) and that the value of the resistor is 470 k. Find the thermal noise across the resistor if the operating temperature of the resistor is 35°C.

$$E_n = \sqrt{4kTR(f_H - f_L)} = \sqrt{(4)(1.38 \times 10^{-23})(35 + 273)(470 \times 10^3)(100 \times 10^3)}$$

$$= 28.3 \ \mu V$$

*Shot (Schottky) noise* is due to the random motion of charge carriers across interfaces of junctions in semiconductor devices, vacuum tubes, and phototubes. As with thermal noise, shot noise is independent of frequency but is related to the dc current through a device. Therefore, an effective technique to reduce shot noise is to limit current through the device which serves as the source of such noise.

Shot noise current is given by the following formula:

$$I_n = \sqrt{2qI_{dc}(f_H - f_L)}$$

where  $q = 1.602 \times 10^{-19}$
$I_{dc}$ = the dc current through the device

### Example 16-5
Assume that the dc current through a diode is 37 mA and that the bandwidth of frequencies that are impressed across the diode is 75 kHz. Find the shot noise current through the diode and noise voltage that results across a 5.1 k resistor as a result of the shot current.

$$I_n = \sqrt{2qI_{dc}(f_H - f_L)} = \sqrt{2(1.602 \times 10^{-19})(0.37)(75 \times 10^3)} = 0.0298 \ \mu A$$

$$V_n = I_n(R) = 0.0298 \times 10^{-6}(5.1 \times 10^3) = 152 \ \mu V$$

*Flicker or 1/f noise* is exhibited by transistors and other solid-state devices, and its name is derived from the fact that its magnitude increases with decreasing frequency. The origin of this excessive low-frequency noise is not well understood, and at frequencies below about 1 kHz it is the predominant cause of internal noise in a system. Therefore, the important conclusion that can be drawn from the characteristics of 1/f noise is that dc measurements should be avoided if sensitive signals are being monitored.

Since the noise introduced at the input of the system is amplified by the succeeding stages, the noise associated with the input stage (often the preamplifier) constitutes the predominant fraction of the total system internal noise signal. Thus, the most important guideline to follow in designing low internal noise measurement systems is to keep the input stage of the system as free of noise as possible. Internal noise in the input section minimized by utilizing the same noise-suppression techniques as those introduced in this section. External noise is kept small by placing the input stage of the system (typically the preamplifier) as close to the signal source as possible. The output signal of the preamplifier can then be transmitted to the main amplifier. Further information on internal noises sources and noise reduction can be found in References 3 and 5.

*Popcorn* or *Burst* noise is the result of a change in the dc current level through a semiconductor. It is probably caused by imperfections near the surface of the semiconductor.

## PROBLEMS

1. Discuss *near-field* versus *far-field* external interference effects.
2. Discuss how capacitive coupling can cause interference to be introduced into a measurement system.
3. How would the following conditions affect capacitively coupled noise signals?
   (a) Frequency
   (b) Input impedance of the measuring instrument
   (c) Magnitude of the capacitive coupling between source and measurement system
4. If the capacitive coupling $C_c$ between a noise source and measurement system is 0.2 pF, the noise frequency is 1 kHz, the amplitude of the noise signal is 15 V, and the input impedance of the measuring instrument is 10 M$\Omega$, find the magnitude of the capacitively coupled noise signal in the measuring system.
5. Discuss how inductive coupling can cause interference to be introduced into a measurement system.
6. Discuss conductively coupled interference signals and steps that can be taken to reduce their magnitude.
7. Explain how properly shielded power transformers can reduce conductively coupled interference in measurement systems.
8. Explain "single-point grounding." Under what conditions can single-point grounding adequately reduce ground-loop problems in measurement systems? Under what conditions must additional measures be employed to adequately reduce ground-loop problems?
9. Define a "ground loop."
10. Explain how ground loops may exist in systems even if only one point in the system is grounded.
11. What is "input guarding"? How does an input guard shield reduce the effects of common-mode noise and ground-loop problems in measurement systems?

12. An unshielded power transformer is used in an attempt to build an isolated (floating) power supply (Fig. P16-1). The capacitance between the $(-)$ output terminal side of this supply (point $A$) and ground is 20 pF. Using an ac voltmeter with a 100-k$\Omega$ input resistance, a student measures the voltage relative to ground at point $A$. He finds a 3-V, 60-Hz since-wave signal. Explain this result. (*Hint*: We can assume in this case that the capacitance between the transformer windings would behave as if a capacitor were connected between the center taps of the two windings as shown in Fig. P16-1. Also, all other transformer impedances can be neglected, to first order.)

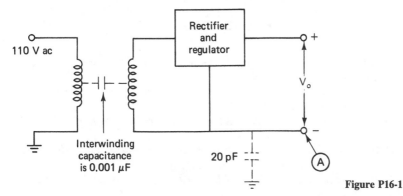

Figure P16-1

13. In attempting to understand the circuit of Fig. P16-1 further, the student uses an oscilloscope with a 1-M$\Omega$, 47-pF input impedance to repeat the measurement at point $A$. This time he observes a 22-V, 60-Hz sine wave. Finally, he uses a 10-M$\Omega$, 4.7-pF attenuating scope probe and observed a 62-V, 60-Hz sine wave. Explain.

14. This problem shows how an improperly isolated power supply for a bridge circuit being used as a thermocouple compensator can introduce conductively coupled interference through the power transformer (Fig. P16-2). The interwinding capacitance is, in effect,

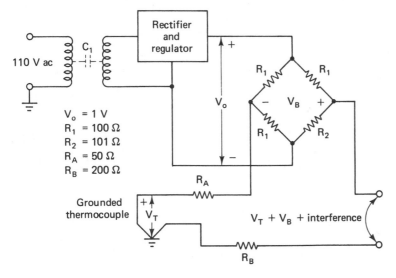

$V_o = 1$ V
$R_1 = 100\ \Omega$
$R_2 = 101\ \Omega$
$R_A = 50\ \Omega$
$R_B = 200\ \Omega$

Figure P16-2

driven with a common-mode voltage of 55 V ac. The rectifier and regulator can be considered a short circuit for all common-mode currents. Show that the amplitude of the 60-Hz interference is $\omega C_1[(R_1/2) + R_a]$ times the power-line amplitude. Evaluate the interference amplitude for

(a) $C_1 = 0.001$ μF (unshielded)
(b) $C_1 = 0.50$ pF (single shield)
(c) $C_1 = 0.1$ pF (properly guarded circuit with triple-shielded transformer)

# REFERENCES

1. *Grounding and Noise Reduction Practices for Instrumentation Systems*. El Segundo, Calif.: Scientific Data Systems, n.d.
2. Malmstadt, H. V., Enke, C. G., and Crouch, S. R., *Instrumentation for Scientists*. Sect. 4-1 and App. 4-A. Menlo Park, Calif.: W. A. Benjamin, 1982
3. Morrison, R., *Grounding and Shielding Techniques in Instrumentation*. New York: John Wiley, 1986.
4. Hieftje, G.M., "Signal-to-Noise Enhancement through Instrumental Techniques," *Analytical Chemistry*, Vol. 44, No. 7, 1972, pp. 81A, 169A.
5. Ott, H. W., *Noise Reduction Techniques in Electrical Systems* 2nd. ed., New York: John Wiley, 1988.
6. Strong, P., *Biophysical Measurements*. Chap. 19. Beaverton, Ore.: Tektronix, 1970.

# 17

# *Introduction to Instrumentation Systems*

Electronic measurement systems are assemblages of instruments and components interconnected to perform an overall measurement function. The system components must not only perform their individual functions properly but must also work effectively with the other components making up the system. This requirement points out the importance of ensuring that proper *interfacing* exists between all components making up the system. *Interfacing is defined as the joining of components in such a manner that they are able to function in a compatible and coordinated fashion.*

A knowledge of interfacing fundamentals has become even more important with the advent of digital electronics. Since both analog and digital instruments are often used in measurement systems, interfacing must be understood in all three of the following situations:

1. Interfacing analog-to-analog instruments
2. Interfacing analog-to-digital instruments
3. Interfacing digital-instruments to other digital devices (printers, computers, etc.)

This chapter is concerned primarily with analog systems and analog-to-digital systems. Digital-to-digital interfacing is treated separately in Chapter 18, which is entirely devoted to that subject.

It should be noted, however, that many of the subjects considered in this chapter have already been discussed elsewhere in more detail in other parts of the book. The treatment in this chapter is meant to serve as an overview of the topics and problems encountered in measurement systems. The reader will therefore be directed to those sections in which the specific topics are explored in more detail.

## ANALOG SYSTEMS

Analog systems are configured entirely of components that measure, transmit, display, or record data only in analog form. In general, such all-analog systems find most use in applications that require wide bandwidth, or when lower accuracy can be tolerated. They also tend to be less complex than analog-to-digital systems but, as a result, cannot handle as great a volume or complexity of input data.

Analog systems typically contain some or all of the following elements:

1. *Signal sources* (of which there are two general classes):
   (a) Elements that produce signals as a result of *direct measurement* of electrical quantities (i.e., voltage, current, resistance, frequency, etc.).
   (b) Devices that convert nonelectrical parameters into an electrical signal (*transducers*). Transducers are discussed in detail in Chapter 14.
2. *Analog signal conditioning elements*. The analog signal from the transducer is rarely in suitable form to be directly displayed or recorded. It must usually first undergo a variety of signal conditioning processes prior to such display. These processes may include some or all of the following: amplification, filtering, linearization, offsetting, and buffering. The signal-conditioning equipment in instrument systems is designed to perform such functions. The process of amplification (and sometimes buffering) is performed by electronic amplifiers. The characteristics of amplifiers and their use in instrument systems was discussed in Chapter 15. The remaining signal condition processes are discussed briefly in this chapter.
3. *Measurement and display instruments*, such as oscilloscopes (Chapter 6), analog meters (Chapters 4 and 9), and so on.
4. *Graphic recording instruments*, such as chart recorders (Chapter 7).
5. *Analog magnetic tape recorders* (Chapter 7).

Figure 17-1 shows a simple example of such an all-analog instrumentation system. A temperature-to-voltage tranducer (i.e., thermocouple) senses the temperature to be measured (analog input) and transforms it into a voltage signal (analog output). The analog voltage signal is transmitted to one or more of the analog measuring/recording instruments shown (oscilloscope, chart recorder, magnetic tape records, or analog meter). In this system, any signal conditioning that amplifies or otherwise modifies the signal of interest is done in an analog manner by the analog measuring/recording instruments as well.

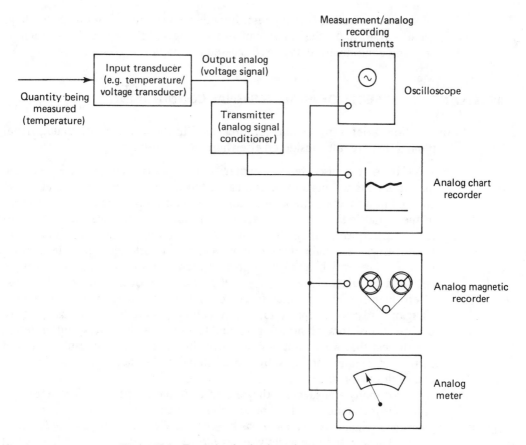

**Figure 17-1**  Example of a simple analog measurement system.

To interface properly all the elements of analog systems, such concerns as impedance matching and loading, susceptibility to noise pickup as data are transmitted from transducer to instrument, proper grounding, ground loops and ground-loop-associated interference, and compatibility of signal source and measurement instruments (i.e., in terms of sensitivity, frequency response, etc.) must all be addressed. Unless the designers of a specific instrument system understand the requirements arising from these interface conditions and ensure that they are all met, it cannot be guaranteed that the measurement system will adequately perform the task for which it was designed. It must also be remembered that the same analog interfacing considerations must be considered in the analog segments of analog-to-digital systems.

Some of the interfacing problems that must be solved in analog systems have been discussed in various other chapters. External noise pickup, and techniques to suppress it, are the subjects of Chapter 16. Instrument sensitivity, frequency response, and other relevant specifications are discussed in the chapters that deal with the specific instruments of interest. Impedance matching, loading, grounding, cables and wiring, and other miscellaneous analog interfacing subjects (such as personnel and equipment safety) are discussed in Chapter 3.

## MISCELLANEOUS ASPECTS OF ANALOG SIGNAL CONDITIONING

In this section, brief introductions are presented to those aspects of analog signal conditioning that are not considered elsewhere in the book.

- *Filtering*. As was discussed in Chapter 16, the environment in which electrical measurements are performed can cause unwanted interference signals to be coupled into the measurement system. The interference may also be present in the signal itself (internal interference). Such interference, despite all attempts to reduce its magnitude, may still cause unacceptably large errors in the measurement data. Often, however, this interference may be further reduced in severity by the use of proper filtering in the system. Filters can be designed to reject signals over specific desired frequency ranges (low-pass filters, high-pass filters, notch filters, etc.). Filter circuits can be implemented using only resistors, capacitors, and inductors (passive filters), or using active devices (transistors, op-amps), gain, and feedback (active filters). Filter design is a subject that we cannot hope to cover in this book, but useful sources of further information on this subject are listed in References 4, 6, 7, and 8 at the end of the chapter.

  For many transducers, the rate of information transfer is rather slow, and their maximum bandwidths are about 10 Hz. Filtering noise from these transducer signals is relatively easily accomplished with the use of *low-pass* filters. These respond perfectly to very low frequency signals and have great attenuation at high frequencies.

- *Linearization*. In many cases, the proportionality that exists between the input variable to the transducer and its output signal is nonlinear. The readouts or recording mechanisms of systems are generally designed to respond to signals as if the relationship between the input and output of the signal sources were linear. Thus the actual nonlinearities cause errors in the measured data. To reduce these errors, the output of the transducer can be linearized as a part of the analog signal-conditioning process. Various linearizing techniques, including modifying the transducer circuitry, or analog processing of the transducer signal, are used to correct (linearize) the transducer output signals.

  For example, if the output voltage of a transducer were to vary exponentially with respect to its input dynamic variable, the transducer output signal could be fed to an amplifier whose output varies as the inverse of the transducer input (i.e., a logarithmic amplifier whose output varies as the natural logarithm of its input). Thereby the output of the amplifier would vary linearly with respect to changes in the transducer input signal.

As a second example, the output voltages of thermocouples vary non-linearly with temperature. The nonlinearities are *undesired but predictable* for each type of thermocouple. A corrective signal that has a quadratic variation can be superimposed on actual thermocouple outputs to linearize them within acceptable error (typically with ±1 percent). This correction can be generated with the use of commercially available *logarithmic multifunction modules* built from integrated circuits.

Additional details of linearizing are beyond the scope of our discussion. References 4, 5, and 9 contain further information on this subject.

- *Offsetting/Level Conversion.* The output of the system transducer might be in one form (i.e., change in voltage, or resistance change), while the readout device might require the signal in another form (i.e., 4- to 20-mA current). The signal might also be a small variation about a large value. These and other situations require some form of signal conditioning which will suitably alter the signal to make it compatable for interaction with subsequent system elements. Three examples of such offsetting/level conversion techniques are (1) a 4- to 20-mA modular voltage-to-current-loop converter (discussed further in the next section) can be used to change signals from voltage to current form; (2) bridge circuits (discussed in Chapter 10) can be used to offset the small output voltage variations from the large fixed value about which they vary; and (3) temperature-monitoring circuits can incorporate instrumentation-amplifier level-conversion elements which offset the signals so that the outputs read, say, in degrees Celsius rather than degrees Kelvin. Many other offsetting/level conversion techniques are used in instrumentation systems as part of the signal-conditioning process.

- *Buffering (impedance matching).* The output impedance of transducers may be the source of problems in certain system applications. For example, high-output-impedance transducers can cause unwanted loading errors or can adversely affect the settling times of system multiplexers (see the following discussion on multiplexers for an elaboration of this point). Unbalanced impedances can lead to the introduction of common-mode interference errors. Both passive and active circuits (the latter usually involving instrumentation amplifiers) are used as impedance-transforming signal-conditioning elements (or *buffers*) to alleviate such problems.

## ANALOG SIGNAL TRANSMISSION

As noted in the previous discussion, analog signals must usually be transmitted to the measuring/recording device from the point of measurement. The data in the signal may be retained in analog form during transmission, or they may be converted to digital before being sent to its destination. In this section we examine analog transmission methods. The transmission of digital data is treated in Chapter 18.

In the example analog system of Fig. 17-1, the analog signal is sent directly from the transducer to the measuring/recording instruments, without any prior conditioning (amplification, filtering, etc.). This transmission method is the simplest but is quite limited in its application. If the signal source and measuring instruments are very close (i.e., within 1–2 m) and the signal levels are not too small (i.e., >100 mV), the technique may be sufficiently effective to yield satisfactory results. That is, the voltage signal may arrive at the instruments without being appreciably degraded by the cable resistance or external noise pickup.

However, if the signal levels are small (<100 mV) and if the instruments must be located at some distance from the measuring point (>5 m), this simple analog signal transmission method will probably no longer suffice. The majority of measurements in process control applications involve the transmission of such low-level signals. As was discussed in Chapter 16, long-distance, low-level, analog transmission is very susceptible to signal degradation due to external noise pickup, common-mode (ground loop) noise problems, and reduction of signal voltage due to cable resistance. Furthermore, if the signal is an ac analog voltage, its alteration by the capacitance of transmission lines (which increases directly with increasing cable length) must also be considered. Therefore, two other data transmission methods are usually employed if it is necessary to send low-level signals over longer distances in analog form. These methods are (1) analog voltage transmission of signals boosted by instrumentation amplifiers (IAs) prior to transmission and (2) analog current transmission, in which the analog signal is amplified and then converted to current form with a value between 4 and 20 mA (4- to 20-mA current loop).

*Analog voltage transmission* employing IAs near the signal source is a popular technique if the distance is less than about 30 m. This method is less expensive than the analog current (4- to 20-mA) method and is equally effective as long as the distance limitation is not exceeded. The low-level signal from the source is amplified by an IA placed as close to the signal source as possible. The output of the IA is a high-level analog voltage signal of 0–5 V or 0–10 V. Because they possess substantially larger amplitudes, in systems that are well designed against external noise pickup such high-level voltage signals do not suffer as much degradation during transmission. For distances greater than 30 ft, however, cable resistance, grounding problems, and so on, reduce the effectiveness of this technique, and the 4- to 20-mA current loop method becomes the better choice. [Note that the characteristics of IAs (instrumentation amplifiers) were discussed at length in Chapter 15.]

The *analog current transmission method* uses an analog dc current signal with a value of 4 mA corresponding to zero signal and 20 mA representing full scale. It has become a favorite data distribution method in many industrial applications. (This is also a fail-safe type of transmission method because the sudden absence of current in the loop immediately signifies an indication of a breakdown in the transmission system.) The current signal can be transmitted for distances up to 2 miles, and this allows measurements of such parameters as temperature, pressure, and so on, to be brought into a control room from remote plant locations. Different loads can be connected to the transmitting circuit because the circuits are designed to work into

**Figure 17-2** The Westinghouse Veritrak model 75 two-wire process transmitter is typical of equipment available from instrument manufacturers. It includes standard RFI/EMI filtering and shielding, the case is explosion proof, and the unit operates in temperatures from −40 to +250°F. (Courtesy of Westinghouse Electric Corp.)

any load from 0 Ω to 1000 Ω. The signal in the current loop is also more immune to noise than is a voltage signal because the circuits of the 4- to 20-mA method consist of low-impedance paths. Furthermore, since the current in a series circuit is constant everywhere along the series path, there is no degradation of the signal with distance as there is when transmitting voltage signals. (Voltage signal degradation with distance occurs because of voltage drops across the resistance of the cable along which the signal is transmitted.)

The 4- to 20-mA transmitter used to implement this method amplifies the analog voltage signals from the transducer, sensor, or signal source, and then converts them to an analog current signal. The transmitters (Fig. 17-2) are powered by a 24- or 48-V dc power supply, and the transmitter outputs are current signals ranging from 4 to 20 mA. The transmitters are enclosed in watertight, explosion-proof housings, which makes them ideal for use in harsh or dangerous environments.

A 4- to 20-mA system in which such a transmitter is used is shown in Fig. 17-3. We see that the current signal can be converted back to the more usual voltage signal that is measured by voltmeters, oscilloscopes, and so on, by use of a 250 Ω dropping resistor placed in series in the 4- to 20-mA path at the instrument location. If it is desired to convert the signal to digital form, the voltage observed across the dropping resistor is in suitable form to be fed to an A/D converter.

## Calibrating Analog Instrumentation Signals (Instrumentation Loops)

An analog instrument signal calibrator (hereinafter referred to as a calibrator) (Fig. 17-4), is commonly used to test and calibrate instrumentation loops. Most calibrators are capable of measuring and outputting both voltage and current signals. They are in essence a precision millivoltmeter/milliameter and millivolt/milliamp power supply. The individual sections are isolated so that simultaneous measurement and signal supplying (commonly called signal sourcing) can occur. The calibrator should have a resolution of 0.01 mV and 0.01 mA and an accuracy of 0.01 percent FS. Analog instrumentation loops (Fig. 17-1) are usually calibrated and tested in sections. For example, the output of a thermocouple is measured before it is connected to the analog signal conditioner (commonly called a transmitter). The output of the

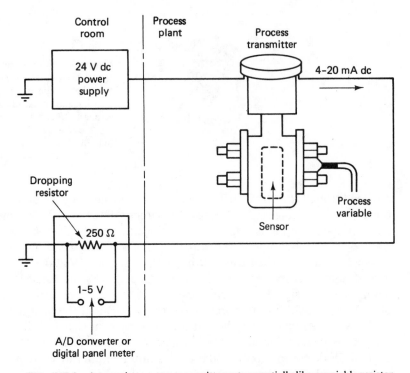

**Figure 17-3**   A two-wire process transmitter acts essentially like a variable resistor. It takes its power from a 24- 48-V dc power supply, converts the input from its sensor to a 4- to 20-mA dc signal, and transmits the output back to the control room. If the signal goes to an A/D converter, a dropping resistor will convert the signal to a voltage.

thermocouple is measured at the end of the wires that connect to the analog signal conditioner, (Fig. 17-5). The transmitter is tested and calibrated by injecting a signal into the transmitter while simultaneously measuring the output signal from the transmitter (Fig. 17-6).

   A calibrator can be used to measure the output of a transducer and compare it to the transducer's specification. The red terminal of the calibrator is positive (+) for both the input and output, whereas the black terminal is negative (−). Let us examine the use of a calibrator to test a thermocouple and the wiring from the thermocouple to a transmitter (Fig. 17-5). Heat the thermocouple to a known temperature. Hold this temperature constant while the test is conducted. Connect the red thermocouple wire to the black terminal (−) on the calibrator (the red wire of a thermocouple is always negative; see Chapter 14). Connect the other thermocouple wire to the red terminal (+) on the calibrator. Select the mV input mode on the calibrator. Read the millivolts on the display, and compare the reading with values in a standard thermocouple reference table (the millivolts in the table must be corrected for the ambient temperature of the calibrator). If the value of the reading

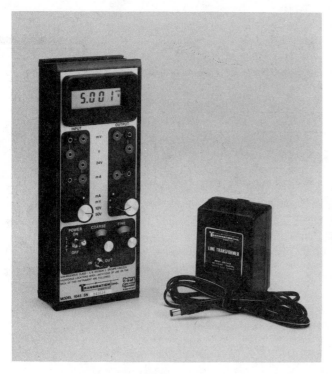

**Figure 17-4** Transmation model 1045 DC (as in *D*irect *C*urrent) signal calibrator for analog instruments. (Courtesy of Transmation, Inc.)

and the corrected reference table value do not agree within a reasonable amount of accuracy, then either the thermocouple or the wiring from the thermocouple is defective. The accuracy is established by the overall system requirements.

As a second example, let us examine the use of a *calibrator* to test a two-wire transmitter (a signal conditioner that converts one signal type into another signal type). The quantities converted do not have to be electrical signals. For example, a two-wire transmitter can convert a mV signal from a thermocouple into a 4–20 mA current signal. A second two-wire transmitter, however, could convert a 3–15 psi pneumatic (air pressure) signal into a 0–5 volt signal. Let us examine a 0–5 volt signal from a potentiometer that is converted to a 4-20 mA signal by a two-wire transmitter (Fig. 17-6). The shaft of the potentiometer is connected to the shaft of a valve so that when the valve is closed the output signal from the potentiometer is 0 volts; when the valve is open, the output voltage is 5 volts. The two-wire transmitter can be tested before the valve and potentiometer are installed by using a calibrator as a potentiometer output simulator or the calibrator can be used to determine if a potentiometer is sending an erroneous signal. First, set the calibrator to mV and then adjust the output by using the Coarse and Fine controls until the display reads 0 volts. The input signal amplitude is read on the calibrator display. If the display does not read 4 mA, the two-wire transmitter must be adjusted for zero offset. The transmitter zero adjusting screw is turned until the transmitter output corresponds to 4 mA. This is called *zero* adjustment. Next, the output from the calibrator is set

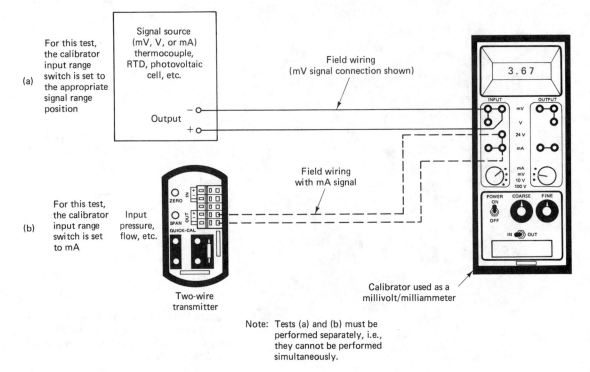

**Figure 17-5**  Block diagram for testing/calibrating a multivolt, volt, or milliampere signal source: (a) testing a millivolt, volt, or milliampere source; (b) testing/calibrating a two-wire transmitter (milliampere source). (Courtesy of Transmation, Inc.)

to 5 volts. The calibrator display should now read 20 mA. Adjust the transmitter output by adjusting the transmitter *span* screw until the display reads 20 mA. Both procedures constitute what is usually called *zero/span* adjustment or test.

## ANALOG-TO-DIGITAL SYSTEMS

Measurement systems in which the measured data are acquired in analog form, but then converted to digital form before display, recording, or transmission, find very widespread use. They are most frequently employed when the electrical signal or physical process being monitored exhibits a narrow bandwidth (one example would be a signal or process that varies slowly with time), and when high accuracy is required. Simple analog-to-digital (A/D) systems may contain only a single channel. More complex A/D systems may contain many input channels. These multichannel systems can measure a large number of input parameters, compare the data against preset limits or conditions, and perform calculations and decisions based on the input data. In general, analog-to-digital systems are more complex than their all-analog counterparts but also possess more enhanced data-handling capabilities.

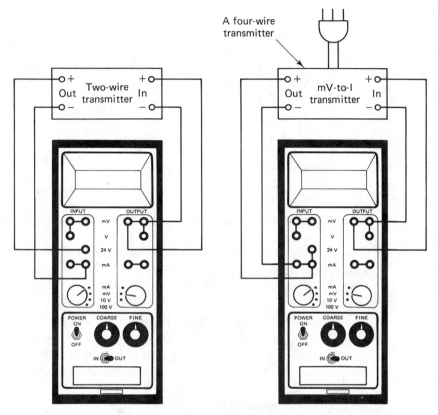

**Figure 17-6** Block diagram for testing/calibrating two-wire and four-wire transmitters. (Courtesy of Transmation, Inc.)

In addition to the superior data-handling characteristics of analog-to-digital systems, they have another very important advantage over all-analog systems: that of the much higher immunity to noise pickup of signals transmitted in digital form. Electrical noise is easily picked up and low-level analog signals can be modified dramatically during transmission even if careful shielding practices are followed. Digital data, on the other hand, can be precisely transmitted because the voltage difference between the "logic O" and "logic 1" states in a digital system usually far exceeds the amplitude of the electrical noise picked up by the transmission line. For example, if the digital signals are transmitted at the widely used levels of "logic 1" = 5 V and "logic O" = O V, a noise pickup of ±1 V superimposed on either logic state would not cause any error in the digital data transmission. Thus, in a practical sense, there is no limit to the precision with which digital data can be transmitted. Furthermore, the cost of the components that perform the conversion of analog signals into digital form has dropped dramatically, so that it has become cost effective to incorporate the use of such conversion techniques in virtually all types of instrumentation.

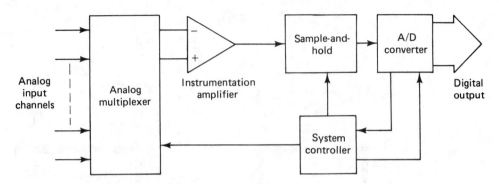

**Figure 17-7**  Simple multichannel analog-to-digital measurement system configuration.

An analog-to-digital system may contain some or all of the following elements:

1. Signal source (direct electrical signal or transducer).
2. *Multiplexer* (found in multichannel systems). Accepts multiple inputs and connects them to a single measurement device (to be discussed later in this chapter).
3. *Signal conditioner* (amplification, filtering, linearization of data, etc.).
4. *Sample-and-hold circuit*. Samples the output of the signal conditioner at a specified time and holds the voltage level at its output until the A/D converter performs its conversion operation (sample-and-hold circuits are discussed in greater detail later in this chapter).
5. *Analog-to-digital converter* (discussed in Chapter 5).
6. *System control device*. Can be a computer, programmable calculator, etc.

Digital voltmeters (a block diagram of which is shown in Fig. 1-8) can be considered as an example of a simple analog-to-digital system. A discussion of the functions of all the various blocks of the DVM has already been undertaken elsewhere in the book (see Chapter 5).

An example of a more complex analog-to-digital system is the data acquisition and conversion system shown in Fig. 17-7. It contains all the elements listed at the beginning of this section. Let us now discuss the elements of this system that are not examined in other parts of our book (the *sample-and-hold circuit* and the *multiplexer*) as well as another important system parameter that must be considered when evaluating the performance of such systems, *settling time*.

## SAMPLE-AND-HOLD CIRCUITS

In some applications of analog-to-digital systems, the signal being monitored varies so slowly that during the time it takes for the A/D converter to perform its conversion, the signal variation is small enough that it does not introduce significant error in the digitized output. In such systems, a direct conversion of the system amplifier output by the A/D converter produces adequately accurate results. The familiar digital panel meter (or DPM) operates in this manner. In the DPM an integrating-type A/D converter with a fixed (and rather long) conversion time converts the analog input signal at a free-running, internally determined rate, into samples of digital data.

If it is required that the A/D converter with an $n$-bit output is able to resolve the input analog data to within one least significant bit (LSB) during the time it takes to make its conversion (full resolution), the maximum permissible rate of change of the input signal is given by

$$\left.\frac{dV}{dt}\right|_{max} = \frac{2^{-n}V_{FS}}{T_{convert}} \tag{17-1}$$

where $n$ is the number of bits in the digital output, $V_{FS}$ the full-scale output voltage of the A/D converter, and $T_{convert}$ the time required for a single conversion.

If the input signal is a full-scale sine wave, Eq. (17-1) can be restated to express the result in terms of the maximum frequency signal that can be applied to an A/D converter and still have it maintain full resolution:

$$f_{max} = \frac{1}{V_{FS}}\frac{1}{2\pi}\left.\frac{dV}{dt}\right|_{max} = \frac{2^{-n}}{2\pi T_{convert}} \tag{17-2}$$

**Example 17-1**

Given an 11-bit integrating-type A/D converter with $T_{convert} = 0.1$s and $V_{FS} = 10$ V, find the maximum rate of change that an analog input signal may have in order that the converter can resolve the input signal into an 11-bit number in a single conversion time.

**Solution.**  Using Eq. (17-1), we find

$$\left.\frac{dV}{dt}\right|_{max} = \frac{2^{-n}V_{FS}}{T_{convert}} = \frac{2^{-11}\times 10}{0.1 \text{ s}} = \frac{1}{2^{11}}\times 100 \approx \frac{1}{20} \approx 0.05 \text{ V/s}$$

Often, the maximum rate of voltage variation far exceeds that given in Example 17-1. If an application calls for the analog-to-digital data acquisition system to maintain its high-resolution capability in the face of faster signal variations, it seems from Eq. (17-1) that a higher-speed (shorter conversion time) A/D converter would need to be specified.

Successive-approximation A/D converters have shorter conversion times (i.e., 1 to 20 μs) than the integration (voltage-to-frequency and dual-slope) types and thus it appears that they would be better suited for applications in which the signals are more rapidly varying. Unfortunately, if the maximum rate of change given by Eq. (17-1) is exceeded in successive approximation converters, they exhibit substantial output linearity errors. This occurs because (as we discussed in the section in Chapter 5 dealing with successive approximation converters), such converters cannot tolerate changes in the input signal amplitude during their conversion (weighting) process. Worse yet, even if the signal variation is apparently slow enough, higher-frequency noise accompanying the signal may induce the same type of errors in the output.

To overcome these weaknesses and to increase the operating speed of successive-approximation converters (while maintaining their high output accuracy), a *sample-and-hold circuit* is introduced at its input. This circuit is designed to acquire (sample) the input signal during conversions and just before conversion takes place, the circuit is placed in the *Hold* mode. During the *Hold* mode, the circuit holds constant the value of the signal that it possessed at the time of the *Hold* command, for the required duration of A/D conversion. Note that this also allows the instant of acquisition to be defined as well. After the hold value is converted, the circuit is switched back to the *Sample* mode to respond again to the input voltage.

The sample-and-hold function is usually carried out by charging a capacitor with the signal value during the sample interval, and the sample-and-hold circuits use voltage followers for isolation purposes as shown in Fig. 17-8.

Although sample-and-hold circuits are used most frequently in data acquisition systems to improve the operating speed and accuracy of successive approximation (and parallel) A/D converters, they can also be used with other converter types if it is desired to establish the precise times at which the signals are being sampled.

Sample-and-hold circuits, however, like all things, are not ideal. The key specifications that describe how well (or poorly) sample-and-hold circuits adhere to their characteristics are described below, together with the assistance of Fig. 17-9.

**1.** *Aperture time.* The time elapsing between the command to *Hold* and the actual opening of the hold switch.

**2.** *Droop.* A decreasing of the amplitude of the *Hold* value, caused by the flow of current from the charged capacitor into the inputs of the amplifier and through the sampling switch.

**3.** *Feedthrough.* Input voltage changes that appear in the output during the *Hold* mode.

**4.** *Acquisition time.* The time it takes to acquire the input voltage, within specified accuracy, after the circuit is switched back from the hold to the sample mode.

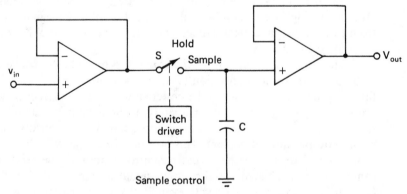

**Figure 17-8**   Basic sample-and-hold circuit.

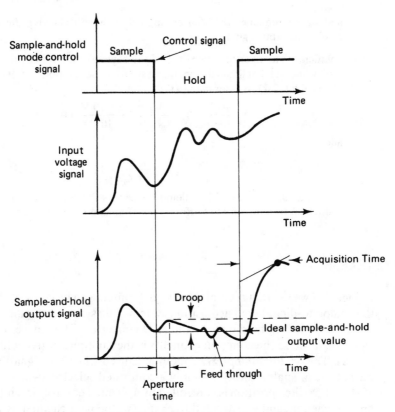

**Figure 17-9**   Characteristics of a sample-and-hold circuit.

**Example 17-2**

Given a 12-bit, 10 V full-scale, successive-approximation A/D converter that has a 20-μs conversion time and is used without a sample-and-hold circuit, find the maximum rate of change of the input signal $(dV/dt)_{max}$, and maximum input frequency (assuming a full-scale sine-wave input signal) that will still allow the A/D to operate at full resolution.

**Solution.**   Using Eqs. (17-1) and (17-2),

$$\left.\frac{dV}{dt}\right|_{max} = \frac{2^{-n}V_{FS}}{T_{convert}} = \frac{1}{4096}\frac{10}{2 \times 10^{-5}\text{ s}} = 125 \text{ V/s}$$

and

$$f_{max} = \frac{1}{2\pi}\frac{2^{-n}}{T_{convert}} \approx 2 \text{ Hz}$$

**Example 17-3**

If a sample-and-hold circuit with an aperture time of 3 ns is used together with the A/D converter described in Example 17-2, calculate the maximum voltage rate of change and signal frequency that can be applied while still allowing the converter to fully resolve the input signal.

**Solution.** The aperture time ($t_{ap}$) of the sample-and-hold circuit is now the cricital timing interval that limits the operating speed of the components. It can be used in Eqs. (17-1) and (17-2) to determine $(dV/dt)_{max}$ and $f_{max}$.

$$\left.\frac{dV}{dt}\right|_{max} = \frac{2^{-n}\,V_{FS}}{t_{ap}} = \frac{\frac{1}{4096}\times 10\text{ V}}{3\times 10^{-9}\text{ s}} = 0.8\text{ V/}\mu\text{s}$$

and

$$f_{max} = \frac{1}{2\pi}\frac{2^{-n}}{t_{ap}} = 14\text{ kHz}$$

Thus, Examples 17-2 and 17-3 demonstrate how the use of a sample-and-hold can radically help to improve the performance of analog-to-digital data acquisition systems.

## MULTIPLEXERS

In Fig. 17-7 we see that several input signal sources can be connected to the inputs of the analog-to-digital data acquisition system. The system contains only a single amplifier, sample-and-hold circuit, and A/D converter. The device that allows all the channels to share the common elements is the multiplexer (sometimes abbreviated MUX). The *analog multiplexer* (one that is used when the signals at its inputs and output are in analog form) is simply a controlled selector switch, as shown in Fig. 17-10. The switch position is controlled by a channel control circuit that can be set to any channel or can be instructed to change to the next channel by external signals. Analog multiplexers are best suited to handle 8–256 input channels.

The multiplexer switch can be an electromagnetically operated rotary-selector switch, a set of relays, or solid-state switches controlled in such a way that only one input channel is connected to the output at a time. The best switching speed of relay-type multiplexers is about 1 ms, and they eventually wear out. Solid-state switches are capable of much higher speed operation (<30 ns) and can have a very long operational life. Field-effect transistors (FETS) are universally used as the switching elements in the solid-state type of multiplexer.

Analog multiplexers are also divided into two other classes: high-level multiplexers and low-level multiplexers. *High-level multiplexers* are designed to handle input signals greater than 1 V without introducing significant error. They are relatively simple and inexpensive. *Low-level multiplexers* must be used when the input signals are in the range 1 mV to 1 V. They are more sophisticated than their high-level counterparts and also considerably more expensive. Differential techniques are employed to reduce the problems of common-mode voltages and cable-impedance unbalance.

The key specifications that characterize analog multiplexers are

**1.** *Settling time* (discussed in detail in the following section)

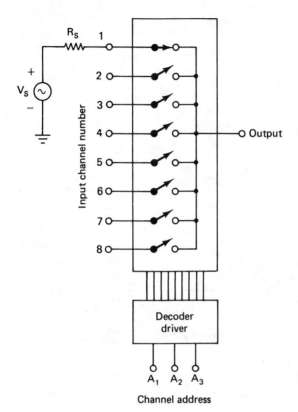

**Figure 17-10** Analog multiplexer circuit.

Channel address

2. *Transfer accuracy.* The input to output error as a percentage of the input

3. *Throughput rate.* The highest rate at which the multiplexer can switch from channel to channel at its specified transfer accuracy (throughput rate is determined by the settling time)

4. *Crosstalk.* Percent of signal transferred from an off channel to the multiplexer output

5. *Voltage range.* Maximum allowable signal above or below ground

### Settling Time

When a multiplexer switch selects a channel, the full signal amplitude does not immediately appear at the multiplexer output. There is a delay in the time following an input change before the output gets to within the specified accuracy of the final value and stays there (Fig. 17-11).

This delay or *settling time* is more formally defined as *the time elapsed from the application of a full-scale step input to a circuit, to the time when the output has entered and remained within a specified error around its final value.*

The importance of settling time in a data acquisition system is that certain

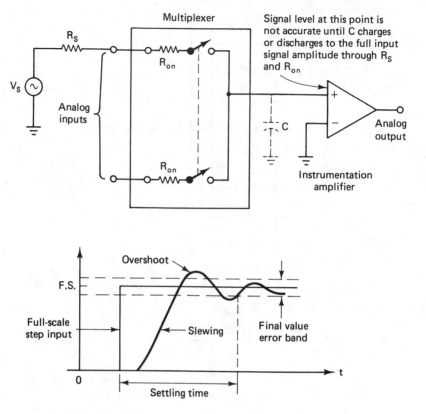

**Figure 17-11**  Why settling time is needed when using an analog multiplexer in a data-aquisition system. The delay in the time that it takes for a multiplexer output to rise to the full input signal amplitude results from the time required to charge the stray input capacitance, $C$, through the multiplexer-switch resistance, $R_{on}$, and the source resistance, $R_s$.

operations must be performed in sequence, and one operation may have to be accurately settled before the next can be initiated. For example, if the output of the multiplexer is fed to an A/D converter, the converted voltage value would not be accurate until the output of multiplexer had reached a value suitably close to the value at its selected input.

As shown in Fig. 17-11, the delay until the multiplexer output signal has settled results from the time required to charge the stray input capacitance $C$ through the multiplexer switch resistance $(R_{on})$ and the source resistance $R_s$. To ensure that the output signal reaches full amplitude, the time allowed for settling should be at least nine times the $RC$ time constant of the input circuit. If the input signals are arriving from sources having different impedances from one another, adequate settling times for each channel are liable to differ. To preserve system accuracy, the system speed might have to be reduced to accommodate the longest settling times.

## ANALOG-TO-DIGITAL DATA ACQUISITION
## SYSTEM CONFIGURATIONS

Analog-to-digital data acquisition and conversion systems are used to acquire analog signals and convert them into digital form for subsequent processing or analysis by computers or for data transmission. They are used in a large and ever-increasing number of applications in a variety of industrial and scientific areas, such as in the biomedical, aerospace, and telemetry industries. They are also used for process control applications in many industries (i.e., chemical, petrochemical, plastic extruder control, power generation, engine test, and environmental monitoring).

The system shown in Fig. 17-7 is the simplest example of a way in which a multichannel analog-to-digital data acquisition system can be configured. This configuration seems appealing in that it appears to yield a low-cost, efficient system. The cost savings arise from the manner in which the multiplexer is connected in the system. It is located so that all the input channels can share a common instrumentation amplifier, sample-and-hold circuit, and A/D converter. However, this cost-saving approach also imposes several major disadvantages. First, since the amplification is performed after multiplexing, an expensive, differential, low-level multiplexer must be used if the system is being used to acquire low-level (i.e., <1 V) signals. Second, because of constraints imposed by the way in which A/D converters operate, wide dynamic ranges between channels are difficult to handle with analog multiplexing. Therefore, unless all the input signals to the system have amplitudes that lie within the same range of values, the common system instrumentation amplifier may have to be a more costly, complex, programmable-gain amplifier (PGA) rather than a simpler fixed-gain type. Finally, the settling times of the input channels may vary widely if the signal sources present differing impedances to the system instrumentation amplifier. Uncertainty in settling times must be dealt with by either reducing the system sampling rate or by accepting a larger error in the sampled data. As a result of these limitations, this simple system configuration is limited to the less demanding data acquisition applications [i.e., those in which all the input voltages are high-level (>1 V) voltages].

The second and more popular configuration of analog-to-digital data acquisition systems is shown in Fig. 17-12. This configuration is both efficient and capable of high performance and is recommended for systems with input signs of <1 V. It is efficient in that the sample-and-hold circuit and A/D converter are shared by all the input channels. Also, the multiplexer can be seeking the next channel input in the time it takes for the A/D converter to convert the signal stored on the sample-and-hold circuit. High performance is possible because the signal conditioning of each channel can be customized to provide the A/D converter with full-scale signal amplitudes that will yield the smallest conversion errors. The premultiplexer signal conditioning also reduces (as well as equalizes) the settling times for all input channels (since the output impedances of the IAs from each channel are small and equal). This also helps maximize system sampling rates. Customization of the signal conditioning of each input channel is implemented by controlling the gain of the

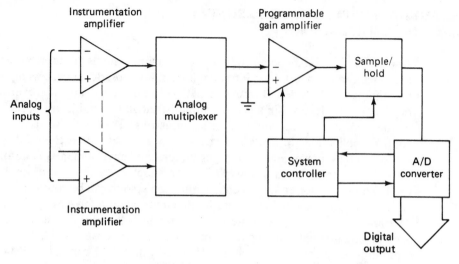

**Figure 17-12** Data-aquisition-system configuration in which the signal condition-ing by instrumentation amplifiers is performed on each individual channel prior to multiplexing. Programmable-gain amplifier is used to customize the gain of each channel to allow optimal operation of the A/D converter.

programmable gain amplifier (PGA) of the system with external signals from the system controller.

Completely assembled (on a single printed-circuit board) analog-to-digital data acquisition systems of this type are sold commercially. Analog Devices' μMac-4000 is an example of such a complete data acquisition system. The μMac-4000 was specifically designed for the large number of process control applications in which the quantities being monitored (temperature, flow, pressure, etc.) vary quite slowly with time. As a result, a slower dual-slope integrating converter is used to perform the A/D conversions and the need for a sample-and-hold circuit is thereby elimi-nated. The μMac-4000 also has the capability of storing the digital data signals and transmitting them (when called upon) to a computer or other digital device.

The third widely used data acquisition system configuration is shown in Fig. 17-13. Each channel applies the signal source voltage to its own instrumentation amplifier, located very near the signal source. The amplifier output is also con-verted near the sensor to a digital signal by a low-cost voltage-to-frequency (V/F) A/D converter. The signal in digital form is then transmitted to a digital multiplexer (MUX), which does not need to be in the immediate vicinity of the sensor. Trans-mitting the data in digital form reduces the possibility of degradation caused by external noise pickup. A multistage counter follows the multiplexer and counts pulses for a given period. Thus this type of configuration finds favor in applications requiring high noise immunity as well as those in which the signal sources cannot be located in the immediate vicinity of the multiplexer. The digital MUX of the system also operates differently from an analog MUX. That is, after a digital signal has

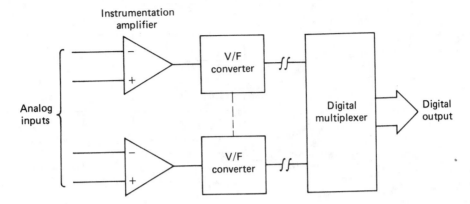

**Figure 17-13** Data-aquisition-system configuration in which the analog inputs are individually converted to digital signals by V/F A/D converters and then transmitted to a digital multiplexer.

been switched and settled, there is no added error at the multiplexer output as there may be when an analog signal is switched and settled. Thus use of a digital MUX eliminates the settling-time error inherent in the two configurations discussed previously. Furthermore, since V/F converters are integrating A/D types, they can also be used without a sample-and-hold circuit. This further reduces system complexity and overall system error. Finally, when applications call for the extraction of a signal in the presence of a high common-mode voltage, the V/F-based system does

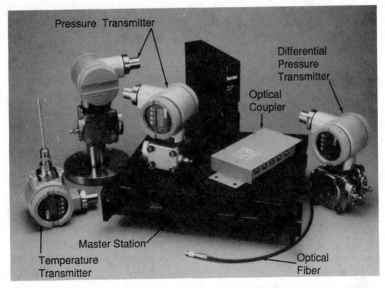

**Figure 17-14** Distributed measurement system hardware. (Courtesy of ITT Barton, A unit of ITT Corporation.)

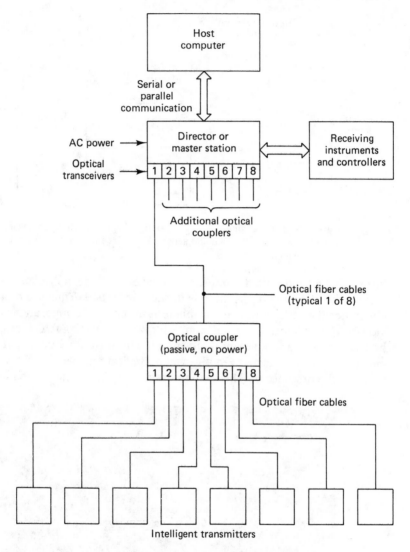

**Figure 17-15**   Distributed measurement system block diagram. (Courtesy of ITT Barton, A unit of ITT Corporation.)

not require the elaborate circuit-protection schemes of the other two data acquisition methods. Instead, each channel requires only a single digital optically coupled isolator, again lowering system cost while maintaining simplicity.

A fourth type of data acquisition system also includes feedback and control signals for both monitoring and controlling an industrial process. Some typical hardware for this type of system is shown in Fig. 17-14, and a system block diagram is shown in Fig. 17-15. The system is controlled and monitored by a director or

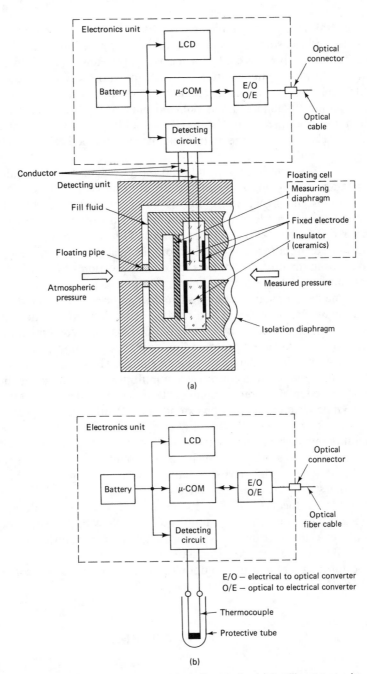

**Figure 17-16** Principle of operation block diagram for (a) intelligent transmitter with pressure transducer and (b) intelligent transmitter with temperature transducer. (Courtesy of ITT Barton, a unit of ITT Corporation.)

master station that communicates to a host computer, receiving instruments, controllers, and transmitters. A system can be configured for a simple monitoring task or for complete process control (such as a petroleum plant). Software in the master station controls communication between the various remote transmitter locations by transmitting and receiving serial data over optical fibers. A main advantage of using optical fibers in an industrial environment is the elimination of ground loops and noise pickup (refer to Chapter 16 for additional details). Installation costs are also reduced because large metal conduits and shielded cables can be replaced by a noise-free optical fiber.

The host computer transmits and receives signals from the director, which, in turn, transmits signals to remote transmitters and controllers. All the control loop calculations and signal levels (signals that open and close control valves and run motors) must be made by a host computer or a distributed control system. A discussion of distributed control systems is beyond the scope of this text. The data can be transmitted either serial or parallel from the host computer since a dedicated microprocessor in the director controls the communication protocol (refer to Chapter 18 for additional details on communication protocols).

The director station contains two microprocessors. One is dedicated to polling (turning on a remote device and reading or sending information to the on device). As many as 64 devices can be polled in 200 ms. A second microprocessor is dedicated to communication protocol (i.e., it makes sure that data transmitted to a device is compatible the device receiving a signal). The director can be located as far as 1 mile from remote transmitters when multiple transmitters communicate through an optical coupler. Direct communication over a single optical fiber can take place over 3 mile distance and can be extended to 10 miles when repeaters are used.

The transmitters, (Fig. 17-16), contain all the electronics necessary for signal conditioning such as linearization, scaling, and amplification. Because the transmitter electronic system uses such small amounts of power (energy consumption is similar to that of a digital wrist watch), a single battery powers the entire transmitter electronics. The transmitters can output signals in any of the five major instrumentation protocols (4–20 mA, 10–50 mA, 1–5 V dc, 0–10 V dc, and 3–15 psi). The detecting circuit in each transmitter is compatible with the type of input signal (such as mV, mA, and resistance). No data are taken or transmitted by a transmitter until polled by the director or master station. The transmitter microprocessor is capable of performing simple troubleshooting on itself (such as battery condition) and transmitting the results to the director.

## PROBLEMS

1. Discuss the differences between *analog systems* and *analog*-to-digital systems.
2. Define the concept of interfacing as it is related to instrumentation systems.
3. Compare *analog voltage transmission to 4- to 20-mA analog current transmission*. Discuss

some of the reasons why the 4- to 20-mA method has become so popular in many industrial process control applications.

4. Discuss the need for the following signal-conditioning functions in instrumentation systems.

   (a) Amplification                         (b) Filtering
   (c) Linearization                      (d) Buffering

5. Discuss the purpose of using sample-and-hold circuits in A/D systems.

6. Define the following terms that are used to specify the capabilities of sample-and-hold circuits.

   (a) Aperture time                     (b) Droop
   (c) Feedthrough                      (d) Acquisition time

7. Given an A/D system with a 10-bit converter that requires 100 $\mu$s to perform a conversion.

   (a) If the full-scale output of the converter is 10 V, find the maximum permissible rate of change of the input signal in order that the converter is able to resolve the analog input data to within 1 LSB.

   (b) For the same converter and a 75-ns aperture time sample-and-hold circuit, calculate the maximum permissible rate of change of the input signal.

8. Discuss the role of the *multiplexer* in A/D data acquisition systems.

9. Explain the difference between a *high-level* and a *low-level* multiplexer.

10. Define the following terms used in specifying the characteristics of multiplexers.

   (a) Settling time

   (b) Crosstalk

11. Explain why variable output-impedance values from the transducers/signal sources feeding the channels of a multiplexer can adversely affect the data-acquisition rate of the system.

## REFERENCES

1. Sheingold, D., ed., *Analog–Digital Conversion Handbook*. Norwood, Mass.: Analog Devices, 1976.

2. Zuch, E., "Principles of Data Acquisition and Conversion, Parts I–V," *Digital Design*, June, July, August, September, and October, 1979.

3. Coombs, C., ed., *Basic Electronic Instrument Handbook*. Chap. 18. New York: McGraw-Hill, 1972.

4. Garrett, P., *Analog Systems*. Chaps. 5 and 6. Reston, Va.: Reston Publishing, 1978.

5. Sheingold, D. H., ed., *Transducer Interfacing Handbook*. Norwood, Mass.: Analog Devices, 1981.

6. *The Application of Filters to Analog and Digital Signal Processing*. West Nyack, N.Y.: Rockland Systems Corp., 1976.

7. Sallen, R. P., and Key, E. L., "A Practical Method of Designing RC Active Filters," *IRE Transactions on Circuit Theory*, Vol. CT-2, March 1955.

8. Tow, J., "A Step-by-Step Active Filter Design," *IEEE Spectrum*, Vol. 6, No. 12, 1969.

9. Sheingold, D. H., *Non-Linear Circuits Handbook*. Norwood, Mass.: Analog Devices, 1974.

# 18

# *Data Transmission in Digital Instrument Systems IEEE-488, CAMAC, and RS/232C Standards*

As discussed in Chapter 17, data must often be transmitted from the point of measurement to some other point in the instrument system. For example, the measured data may need to be sent to display devices, recording devices, computers, or other process controllers. In many systems the destination may be quite far from the measurement point. Thus the problem of how to send the data over the transmission path (at the required rate while preserving the desired accuracy) must be addressed.

We noted in Chapter 17 that measurement data may be transmitted in either *analog* or *digital* form. We discussed that *analog* data transmission is usually performed in one of two ways: (1) transmission of an analog voltage signal (often the output of an *instrumentation amplifier*) or (2) transmission of the signal of an analog current, typically from 4 to 20 mA. Analog data transmission, however, suffers from the following limitations: First, the transmission of *analog voltage* signals is limited to about 30 m and *analog current-form* transmission to about 3000 m. Second, we saw in Chapter 16 how analog transmission is prone to degradation by interference signals. Third, digital computers and recording devices must have the analog data converted to digital form before the data can be handled by these devices. Therefore, analog transmission is not always the best choice for distribution of data.

Measured data can also be converted to and transmitted in *digital* form. By utilizing various transmission techniques, digital data can be sent to destination points of virtually unlimited distance. This can overcome the distance limitations of analog transmission. In addition, the data transmission can be performed in a highly efficient and virtually errorless manner, eliminating the problem of reduced accuracy that results from interference. Finally, digital data can be transmitted in the format required by the receiving computer or digital display device inputs. For all of these reasons cited, it is important to be familiar with the techniques of digital data transmission and interfacing in instrument systems.

The general subject of digital data transmission and interfacing, however, is a very broad one. It encompasses techniques of data transfer within and between computers, from computer to computer peripherals (i.e., to such devices as printers, CRTs, floppy disks, and tape cassettes) and between instruments and computers. We are interested primarily in the latter subject. Although much of what we discuss will also apply to digital data transmission in general, we intend to restrict the scope of the discussion to a brief and introductory overview of data transmission and interfacing in instrument systems (i.e., between two or more digital instruments or between digital instruments and computers). The goal will be to present sufficient information to allow readers to understand the specifications and instructions that are encountered when interfacing digital instruments with other digital devices.

Data transmission in instrumentation systems is usually performed with one of two goals in mind: (1) real-time data logging and (2) real-time data acquisition and control. In real-time *data logging*, there is essentially no control (e.g., by a computer) of the instrument, process, or system that provides the data. The information is simply continuously sent to the data logger (i.e., recording device, printer, or computer) and stored or displayed.

In contrast, real-time *data acquisition and control* implies that a computer not only acquires data and stores them in memory, but also mathematically "processes" such data and then transmits control signals back to the instrument, process, or system. We shall discuss how both data logging and data acquisition and control are achieved in digital instrument systems.

## LANGUAGE OF DIGITAL DATA TRANSMISSION

Five types of interface are most commonly encountered when transmitting digital data between instruments and other digital devices. These are

1. Parallel
2. Binary-coded decimal (BCD)
3. IEEE-488 bus
4. CAMAC
5. Serial, asynchronous

The *parallel interface* is a general interface method. In fact, the BCD, IEEE-488 bus and parallel CAMAC are all interfaces that come under the category of parallel interfaces. For this reason, we open our discussion with a brief introduction to the parallel interface, and then move on to a more detailed exploration of the other four types.

Digital data are transmitted along paths that may physically consist of wires, radio waves, microwaves, and so on. Such paths (also often referred to as data buses) usually consist not only of lines for carrying measured data but additional lines whose function is to carry control signals between instruments and digital devices. (A *bus* is more formally defined as a signal line or set of signal lines used by a system and over which messages are carried.) The control lines make it possible for computers to manage the measurement systems and dictate valid data transfers within the system.

The digital data that are carried on such buses are encoded in a digital format such as BCD, ASCII, or 8-bit words (note that these codes are introduced and discussed in Chapter 1). If all the bits that make up a digital word are transmitted simultaneously, this is known as *parallel transmission*. In parallel transmission, each bit of the data word requires its own data line, and together with the control lines, appears as the interface shown in Fig. 18-1. Parallel interfacing can be an efficient method of transmitting digital data if the pieces of equipment are in proximity. For long distances, the cost of providing an individual line for each bit in a word of transmitted information can become cost prohibitive. Thus for long distances a method that requires only one data path, and on which data are transmitted *serially* (i.e., one bit at a time) is used.

The data in all (parallel or serial) systems can be transmitted either synchronously or asynchronously. Synchronous transmission means that the data being transferred are in synchronization with a timing or control pulse. In *parallel synchronous* systems, a clock (or strobe) pulse is also transmitted in parallel with the data. The clock pulse notifies the receiving device that valid data have arrived. Parallel synchronous transmission rates can be very high. The distance of transmission, however, is limited to about 12 ft, and synchronization is costly and complex to implement. Therefore, parallel, synchronous transmission is employed primarily for data transfers within computers and from computer to computer. It is rarely implemented in instrument systems.

Asynchronous data transmission is performed without the use of synchronizing clock pulses. Instead, *asynchronous* transmission (either serial or parallel) requires a *handshake* between the transmitting and receiving devices to ensure that valid data transfers are made. Handshaking is a technique in which the transmitting device sends a pulse that says "data ready." In response the receiving device replies

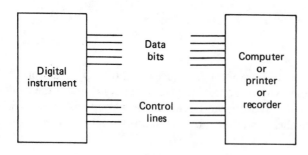

**Figure 18-1**  Parallel interface.

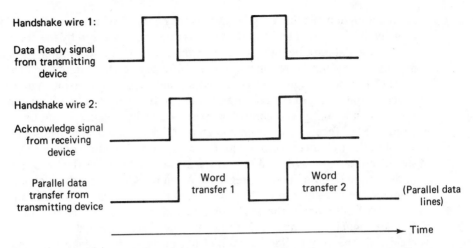

**Figure 18-2** Parallel handshaking. Handshaking in asynchronous systems allows the sending device to synchronize data transfers with the receiving device. In the parallel asynchronous interface example of this figure, the sending device sends a *Data Ready* signal to the receiving device. When the receiver sends back an *Acknowledge* signal, a data byte can be transferred between devices.

"acknowledged." Upon receipt of the "acknowledged" signal, the transmitting device is sure that the receiver is in a state ready to accept data, and so the data word can be transferred. Rudimentary *parallel* asynchronous systems use two wires to perform the handshaking event, as shown in Fig. 18-2.

Handshaking in *serial, asynchronous* systems is performed in a different manner; this technique is examined later in the section that deals with the serial, asynchronous interface.

Parallel, asynchronous interfaces are used in many digital instrument systems, usually in the form of the BCD, IEEE-488 bus, or CAMAC interface. Many other parallel, asynchronous interfaces are also used as methods of transmitting data between computers and their peripheral devices (i.e., between computers and printers, CRTs, or floppy disks, etc.). Most such parallel interfaces are more or less private affairs between the computer manufacturers and the peripheral devices they manufacture. As such, outside of the BCD, IEEE-488, and CAMAC, very little standardization exists in parallel interfacing. However, since the operation of the nonstandard interfaces does not affect (or illuminate) the operation of data transfer in instrument systems, we will not discuss their operation further from this point onward.

Within computers, data are transferred as parallel digital signals along the central processing unit (CPU) bus. (As an example, in many microcomputers the CPU bus is an 8-bit parallel bus.) However, for the computer to be able to acquire and process data from the outside world and to control external devices, its CPU bus must be made accessible. The *input/output (I/O)* ports of the computer perform the function of providing an accessway between the CPU bus and the outside world. Usually, both parallel and serial I/O ports must be possessed by the computer.

If the data are being received or transmitted by the computer from a *parallel* interface, they pass through the *parallel I/O port* of the computer. The simplest parallel I/O port (Fig. 18-3) contains parallel *register* (an *n*-bit digital circuit on which digital data can be temporarily stored) together with lines that can send and receive handshaking and control signals. When the register in the parallel port is being used to receive data from an external device, the device sends a Data Ready signal to the I/O port together with the data. This procedure enables the data to be stored in the register of the I/O port. Next, the port notifies the CPU that new data have been entered in its input register. When the CPU is ready, it collects the data from the register and routes them to the desired location within the computer along the CPU bus. Finally, the port sends to the transmitting external device a Data Request signal, informing it that the data have been received and that it is ready to receive the next byte. Simple parallel I/O ports act in a reverse fashion when the computer is transmitting data to the outside world.

When data are sent serially to a computer, they enter through the serial I/O port. They are converted by the serial port into the parallel data required by the CPU bus. The CPU bus can then accept the data. The port also collects parallel data from the CPU bus and serializes it into a suitable format (including handshake bits) for serial transmission. The devices within the serial I/O ports that perform the serial-parallel conversions are LSI chips called UARTS (universal asynchronous

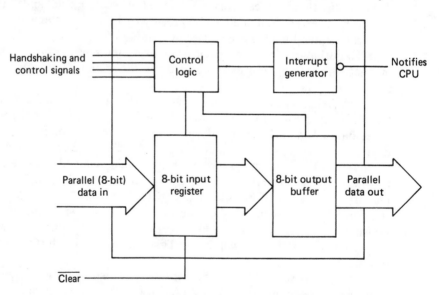

**Figure 18-3** Simple parallel computer input/output port. When Data In comes from external device, Data Out goes to CPU bus. Port then acts as Input Port. When Data In comes from the CPU bus, Data Out goes to external device. Port then acts as an Output Port.

receiver-transmitters) or USARTS (universal synchronous/asychronous receiver-transmitters). We shall discuss UARTS in more detail in the section dealing with asynchronous, serial interfacing (see Fig. 18-19).

## BINARY-CODED-DECIMAL INTERFACE

The BCD interface is a parallel, asynchronous interface that originated when attempts were first being made to connect digital measurement instruments to other digital devices. Thus, it is not surprising to learn that this interface method is rather primitive. Nevertheless, since many digital instruments are equipped for BCD interfacing and are still in use, it is useful to understand the interface and the basis of its operation.

To help analyze the characteristics of the BCD interface, let us review the means by which a digital instrument typically displays its measured data (Fig. 18-3); the output of the analog-to-digital converter in the instrument is fed to a series of 4-bit counters. These counters encode the digital data into binary-coded-decimal (BCD) words. The output of the counters is used to activate display devices that exhibit a decimal digit in response to signals from each of the 4-bit counters.

If the digital outputs of the counters are also made available (Fig. 18-4) on a set of signal lines (usually accessible from connector pins located on the rear panel of the instrument), together with handshake and control lines, the basis for a

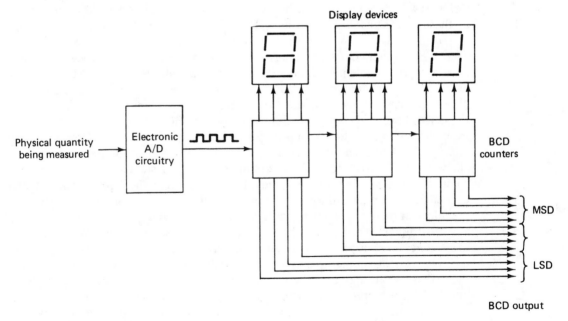

**Figure 18-4**   BCD output from a digital instrument.

parallel interface based on 4-bit binary groupings (that each represent a decimal number from 0 through 9) is established.

The first application that engineers considered for such BCD interfaces was the connection of digital instruments to printers. Successful interfacing between digital instruments and printers would allow continuous and unattended logging of data by the instruments. In the printers available at the time, each decimal digit had its own print wheel, and signals from the BCD counters controlled the position of the wheel when it hit the paper. Only two other signals were necessary to complete this interface: one to tell the printer when the data on the BCD lines was valid (a PRINT command), and one to allow the printer or other external device to control the rate at which readings were made (an External Trigger). Each of these two control signals requires its own wire, and thus with two wires and appropriate control signals, a handshake mechanism between the instrument would be formed. Today, the BCD interface allows digital signals to be sent from digital instruments to other devices as well as just printers (computers, terminals, etc.). However, the BCD interface still operates between these devices in the same manner as if the instrument were communicating with a printer.

It should also be noted that for each decimal digit of resolution provided by the digital instrument, four parallel data lines must be provided. For a BCD interface from an instrument with five-digit resolution, a minimum of 23 parallel lines (four each per digit, two control lines, and one ground line) would have to be provided by the interface. Thus, a connector with 23 pins would need to be available on the back of the instrument. Higher-resolution instruments with additional control lines would require a connector with even more pins. The logic levels of the bits on these BCD signal lines are usually TTL compatible ($1 = 2.0$ V $<$ TTL Hi $< 5.0$ V and $0 = 0.0 <$ TTL Low $< 0.8$ V).

When a BCD output of an instrument is sent to a computer, the computer is faced with the problem of simultaneously accepting all the parallel data bits being presented by each reading, regardless of their number. There is no standard procedure by which a computer is designed to perform this acquisition. But the general sequence of events does go something like this: The BCD information is transmitted from the instrument to a group of parallel registers (a kind of "interface to an interface"). An assembly language program is written for the computer which allows the BCD words to be read from the group of registers one at a time into the computer. The BCD words are usually then translated by the computer into ASCII characters (a digital word format that most computers can recognize). Finally, with the information in recognizable form, the computer can digest it.

Instruments equipped with BCD outputs are designed only to send messages to printers, computers, or other digital devices. That is, they are not capable of receiving messages other than handshaking and simple control commands (such as those that set the range of the instrument). The instruments equipped for BCD interfacing are therefore just slaves, and the flow of data is unidirectional. Because of this limitation, the BCD interface is, as we said, quite simple and primitive. As such, it is suited primarily for data logging. In contrast, the IEEE-488 is a more modern instrument-interfacing standard. It is much more flexible and complex, and can be used in either data logging or control applications.

## IEEE-488 BUS

The IEEE-488 standard defines a byte-serial, 8 bit-parallel, asynchronous type of instrument interface. It has been adopted by the IEEE in 1975 as an international standard for instrument interfacing and is designed to allow instruments made by any company in the world to link up if they are configured into systems in which all the pieces of equipment and elements conform to the standard. Hundreds of *IEEE-488*-compatible products (including several minicomputers) are already being manufactured. References are occasionally made to the *General Purpose Interface Bus* (GPIB) and the *Hewlett-Packard Interface Bus*. These, however, are merely other names for the IEEE-488 bus. The Hewlett-Packard Company originally developed this bus so that it could interconnect its own instruments to each other. The interface proved to be so popular that the IEEE picked it up as a standard.

It is also useful at this point to make the distinction between the IEEE-488 *standard* and the *IEEE-488 bus*. The *standard* is a *document* that spells out the rules, specifications, timing relationships, physical characteristics, and so on, of an interfacing technique that allows digital instruments and devices to be interconnected. The *IEEE-488 bus* is the hardware (wires, connectors, etc.) that is used to implement the Standard. Figure 18-5 shows an example of a system of instruments interconnected with the IEEE-488 bus.

The IEEE-488 Standard more specifically defines the electrical, mechanical, and functional specifications of the instrument interface. The *electrical specifications* discuss the electrical parameters of the digital signals transmitted on the bus (i.e., voltages and currents corresponding to the logic levels of the transmitted signals). The *mechanical specifications* define the physical makeup of the bus (number of wires, connector type, and pin designations, etc.). The *functional specifications* determine the precise use of each of the signal lines, the rules (protocol) that must be obeyed in order to properly transfer messages across the interface (including a handshaking procedure), timing relationships between signal lines, and the repertoire of messages that may be carried between devices. The most important of these specifications are discussed in this section.

Sixteen signal lines comprise the complete IEEE-488 bus structure as shown in Fig. 18-6. Eight of the lines are assigned to carry data (measurement data, program data, addresses, and universal commands), three lines are for the handshaking function (required since the bus operates asynchronously), and five lines are for the bus management functions that effect an orderly flow of messages across the interface.

Up to 15 instruments can be interconnected in almost any fashion to the bus, provided that the total cable length does not exceed 20 m (66 ft). Data can be transferred at rates up to 1 megabyte per second, but the actual rates are controlled

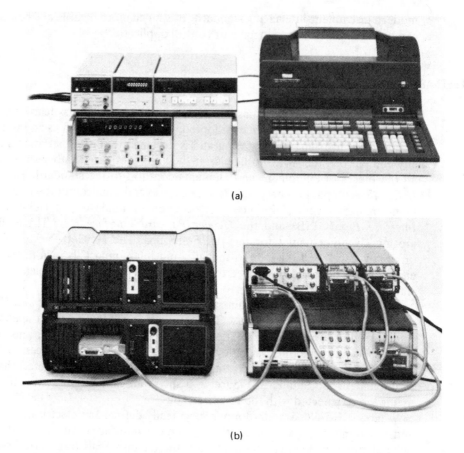

(a)

(b)

**Figure 18-5** Instrument system interconnected by the IEEE-488 bus: (a) front view of an assembled system; (b) rear view of an assembled system. (Courtesy of Hewlett-Packard Co.)

by the limitations of the instruments comprising each measurement system. The signals transmitted over the bus are TTL compatible, although they employ a negative logic convention (logical $0 =$ TTL high state [$\geq 2.0$ V], and logical $1 =$ TTL low, [$\leq 0.8$ V]). Although the coding format of digital words on the eight data lines is not mandated by the standard, many instruments code the data in ASCII format because most computers or programmable calculators that serve as the controllers of the IEEE-488 based systems read and write ASCII strings and numbers. For example, a voltmeter in a IEEE-488 bus system may wish to tell the controlling computer that it is reading $+1.433$ V at its input. It would be convenient if the DVM sent the data encoded in ASCII. Furthermore, since most computers prefer most-to-least significant byte ordering of the data, it would be best if the volmeter would send $+$, 1, ., 4, 3, 3, CR, LF. (The two characters CR and LF stand for Carriage

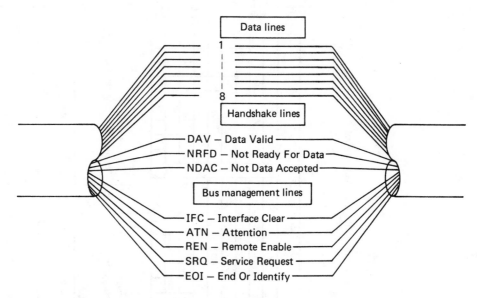

**Figure 18-6**  Sixteen lines make up the complete IEEE-488 bus structure.

Return and Line Feed and are used to terminate a transmission.) *However, since it is not guaranteed by the standard that instruments will send information coded in this suggested manner, two IEEE-488 interconnected instruments may always be able to talk to each other, but they may not always be able to understand one another.*

The instruments connected to the bus are grouped into four categories: controllers, talkers, listeners, and talker/listeners (Fig. 18-7). *Controllers* (computers or programable calculators) manage the operation and direct the flow of data on the bus, mainly by designating which instruments are to send data and which are to receive data. *Listeners* are capable only of receiving data (i.e., printers and recorders), and *talkers* (i.e., digital voltmeters) can merely send data over the bus to listeners. *Talker/listeners* (i.e., digital multimeters and network analyzers) have the capability of either receiving or sending data (controllers must also have the capability of functioning as both talkers and listeners). A minimal system does not actually need a controller, but may consist of just one talker and one listener (i.e., a counter and a printer). However, this type of system is limited to operating in the data logging mode. But more typical systems usually include a controller and a variety of other elements that would be talkers or listeners or both. In such systems, the function of *data acquisition and control* is quite readily exercised.

The IEEE-488 standard requires that certain strict rules or "protocol" be followed during its operation. Only one talker at a time is allowed to transmit data onto the bus, but there may be several active listeners which can simultaneously receive data. When there are several listeners, it is the slowest of them that dictates the rate of transmission to ensure that data are not lost.

To order talkers and listeners to become active, the controller uses a technique

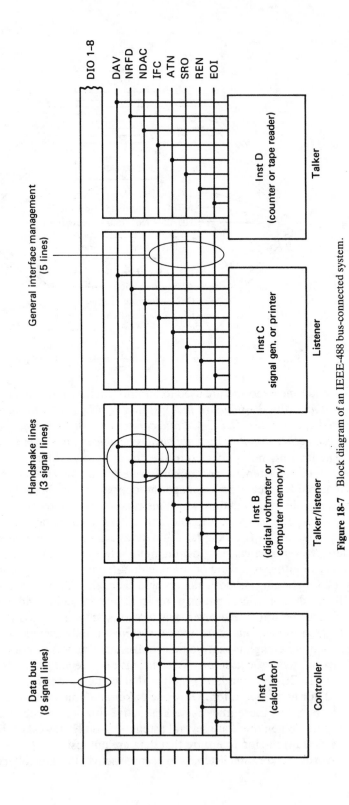

**Figure 18-7** Block diagram of an IEEE-488 bus-connected system.

called "addressing." Each of the instruments has a unique address for its talker or listener functions. The controller can cause a talker to become active by "addressing" the device to talk. Similarly, the controller can address one or more listeners on the bus to listen. The handshake sequence is also performed according to a highly ordered procedure.

The three-wire handshake used in the IEEE-488 bus is used because it has important characteristics that give the interface system wide flexibility. By this we mean that it allows interconnection of multiple devices which may operate at different speeds. The slowest active device controls the rate of data transfer and more than one device can accept data simultaneously.

The timing diagram in Fig. 18-8 helps to illustrate the sequence in which the handshake and data transfer is performed:

**1.** All active listeners use the Not Ready for Data (NRFD) line to indicate their state of readiness to accept a new piece of information. Nonreadiness to accept data is indicated by the NRFD line being held at zero volts. If even one active listener is not ready, the NRFD line of the entire bus is kept at zero volts and the active talker will not transmit the next byte. When *all* active listeners are ready, and they have released the NFRD line, it now goes high.

**2.** The designated talker drives all eight data input/output lines, causing valid data to be placed on them.

**3.** Two microseconds after putting valid data on the data lines, the active talker pulls the Data Valid (DAV) line to zero volts (this is called *asserting* the DAV line) and thereby signals the active listeners to read the information on the data

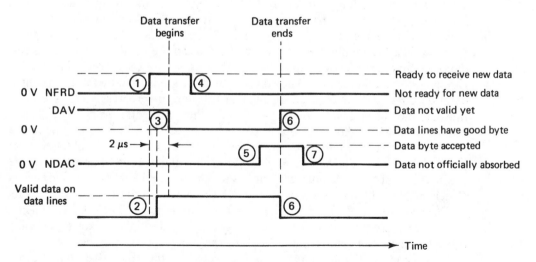

**Figure 18-8** Timing diagram showing the sequence of events during an IEEE-488 bus handshake.

bus. The 2-$\mu$s interval is required to allow the data put on the data lines to reach (settle to) valid logic levels.

**4.** After the DAV is asserted, the listeners respond by pulling the NRFD line back down to zero (asserting the NFRD line). This prevents any additional data transfers from being initiated. The listeners also begin accepting the data byte at their own rates.

**5.** When each listener has accepted the data, it releases the Not Data Accepted (NDAC) line (which until now has been held at zero volts by the listeners). Only when the last active listener has released its hold on the NDAC line will that line go to its high-voltage level state.

**6.** (a) When the active talker sees that NDAC has come up to its high state (thus signifying that all active listeners have accepted the data byte), it stops driving the data lines. (b) At the same time the talker releases the DAV line, ending the data transfer. The talker may now put the next byte on the data bus.

**7.** The listeners pull down the NDAC line back to zero volts and put the byte "away."

Note that the handshake operation is effected by the active talker and active listeners; once the controller has set up the bus for operation (by specifying active talker and listeners), it takes no part in the data transfer. Note also that the DAV line is controlled by the talker to indicate that data are ready to transmit. The NFRD and NDAC lines are controlled by the listeners to indicate readiness to receive data and receipt of data, respectively.

The five interface management lines help supervise the data lines. The primary management line, *Attention* (ATN), determines how many data lines are processed. When ATN is true, data lines are interpreted as containing addresses or universal commands by all the bus devices. In addition, only the controller talks. When ATN is false, only those devices addressed can use the data lines; in this case data transmitted are device dependent. With the *Interface Clear* (IFC) line, the bus is returned to a quiescent, or known state. Such action is used by the controller to override all activity and to abort all data transfers in progress. IFC is used to initialize the bus and clear the bus when something has gone wrong. The *Remote Enable* (REN) line allows the bus to control a device. That is, it makes the instruments of the system programable by the bus, rather than under control by the front panel of the device. The *Service Request* (SRQ) line can be used by any device to get the attention of the controller when it has data to send (talker) or needs to receive (listener). Note that this is a request and not a command, and can be ignored by the active controller until there is time to service the request. The *End Or Identity* (EOI) line is used in two ways: It can be asserted by the active talker to designate the end of a message, and it can be used by the controller as a polling line.

A standard 24-pin connector is prescribed by the IEEE-488 standard and is shown in Fig. 18-9. An illustration of the connector and cables is shown in Fig. 18-10.

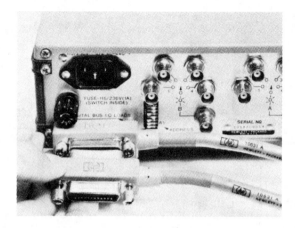

**Figure 18-9**   Connecting bus cables.
(Courtesy of Hewlett-Packard Co.)

**Figure 18-10**   Bus cable. (Courtesy of
Hewlett-Packard Co.)

Sometimes it is necessary to extend the instrument interconnection beyond the 20-m limitation of the bus. *Bus extenders* are available to allow this distance limitation to be overcome. With bus extenders an IEEE-488 bus system may be split into two isolated groups of instruments that may be separated by distances up to thousands of miles. This allows the IEEE-488 bus to be used in such new applications as industrial process control and remote sensing. When using bus extenders, the remote instruments still appear functionally as if they were directly cabled to the local bus. The extenders are essentially transparent. That is, the interface behavior of the system is operationally identical with and without extenders.

The extenders function by serializing the information on the 16 signal lines of the bus and transmitting the information serially in the RS-232 manner (serial interfacing and RS-232C are discussed in subsequent sections). Communications for short distance of up to 1000 m can be implemented by connecting twin-pair cable (four wires) directly to the extenders (Fig. 18-11). For longer distances the RS-232C interface of the extender allows the data to be sent over telephone lines, with the assistance of modems (Fig. 18-20).

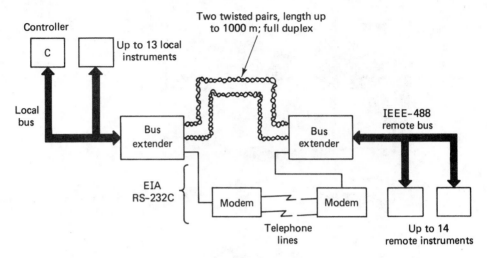

**Figure 18-11**   Bus extenders can be used to increase transmission distances, either through dedicated lines or through phone lines.

Use of the extender, however, does decrease the maximum bit transmission rate. For short distances (up to 1000 m), using twin-pair cables, a fixed bit rate of 20 kilobits/s is employed. This gives a useful transmission rate up to 775 data bytes per second. For transmissions linked by telephones lines, both synchronous and asynchronous modems are supported by the bus extender. This means that transmission rates of 19.2 kilobits/s (synchronous) and 150, 300, 600, or 1200 bits/s (asynchronous) can be achieved.

## CAMAC INTERFACE

The CAMAC interface can be configured to transmit data in parallel, serial, or byteserial manner. It was originally designed to meet the requirements of nuclear instrumentation labs, where instrumentation modules were constantly being swapped between systems. The basic word length in CAMAC systems is 24 bits, but when all the control lines are added, a parallel CAMAC system requires a 66-wire interface. However, when data are transmitted by a serial CAMAC interface just two data lines are required and only a nine-wire cable is needed.

The process I/O equipment built for use in CAMAC systems (i.e., A/D converters, microprocessor controllers, CRT drivers, etc.) is designed in a standard modular form so that, theoretically, the modules can simply be plugged into any of the remote CAMAC I/O boxes, called CAMAC *crates* (Fig. 18-12). The module design, as well as the method of connection between module and crate, are defined in one part of the CAMAC standard. The *crate* can contain a power supply and up to 25 plug-in modules, the right-most of which is the *crate controller*. This controller

**Figure 18-12** CAMAC crate. (Courtesy of KineticSystems, Inc., Lockport, Ill.)

interfaces the crate to the system bus (parallel or serial, depending on the interface), and also controls communication within the crate itself.

The CAMAC standard (now also adopted by the IEEE as Standards 583, 595, and 596) defines the mechanical characteristics of the crates, connectors, and modules; data highways; and the command structure of protocols that control communications in the system.

Five types of crate controllers are available, depending on the type of data transmission to be used by the specific CAMAC interface: parallel, serial, byte-serial, direct parallel transmission to a computer I/O bus, or stand-alone operation (i.e., operating without any connection to any highway or supervisory computer). Except for stand-alone operation, CAMAC interfaces are ultimately connected to a supervisory computer. Most minicomputers have CAMAC interface capability and communicate with CAMAC systems using versions of FORTRAN or BASIC.

The fastest data transmission rates in CAMAC systems are achieved using parallel interfaces. By tying crates directly to the computer I/O bus (parallel) up to 1 million words per second can be transmitted. However, such applications are limited to very short distances and only a few crates can be tied directly to a computer bus. Longer-distance transmission (up to 300 ft) at equally high speeds can be performed using the parallel CAMAC highway (the 66-wire cable, called the *Branchway*). The Branchway connects the crate controllers with a Branchway driver that connects to the computer I/O bus (and which can manage up to seven crate controllers). For still-longer-distance transmission, either of the two serial transmission options can be used. Data rates of up to 5 megabaud/s along CAMAC serial highways are possible, making this one of the fastest serial interfaces available. However, if telephone lines are used for transmission, much reduced data rates are encountered.

The chief advantages of CAMAC systems over other interfaces are their speed and ease of use. No engineering design is required. The user decides on the number of crates needed, whether to use parallel or serial transmission, and then orders the crates, control devices, and modules. But this top-of-the-line interface is also quite expensive, and is used primarily in large industrial process systems. However, stand-alone CAMAC configurations that use a microcomputer as the supervisory computer, together with economy-priced crates (13 module capacity), allow users to have a CAMAC system at a lower price.

## SERIAL, ASYNCHRONOUS INTERFACING

As mentioned earlier, transmission of data over long distances becomes prohibitively expensive if attempted in parallel fashion. However, if the data are transmitted serially, only one data path is required since the data are sent only one bit at a time (Fig. 18-13). In view of the fact that the single serial data path requires just two wires, cabling costs are kept low. In addition, the path needs only one transmitting processor to log the data out and one receiving processor to log it in. Hence, when data must be transmitted over long distances, serial interfacing is the obvious choice.

Serial transmission interfaces can be built to operate in either simplex, half-duplex, or full-duplex modes. *Simplex* transmission is unidirectional. A system designed to operate in a simplex mode has only a single transmitter at one end and a receiver at the other. If, however, the interface contains both a transmitter and receiver at each end, it can be designed to operate in either the half-duplex or full-duplex mode. If information can only be sent in one direction at a time, the path functions in the *half-duplex* mode (transmission in either direction, but not both directions at once). If both transmitters can simultaneously and independently transmit data on the path, the operation is called *full-duplex* (simultaneous transmission in both directions). Full-duplex serial transmission most often uses four wires for its implementation.

Serial transmission methods are also often characterized by how many bits per second they can transmit (1 bit/s is known as a *baud*). Therefore, a system that is rated at 110 baud can transmit, for example, ten 11-bit characters per second. Standard bit rates in serial systems are 50, 75, 110, 134.5, 150, 300, 600, 1200, 1800, 2400, 3600, 4800, and 9600 baud. (We shall see later why standard bit rates are a necessity for carrying out serial transmission.)

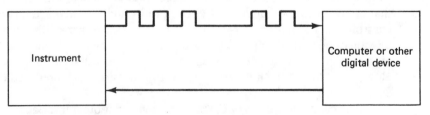

**Figure 18-13**   Serial data transmission.

Most *serial* interfacing in instrumentation systems is done in an *asynchronous* manner (rather than a synchronous one). Although serial synchronous transmission rates can be higher (up to 9600 baud) than asynchronous serial rates, the speed is achieved at the expense of greater system complexity. Since the lower transmission rates provided by serial, asynchronous methods are usually adequate for most instrumentation applications, there is no need to complicate such systems by requiring them to adopt synchronous transmission methods.

Asynchronous methods are relatively slow because they require a handshake for each character of data transfer. (Synchronous methods do not.) In *serial asynchronous* systems, handshaking is performed by using start and stop bits at the beginning and end of each character that is transmitted. For example, an 8-bit character is usually preceded by a single start bit and followed by one or two stop bits. Let us examine the details of such data transfer in asynchronous, serial systems more closely at this point.

Conventions governing serial, asynchronous data transmissions stem from the time when teleprinters were used to transmit information over long distances. Because of the mechanism design in early teleprinters (and to ensure their fail-safe operation), it was decided that an idle line (i.e., no data are being sent) is one in which current is flowing. Data transmission begins when the current in an idle line is interrupted in a specified fashion. This convention is still adhered to in modern asynchronous serial systems. In addition, the idle state of the line (current flowing) was named the "1" state or MARK condition, while the lack of current on the line was named the "0" state or SPACE condition (Fig. 18-14).

The beginning of the transmission of a serial, asynchronous word is announced by the "start" bit, which is always a "0" state bit. This convention was also dictated by teleprinter design insofar that it was necessary to bring the line from the idle (MARK) state to the "0" state in order to start the mechanism of the receiving teleprinter. For the next five to eight successive bit times (depending on the code

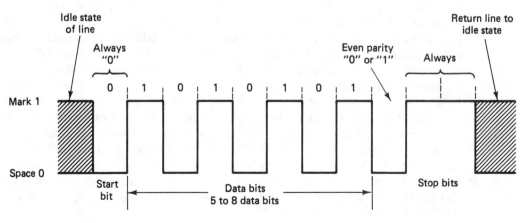

**Figure 18-14**   Serial data word.

and the number of bits that specified the word length in that code) the line is switched to the "1" states or "0" states as required to represent the character being sent. Following the last data bit in the word and the parity bit (about which more will soon be explained), it was necessary that the mechanism of the receiving teleprinter be allowed to coast back to a known position, in time for the beginning (start bit) of the next character. This is accomplished by transmitting one or more bits of the "1" (idle) state. The time period associated with this transmission is called the "STOP" bit interval.

Some time after transmission, the data words arrive at their destination (i.e., the receiver). The receiver, having been alerted to the incoming character by the start bit, times the incoming signal and samples the data line as close to the centers of each of the bits as possible. This illustrates the importance of the requirement that the receiver and the transmitter must both agree on the length of time each bit will be held on the line. Otherwise, the transmission would be garbled by the receiver taking samplings at the wrong times. The bit time then determines the maximum rate at which bits can be transmitted and thus defines the bit rate at which the serial interface runs. The data rate is controlled by interval timing devices called *clocks* in both the transmitting and receiving terminals. In asynchronous systems, the receiver restarts its clock for each transmitted digital word, and there may be a gap of any duration between words.

If noise pulses should somehow affect the transmission line at the wrong times, it is possible that a bit in the transmission could be misread. Therefore, following the 5–8 transmitted data bits, there may be a parity bit (see again Fig. 18-14), which is used for error detection. The parity bit is used in this manner: If the transmitter keeps track of the number of "1"s in the word being sent, it can send a parity bit— either a "1" or a "0"—so that the total number of "1"s sent is always even (even parity) or always odd (odd parity). Similarly, the receiver can keep track of the "1"s received, and so determines whether the transmission was received without error. If an error is detected, retransmission of the corrupted word can be requested.

The most widely used code for formatting the data words in serial, asynchronous transmission is ASCII (defined and discussed in Chapter 1). It is a 7-bit code and is almost always transmitted with a parity bit, which makes a total of 8 bits per character. [There are also other codes that can be used to format serial, asynchronous words. These other codes, however, are being used less and less frequently with time. They do, however, include Baudot code (5 bits per character), IBM's 6-bit correspondence code, and the 8-bit EBCDIC code].

The serially coded data words are transmitted according to one of two electrical conventions: RS-232C or 20-mA current loop. Virtually all computers produced have an RS-232C, serial I/O port and many also have a 20-mA current loop serial I/O port. Thus we need to study these transmission methods and the specifications that govern their characteristics. (Note that two newer standards RS-422 and RS-423 have also been adopted, but their acceptance is growing very slowly. RS-232C is still preeminent).

## DATA LINE MONITORS (DATA COMMUNICATIONS ANALYZERS)

Data line monitors permit us to diagnose and resolve problems in communication networks. Transmitting data between remote locations can result in lost data and error injection. Lost data can result from an intermittant open or broken transmission line. The open could result from a loose or broken connection. Vibrations in a building are random, and these vibrations can make and break a loose connection. This type of defect in a communications system is difficult to isolate because it is random; therefore, the transmission line must be continually monitored until an error occurs. By locating a data line monitor at different locations in the communication network, the defective connection can be isolated and then repaired. Error injection (adding unwanted data) can result from noise being superimposed on the communication transmission line (see Chapter 16 for additional details). The noise source can also be located and isolated by locating a data line monitor at different locations within the communication network. A common cause of noise in a communication transmission line is the large currents in power lines located near communication transmission lines. Coaxial cables will not prevent the induction of noise on a communication transmission line. Magnetic shielding can only be effected by installing the communication transmission line in an iron conduit or by locating the line a large distance from any power or lighting cables. Also, whenever a communication and power line must cross, make sure that they cross at 90°. This will substantially reduce noise as a result of magnetic coupling (see Chapter 11 for a discussion on *magnetic coupling*).

Another common use of a data line monitor is locating and isolating defective software. As an example of defective software, consider the following example that occurred in a major industrial plant. The velocity of a conveyor is monitored by a pulse generator (a device that generates a squarewave that has a frequency that is proportional to speed). The pulses are transmitted over a communications line to a computer that controls the starting and stopping sequence of the conveyor. By counting the number of pulses, the computer can calculate both the distance that a conveyor has traveled and its velocity. The error in counting the pulses resulted from the computer counting too many pulses (i.e., it would sometimes count a single pulse twice). A data line monitor was used to transmit a known series of pulses to the computer while monitoring the distance and velocity calculated by the computer. When the transmitted data and computer calculated results did not agree, the software was isolated as the cause of the discrepancy. Modifying the "read input data" routine in the software solved the problem.

A typical data line monitor is shown in Fig. 18-15. Because these devices are often used in the field, there are both portable and stationary models available. There is an initial setup procedure that is required. The actual procedure depends upon the manufacturer and the model being used. As an operator, first install the monitor in a communications network depending upon the end purpose. Figure 18-16 shows four commonly used locations for installing a monitor in a communica-

**Figure 18-15** Data communications analyzer-data line monitor. (Courtesy of Tektronix, Inc.)

tions network (refer to the next section in this chapter for a discussion of DTE and DCE). Second, we have to set-up the monitor by entering the communications protocol such as the following: Baud rate, parity (even or odd), data code (ASCII, EBCDIC, HEX, etc.), bits/character, stop bits, and timing. Third, we have to enter an operating sequence such as read or monitor incoming data and stop (called triggering) when a particular data string is read (data are stored in a *capture buffer*), trigger on an error, or transmit a particular data string. Once data have been stored in the capture buffer, it can be read as it is scrolled across the display or it can be retransmitted to a printer for outputting a permanent record.

## RS-232C STANDARD

When serial data are transmitted according to the RS-232C standard, *voltage* pulses transmitted at the selected baud rate are used to represent the digital data (RS, recommended standard). Logical "1" is represented by a transmitted voltage level in the range $-3$ V to $-15$ V, while logical "0" is represented by a transmitted voltage level of $+3$ V to $+15$ V. (Note that RS-232C therefore employs a negative logic convention, which is also not TTL compatible, for representing its data signals.) Digital data can be transmitted serially for distances up to 50 ft according to the RS-232C standard. It is the more widely used of the two serial transmission methods. RS-232C has also been adopted by the Electronic Industries Association (EIA) as an official standard for serial communication. In addition to specifying the voltage levels and maximum transmission distance of the digital signals, the standard defines many other characteristics required of a conforming data interface,

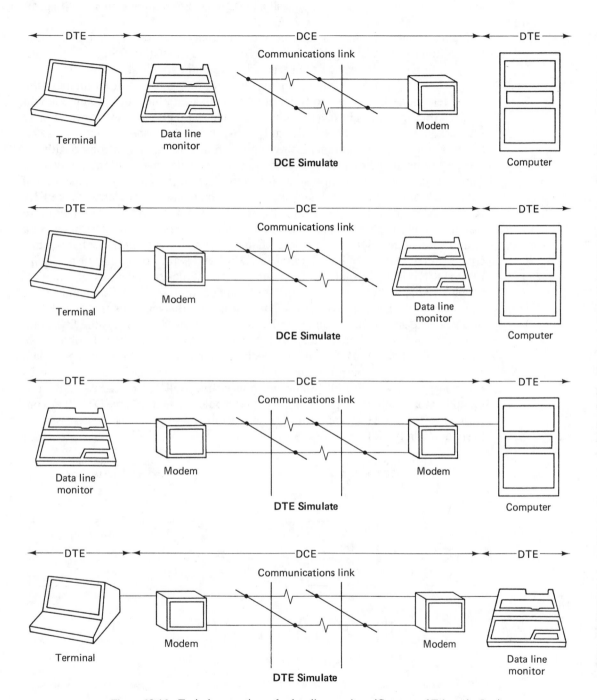

**Figure 18-16** Typical connections of a data line monitor. (Courtesy of Tektronix, Inc.)

including connector pin designation and other electrical considerations for the transmitting and receiving devices. Let us discuss how the RS-232C standard evolved; this will help explain the basis of the resultant standard specifications.

When serial, asynchronous digital data transmission was in its infancy, the public telephone lines were considered as potential data links of indefinite length for such transmission. However, it was noted that these phone lines were originally established to carry voice signals, not high-volume, high-speed digital data. In addition, the telephone companies were not enthusiastic when confronted with the prospect of having to deal with all kinds of strange digital signals in their networks. Thus it became obvious that a standard would have to be created to make digital data compatible with phone-line transmission. The resultant standard was the EIA RS-232C (Fig. 18-17). More specifically, the standard was formulated so that there would be a uniform set of electrical and mechanical interface characteristics as well as a functional description of the interchange circuits for the interface that permits connection of Data Terminal Equipment (DTE) and Data Communications Equipment (DCE). Data Terminal Equipment (DTE) includes such devices as computer terminals, digital instruments, teleprinters, and computers. Data Communications Equipment (DCE) includes the category of devices that encodes the digital data into voicelike signals permissible on the telephone system. Thus it is the DCE that actually converts the digital data into signal waveforms compatible with phone-line transmission. The RS-232C standard specifies the manner in which the digital data from the DTE must be presented to the DCE so that the DCE can perform its function properly. The Bell System introduced the Bell 103 *Modem* (modulator/demodulator) as the first piece of DCE that would respond to any DTE that conformed to the RS-232C standard (The characteristics of modems are discussed in more detail in the following section.)

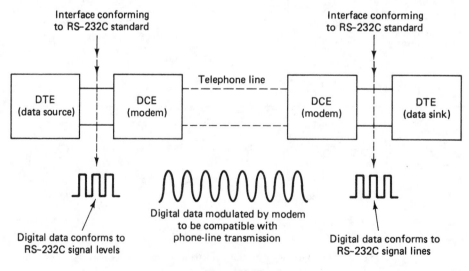

**Figure 18-17**

Although this bit of history helps explain the origin of RS-232C, it does not fully explain why the standard has been adopted for use in transmitting serial, asynchronous data in many other digital systems (most of which never incorporate the phone system in the transmission paths). The answer is that RS-232C can work just as well for transmitting digital data in serial form between any two digital devices, regardless of whether one of them is a DCE or not. As long as both pieces of digital equipment are receptive to transmitting and receiving according to the RS-232C specifications, digital data can be sent serially between them by obeying the standard.

We can, therefore, summarize RS-232C as a *standard* that defines the physical and electrical links of systems that allow serial, asynchronous data transmission over (1) telephone lines equipped with modems and (2) in digital systems in which the cabling distance is less than about 50 ft.

The most important specifications and characteristics that define RS-232C transmissions include the following (refer to Fig. 18-18):

**1.** Voltage signals employing a negative logical convention define the *data signal levels*. That is, data signals are considered Logic "1" = MARK, when the voltage, $V_1$, is $-3$ to $-15$ V. Data signals are considered Logic "0" = SPACE, when the voltage, $V_1$, is $+3$ to $+15$ V.

**2.** *Control signal levels*, on the other hand, use a positive true logic (control signal logical "1" = $+3$ to $+15$ V and control signal logical "0" = $-3$ to $-15$ V). The mixing of negative and positive logic in the same interface can be a point of confusion unless the distinction is clearly discerned by the user.

**3.** The shunt capacitance ($C_L$) of the terminator (load) must not exceed 2500 pF, including the capacitance of the cable. Since cables typically exhibit capacitances of 40 to 50 pF/ft, this specification limits the maximum transmission distance to about 50 ft.

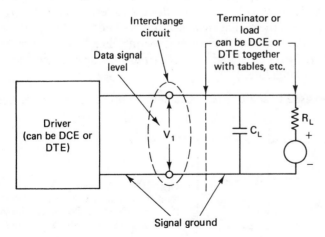

**Figure 18-18** Equivalent circuit of the interface that is used when defining the specifications required by the RS-232C standard.

**4.** The dc impedance of the load, $R_L$, must be between 300 and 3000 $\Omega$.

**5.** There are three wires used for data transmission: two are data-carrying wires (one each for the transmitted and received data) and the third is a signal ground wire that serves as a signal current return path (Fig. 18-19). There are also 22 other wires that serve as control wires between the DTE and DCE. The standard pin connections called out by RS-232C are shown in Fig. 18-20. Pins 2, 3, and 7 access the lines used for data transmission; the remaining pins access the control signal lines. RS-232C specifies that the male part of a connector is on the DTE, and the female part on the DCE.

**6.** The nomenclature of the pin connections shown in Fig. 18-19 is derived from the convention that according to RS-232C, the DTE transmits on the *Transmit* line and the DCE receives on it. Similarly, the DTE receives on the *Receive* line and the DCE transmits on it. If the system actually contains DTE and DCE, there may be no problem with plug-to-plug compatibility. However, in systems in which there is no DCE (e.g., in a system containing only a computer and printer, neither one is in fact DCE or modem) a problem therefore arises as to which piece of equipment is considered to be the DTE and which the DCE. The question that must be answered is this: Which device will transmit data on the Transmit line and which will receive on it? The answer is that some equipment manufacturers may offer cables to allow their gear to appear as either DTE or DCE. But in many cases, when the RS-232C connector is mounted on the rear panel, no choice is possible. In the case when two instruments both appear to be the same type, a cross-wire cable may need to be assembled to get signals on the correct lines.

**7.** Up to 50 ft, the bit rates may be any of the standard rates, from 50 bits/s to 19.2 kilobits/s. The 50-ft-distance limitation of RS-232C transmission can be extended up to about 10,000 ft, at maximum transmission rates of 600 baud, by using *line drivers* and twisted-pair lines. Line drivers are signal converters that amplify digital signals to ensure reliable transmission beyond the 50-ft EIA limit. Regular RS-232C interface cables are used to connect line drivers to terminals and ports. The line driver does not convert the digital signals into modulated audio signals (as does a modem) but retains the pulses in their original form. Therefore, such transmission is sometimes referred to as *baseband signaling* (transmission of a signal at its original frequencies, i.e., unmodulated).

**8.** Any code can be used to send serial characters according to RS-232C. As

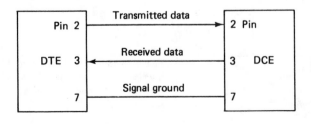

**Figure 18-19** RS-232C uses three wires for data transmission between DTE and DCE. All other wires associated with the interconnection are control signal lines.

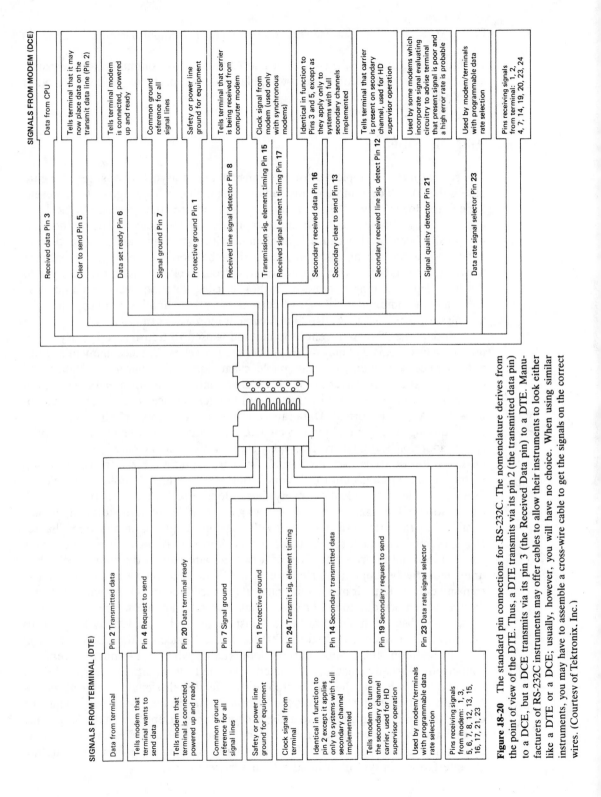

**SIGNALS FROM MODEM (DCE)**

|  |  |
|-----|-------------|
| Received data Pin 3 | Data from CPU |
| Clear to send Pin 5 | Tells terminal that it may now place data on the transmit data line (Pin 2) |
| Data set ready Pin 6 | Tells terminal modem is connected, powered up and ready |
| Signal ground Pin 7 | Common ground reference for all signal lines |
| Protective ground Pin 1 | Safety or power line ground for equipment |
| Received line signal detector Pin 8 | Tells terminal that carrier is being received from computer modem |
| Transmission sig. element timing Pin 15 | Clock signal from modem (used only with synchronous modems) |
| Received signal element timing Pin 17 | Identical in function to Pins 3 and 5, except as they apply only to systems with full secondary channels implemented |
| Secondary received data Pin 16 | |
| Secondary clear to send Pin 13 | |
| Secondary received line sig. detect Pin 12 | Tells terminal that carrier is present on secondary channel, used for HD supervisor operation |
| Signal quality detector Pin 21 | Used by some modems which incorporate signal evaluating circuitry to advise terminal that present signal is poor and a high error rate is probable |
| Data rate signal selector Pin 23 | Used by modem/terminals with programmable data rate selection |
| | Pins receiving signals from terminal: 1, 2, 4, 7, 14, 19, 20, 23, 24 |

**SIGNALS FROM TERMINAL (DTE)**

|  |  |
|-----|-------------|
| Pin 2 Transmitted data | Data from terminal |
| Pin 4 Request to send | Tells modem that terminal wants to send data |
| Pin 20 Data terminal ready | Tells modem that terminal is connected, powered up and ready |
| Pin 7 Signal ground | Common ground reference for all signal lines |
| Pin 1 Protective ground | Safety or power line ground for equipment |
| Pin 24 Transmit sig. element timing | Clock signal from terminal |
| Pin 14 Secondary transmitted data | Identical in function to pin 2 except it applies only to systems with full secondary channel implemented |
| Pin 19 Secondary request to send | Tells modem to turn on the secondary channel carrier, used for HD supervisor operation |
| Pin 23 Data rate signal selector | Used by modem/terminals with programmable data rate selection |
| | Pins receiving signals from modem: 1, 3, 5, 6, 7, 8, 12, 13, 15, 16, 17, 21, 23 |

**Figure 18-20** The standard pin connections for RS-232C. The nomenclature derives from the point of view of the DTE. Thus, a DTE transmits via its pin 2 (the transmitted data pin) to a DCE, but a DCE transmits via its pin 3 (the Received Data pin) to a DTE. Manufacturers of RS-232C instruments may offer cables to allow their instruments to look either like a DTE or a DCE; usually, however, you will have no choice. When using similar instruments, you may have to assemble a cross-wire cable to get the signals on the correct wires. (Courtesy of Tektronix, Inc.)

we discussed earlier, ASCII characters and ASCII control commands for the protocol of the serial system are the most popular code and protocol choices. But other codes and protocols can be used.

## 20-mA CURRENT LOOP

The second method employed for the transmission of asynchronous, serial digital data is the 20-mA current loop. Instead of using digital voltage signals to represent logic 0 and logic 1 levels, the current-loop method uses the presence (logic 1) or absence (logic 0) of a 20-mA current to represent the logic levels. The method has historically been used because it originated with data transmission to and from Teletypes. In the mechanical teleprinter such as the Model 33 Teletype, line current is switched on and off to transmit information from the keyboard of the Teletype to the line. The switching is done by carbon brushes rotating over a commutator. Generally, a minimum of 18 mA of current must be maintained to keep the contacting surfaces clean; thus, the 20-mA current loop was born.

Today, the 20-mA-current-loop method is used to transmit serial, asynchronous data in systems that may or may not contain a Teletype. Since many minocomputers, microcomputers, programmable calculators, time-sharing terminals and CRTs possess 20-mA-current-loop ports, digital information from data-acquisition sub-systems can be received or transmitted by these devices in a serial, asynchronous 20-mA-current format. Data rates are typically 110, 150, or 300 bits/s.

Although the 20-mA loop was created for Teletype communications, its widespread adoption beyond mere Teletype systems arose from the natural advantages it possesses for the transmission of serial, asynchronous data. First, the current loop is, in fact, a continuous unbroken conducting path in which 20 mA of current can flow. Since current is identical everywhere along a series path, there will be no degradation of the signal with distance along the loop. Second, the total impedance exhibited by the path of the loops is low. This property (as we discussed in Chapter 16) is an important factor in reducing the susceptibility of a circuit to capacitively coupled external interference. Use of a twisted-wire pair as the transmission line limits inductively coupled interference as well. Thus current loops can be used to transmit data up to their maximum practical distance (about 3000 m) with no degradation of accuracy due to either loss of signal or external interference. [Note that the 3000-m distance limit allows data transmission over considerably longer runs than RS-232C (which is limited to 50 ft for direct-connection data transmission).]

In addition, the 20-mA current loop (along which the data receivers are connected) is usually electrically isolated from the transmitters of the digital data. This is done by having the digital transmitters feed their signals to an *optical isolator*, the output of which controls the current flowing in the 20-mA loop. Thus the digital transmitters are allowed to actively control the current flow in the loop while remaining electrically isolated from it. The complete electrical isolation between transmitters and the receiver results in elimination of ground loops and their attendant interference in systems using the 20-mA current loop.

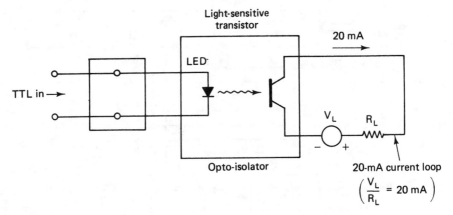

**Figure 18-21**   Opto-isolator used in a 20-mA current loop.

As shown in Fig. 18-21, digital signals in the form of TTL logic levels are fed to the input of the digital transmitter. The digital transmitter, in response to a *TTL high* level input (logic 1), drives current through the light-emitting diode (LED) of the opto-isolator. The light-sensitive transistor responds to the LED light output and turns on. The collector-emitter resistance of an "on" transistor drops to a very low value, and in the circuit of Fig. 18-21, this permits 20 mA to flow through the loop (logic 1 state in loop). When the logic level input at the digital transmitter is *TTL low*, it will provide no current to the LED, and the light-sensitive transistor will be turned off. In the "off" state its collector-emitter resistance becomes very large. Thus the current flowing in the loop decreases to a value near 0 mA (logic 0 state of the loop). Receivers of digital information (computers, Teletypes, terminals, etc.) are attached passively to this loop through this 20-mA current loop port and simply detect the logic state in the loop. If it is necessary that digital information be transmitted back to the digital devices from Teletypes or computers, a circuit must also exist that converts 20-mA logic signals back to TTL logic signals. Again, the opto-isolator plays a major role in such circuits.

The 20-mA current loops do, however, suffer from the problem that they are liable to introduce inductively coupled crosstalk interference in nearby circuits. Thus if adjacent wires are suspected of being sensitive to such interference, steps must be taken to reduce the degrading effects. The RS-232C standard is also becoming more pervasive with time, and some computer manufacturers are ceasing to offer the 20-mA current loop input-output capabilities.

## UNIVERSAL ASYNCHRONOUS RECEIVER-TRANSMITTERS

The devices used to convert parallel data words of 5, 6, 7, or 8 bits into asynchronous serial data words (with one start bit, one parity bit, and one or two stop bits) are called *universal asynchronous receiver-transmitters* (UARTs). That is, they can also receive serial data and convert it back to parallel form. The UARTs are

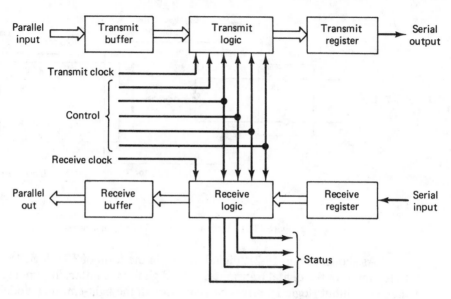

**Figure 18-22**   Block diagram of a standard UART.

typically 40-pin integrated circuits which are programable to allow selection of the number of bits per transmitted serial word. They are double buffered at both the transmitter and receiver and can thus permit the storage of a single 8-bit word while another 8-bit word is being either transmitted or received. UARTs can be transmit data at rates over 100 kilobits/second. They also usually provide their outputs in tri-state form, which allows use in shared-line situations. An external clock is synchronized with the serial signal to establish the bit period. Figure 18-22 shows a block diagram of a standard UART.

## PREASSEMBLED INTERFACE SUBSYSTEMS

Entire subsystems, which can acquire analog data from a variety of sources, convert the data to digital form and then make them available in the form of ASCII-coded characters ready for serial transmission in either RS-232C or 20-mA-current-loop format, are available in preassembled, ready-to-use form. One such subsystem is Analog Devices μMac-4000. Such microcomputer-based measurement systems come complete on a single board. They are typically capable of accepting analog data from tens of input channels, including such sources as thermocouples, strain gauge transducers, resistance-temperature detectors, analog dc current loops (0.1 mA, 0 to 20 mA, or 4 to 20 mA), and dc voltage sources (with full-scale ranges from ±25 mV to ±10 V). The input signals from these sources are first amplified to an optimum voltage level. The gain required for particular amplifications is achieved by combining the gain characteristics of the IAs in each individual channel together with a selectable gain, the *programable gain IA* (known as a PGA). Note that a PGA is an amplifier capable of having its gain programed by remotely generated logic-level signals. The amplified analog data are converted to digital form by a dual-slope A/D converter.

A full-duplex UART is used to convert the parallel digital output of the A/D converter to serial ASCII-formatted characters. The data are available for serial transmission from a communications port in either the RS-232C or 20 mA full-duplex current-loop manner. Even or odd parity and baud rates up to 9600 (maximum transmission distance then limited to 600 ft) are selectable. Use of the 20-mA current loop also optically isolates the interface from the host computer and allows it to withstand common-mode voltages up to 300 V. Built-in intelligence (from a microprocessor) unburdens the host computer by allowing on-board supervisory control.

## LONG-DISTANCE DATA TRANSMISSION (MODEMS)

When digital data must be transmitted over distances that exceed the distance limits of direct serial transmission methods (i.e., 10,000 ft with the 20-mA current loop, or RS-232C with line drivers), telephone lines with a *modem* at each end become the data links over which digital information can be sent. Such use of the phone lines allows data to be transmitted over essentially unlimited distances.

The digital data originate in a digital device (i.e., and instrument, computer, or computer peripheral) and is outputted according to the RS-232C standard to the input of the modem (Fig. 18-14). The *modem* (also known as a *data set*) accepts and converts the digital pulses into audible tones suitable for transmission over the telephone network and then another modem does the reverse at the other end. That is, modems perform the functions of the data communications equipment (DCE) devices described in the section that introduced the RS-232C standard. The transmitting modem *modulates* the digital signal into an audio tone suitable for transmission over a voice-channel communications facility, while the receiving modem at the other end *demodulates* the audio tone back into digital form, hence the name "modem" (modulator/demodulator).

A modem and the communications (telephone) line can be connected directly (hardwired) or indirectly (acoustically coupled). Acoustically coupled modems have an acoustic connection to the telephone network via transducers in an acoustic pad located in the acoustic modem and also in the telephone handset. Since acoustic modems have no hardwire connection to the telephone lines, they are portable. Such acoustic modems are more commonly known as *acoustic couplers* and are widely used when the regular dial-up telephone network is used to transmit data. *Hardwired* modems, on the other hand, are usually chosen when a user leases telephone lines from the phone company and uses these leased lines exclusively to transmit digital data.

Modems use one of (or a combination of) three basic modulation schemes

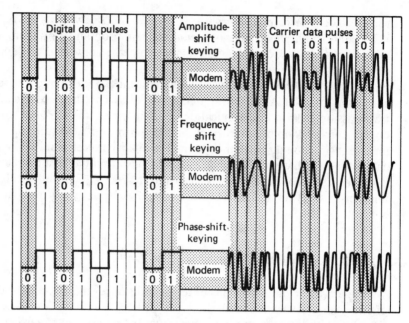

**Figure 18-23** Three basic modulation techniques are available for translating digital data into a form more compatible with telephone system transmission facilities.

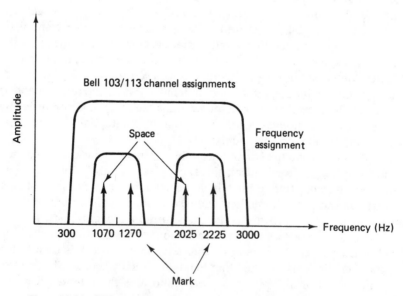

**Figure 18-24** FSK modems use frequency modulation. The Bell 103/113 can operate full duplex on two-wire lines.

to convert digital signals into audio tones: the amplitude-modulation, frequency-modulation, or phase-modulation schemes shown in Fig. 18-23. In *amplitude modulation*, the carrier signal amplitude is varied according to the bit pattern. In *frequency modulation*, the frequency is varied and in *phase modulation*, the phase is varied. Many variations are possible within these three main types of modulation.

Most asynchronous modems use *frequency-shift-keyed* (FSK) modulation (a simple frequency modulation scheme), in which a signal of one frequency indicates a space, and a signal of another frequency indicates a mark (Fig. 18-24). The most popular FSK modems are the Bell 103/113 and 202 types. The Bell 103/133 modems can transmit up to 300 baud full-duplex, and the 202 operates at speeds up to 1200 baud full-duplex. Higher-speed synchronous modems employ more sophisticated modulation and encoding speeds to allow data transmission rates up to 9600 baud.

## PROBLEMS

1. Discuss the advantages of digital data transmission to analog data transmission.
2. Define the following terms.
   (a) Handshaking
   (b) Baud
   (c) Baseband transmission
3. Describe the BCD interface.
4. Describe the method of handshaking used in serial transmission.
5. Draw the digital waveforms for the following ASCII-coded, serial words.
   (a) @                               (b) >
6. Differentiate between the concepts (1) IEEE-488 standard and (2) IEEE-488 bus.
7. Explain why three wires are required for the handshaking technique employed in IEE-488 bus systems.
8. Explain why synchronous data transmission can be faster than asynchronous data transmission.
9. How does the receiving device of an asynchronous serial system know the proper times at which to sample the data line to ensure that it is sampling when valid data exist on the line?
10. Discuss the RS-232C standard.
11. Define the following terms.
    (a) Modem
    (b) Acoustic coupler

## REFERENCES

1. McNamara, J., *Technical Aspects of Data Communication*. Maynard, Mass.: Digital Equipment, 1979.

2. Liebson, S., "Computer I/O Course," *Instruments and Control Systems*, Parts I–VI, October 1979–April 1980.

3. Washburn, J., "Communications Interface Primer," *Instruments and Control Systems*, Parts I and II, March–April 1978.

4. Garret, P. H., *Analog Systems for Microprocessors and Minicomputers*. Reston, Va.: Reston Publishing, 1978.

5. *HP-IB (Hewlett-Packard Interface Buss): Improving Measurements in Engineering and Manufacturing*. Hewlett-Packard Company, no date.

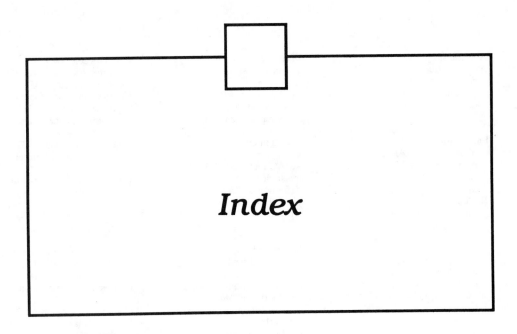

# Index